Energy Conservation
and
Thermal Insulation

PROPERTIES OF MATERIALS

SAFETY AND ENVIRONMENTAL FACTORS

Edited by

S. S. Chissick

University of London, King's College, The Strand, London

and

R. Derricott

The National Building Agency, Wales

Titles in Series

Asbestos Volume 1

Published 1979

Occupational Health and Safety Management

Published 1981

Energy Conservation and Thermal Insulation

Published 1981

in preparation

Asbestos Volume 2

Rebuild

Energy Conservation
and
Thermal Insulation

Edited by

R. Derricott

The National Building Agency, Wales

S. S. Chissick

King's College, London

A Wiley–Interscience Publication

JOHN WILEY & SONS

Chichester · New York · Brisbane · Toronto

British Library Cataloguing in Publication Data:

Energy conservation and thermal insulation.—
  (Properties of materials: safety and environmental
  factors).
  1. Architecture and energy conservation
  I. Derricott, R.
  II. Chissick, Seymour S.
  III. Series
  658.2'6   TJ163.3   80-41587

  ISBN 0 471 27930 7

Typeset by Preface Ltd, Salisbury, Wilts. and printed
by The Pitman Press, Bath, Avon

Contributors

G. APPLEGATE — *Curwen & Newbery Ltd, Alfred Street, Westbury, Wilts BA13 3DZ*

A. BARNATT — *Lankro Chemicals Ltd, Urethane Division, Eccles, Manchester M30 0BH*

J.-P. BARTHELEMY — *Aerospatiale, Établissement des Mureaux, Route de Verneuil, 78130 Les Mureaux, France*

P. BASNETT — *Electricity Council Research Centre, Capenhurst, Chester CH1 6ES*

A. M. BERMAN — *Scientific Branch, Greater London Council, County Hall, London SE1 7PB*

E. G. BUTCHER — *Fire Check Consultants, 6 Welbeck Street, London W1*

S. S. CHISSICK — *Department of Chemistry, King's College, Strand, London WC2R 2LS*

D. CROGHAN — *Anglian Architects, 39 Newton Road, Cambridge CB2 2AL*

R. DAFTER — *Center for International Affairs, Harvard University, 1737 Cambridge Street, Cambridge, Massachusetts 02138, USA*

R. DERRICOTT — *National Building Agency, Caerwys House, Windsor Lane, Cardiff CF1 3DE*

R. GRADIN — *International Energy Agency, Organization for Economic Co-operation and Development, 19 Rue de Franqueville, Paris 16e, France*

C. GREEN — *Department of Architecture, University of Sheffield, The Arts Tower, Sheffield S10 2TN*

J. H. HAMPSHIRE	*TAC Construction Materials Ltd, Gorsey Lane, Widnes, Cheshire WA8 0RL*
V. I HANBY	*Department of Architecture, University of Nottingham, University Park, Nottingham NG7 2RD*
A. C. HARDY	*Building Science Section, School of Architecture, The University, Newcastle upon Tyne NE1 7RU*
L. M. HOHMANN	*LMH Design, 17 West Grove, London SE10*
P. J. JONAS	*Department of Energy, Thames House South, Millbank, London SW1P 4QJ*
R. JONES	*British Gas Corporation, 326 High Holborn, London WC1V 7PT*
A. R. D. LAMBERT	*Bayer UK Ltd, Bayer House, Paradise Road, Richmond, Surrey TW9 1SJ*
W. LAWSON	*National Coal Board, Stoke Orchard, Cheltenham, Glos. GL52 4RZ*
S. J. LEACH	*Department of the Environment, Building Research Establishment, Garston, Watford WD2 7RJ*
W. D. MCGEORGE	*Environmental Design Services, 1 Lairg Drive, Priory Bridge, Blantyre, Scotland*
A. C. PARNELL	*Fire Check Consultants, 6 Welbeck Street, London W1*
R. C. PAYNE	*National Coal Board, Stoke Orchard, Cheltenham, Glos. GL52 4RZ*
J. RICH	*British Gas Corporation, 326 High Holborn, London WC1V 7PT*
H. ROBERT	*Pittsburgh Corning Europe N.V., Tervurenlaan 32–38, 1040 Brussels, Belgium*
R. SHOTTON	*Commission of the European Communities, DC XII, Rue de la Loi 200, B-1040 Brussels, Belgium*
A. STRUB	*Commission of the European Communities, DC XII, Rue de la Loi 200, B-1040 Brussels, Belgium*
S. V. SZOKOLAY	*Architectural Science Unit, University of Queensland, St Lucia, Queensland, Australia 4067*
R. W. TODD	*National Centre for Alternative Technology, Llwyngwern Quarry, Machynlleth, Powys, Wales*

G. R. WINCH   *School of Architecture, University of Manchester, Manchester M13 9PL*

P. P. YANESKE   *Environmental Design Services, 1 Lairg Drive, Priory Bridge, Blantyre, Scotland*

Contents

Foreword

We expect oil to become scarcer and dearer. By the year 2000, its price in real terms is likely to have more than doubled. The prices of other fuels will increase in its wake. Buildings designed today will still be in use well into the twenty-first century when the recent era of cheap and plentiful energy will have passed into history. As space heating, lighting, and provision of hot water in buildings now account for something like half the United Kingdom's total energy consumption, the need, and indeed the scope, for more energy-efficient buildings is self-evident. I welcome this collection of essays as a valuable contribution to assist architects and others concerned with the design of buildings to meet this need.

JOHN MOORE
Parliamentary Secretary
Department of Energy

Introduction

THE NECESSITY FOR ENERGY MANAGEMENT

The timing is just right for emphasizing the role that managers in industry, business, and commerce have to play in implementing energy management. Some bemused eyebrows may be raised at the idea of the Chairman of one of the world's largest producers and marketers of oil extolling the virtues and hard-headed business sense of efficient energy use, particularly at a time when most European oil refineries are under-utilized and operating relatively uneconomically. However, as a major supplier of an exhaustible and probably increasingly expensive resource, British Petroleum has a responsibility to its customers, and in a wider sense to society, to encourage its efficient utilization.

Complacency over the past two years of slack demand and relatively depressed oil prices will only stoke up problems for the future. Unless one believes that the world will be unable to pull itself out of the current recession in the foreseeable future, it would be short-sighted not to see that a revival in economic growth, and consequently strongly rising energy demand, is likely to have some effect on energy prices. This refers to long-run effects, but even in the short term the response of price to demand can be significant, as the increase in oil prices in many European markets in recent months has shown.

Now BP has no special claim to foretelling the future; neither should it attempt to. However, there is some merit in extrapolating trends in energy demand and supply to see where they might lead. If we take the officially declared economic aspirations of the OECD countries, namely a GNP growth rate of 4% per annum, our studies show that potential oil demand in the non-Communist World doubles by 1995, in spite of a substantial allowance for more efficient use of energy. However, oil supply could fail to rise with potential demand by 1990. World oil production would then be in decline as our finite resources are slowly exhausted. Note that this does not mean that 'oil is going to run out in 1990'. It is to be expected that the world will be producing—and consuming—more oil in 1990, and even in the year 2000, than it does today. But not twice as much, or three times as much as past trends would suggest. However, it is unrealistic to think that this maximum potential oil production rate will actually be obtained. Problems could develop much sooner, as soon as the early 1980s. This is so because many of the OPEC countries have indicated that they will not increase production beyond certain self-imposed levels. For example, Kuwait has declared that it will not increase output beyond 2 million barrels per day, though its oil reserves are capable of much more. The effect of such policy limits on oil production is that supply would fail to meet potential demand by the early 1980s. Even then, the question of whether the projected growth in oil demand will be met depends to a large extent on the oil production policies of only one country, Saudi Arabia, which by virtue of its vast oil reserves and small population has great flexibility in choosing its oil production rate.

This scenario therefore suggests that sometime in the 1980s oil supply would tighten, and that oil demand would be balanced with supply presumably through price. By that is meant higher price in real terms. With the non-Communist World depending on oil to meet 50% of its energy demand in this scenario, the consequences for the oil-importing nations are clear. One recent example is the effect that a yawning balance-of-payments deficit derived from oil-import dependence has done to the mighty dollar. Once again, this is not a precise BP forecast of future events, but it illuminates a few potential pitfalls. It also highlights the need for vigorously pursuing energy efficiency as well as developing non-oil energy supplies.

If this appears to be a rather gloomy scenario, perhaps a note of optimism can be added by recalling what Dr Lantzke, Executive Director of the International Energy Agency, said in September 1978 at a conference on 'Oil and Economic Power' at the Royal Institute of International Affairs. His view was that there is more scope for reducing energy growth relative to economic growth than has been allowed for in this scenario. However, it is clear that it will require positive and consistent pursuit if the potential is to be realized. The responsibility for that will lie with private industry, business, and commerce as well as with national governments and the public sector.

Let us consider how Britain fits in the global energy scene. Britain will soon cease to be an oil-importing nation, with the benefit of North Sea oil production. But it would be quite mistaken to see our oil self-sufficiency in the 1980s as grounds for complacency in tackling energy conservation now. This is because major projects required to develop alternative energy supplies have very long lead times. For example, the installed nuclear capacity of the late 1980s is already defined by today's order book. The development of non-oil energy supplies is therefore unlikely to avoid the need to import oil again in the early 1990s. One could argue about whether Britain should aim for an export surplus during the 1980s to earn the foreign currency necessary to improve our economy for the future, or whether one should invest that surplus oil in the ground for consumption after 1990. Whatever we do, the marginal cost of energy to the economy is the international oil price—we will or could be earning it in the 1980s or paying it in the 1990s. That is the test against which the economics of energy conservation should be measured.

The foregoing should place the need for energy management in a national and international context of future supply, demand, and price. Depending, as we do, and shall continue to do, on a relatively scarce and probably increasingly expensive exhaustible resource to meet our energy requirements, the need for efficient use of our energy resources is imperative. Indeed, the events of 1973 and the subsequent oil price rise emphasized the need for action five years ago. In this respect, the national interest and the interest of the domestic and industrial energy consumer coincide. The cost of energy today is such that it makes good financial sense to examine critically how one's company consumes energy; energy conservation measures can be profitable investments now, as well as prudent insurance for the future. This can be stated confidently, on the basis of the results of BP's in-house energy programme. The following describes this in more detail as an example of how BP set about monitoring and improving its own 'in-house' energy consumption.

BP spent about £114 million in 1973 on the energy consumed in operating its main companies. In today's money that would be about £240 million. That is a significant sum, accounting for 10% of the operating costs at that time. It would have been very foolish indeed not to devote serious effort to determining how it might be reduced, particularly as oil prices started to rise dramatically in the latter part of that year. In fact, at today's energy price, that £114 million would now be £684 million.

The major consumers were shipping and refining, accounting for more than two-thirds of total energy consumption. At the time, the chemical company already had a detailed energy monitoring programme. The reason for this lies in the very high usage per ton of product in chemical operations by comparison with oil refining and shipping. In the summer of 1973 were also started energy investigations in certain of the BP refineries, and identifying where it would be possible to reduce energy consumption. However, it was after the oil price rises of 1973 and 1974 that the need to direct new and greater efforts to reducing energy consumption became clear.

In late 1973, the previous Chairman of BP, Sir Eric Drake, requested all the subsidiary companies to review their energy consumption and take what steps were appropriate to reduce it. However, to ensure a sustained follow-through of effort throughout such a large group of companies spread around the world, it was later recognized that an initiative had to be taken at the centre, and at a senior management level. At the Head

Office, Dr Birks, then an Executive Director on the Board of BP Trading, the principal operating company, and now an Executive Director on the Main Board of BP, set up a committee comprising representatives of the functions which are principal energy consumers—such as refining and shipping. This was augmented with representatives from other departments who could advise on technical matters, such as engineering, and research and development. The BP Energy Conservation Committee, as it was called, was chaired by the General Manager of the Refining Department, and reported to Dr Birks. Mr Clement, Manager of BP's Refinery Design Division, was Secretary to the Committee.

The Committee had four principal functions. The first was to organize energy surveys and seminars to demonstrate clearly the need for action. The second was to encourage the subsidiaries to pursue more vigorously energy conservation measures. The third was to collect data on what was being done and planned by the various subsidiaries. The fourth function was to present an annual report on what had been achieved. This report went to the Executive Directors, the relevant Head Office Department General Managers, and to the General Managers of the refining and marketing subsidiaries.

Thus the committee provided a coordinating, advisory, and reporting role at a senior management level. The responsibility of carrying out energy conservation measures was devolved to the operating subsidiary companies. It is of course only the people with close working knowledge of the energy-consuming operations who are in a position to put energy-saving measures into effect quickly and effectively. But the attention devoted at Head Office to energy conservation, and the interest of the senior management, were important factors in motivating the operating companies to examine critically how they used energy, and to take the appropriate action.

As refining accounted for a third of the total energy consumption, one of the first actions was to set up a study team in 1973 to visit refineries and carry out energy surveys. Major energy conservation schemes were identified and a target set for 15% saving by 1980. Priorities and investment criteria were defined. Priority naturally focused on those measures which required no capital investment, and could be immediately implemented. Thus the main emphasis was initially placed on improved housekeeping and operational efficiency. In order to limit demand on capital at a time when BP was heavily committed with the Alaska and North Sea projects, and at the same time to ensure that the most cost-effective energy-saving schemes were adopted quickly, an investment guideline of two-year payback was adopted. This was relaxed in 1976, such that energy-saving schemes are now judged by the same economic criteria as other potentially profit-making investment. For every barrel of fuel oil saved through efficient energy use, there is an additional barrel of fuel oil to sell or process. In that respect it is a profit-earning operation.

Consider the position in 1977. BP's energy consumption cost £500 million, and was just under a quarter of our operating costs. This is a greater proportion of the operating costs than it was in 1973, reflecting the real increase in the cost of energy supplies. Refining accounted for nearly half, shipping a quarter, chemicals a sixth, and marketing a tenth. The relative proportions are quite different from what they were in 1973, but to provide a sound comparison, one should look at each sector individually.

In refining, a 14% saving has been achieved in 1977 over 1973 levels of consumption, and it is expected to attain 15% in 1978—two years ahead of target.

The chemical company has had a detailed energy monitoring system since before 1973. This has of course been continued, and as in refining, savings have been made through a continuation of improved operational efficiency, and investment in new equipment. Energy consumption in chemicals production is particularly susceptible to changes in feedstock quality, product mix, and plant throughput. Energy savings have consequently fluctuated from year to year, but 10% was achieved in 1976 over the 1973 level.

Energy consumption in shipping in 1977 was 37% less than the 1973 level. However, this is substantially due to a complex series of factors not related to energy conservation *per se*. For example, the replacement of older ships with new larger vessels, together with changes in the shipping pattern have reduced specific energy consumption. However, a contribution is the interesting development of the use of biocidal wax coatings in ship hulls to minimize marine growth and hence frictional drag, as one of several energy-saving measures.

In marketing and distribution, careful attention has been paid to equipment maintenance and vapour emissions, and reorganization of distribution networks to improve vehicle mileage. These measures have resulted in a 10% reduction in energy consumption compared to 1973.

It has not been possible to quantify energy savings in other activities, such as exploration and production. However, although the Head Office complex uses relatively small quantities of energy, the attention paid to better control of heating, ventilation, and lighting has resulted in a 50% reduction of fuel consumption compared to 1973. Yet the staff are not working by candlelight, wrapped in overcoats!

Expressing these savings in monetary terms, it is estimated that BP has saved about £94 million in 1977 compared to what might have been incurred had operations continued with the same degree of energy efficiency as in 1973. Excluding the savings in shipping which arose primarily for reasons other than energy conservation measures alone, the figure is about £50 million. This is a saving of 10% of the energy bill.

But what have been the financial implications of all this effort? Consider refining, as most of the savings have derived from this sector. Determining priority was relatively simple; initially attention was directed to those actions which would have immediate effect with little or no capital expenditure. Following in priority were those schemes whose capital investment was significant. At first a limit of a two-year payback period was imposed. This was subsequently relaxed when the most cost-effective measures were implemented. This approach was not particularly limiting in the scope for action, and the results have been quite revealing.

BP saved about £23 million in 1977 compared to 1973, from improvements in operational efficiency alone. A further £7 million per annum is being saved as a consequence of a capital investment of £8 million on energy conservation.

Schemes are now being studied with significant capital investment to achieve further energy savings. Planned investment and schemes still being evaluated could lead to expenditure on energy conservation in refining operations of more than £30 million by 1980. The potential additional savings would be about £33 million per year from 1980 onwards.

For the UK, expenditure on energy conservation could amount to nearly £15 million by 1980 if schemes currently being evaluated go ahead. The potential savings would be almost £8 million per annum. Energy conservation makes good business sense, and these figures clearly demonstrate this.

How relevant is the BP Group's experience to the average UK industrial company? The BP Group is composed of a large number of associates covering more than 60 countries throughout the world, some of quite significant size, others very small companies indeed. Responsibility for energy conservation has been devolved to each and every one of them, and the Group's achievement is only the sum of their individual achievements.

It is now felt that the essential role of the BP Energy Conservation Committee at Head Office, to motivate the subsidiaries in pursuing energy conservation measures, has been fulfilled and it has now formally stepped down. After having the active responsibility of carrying out conservation measures delegated to them, the operating companies have generated their own momentum, as the forward programme of investment illustrates. However, continuity in energy conservation matters is maintained through nominated personnel in each of the key departments.

In summary, BP takes energy conservation seriously. This derives both from the view of future energy supply and demand, and the fact that there are clear financial incentives in doing so. This will increasingly be the case in the next decade, when energy prices are likely to increase in real terms.

The question is not when to tackle energy consumption but how. The Government's 'Save-it' campaign had BP's support, but, for it to have the greatest effect, private industry must respond, as well as the domestic and public sectors. The Government exhorts, but ultimately it is industry who must act.

The responsibility of investigating and evaluating energy-saving measures, and of implementing them, must of course be delegated to those with detailed working knowledge of the energy-consuming systems. But in delegating responsibility, senior management must not abdicate it. Energy, like raw materials, equipment, and manpower, is a resource to be utilized efficiently and effectively. The responsibility for insisting and ensuring that it is done lies with a company's senior management.

Based on a speech by Sir David Steel,
Chairman of the British Petroleum Company,
to the National Energy Management Conference,
Birmingham Conference Centre, 10 October 1978

Energy Conservation and Thermal Insulation
Edited by R. Derricott and S. S. Chissick
© 1981 John Wiley & Sons Ltd.

CHAPTER 1

International energy management

Rolf Gradin

Head of Energy Conservation Division, International Energy Agency, Paris

INTRODUCTION

Energy conservation became an element of energy policy in most countries at the time of the oil crisis of 1973–74. The oil embargo led to the introduction of emergency measures, such as a ban on Sunday driving, which were intended to produce a drop in energy demand. The need for such measures was well understood by the man in the street; and public perception of the requirement for conservation was sharpened by the spectacular rise in oil prices.

Now the main factors have changed. The rate of increase in oil demand has dropped significantly, principally as a result of the economic recession which followed the events of 1973–74, and, prior to the Iranian crisis, oil was in plentiful supply. In addition, the price of oil had stabilized, or even, in certain industrial countries, dropped in real terms before OPEC decided in December 1978 to increase oil prices. Within national administrations, energy has ceased to be a matter for emergency action, and has taken its place alongside other concerns, such as stable economic growth, environment protection, and control of inflation, which must be balanced against each other. The public no longer perceives energy to be the major problem; and there are some who think that the unfettered market will provide sufficient incentives for consumers to take adequate measures to avoid any further shortfalls in energy supply.

ENERGY CONSERVATION—A CORNERSTONE IN ENERGY POLICIES

General

There has been a growing consensus in the Western World that world demand for oil will exceed the capacity or willingness of producers to sup-

1

ply oil during the 1980s and that prices then—or even earlier—will increase sharply. There has also been a slippage in nuclear power programmes throughout the world, the wider use of coal is difficult, and a large contribution from renewable energy sources is still not envisaged.

Although the difficulties are experienced in the areas of energy supply, it has however been understood that policies to strengthen energy supply alone cannot solve the problem. Furthermore, in many countries the public in general cannot be persuaded to accept policies promoting energy supplies unless it can be demonstrated that equally strong efforts are made in the area of conservation.

For the IEA area*, the crucial importance of energy conservation has been clearly demonstrated. When the IEA Energy Ministers met in October 1977 and decided to help alleviate the potential energy imbalances, to hold down oil imports to 26 million barrels a day (mb/d), approximately 1300 million tons of oil equivalent (mtoe), in 1985, the 12 principles of energy policy that they adopted included a Principle 4 on conservation, which states that member countries should strive for:

> Strong reinforcement of energy conservation, on a high priority basis with increased resources, for the purpose of limiting growth in energy demand relative to economic growth, eliminating inefffi-cient energy use, especially of rapidly depleting fuels, and encouraging substitution for fuels in shortest supply, by implementing vigorous conservation measures in various sectors along lines which include the following elements:
> — pricing policies (including fiscal measures) which give incentives to conservation;
> — minimum energy efficiency standards;
> — encouragement and increase of investments in energy saving equipment and techniques.

Considerations, of an economic, environmental, social, and safety character, have often spoken in favour of energy conservation. It has, in particular, been pointed out that energy saved does not, as a rule, pollute the water, the air or the soil, and that energy conservation thus contributes to the quality of life; and also that energy conservation can lead to employment opportunities and to a more energy-efficient and, therefore, more competitive economy.

---

*Austria, Belgium, Canada, Denmark, West Germany, Greece, Ireland, Italy, Japan, Luxembourg, the Netherlands, New Zealand, Norway, Spain, Sweden, Switzerland, Turkey, the United Kingdom and the United States of America are the member countries of the IEA treated in this report. Australia has recently joined the Agency but is not yet included in the statistics.

Energy conservation measures

In most IEA countries, combinations of 'sticks and carrots' are a main feature of conservation programmes. Due to differences in underlying philosophies and socioeconomic climates, however, some countries place the major emphasis on voluntary measures for the promotion of conservation, whereas others employ primarily regulatory measures.

The most important measures that have been adopted so far in IEA countries to promote energy conservation in *industry* are:

(a) information and motivation—there have been campaigns and advice services, especially for small- and medium-sized industries;

(b) incentives, such as grants and tax reductions for energy-saving equipment and processes; and

(c) target setting and reporting for energy-intensive industries.

In the *transport* sector, the most important conservation measures so far adopted are:

(a) heavy taxes on petrol and progressive taxation on cars according to weight or engine size in most IEA countries; and

(b) mandatory fuel-efficiency targets.

Promotion of public transport by increased public funding, publicity campaigns, and speed limits on roads also help to save energy in this sector.

The important measures taken in the *residential/commercial* sector include:

(a) building codes laying down minimal thermal efficiencies for new buildings; and

(b) incentive schemes for retrofitting existing buildings.

Publicity campaigns, improvement of heating and cooling equipment, and the labelling of appliances also contribute to the saving of energy in this sector.

Except in a few countries, district heating and combined production of heat and electricity do not yet play a major role. The measures so far adopted to promote district heating are subsidies, loans or tax reduction to help finance district heating networks; and urban planning, designating particular areas for district heating.

Energy conservation results

IEA total consumption of primary energy in 1977 was 3348 mtoe. This corresponds to an increase of 0.8% per annum between 1973, 3247 mtoe, and

Table 1 Energy consumption in IEA member countries, 1960–77, in million tons of oil equivalent (mtoe), total primary energy (TPE), total final consumption (TFC), and final consumption in the three consumption sectors. (Source: OECD, 1979)

	1960	1968	1973	1974	1975	1976	1977	Annual growth (%) 1960–73	Annual growth (%) 1973–77	Share of TFC (%) 1960	Share of TFC (%) 1973	Share of TFC (%) 1977
Austria												
TPE	12.38	17.37	23.54	23.93	23.27	24.27	24.80	5.1	1.8	100	100	100
TFC	8.41	12.69	18.05	17.90	17.51	18.52	17.90	6.1	-0.3	51	49	41
Industry	4.32	6.10	8.77	8.75	7.69	8.15	7.44	5.2	-5.3	18	20	21
Transport	1.48	2.47	3.68	3.39	3.46	3.54	3.70	7.3	0.2	31	31	38
Res/Comm	2.61	4.12	5.60	5.76	6.36	6.83	6.76	6.1	6.5			
Belgium												
TPE	25.38	36.42	46.74	45.38	41.83	44.76	44.54	4.8	-1.2	100	100	100
TFC	18.30	28.29	37.48	35.89	33.23	35.21	34.78	5.7	-1.9	48	51	47
Industry	8.73	13.62	19.27	19.02	15.32	17.31	16.16	6.3	-4.3	14	14	16
Transport	2.55	3.77	5.05	4.77	5.06	5.34	5.61	5.4	2.7	38	35	37
Res/Comm	7.02	10.90	13.16	12.10	12.85	12.56	13.01	4.9	-0.3			
Canada												
TPE	96.05	139.03	185.70	189.50	193.31	197.02	203.36	5.2	2.3	100	100	100
TFC	62.67	96.99	125.54	130.57	125.95	132.87	135.85	5.5	2.0	39	37	37
Industry	24.32	35.85	46.05	49.63	44.00	46.94	49.81	5.0	2.0	25	27	29
Transport	16.01	24.12	33.85	35.11	36.15	38.91	39.49	5.9	3.9	36	36	34
Res/Comm	22.34	37.02	45.64	45.83	45.80	47.02	46.55	5.6	0.5			
Denmark												
TPE	8.99	16.34	19.72	17.82	17.80	19.37	19.92	6.2	0.3	100	100	100
TFC	7.11	13.28	16.23	14.02	14.29	15.68	15.80	6.5	-0.7	30	25	23
Industry	2.11	3.03	4.08	3.29	3.15	3.49	3.66	5.2	-2.7	18	22	23
Transport	1.29	2.72	3.52	3.20	3.49	3.55	3.62	8.0	0.7	52	53	54
Res/Comm	3.71	7.53	8.63	7.53	7.65	8.64	8.52	6.7	-0.3			

										%	%	%
West Germany												
TPE	145.78	201.81	265.82	260.26	243.12	263.73	261.09	4.7	-0.4		100	100
TFC	102.10	150.57	203.78	196.63	184.03	191.03	191.78	5.5	-1.5	100	51	45
Industry	51.75	72.10	92.23	92.81	79.85	80.42	79.94	4.5	-3.5	42	15	16
Transport	15.84	23.40	32.88	31.16	32.96	33.15	35.10	5.8	1.7	18	34	39
Res/Comm	34.54	55.07	78.67	72.66	71.22	77.46	76.74	6.5	-0.6	40		
Greece												
TPE	2.76	6.76	12.24	11.63	12.11	13.11	14.26	12.1	3.9		100	100
TFC	2.28	4.78	9.02	8.11	8.46	9.43	9.83	11.2	2.2	100	37	41
Industry	0.84	1.76	3.70	3.47	3.69	3.97	4.03	12.0	2.2	41	39	29
Transport	0.88	1.89	2.62	2.14	2.13	2.91	3.31	8.8	6.0	34	24	30
Res/Comm	0.56	1.13	2.70	2.50	2.64	2.55	2.49	12.9	-2.0	25		
Ireland												
TPE	4.24	5.88	7.62	7.45	7.00	7.17	7.59	4.6	-0.1		100	100
TFC	2.66	4.15	5.46	5.71	5.31	5.43	5.82	5.7	1.6	100	27	28
Industry	0.73	1.24	1.53	2.20	1.59	1.54	1.86	5.9	5.0	32	17	26
Transport	0.44	0.82	1.41	1.43	1.44	1.51	1.58	9.4	2.9	27	56	46
Res/Comm	1.49	2.09	2.52	2.08	2.28	2.38	2.38	4.1	-1.4	41		
Italy												
TPE	49.78	94.15	132.59	133.62	127.39	136.41	138.58	7.8	1.1		100	100
TFC	35.05	72.47	100.97	101.66	98.34	106.85	103.54	8.5	0.6	100	53	49
Industry	18.42	37.83	49.29	50.70	46.66	52.23	50.09	7.9	0.4	48	19	19
Transport	6.65	14.76	19.56	19.09	19.20	18.42	20.92	8.7	1.7	20	28	32
Res/Comm	9.98	19.88	32.12	31.87	32.48	36.20	32.53	9.4	0.3	32		
Japan												
TPE	94.66	216.75	337.77	337.45	331.95	343.08	350.46	10.3	0.9		100	100
TFC	58.60	148.90	243.16	238.04	230.41	252.99	254.21	11.6	1.6	100	62	57
Industry	36.22	90.39	137.74	138.86	129.81	151.67	149.21	10.8	2.0	59	20	16
Transport	11.70	26.53	39.67	39.83	41.44	41.63	43.22	9.8	2.2	17	18	27
Res/Comm	10.68	31.98	65.75	59.35	59.16	59.69	61.78	15.0	-1.6	24		

Table 1 (Continued)

	1960	1968	1973	1974	1975	1976	1977	Annual growth (%)		Share of TFC (%)		
								1960–73	1973–77	1960	1973	1977
Luxembourg												
TPE	3.27	3.94	4.76	5.06	3.99	4.09	3.84	2.9	−5.2			
TFC	2.72	3.41	4.10	4.43	3.59	3.61	3.46	3.2	−4.1	100	100	100
Industry	2.33	2.83	3.22	3.62	2.72	2.67	2.47	2.5	−6.4	86	79	71
Transport	0.13	0.17	0.29	0.29	0.35	0.37	0.38	6.4	7.0	5	7	11
Res/Comm	0.26	0.41	0.59	0.52	0.52	0.57	0.61	6.5	0.3	9	14	18
Netherlands												
TPE	21.94	39.47	61.67	60.98	59.09	65.03	63.51	8.3	0.7			
TFC	15.61	30.35	49.67	48.92	48.33	53.70	52.44	9.3	1.4	100	100	100
Industry	6.12	12.88	21.80	22.95	20.23	22.83	22.83	10.3	1.2	39	44	44
Transport	2.81	5.23	7.45	6.98	7.44	8.24	8.67	7.8	3.9	18	15	16
Res/Comm	6.68	12.24	20.42	18.99	20.66	22.63	20.94	9.0	0.6	43	41	40
New Zealand												
TPE	5.41	7.42	9.45	10.02	10.14	10.75	11.30	4.4	4.6			
TFC	3.52	4.83	6.52	6.55	6.50	6.72	6.83	4.8	1.2	100	100	100
Industry	1.20	1.17	1.89	2.64	2.29	2.28	2.40	3.6	6.1	34	29	35
Transport	1.12	1.83	2.25	2.30	2.15	2.24	2.49	5.5	2.5	32	35	37
Res/Comm	1.20	1.83	2.38	1.61	2.06	2.20	1.94	5.4	−5.0	34	36	28
Norway												
TPE	8.98	15.93	19.74	19.59	19.83	20.70	19.82	6.2	0.1			
TFC	6.64	11.26	13.82	13.57	13.73	14.23	13.82	5.8	±0	100	100	100
Industry	3.24	5.83	7.12	7.31	7.05	7.14	6.86	6.2	−0.9	49	52	50
Transport	1.33	2.17	2.52	2.44	2.85	2.80	2.59	5.0	0.7	20	18	19
Res/Comm	2.07	3.26	4.18	3.82	3.83	4.29	4.37	5.5	1.1	31	30	31

Spain

TPE	19.90	36.83	56.61	59.87	62.10	65.73	67.90	8.4	4.6	100	100	100
TFC	13.27	23.88	43.42	42.16	43.63	47.02	49.22	9.5	3.2			
Industry	5.85	13.26	24.61	22.90	23.11	24.28	25.26	11.7	0.6	44	57	51
Transport	3.96	6.12	11.26	11.16	12.26	13.66	13.86	8.4	5.3	30	26	28
Res/Comm	3.46	4.50	7.55	8.10	8.26	9.08	10.10	6.2	7.5	26	17	21

Sweden

TPE	27.24	40.14	47.10	44.91	47.15	50.25	50.30	4.3	1.6	100	100	100
TFC	19.25	30.04	36.03	33.65	35.17	36.85	35.33	5.0	-0.5			
Industry	9.93	13.57	16.72	16.48	16.34	16.24	14.92	4.1	-2.8	52	46	42
Transport	2.78	4.41	5.22	5.23	5.46	5.84	6.07	5.0	3.9	14	15	17
Res/Comm	6.54	12.06	14.09	11.94	13.37	14.77	14.34	6.1	0.4	34	39	41

Switzerland

TPE	11.80	18.00	23.39	22.32	22.60	22.32	24.50	5.4	1.2	100	100	100
TFC	6.83	12.29	17.36	16.23	15.60	16.21	16.60	7.4	-1.1			
Industry	2.60	3.82	4.80	4.51	4.16	4.03	4.24	4.8	-2.9	38	28	26
Transport	1.54	3.04	4.29	4.12	3.93	3.94	4.20	8.2	-0.5	23	25	25
Res/Comm	2.69	5.43	8.27	7.60	7.51	8.24	8.16	9.0	-0.3	39	47	49

Turkey

TPE	9.0	17.8	24.49	26.26	27.15	29.43	32.00	8.0	6.9	100	100	100
TFC	4.0	13.7	20.59	21.04	22.40	24.14	26.24	13.4	6.3			
Industry	1.1	4.0	4.14	4.15	4.70	5.66	6.04	10.7	9.9	27	20	23
Transport	1.5	4.8	4.52	4.65	5.36	5.80	6.87	8.9	11.0	38	22	26
Res/Comm	1.4	4.9	11.93	12.24	12.34	12.68	13.33	17.9	2.8	35	58	51

United Kingdom

TPE	169.69	201.05	223.96	214.52	203.06	206.51	211.50	2.2	-1.4	100	100	100
TFC	125.03	141.14	154.32	147.32	139.81	143.53	146.72	1.6	-1.3			
Industry	55.00	65.03	69.67	64.57	57.97	60.34	60.46	1.8	-3.5	44	45	41
Transport	21.11	24.76	30.84	29.73	29.37	30.46	31.26	3.0	0.3	17	20	21
Res/Comm	48.92	51.35	53.81	53.02	52.73	52.47	55.00	0.8	0.6	39	35	38

Table 1 (Continued)

	1960	1968	1973	1974	1975	1976	1977	Annual growth (%)		Share of TFC (%)		
								1960–73	1973–77	1960	1973	1977
USA												
TPE	1014.18	1412.78	1744.14	1704.26	1666.92	1761.07	1799.22	4.3	0.8			
TFC	810.15	1108.17	1306.62	1263.09	1213.16	1293.01	1309.17	3.7	0.1	100	100	100
Industry	309.80	429.90	467.47	445.28	393.05	429.83	429.84	3.2	−2.1	38	36	33
Transport	234.26	324.23	406.48	399.86	412.13	423.46	437.14	4.3	1.9	29	31	33
Res/Comm	266.09	354.04	432.67	417.95	407.98	439.72	442.19	3.8	0.6	33	33	34
IEA total												
TPE	1731.43	2527.87	3247.05	3194.83	3114.81	3282.80	3348.49	5.0	0.8			
TFC	1304.21	1905.80	2412.14	2345.49	2259.45	2407.03	2429.34	4.8	0.2	100	100	100
Industry	544.58	812.59	984.10	963.14	863.38	941.02	937.52	4.6	−1.2	42	41	39
Transport	327.42	475.42	622.36	606.88	626.63	645.77	670.08	5.1	1.9	25	26	27
Res/Comm	432.21	617.79	810.68	775.47	769.44	820.24	821.74	5.0	0.4	33	33	34

1977. Compared with an historical growth of 5.0% per annum between 1968 and 1973, there has thus been a dramatic slowdown of energy demand growth in IEA countries.

There are many reasons for this development, e.g. the economic recession. GDP growth decreased from 4.5% per annum between 1968 and 1973 to 2.1% per annum from 1973 to 1977. Because of the relationship between GDP and energy demand, this factor alone had a strong dampening effect on energy consumption.

It can, however, also be noted that total primary energy (TPE) requirements grew less quickly in relation to GDP since 1973 than in the past. The TPE/GDP ratio (tons of oil equivalent per thousand US dollars), which increased from 1.45 in 1960 to 1.48 in 1973, has now started to decline, to 1.41 in 1977. The TPE/GDP elasticity (ΔTPE/ΔGDP, that is percentage change in TPE for a 1% change in GDP), which was more than unity before 1973 has thus dropped to below unity for the period 1973–77. This can be interpreted as an overall improvement in energy efficiency, that is as a response to higher energy prices, and to specific energy conservation measures, which have been introduced.

From a sectoral analysis (Table 1), it can be seen that the industrial energy consumption in the IEA was in fact lower in 1977 than in 1973, 938 mtoe compared to 984 mtoe. The industrial sector, of course, has been influenced by the overall economy; the industrial output, characterized by the total industrial production index, 1970=100, only grew from 119 in 1973 to 124 in 1977. The energy consumption in the residential/commercial sector was slightly larger in 1977 than in 1973, 822 mtoe compared to 811 mtoe. The highest growth in energy consumption has been experienced in the transport sector, growing from 622 mtoe in 1973 to 670 mtoe in 1977.

Industry is the most important energy-consuming sector in the IEA. Due to the actual decrease of consumption in this sector since 1973, compared to a continued increase in the other sectors, in particular for transport, the share for industrial consumption has, however, decreased from 42% in 1973 to 39% in 1977. The residential/commercial sector's share has remained constant, 33–34%.

It can further be seen from the detailed analysis that total primary energy has been growing faster than total final consumption, especially for the period after 1973. This means that energy conversion losses, in the production of electricity and gas from primary energy, and in refineries, have in fact grown from 24.7% in 1960 to 25.7% in 1973 to 27.6% in 1977. This follows from a gradual shift to more refined energies at the users' end.

Table 2 Energy demand projections for IEA member countries 1985 and 1990, in mtoe. (Source: IEA, 1978)

	1977	1985	1990	Annual growth (%)		Share of TFC (%)		
				1985–77	1990–85	1977	1985	1990
Austria								
TPE	23.3	32.8	38.1	4.4	3.1	100	100	100
TFC	17.0	23.5	27.4	4.1	3.1	40	38	37
Industry	6.8	9.0	10.0	3.5	2.2	24	22	20
Transport	4.1	5.1	5.6	2.8	1.9	36	40	43
Res/Comm	6.1	9.4	11.8	5.6	4.7			
Belgium								
TPE	45.2	57.0	64.0	3.0	2.3	100	100	100
TFC	35.8	43.9	49.3	2.7	2.3	51	48	49
Industry	18.2	21.8	24.9	2.3	2.7	15	18	20
Transport	5.3	7.3	8.9	4.1	3.9	34	34	31
Res/Comm	12.3	14.8	15.5	2.4	0.9			
Canada								
TPE	201.8	262.0	297.5	3.3	2.6	100	100	100
TFC	138.5	174.0	193.7	2.9	2.2	38	42	43
Industry	52.6	73.8	82.7	4.3	2.3	28	26	23
Transport	39.0	45.3	48.1	1.7	1.2	34	32	34
Res/Comm	46.9	54.9	62.9	2.0	2.8			
Denmark								
TPE	20.4	20.7	23.9	0.2	2.9	100	100	100
TFC	16.4	17.2	19.7	0.6	2.8	22	26	26
Industry	3.6	4.4	5.1	2.5	3.0	21	24	25
Transport	3.5	4.2	4.9	2.3	3.1	57	50	49
Res/Comm	9.3	8.6	9.7	-0.9	2.4			

West Germany								
TPE	260.7	337.8	371.0	3.3	1.9	100	100	100
TFC	195.8	243.4	259.4	2.8	1.3			
Industry	82.9	105.4	114.5	3.1	1.7	42	43	44
Transport	35.8	39.6	42.0	1.3	1.2	18	16	16
Res/Comm	77.1	98.4	102.9	3.1	0.9	40	41	40
Greece								
TPE	13.8	24.4	31.7	7.4	5.4	100	100	100
TFC	9.7	16.2	20.2	6.6	4.5			
Industry	4.1	7.2	8.9	7.3	4.3	42	44	44
Transport	3.1	4.5	5.3	4.8	3.3	32	28	26
Res/Comm	2.5	4.5	6.0	7.6	5.9	26	28	30
Ireland								
TPE	7.4	13.7	17.0	8.0	4.4	100	100	100
TFC	5.0	9.3	11.3	8.1	4.0			
Industry	1.5	4.8	6.4	15.7	5.9	30	52	57
Transport	1.6	2.3	2.7	4.6	3.3	33	25	24
Res/Comm	1.9	2.2	2.2	1.9	± 0	37	23	19
Italy								
TPE	134.9	193.0	235.0	4.6	4.0	100	100	100
TFC	101.6	138.1	160.0	3.9	3.0			
Industry	46.8	64.3	75.0	3.6	3.1	46	47	47
Transport	21.3	27.9	31.9	3.4	2.7	21	20	20
Res/Comm	33.5	45.9	53.1	4.0	2.9	33	33	33
Japan								
TPE	348.1	560.5	674.7	6.1	3.8	100	100	100
TFC	248.8	391.6	466.4	5.8	3.6			
Industry	139.6	228.1	271.9	6.3	3.6	56	58	58
Transport	52.2	72.0	83.2	4.1	2.9	21	19	18
Res/Comm	57.0	91.5	111.3	6.1	4.0	23	23	24

Table 2 (Continued)

	1977	1985	1990	Annual growth (%) 1985–77	Annual growth (%) 1990–85	Share of TFC (%) 1977	Share of TFC (%) 1985	Share of TFC (%) 1990
Luxembourg								
TPE	3.77	4.55	4.99	2.4	1.9			
TFC	3.45	4.25	4.67	2.6	1.9	100	100	100
Industry	2.46	2.96	3.18	2.4	1.5	69	70	72
Transport	0.36	0.51	0.58	4.4	2.6	11	12	11
Res/Comm	0.63	0.78	0.91	2.7	3.1	20	18	17
Netherlands								
TPE	63.3	90.1	98.8	4.5	1.9			
TFC	51.5	71.6	80.2	4.2	2.3	100	100	100
Industry	21.6	32.1	37.5	5.1	3.2	42	45	47
Transport	8.0	11.0	11.5	4.1	0.9	15	15	14
Res/Comm	21.9	28.5	31.2	3.3	1.8	43	40	39
New Zealand								
TPE	11.1	15.4	18.2	4.2	3.4			
TFC	7.0	9.1	10.9	3.3	3.7	100	100	100
Industry	2.7	3.9	4.8	4.7	4.3	39	43	44
Transport	2.5	2.8	3.2	1.4	2.7	36	31	29
Res/Comm	1.8	2.4	2.9	3.7	3.8	25	26	27
Norway								
TPE	20.6	27.5	31.0	3.7	2.4			
TFC	14.1	20.3	22.9	4.7	2.5	100	100	100
Industry	6.1	9.3	10.6	5.4	2.7	43	46	46
Transport	3.0	4.2	4.4	4.3	0.9	21	21	19
Res/Comm	5.0	6.8	7.9	3.9	3.0	36	33	34

Spain								
TPE	69.4	94.0	114.2	4.0	3.9	100	100	100
TFC	50.0	65.8	79.3	3.5	3.8	100	100	100
Industry	27.0	34.7	41.7	3.2	3.7	54	53	53
Transport	13.6	16.7	18.5	2.6	2.1	27	25	23
Res/Comm	9.4	14.4	19.1	5.5	5.8	19	22	24
Sweden								
TPE	48.7	58.4	63.2	2.3	1.6	100	100	100
TFC	33.9	39.6	41.0	2.0	0.7	100	100	100
Industry	14.1	19.2	20.6	3.9	1.4	42	48	50
Transport	6.0	6.7	6.6	1.4	-0.3	18	17	16
Res/Comm	13.8	13.7	13.8	-0.1	0.1	40	35	34
Switzerland								
TPE	23.4	26.7	30.1	1.6	2.4	100	100	100
TFC	16.4	18.5	20.5	1.5	2.1	100	100	100
Industry	4.9	5.0	5.7	0.3	2.6	30	27	27
Transport	4.0	4.9	5.2	2.6	1.2	24	27	26
Res/Comm	7.5	8.6	9.6	1.7	2.2	46	46	47
Turkey								
TPE	31.9	69.4	91.2	10.2	5.6	100	100	100
TFC	26.2	57.5	76.4	10.3	5.6	100	100	100
Industry	6.0	23.2	33.6	18.4	7.7	23	40	44
Transport	6.9	9.6	14.3	4.2	8.3	26	17	19
Res/Comm	13.3	24.7	28.5	8.0	2.8	51	43	37
United Kingdom								
TPE	211.4	245.0	269.0	1.9	1.9	100	100	100
TFC	150.1	170.0	185.0	1.6	1.7	100	100	100
Industry	64.5	81.0	92.0	2.9	2.6	43	48	50
Transport	31.3	36.0	38.0	1.8	1.1	21	21	20
Res/Comm	54.3	53.0	55.0	-0.3	0.7	36	36	30

Table 2 (Continued)

	1977	1985	1990	Annual growth (%) 1985–77	Annual growth (%) 1990–85	Share of TFC (%) 1977	Share of TFC (%) 1985	Share of TFC (%) 1990
USA								
TPE	1795.7	2263.3	2605.3	3.0	2.9	100	100	100
TFC	1311.5	1536.5	1739.3	2.0	2.5	33	38	41
Industry	421.9	591.2	719.2	4.3	4.0	34	31	30
Transport	443.6	471.9	521.4	0.8	2.0	33	31	29
Res/Comm	446.0	473.4	498.7	0.7	1.1			
IEA total								
TPE	3334.8	4396.3	5078.9	3.5	2.9	100	100	100
TFC	2432.8	3050.4	3467.5	2.9	2.6	38	43	45
Industry	927.3	1321.4	1568.4	4.5	3.5	28	25	25
Transport	685.2	772.5	856.2	1.5	2.1	34	32	30
Res/Comm	820.3	956.5	1042.9	1.9	1.7			

Energy forecasts

Forecasts submitted for the 1978 review of IEA countries' energy programmes (Table 2) indicate that total primary energy demand for IEA countries is expected to increase by an average of 3.5% per annum from 1977 to 1985 and by 2.9% per annum between 1985 and 1990. The key factor for these forecasts is the underlying assumption concerning the economic growth. On average, GDP has been expected to grow by 4.2% per annum in the IEA between 1977 and 1985, and by 3.6% per annum between 1985 and 1990. It is thus believed that there will be a further decrease of the TPE/GDP ratio, that is a continued improvement of overall efficiency in energy use. The TPE/GDP ratio (1.41 in 1977) will decrease to 1.40 in 1985 and further to 1.38 in 1990.

Industrial energy demand is expected to grow faster than demand in other sectors, an average of 4.5% per annum between 1977 and 1985 and of 3.5% per annum between 1985 and 1990 (Table 2). Energy demand in the transport sector is expected to increase by 1.5% per annum from 1977 to 1985, and then by 2.1% per annum up to 1990. The residential/commercial sector is expected to increase by only 1.9% per annum between 1977 and 1985 and by 1.5% per annum from 1985 to 1990. As a consequence, the share of the residential/commercial sector in total final consumption will decrease from 33.7% in 1977 to 31.3% in 1985 and to 28.5% in 1990.

Most IEA countries have in their energy forecasts taken account of energy conservation measures in place. However, they lack the experience in assessing the effects of conservation measures, which might therefore have been underestimated. Also, many countries have not specifically evaluated the effects of higher energy prices on energy demand, or just simply assumed energy prices to remain constant in real terms, at least up to 1985. As a conclusion, there has probably been a tendency to submit demand forecasts that are too high. This is underlined by the fact that GDP growth targets often seem to be unrealistically high.

ENERGY CONSERVATION IN THE RESIDENTIAL/COMMERCIAL SECTOR

The main responsibility for energy conservation, as well as for all other aspects of energy policy, rests with national governments. It is, however, recognized in the IEA that energy conservation can be strengthened by international cooperation, because:

(a) individual country's programmes can gain from international exchange of information and practical experience made in other countries;

(b) all countries do have a considerable potential to save energy which can be harnessed on economically profitable terms;

(c) all countries need to take the same measures, broadly speaking, to promote energy conservation;

(d) energy conservation is an area with less controversial views than energy supply; and

(e) international cooperation could prove particularly rewarding in this area, as strong measures may be more palatable to the extent they are adopted by many countries.

In the IEA, energy conservation is developed within the framework of the Long-Term Co-operation Programme. A Sub-Group on Energy Conservation, reporting to the Standing Group on Long-Term Co-operation, provides the opportunity for exchange of information on conservation and also acts to guide the IEA Secretariat's work with analysis of conservation potentials and results.

For detailed work on energy conservation, five Expert Groups have been established under the aegis of the Conservation Sub-Group to study energy conservation in major demand sectors (industry, transport, building sector) and in the areas of district heating and public information.

The findings of the Expert Group on Energy Conservation in Buildings, with Canada acting as lead country, form a basis for action on energy conservation in the Agency and in the Agency countries. The Expert Group on Energy Conservation in Buildings has *inter alia* looked into the constraints to energy conservation and identified the measures that could be taken to overcome these constraints and enhance energy conservation in the building sector.

The Expert Group on District Heating and Combined Production of Heat and Electricity (lead country, Denmark) is also involved in questions of energy conservation relevant to the residential and commercial sector.

Constraints to energy conservation

Energy conservation in the residential and commercial sector is hampered by financial, technical, and institutional difficulties. A lack of general public awareness of the energy situation and of the potential for energy conservation is also a serious problem. The economic recession has furthermore hindered a rapid development of conservation programmes in most countries.

Retrofitting the housing stock of a nation requires considerable financing. In many countries, governments have problems to provide incentives for retrofitting existing houses, particularly where firm commitments beyond any one fiscal year are difficult. Financial problems also exist for countries with moderate public fundings, but which have programmes where matching funds are needed from local authorities. Existing tax systems, rent control regulations, and building codes very often conflict with current goals of energy conservation. Many of these regulations were established at a time when energy efficiency was not a primary concern, and must now be reviewed in detail as a prerequisite for progress in this area of conservation.

Trained advisers and installers are often required if actions are to be wisely chosen and properly implemented. There is in some instances an inadequate number of these individuals, and training programmes need to be established. In some countries the quality of installation material has not always been high enough to meet the requests.

Different patterns of house ownership have in particular been identified as a major constraint, as the burden of the retrofitting cost does not necessarily fall on the consumer. For example, public sector housing costs are attributable to housing authorities, but the benefits of retrofitting go to the tenants. In the private sector, problems arise when the expense of retrofitting may not be recovered in the selling price.

Owners of apartment houses might not be interested in investing in retrofitting, when tenants pay the heating costs. Statutory or tax measures are therefore needed. Low-income groups do not have the money to make energy-saving investments, even if they would benefit most from conservation.

The development of district heating and combined production of heat and electricity is hampered by economic, institutional, and environmental problems.

The establishment of extensive district heating networks, and the corresponding heat production plants, calls for large amounts of investment money. This might cause problems, especially as district heating is usually looked upon as an industrial matter, with no access to special funds devoted to residential buildings. In addition, district heating has to compete with other heating systems. The consumer himself, when making his choice, normally will take only his own costs into consideration. Common costs for the society, e.g. environmental costs and costs for energy supply in emergency situations, will be observed only if they are included in the energy costs—which normally is not the case.

Conservation measures already implemented

Public awareness and the availability of information on energy conservation measures is recognized by all IEA countries as vital to the success of an effective energy programme in the residential/commercial sector. Information alone is, however, often not enough to motivate the consumer

Table 3 Implementation in the IEA of energy conservation measures in the residential and commercial sector

	Fiscal/financial incentives					Building codes								Appliance efficiency					
	Taxes	Removal of sales tax	Discount on taxable income	Subsidy/grant	Loan	New homes: federal	New homes: local	Existing homes: federal	Existing homes: local	Federal/public buildings	Max. temp: air	Max. temp: water	Prohibition of bulk meter	Energy label: mandatory	Energy label: voluntary	Standard: federal	Standard: local	Maintenance	Others
Austria	—	—	—	X	—	X	—	—	—	X	—	—	—	—	X	—	—	—	—
Belgium	—	X	—	X	—	X	—	—	—	X	X	—	—	X	—	—	—	X	—
Canada	—	—	P	X	X	X	—	X	—	X	—	X	X	X	—	P	—	P	—
Denmark	—	—	X	X	—	X	—	X	—	—	X	—	—	—	X	P	—	—	—
West Germany	—	—	X	X	X	X	—	X	—	X	—	—	—	—	P	P	—	—	—
Greece	—	—	—	—	—	—	—	—	—	—	X	—	—	—	P	—	—	—	—
Ireland	—	—	—	—	—	X	—	X	—	—	X	—	—	—	P	—	—	—	—
Italy	—	—	—	—	—	X	—	X	—	P	X	—	—	—	X	—	—	—	—
Japan	—	—	—	X	—	—	—	—	—	—	—	—	—	—	—	—	—	P	—
Luxembourg	—	—	—	—	X	X	—	—	—	—	—	—	—	—	—	—	—	—	—
Netherlands	X	—	—	X	X	P	—	—	—	—	—	—	—	—	—	—	—	P	—
New Zeland	X	—	—	—	—	X	—	X	—	P	—	—	—	P	—	—	—	—	X
Norway	—	—	—	X	X	X	—	X	—	X	X	—	—	—	—	—	—	P	—
Spain	—	—	X	—	—	P	—	—	—	P	P	—	P	P	—	—	—	—	—
Sweden	—	—	—	X	—	X	—	X	—	X	—	—	—	—	—	—	—	X	—
Switzerland	—	—	—	X	—	X	X	—	X	—	—	—	—	—	—	—	—	—	—
Turkey	—	—	—	X	—	P	—	P	—	—	—	—	—	—	—	—	—	—	—
United Kingdom	—	—	—	X	X	X	—	—	—	X	—	—	—	X	—	P	—	—	—
USA	—	—	P	X	P	X	P	—	—	—	—	—	—	—	—	P	—	—	—

X existing measures; P planned measures; — no measures.

to conserve energy, and financial or fiscal incentives in the form of taxes, discount on taxable income, subsidies, grants or loans are provided in most countries. Programmes vary considerably in scope and funding between the various IEA member countries (Table 3).

Financial and fiscal incentives

Among financial and fiscal incentives, subsidies and grants are most commonly used though of widely varying scope. It has, however, been noticed in many countries that grant (or loan) programmes can result in very high costs of energy savings. Grants and subsidies range from very limited budgets and small incentives to considerable amounts of public funds made available (e.g. in Denmark, West Germany, the Netherlands, Sweden, and the United Kingdom). In general, to be sufficiently attractive, a subsidy level of at least 20% appears to be required.

Loan schemes are employed by about half of those countries which provide incentives. The payback period for loans ranges from four years interest-free for the purchase of appliances to 15 to 20 years with, for most countries, an interest rate of about $4\frac{1}{2}$% for longer-term projects.

Tax credits or discounts on taxable income for the purchase of energy-conserving equipment or insulation materials is, in general, preferred to using a direct tax applied to high energy-consuming appliances or equipment.

The effectiveness of all these programmes often is still uncertain. The programmes are, in general, new and there has not been sufficient time to receive strong feedback on their impact. A long lead time is often required before significant changes can be achieved in this sector; for an effective penetration of the market, at least five to ten years are necessary. In some countries, e.g. the United States of America, Canada, West Germany, and Sweden, incentives are also provided to promote the use of renewable energy, e.g. solar collectors and heat pumps.

Building codes

Building codes for all new homes with stipulations also for energy are very common for new homes in the IEA. The codes generally apply to insulation of outer ceilings, outer walls, ground floors, and windows. They are mandatory only in Denmark, West Germany, Italy, the Netherlands, Norway, Sweden, and the United Kingdom. Six countries have guidelines or voluntary standards also for existing homes.

As an alternative to the traditional formulation of building codes, with minimum requirements for the various assemblies of the building envelope, a stated combined energy efficiency for the whole house is also possible.

With this energy budget-type code, the consumption for heating, ventilation, hot water, cooling, electrical appliances, lighting, and communications would all be taken into account. The limits would be designed according to the size, presupposed use, and, possibly, the shape of the building.

Appliance efficiency and energy labelling

Information by means of labelling of household appliances, air conditioners, etc., is regarded as an important element of consumer policy and as an energy conservation measure both to encourage consumers to buy energy efficient products and to encourage producers to manufacture appliances above the minimum standard. Energy consumption labels must, however, be based on adequate internationally tested methods which measure performance under agreed standardized conditions. Energy efficiency standards now exist in seven countries, Canada, Denmark, West Germany, Greece, Italy, the United Kingdom, and the United States of America. Some other countries are also considering their introduction. Energy efficiency labelling for consumer appliances exists or is planned in seven countries. Canada, Spain, and the United States of America have enforced mandatory labelling schemes for major consumer appliances. In Austria, West Germany, Italy, Japan, and the United States of America voluntary labelling on a limited number of electrical appliances has been carried out. The EEC is giving consideration to Community-wide action in this field.

District heating and combined production of heat and electricity

So far, district heating covers an essential part of total residential and commercial energy use in only two IEA countries, in Denmark about 30% and in Sweden about 15% in 1977. In many countries, there is evidently no district heating worth mentioning. This is mainly due to the geographical position and climate, e.g. in Greece, Italy, and Spain. In three countries, district heating has not come into great use because of competing energy distribution systems, natural gas in the Netherlands and the United Kingdom, and electricity in Norway. In the remaining countries, there is, so far, limited use of district heating, often only a few local networks in densely populated areas.

About half of the heat for district heating in Denmark and in Sweden is produced in co-generation plants. Combined production of heat and electricity is also widely used in industry in several countries.

District heating and combined production of heat and electricity can be based on practically any fuel and thus constitutes an effective method to substitute oil by other fuels, especially for space heating.

Renewables and new technologies

So far, work on renewable energies, e.g. solar heat, geothermal energy, wind energy, and biomass, in the IEA has mainly been a question of R&D. The same applies to certain other nonconventional technologies, such as heat pumps and use of waste heat. There is, however, a growing amount of feasibility studies and demonstration projects.

In a few cases, consumers are given direct support to use these energies and technologies.

Experience in many IEA countries is that the economics of many conventional conservation measures are for the time being better, and often much better, than the economics of conservation by renewables or by nonconventional techniques. In the short term, therefore, the potential for energy savings by measures of the first type is greater. In the long term, and in particular following the comprehensive R&D programmes on renewables and nonconventional techniques, the prospects for the latter kind of conservation are also better. Canada, for instance, has reported on a study concluding that in the year 2000 as much as 10% of projected energy demand could be supplied by renewable energies.

Energy price development

Energy consumption is strongly influenced by energy prices. Energy pricing thus has to be considered as one of the most important tools of energy conservation policies. Energy price increases have a dampening effect on energy demand, both in the short and in the long term, since the implementation of energy conservation measures often depends on a comparison of investment costs with anticipated savings in terms of energy costs. To enable the efficient allocation of resources, energy prices should be based on long-run marginal costs.

Furthermore, energy taxes can be used as a flexible energy policy tool, both to influence demand and, as has been proposed in several countries, to finance incentives for energy-saving investments, especially in the residential and commercial sector.

Two aspects have been considered when looking at energy prices in IEA countries: the absolute price level, e.g. on 1 January 1978, and energy price increases since 1973. The former gives an idea of where a country stands with respect to pricing policies compared with other countries, while the latter shows pricing policy performance since the oil crisis.

Energy prices have sharply increased in 1973–74, followed, in most member countries, by a slow decline, especially in 1976 and 1977, mainly due to currency appreciations. Broadly speaking, increases have been higher in industry than in the residential and commercial sector, with transport showing the smallest price rises.

The development, in real terms, of the 'fuel and light' component of the consumer price index (Table 4) confirms this for the household sector. Except in Canada, the United States of America, the Netherlands, Sweden, and the United Kingdom, where a steady rise has been observed until January 1978, the 'fuel and light' price index reaches a maximum between January 1974 and 1976 and declines slightly thereafter.

If prices, especially for oil products, were to continue to decrease in the future, thus giving consumers the wrong signals, energy conservation efforts of IEA member countries could be seriously jeopardized.

In the residential and commercial sector, the lowest fuel prices are offered in Canada, and for natural gas also in the United States of America. Natural-gas prices are less than half of the cheapest prices in Europe (the United Kingdom and the Netherlands), and gas diesel oil prices in Canada are only half of the prices which are paid in Japan and Austria. Electricity prices, again, are much lower in Canada than in all other countries; 40% below those of the United States of America, 60% below those of Norway and 75% below those of West Germany.

Conservation potential

While the absolute and percentage reduction in energy demand that can be achieved through conservation actions in the building sector varies from country to country, the overall potential is great. In all countries, the building sector accounts for a significant percentage of total energy consumption (over 50% in some) and the potential for economically justified energy sav-

Table 4   Fuel and light price index in households, in real terms, 1972 = 100 (Source: Internal unpublished OECD price statistics; available only for some countries)

| | Prices as at January of each year | | | | | |
	1973	1974	1975	1976	1977	1978
Austria	99.0	100.9	109.1	109.0	109.5	107.0
Canada	101.9	104.8	106.6	113.4	122.4	127.0
Denmark	99.5	145.3	142.5	140.2	133.3	138.4
West Germany	102.4	125.2	122.0	128.9	125.6	124.0
Italy	93.6	99.3	112.8	94.9	102.5	104.3
Japan	98.4	94.6	100.9	95.3	99.3	96.9
Luxembourg	97.6	99.1	100.3	105.9	102.7	100.2
Netherlands	98.6	102.2	110.4	115.5	115.6	119.5
Norway	99.5	105.2	104.5	103.1	103.3	102.7
Sweden	101.7	136.5	148.9	156.2	150.2	158.1
Switzerland	103.5	158.3	116.9	119.8	121.0	113.6
United Kingdom	98.6	93.2	97.0	106.3	107.4	108.1
USA	100.3	105.1	107.2	110.4	115.9	116.3

ings has, in some cases, been estimated at 40% in existing buildings and very often over 50% in new buildings.

PUBLIC INFORMATION

The success of an effective energy policy depends on the extent of public support and cooperation. The need to convince the public of the seriousness of the long-term energy situation and the need for public support of governments in their efforts to introduce more stringent energy policies have often been stressed as essential to the development of effective energy conservation programmes. At the IEA Ministerial meeting in October 1977, Ministers expressed 'the importance of increased public awareness of the gravity of today's energy situation and the need for public support in accepting sacrifices now to avoid fundamental difficulties later.' Events over the last years have reinforced the need for effective promotion of energy conservation and this has been recognized in all IEA countries.

Energy conservation promotional campaigns are now an established feature of most member countries' policies and the indispensable role within these policies of public information/education and motivation campaigns has also been firmly recognized.

The effectiveness of campaign efforts is increased by directing programmes towards specific economic and social groups. The main target groups for the residential sector are single-unit households and apartment households. Most countries with well developed publicity campaigns are likely to put the major emphasis in campaigns on more specific segments of the public or the economy, e.g. the thermal insulation of houses.

THE WAY AHEAD

The work in the Expert Group on Energy Conservation in Buildings, and then in the Sub-Group on Energy Conservation, has in particular aimed at indicating possible lines of action which countries might take to promote energy conservation in the residential/commercial sector, and by describing the future work programme of the Agency in this area.

While virtually all IEA countries have developed or are in the process of developing energy codes, there is a strong feeling that these codes are merely a first step in the direction of developing standards which are sufficiently stringent to reflect current and expected energy prices. It appears that codes rarely go far enough in prescribing justified measures.

In addition, the scope of codes is frequently limited to very obvious conservation measures (e.g. insulation in housing) rather than covering all conservation measures or all types of building in a comprehensive manner. This approach is often adopted because of real or perceived difficulties in enforcing the codes if they become too complex and because of the time required to develop comprehensive standards. In the longer term, it appears that performance or energy budget-type codes are the most desirable, provided that adequate preconstruction approval and postconstruction procedures can be developed. Information exchange in this regard is most desirable.

Where energy conservation in existing buildings is concerned, there is strong agreement that, while higher energy prices are required to support existing building retrofitting activities, the market system alone, even with adequate information programmes, is not sufficient to lead to the implementation of retrofitting measures. Some IEA countries believe that voluntary subsidy programmes can achieve the desired result if the percentage of subsidy is large enough; others express the view that mandatory programmes alone, complemented with government subsidies, will be successful. As most experience is gained by other countries with voluntary subsidy programmes, if the experience is unsatisfactory, mandatory programmes or semi-mandatory programmes (e.g. limited government financing to upgraded buildings) may be seriously considered on a more widespread basis.

It has often been stressed that most consumers are concerned about the initial cost of an appliance, rather than its operating cost. That is why energy labelling is extremely important where competition between manufacturers results in low selling prices to attract buyers.

Considerable savings can be derived from proper maintenance of furnaces at relatively small costs compared to investments in alternative equipment. A basic programme of maintenance could be designed; Canada, for instance, is already developing a programme along these lines.

Maximum air and water temperatures, of course, are a question of 'lifestyle' and comfort. However, there must be a moderate range that each country could agree to in order to avoid excessive waste.

Finally, experiences show that federal and public buildings should be used as model examples for the private residential sector, at least for all new buildings planned or under construction. If government cannot control the buildings in which it operates, then modifying a nation's housing stock would seem to be an impossible task. Besides, consumers are quick to point to the excesses of government whenever they are asked to make any kind of 'sacrifice'.

Following discussions of this kind, the Sub-Group on Energy Conservation and its Expert Groups have developed a series of measures which could be considered in order to strengthen energy conservation policies in the industrial, transport, and residential/commercial sectors, in district heating, and in the area of public information/education and motivation.

The possible lines of action, affecting the residential/commercial sector, are the following:

(a) oil and gas prices at a level which encourages energy conservation;

(b) electricity prices covering long-term marginal costs of production;

(c) significant taxes on certain fuels to reinforce the effects of market prices where these prices are judged to provide inadequate incentives for conservation;

(d) consideration of energy use in urban planning, given that strong relationships exist between physical land use development and the transportation system, and between the transportation system and energy use;

(e) building codes with mandatory minimum thermal and lighting efficiency standards for all new buildings;

(f) periodic review and strengthening of building codes;

(g) financial incentives for encouraging local governments to adopt building codes;

(h) a building design advisory service provided by utilities and by architectural, engineering or similar institutions;

(i) incentives and other measures for promoting retrofitting of existing buildings, including:

    (i) government subsidies for feasibility studies and for retrofitting activities for which the payback periods would otherwise be commercially unattractive,

    (ii) reduction of sales taxes and import duties on energy-efficient equipment and insulation material,

    (iii) inspection of existing buildings leading to subsidized retrofitting activities,

    (iv) rent control programmes, if implemented, which include special provisions to avoid discrimination against retrofitting, and

    (v) equitable sharing of costs and benefits by tenants and landlords;

(j) regulations for individual metering of energy used for space heating and for hot water in all new buildings, as well as of gas and electricity in new and existing buildings;

(k) mandatory standards for central heating equipment, and mandatory maintenance programmes for all major heating and cooling devices;

(l) energy efficiency labelling for all major consumer appliances (e.g. water heaters, heating equipment, air conditioners, refrigerators, freezers);

(m) incentives and regulations to encourage district heating schemes and combined production of heat and power;

(n) encouraging the use of industrial waste heat and other sources of surplus heat;

(o) comprehensive public information, education, and motivation programmes with a conservation message, including programmes specifically directed at schools;

(p) specialized energy conservation education and/or training for such personnel as architects, engineers, and building contractors, supervisors and inspectors; and

(q) an exemplary and effective effort to reduce all central government and local government energy use.

Each of these measures is in place in one or more IEA member country. They are believed to have been effective, and there will be a continuous accumulation of further experience in energy conservation in member countries. It is now hoped that, insofar as they are not already being applied, IEA member countries may wish to consider such measures, as appropriate to their national circumstances, for possible inclusion in national conservation programmes.

REFERENCES

IEA, 1978, *Review of Member Countries' Energy Programmes,* International Energy Agency, Organization for Economic Co-operation and Development, Paris.

OECD, 1979, *Energy Balances of OECD Countries*, Organization for Economic Co-operation and Development, Paris.

Energy Conservation and Thermal Insulation
Edited by R. Derricott and S. S. Chissick
© 1981 John Wiley & Sons Ltd.

CHAPTER 2

Energy conservation and the European Communities' energy policy

A. Strub and R. Shotton

Commission of the European Communities, Brussels

1. THE EUROPEAN COMMUNITY IN THE WORLD ENERGY MARKET

With about one-fifth of the world's gross national product, the European Community's energy requirements amount to about one-sixth of world requirements, but are expected to account for only about one-tenth of the forecast increase in world energy requirements between now and the year 2000.

However, the European Community is still the largest single purchaser of oil in the world market today: in 1977, the European Community imported 483 mtoe, the United States of America 447 mtoe, and Japan 274 mtoe. Although the European Community's oil import requirements are expected to remain nearly constant at least up to the year 1990[1], those elsewhere may increase dramatically.

One of the main objectives of the European Communities' energy policy is to reduce its dependence on energy imports by developing indigenous energy sources and by moderating the future growth of energy demand, without adversely affecting economic and social aims. But the European Community, together with the other main energy importers, should also make sure that the demands of the industrialized Western World leave room for the newly industrialized countries in the Third World to satisfy their legitimate ambitions for economic and social development, without excessive and destabilizing pressures being put upon the principal oil exporters.

The authors work, respectively, for the Research, Science and Education and Energy Departments of the Commission. The views expressed in this article are, however, their own, and not necessarily those of the Commission. The work represents a viewpoint taken in 1979/80.

Indeed, the European Community, as the foremost trading group in the world, has a central role to play and interest in the future health and stability of the world trading system.

2. ENERGY'S IMPACT IN THE EUROPEAN COMMUNITY

Within the European Community, it is clear that individual member states are facing different energy situations which sometimes lead to differing attitudes to energy policy.

The direct economic burden of energy supply, measured by the sum of the resources required for energy investment and for energy imports as a percentage of gross national product, is expected to range from nearly 9% for Italy through a community average of nearly 5%, to a minimum of approximately 3% for the Netherlands and 2% for the United Kingdom, in the period 1981 to 1985.

Some member states, for example Ireland and Italy, will have to invest heavily without seeing any appreciable reduction in their external dependence, while others, such as the United Kingdom, should be rewarded by decreasing import requirements and increased internal production.

Those member states with the strongest currencies against the dollar have obtained gains in their terms of trade equivalent to nearly 30% of the dollar price of oil, compared to member states with relatively weak currencies such as Italy or Ireland.

The political and economic benefits of cheap and secure energy supplies are felt to be very important. Economically weaker countries are tempted to protect themselves against the economic power of stronger economies by erecting barriers to free competition and to the free movement of energy products. These barriers may come into conflict with the provisions of the Treaties establishing the Community, which deal with such questions. Similarly, pricing energy below cost, or subsidizing in other ways the purchase and use of energy by the industrial and commercial sectors, raises questions of fair competition.

In short, energy is a powerful factor determining economic integration in the European Community. The situation described above tends to work against greater economic convergence.

In times of severe market tension, such as in the 1973 crisis, these underlying differences can also endanger the political solidarity of the European Community.

One part of a Community energy policy must therefore be to minimize these problems. But there are important positive aims too.

The first aim is to give the member states the opportunity, by acting together, to determine with more degrees of freedom their own conditions for the supply and use of energy. Member states individually still play a

Table 1 Importance of intra-Community trade in energy

	Dependence on imported energy from non-member countries (%)[a]		Total energy dependence (%)[b]	
	1977	1990	1977	1990
Belgium & Luxembourg	53	67	85	83
Denmark	58	27[c]	100	63
West Germany	42	43[c]	58	54
France	67	59[c]	77	64
Ireland	42	31	83	81
Italy	78	69[c]	83	70
Netherlands	43	60	0	50
United Kingdom	27	16[c]	25	20–0
EEC			56	54–49

(Source: Commission of the European Communities[11]).
[a]*Imports–exports* (Indigenous consumption + bunkers) + EEC deliveries.
[b]Imports from non-member countries + EEC deliveries.
[c]The breakdown of imports by origin was estimated by the Commission's services.

major part in determining market conditions, and especially price, but they are restricted by the wider economic and social impact of their decisions.

The second aim is to enable member states to obtain the advantages from the potential for intra-Community trade in energy which exists. To the extent that energy imported from another member state could be treated as virtually synonymous with indigenous energy, the dependence of individual member states is transformed, as Table 1 illustrates.

The third aim is to help all member states to achieve a smooth transition into the post-oil era. Industrial structures must be adapted not only to meet new trading conditions in external markets but also to reflect new social values and forms of economic growth within the Community. Not only will the energy industry itself be altered by the new emphasis on a more rational use of primary energy and on the development of new energy sources but industrial activity in general, and patterns of consumption too. The free flow of technical experience and of ideas can contribute greatly to successful adaptation, and this must therefore also be an important objective for Community energy policy.

3. ENERGY CONSERVATION AND THE EUROPEAN COMMUNITIES' ENERGY POLICY

Energy conservation, together with coal and nuclear energy, determines the essential factor of the margin of choice that policy-makers have between now and the year 2000[2].

In this context, it might be of interest to show the present estimates of energy losses in the four main sectors of energy consumption:

Domestic and commercial	55%
Transportation	85%
Industry	45%
Energy transformation	60–70%

Although, because of the laws of thermodynamics, not all of these losses can be avoided, the above figures constitute, to a certain extent, an indication of the considerable potential for energy conservation.

Detailed studies[3] of the technical potential for saving energy and for renewable sources, assuming the use of technologies presently available and cost-effective at current energy prices, suggest that there may be a long-term potential for more than doubling standards of material welfare with very small increases in primary energy requirements, in advanced industrialized societies.

If it was imagined, as a theoretical exercise, that a substantial dissociation (or 'decoupling') between energy requirements and economic growth were possible in practice, it could mean that the European Community could contemplate a situation by the year 2000 in which as little as one-third of its requirements would need to be imported, which in volume terms would be substantially less than the present level of imports. And if other industrialized nations played their part, world energy markets would have the flexibility to absorb disruptions and shocks without massive price increases.

Is such a vision unrealistic and utopian—or are there conditions in which such a transformation would be economically and socially feasible?

While there is general acceptance of a 'natural' trend towards dissociation which is likely to continue and even strengthen up to the year 2000, for the European Community as a whole, each unit increase in gross national product is still expected to result in a 0.8 unit increase in primary energy requirement at least up to the year 1990. Governments do not yet believe that it would be prudent to base their plans on a more ambitious achievement than this, although it does not reflect much more than the result of the market forces and expectations presently at work, and the measures already adopted.

By and large, energy demand policy consists of encouraging a myriad of projects and measures of very diverse character. Responsibility is dispersed amongst individuals and organizations of equally diverse character, all of whom need to be motivated, and to have the resources and the information to do the job. The costs and benefits of energy conservation are difficult to predict: there is no easy or generalized access to an empirical experience on which people can determine their actions. While it is partly a matter of new technologies and equipment, it is also, and very important-antly, a problem of human behaviour, and especially of human interaction with instruments that measure and control the use of energy.

For all these reasons, energy conservation requires interventions that are different from those traditionally employed on the supply side. The only but a major compensating advantage is that all the actions concerned are within the European Community, and hence under direct influence and jurisdiction of the governments concerned.

Nevertheless, despite its difficulty, there is growing acceptance that the member states are barely beginning to tap the potential of energy conser-vation as a cheap 'source of energy'. The large number of investment pro-jects still to be undertaken with payback periods of less than two to three years is testament to this. An economic and institutional framework is gradually being established to ensure that the economies of every member state take full advantage of this potential.

4. THE EUROPEAN COMMUNITY'S ACTIONS IN THE FIELD OF ENERGY CONSERVATION

The European Community as well has to steer a difficult path between the need to respect specific local circumstances, and the need to encourage every member state to reflect the concerns and experience of the Commun-ity as a whole in the formulation and application of national programmes.

The European Community's action has therefore concentrated on the following four items.

(1) An annual review of national energy programmes and objectives

This is conducted in terms of broad policy guidelines, and specific recom-mendations or directives, adopted by the Council of Ministers. The inten-tion is to promote the convergence and coherence of national policies in the framework of Community guidelines. An annual report drawn up by the Commission is discussed by the Energy Council of the European Communities.

In 1974, the Council of Ministers adopted an objective of a 15% reduc-tion in energy requirements in 1985 compared to forecasts based on pre-1973 trends. This reduction should mainly be achieved by increased energy conservation. The Heads of State and Government meeting at Bremen in July 1978 further adopted the objective of a ratio between energy requirements and economic growth of 0.8 to 1 between 1978 and 1985. They also confirmed the objective adopted in 1974 of reducing the dependence of the Community on imports from 63% of energy requirements to 50% or less by 1985.

By 1977, the Community had saved about 8% of its originally forecast energy requirements and reduced its import dependence to 56% of energy requirements. It is expected that, on the basis of present policies, the above targets for 1985 will be achieved, though in the context of much lower rates of economic growth than had been supposed in 1974.

The Council of Ministers will shortly be examining broad policy guidelines at the horizon 1990.

In addition, partly as a result of the work of groups of experts chaired by the Commission of the European Communities, a number of specific legislative acts aiming at a more rational use of energy have been adopted or are in the process of being adopted by the Council of Ministers. Others will be proposed.

A number of recommendations[4] have been addressed to the member states by the Council of Ministers in the fields of buildings, motor vehicles, domestic appliances, the combined generation of heat and power and the use of waste heat, and energy saving in industry. Periodic reports will be made to determine the extent to which national legislation reflects these recommendations.

In addition, a directive[5] has been adopted by the Council of Ministers in February 1978, requiring member states to establish minimum performance standards for heat generators produced in series, and appropriate testing and information procedures for purchasers and users. A further proposal for heat generators designed to an individual requirement is expected shortly. The Commission of the European Communities also chairs a working group which exchanges information about standards for heat generators adopted by member states, and the testing procedures used.

One further proposal for binding legislation is currently being discussed by the Council of Ministers. It requires all energy labels for domestic appliances to conform to a standard European label[6].

The intention in preparing such legislation is not to try to impose a single approach on member states, which may not be adapted to local circumstances. Rather it is to ensure that, in a limited number of priority areas, all member states contribute at least a minimum effort to the achievement of the policy guidelines laid down by the Council of Ministers, while leaving to them the freedom of choice they need properly to reflect variations in local circumstances.

(2) Policy studies and analysis

The Commission undertakes a number of background studies in support of themes of general interest to all member states. For example, the Commission has a strong interest in pricing-policy. Most member states insisted after 1974 that the energy industries restore their pricing patterns to a strictly commercial basis, with the removal of various subsidies. However, in some cases, the real price of energy to consumer has begun to decline relative to 1975–76 levels and this is a matter of concern, since consumers will be getting misleading signals from the market about long-term fuel scarcities. The Commission has begun fact-finding studies, concentrating in this first stage on prices for electricity and for gas. A report will soon be made to the Council of Ministers.

The Commission is engaged in short- and long-term market forecasts. A study on scenarios for the year 2000 is now well advanced. This examines in detail the demand for energy, and is designed to explore the margins of choice with due regard both to the social and economic contexts and to technical and engineering capabilities.

The Commission is also sponsoring a study entitled *Low Energy Growth Societies* which will examine the social, economic, and institutional barriers to the rapid introduction of energy-saving behaviour and technologies. The aim is to identify ways in which society's economic, social, and environmental goals might be achieved without large increases in requirements for primary energy. It is the responsibility of a panel of eminent personalities drawn from the member states of the Community, and a first report is expected for May 1979.

(3) Technology: Research, demonstration, and design

The Community intends to help promote the use of best-practice technology and design, and to develop an industry specializing in energy-saving equipment and materials at a European level.

As regards research and development, the Commission organizes regular meetings of those responsible for research and development policy to encourage better coordination between national research and development programmes, and to avoid overlapping. It also manages directly a research and development programme at the European level. Within this programme, about 150 projects dealing with energy conservation are carried out by research institutions, industrial firms, and universities located in the member countries. The programme covers the following sectors:

(a) insulation of buildings,
(b) heat pumps,
(c) urban transport,
(d) residual heat recovery,
(e) materials recycling,
(f) energy from waste,
(g) energy consumption in industrial processes, and
(h) energy storage.

Projects are chosen from an open call for tenders which is made periodically. Some examples of realizations in the four-year programme adopted in 1975 and now terminating are:

(a) a large-scale air/water heat pump operated by gas for heat and hot water supply to a building with 60 apartments;
(b) prototypes for absorption heat pumps;
(c) work on advanced electrical batteries for energy storage; and
(d) computer software for building design.

A total budget of 13.3 million European units of account (MEUA) was approved for this programme. As these funds are used on a cost-sharing basis, the total cost of the projects supported is roughly 25 MEUA. The results of this ongoing programme are published by the normal procedures of presenting papers in scientific reviews, or at conferences, and also by making the final reports available to all interested parties within the member countries.

A second four-year programme[7] has just been proposed to the Council of Ministers, for which a total budget of 125 MEUA is asked, of which 25 MEUA will be for energy conservation in the years mid-1979 to mid-1983.

The demonstration of successful developments is a natural follow-up to research and development. Here again the Community has adopted a scheme (Regulation 1303/78)[8] which complements existing national programmes. As part of the management process of the Community scheme, information is exchanged about national schemes.

Financial assistance in the form of grants of up to 49% of allowable costs can be offered to projects which establish references on a European scale for new technologies or equipment, or new combinations of existing technologies for energy saving. The objective is to accelerate their widespread introduction throughout the European Community. A budget envelope of 55 MEUA was approved in principle in December 1978.

A first call for tenders was made in July 1978. Over 300 projects were received, which illustrates the great interest this scheme has aroused.

The Commission proposed, in December 1978, a first set of projects as worthy of support, in which there were, for example:

(a) combined heat and power production from diesel engines;
(b) electric heat pumps (air–air) in housing;
(c) straw-firing plants;
(d) small hydroelectric plants on rivers with low head falls;
(e) housing with new types of construction, advanced methods of insulation, and heating systems based on non-oil-consuming techniques; and
(f) a series of projects in the industrial and agricultural sectors.

Further proposals will be made as the programme develops.

Looking to the future, it may be decided to supplement these initiatives by a European databank to pool the experience obtained in real-life situations, beginning with buildings and selected industrial processes. In each member state, an inventory of such case studies is gradually being built up, and pooling the experience at European level might provide a better basis for design manuals or even computer software to assist engineers and architects. They could check quickly, simply, and cheaply their situation against comparable situations elsewhere, to see how far they reflect the best practice, and aginst performance standards or targets.

(4) Investment

The level of investment is a critical factor in determining the penetration of better energy-saving technologies and equipment. However, a means has to be found of overcoming the very short time horizons and high discount rates used by the individual investor. Many worthwhile investments from society's point of view are not undertaken because of the private investor's reluctance to accept long payback periods or life-cycle costing.

A certain amount of money could be available from Community sources. Community loans for energy purposes account for only around 4% of the total energy investment by member states. However, in absolute terms, the volumes are important—currently running at over 600 MEUA per annum—and can make a major impact if concentrated in particular sectors or member states. New loan schemes are proposed which would provide for a major increase in lending, and the possibility of interest-rate rebates in some member states.

Up to now, these loans have been mainly given for large projects to develop energy supplies, but the Commission is now seeking, in cooperation with the European Investment Bank, ways of increasing the share for investments in energy saving.

Much will depend upon the general economic context, but, without waiting for a return to rates of economic growth comparable to those of the 1960s, the Commission is exploring the possibilities offered by combining loan schemes with grant, interest rebate, and tax credit schemes operated by the Community or the member states, and the possibility of developing specialized financial or leasing institutions in the field of energy saving. It is hoped that the progress towards a European zone of monetary stability will help to improve access to such sources of finance where there has been a significant exchange risk in the past, and indeed that such a zone will help to contribute to a more stable planning basis for investment of all kinds and hence a better acceptance of long payback periods.

In addition to such incentives, there is the possibility of adopting performance standards. A start has been made with heat generators. The

Commission is examining what further action might be appropriate at the level of the Community in other areas, such as buildings, and selected consumer durables.

5. ENERGY CONSERVATION IN BUILDINGS

Some of the themes which have been developed in the previous section are reflected, of course, in work specifically related to the task of energy conservation in buildings.

A first theme is that of examining and comparing national programmes with a view to establishing a shared view of the right degree of effort and the best way to tackle the problem. But it is by no means an easy task to determine the scope for cost-effective action to save energy in buildings.

One of the most immediate and most intractable problems is to establish a common language for discussing such problems. It is a problem that the Community has now been wrestling with for four years with the aid of experts nominated by the member states, and has not finished yet.

Conventions and simplified calculation methods have to be established so that architects and builders can make rapid assessments of the effect of design or structural change on such key factors as:

(a) the recoverable free heat gains,
(b) the levels of ventilation,
(c) the thermal inertia of the building,
(d) the overall efficiency of the heating or cooling system (and not just the heat generator), and
(e) the thermal conductivity of the external envelope.

Climate and user requirements or behaviour must be taken into account also.

This is a completely new field. The empirical evidence for evaluating most of these factors only exists for a very limited number of case studies, difficult to generalize. There is no clear definition of the performance required of a building and its systems in terms of the comfort and convenience of the user. In addition, there are a number of important side-effects which must be introduced as constraints—for example, the inconvenience and eventually health risks associated with low levels of ventilation, the danger, particularly for existing buildings, of structural damage due to condensation or rising damp. There is also the level of competence which can reasonably be expected of the construction industry, and the behaviour of the user of the building, to be taken into account.

Four years in that thicket of unknowns has convinced the Commission that, however necessary it is to press on with the development of a comprehensive approach to building design and renovation, policy and action for some years to come must depend on an essentially empirical approach.

For new buildings, an obvious and attractive approach is simply to collect case studies of best-practice buildings and carefully monitor their use and behaviour over a number of years, to establish whether there is any obvious reason why the measures should not be taken to bring the average closer to the existing best practice. The limited information so far available suggests that the range in energy requirements between best and worst for a given floor space and use is so large, depending upon the quality of the design and construction, that it seems superfluous to worry unduly whether the best practice could be improved further. How much more important it is to get on with eliminating the worst practice!

Today all member states except one have adopted more stringent standards for new buildings since 1974. For this reason, and because of the relatively slow rate of replacement of the existing built stock, the Community has recently concentrated its attention on an initiative in the field of existing buildings.

Thus, in December 1978, the Council of Ministers adopted a recommendation[9] to member states to adopt, where they had not already done so, a four-year programme for the renovation of existing buildings with a view to saving energy. The priority was to be given to buildings in residential use and to buildings in public ownership, without excluding buildings in other categories of use. National programmes should be communicated to the Commission by January 1980, and a first review report, based above all on the practical experience of countries with major programmes such as the Federal Republic of Germany, Denmark, the Netherlands, and the United Kingdom, is to be prepared by the summer of 1980. It is hoped that this report can contain an international comparison of expectations and results of great interest.

For existing buildings the pragmatic approach is similar to that described for new buildings, but with the nuance that it should be possible to rank for each building, with a high degree of certainty, some of the most obvious priorities for action in terms of cost-effectiveness. For example, if a building has no roof insulation, and a fireplace with no damper, or the ventilation rate, through general leakiness, equals several volumes an hour, the priorities are evident.

Many countries which have costed programmes to save energy in existing buildings, especially those in residential use, have come up with total programme costs that seem high relative to the value of the energy saved. Some countries which have introduced incentives have been disappointed by the energy savings achieve in practice. The answer lies, we believe, in the absence of any insistence that a proper evaluation be made for each house of the most cost-effective solution, and that the initial programme should limit itself to solutions that are cheap, and highly cost-effective, leaving aside marginally advantageous investments.

Fundamental therefore to the success of any such programme is the availability of cheap, impartial, and credible expertise to the owner or investor. And that expertise would need to be steadily refined by the testing of results against expectations in representative samples of renovated buildings.

In Denmark, the government has adopted a scheme by which an approved list of experts is published in every local area, and their fees are controlled and widely publicized. In the United Kingdom, the initiative has been taken by the private sector. Energy conservation advice services for the householder have been launched, which set out to be impartial in their assessment of the right solution for an individual home. In most member states, however, the lack of availability of cheap comprehensive and impartial expertise to the householder remains a serious obstacle to action. The Commission is examining the role that the energy utilities could play to fill this gap.

Finally, a part of the European Community's research and development effort and of the demonstration programme is devoted to the buildings sector. The demonstration projects that the Commission has proposed to support so far were mentioned previously.

In the research and development programme, quite a number of the projects[10] under way deal directly or indirectly with better energy use in buildings.

The thermal insulation of windows and walls is, of course, in the forefront of our interest. Several projects aim at improving the insulation characteristics of windows by developing or optimizing optical- and heat-reflecting coatings, as well as by using gas fillings between the two glass layers. Wall insulation, in itself an already well known technique, is looked into by developing low-cost foam materials, with improved mechanical properties and better fire resistance. Other work aims at giving assistant to architects who have to design low-energy-consumption houses, without using costly techniques and materials. The detection of heat leaks by infrared thermography, a very useful but rather expensive technique, might receive support from a project within which typical heat-leak profiles, taking into account current building technologies, are identified and collected in a catalogue.

Other work under way in this programme, dealing with heat pumps, low-temperature heat storage, heat recovery, and the improvement of water heating appliances and house heating systems, is also of direct interest to energy conservation in buildings.

This effort in the field of energy conservation in buildings is clearly worthwhile, given that about 40% of the final energy consumption within the European Community is attributable to the domestic sector, the lion's share being heat produced from oil.

NOTES AND REFERENCES

1. Commission of the European Communities, 1978, *Energy Objectives for 1990 and Programmes of the Member States*, Doc. COM(78)613 final, 16 November.
2. New energy sources or technologies (such as thermonuclear fusion and solar energy) might come into play after the turn of the century. But, because of their long lead times, such technologies need to be strongly supported now.
3. For example: Fichtner, *et al.*, 1977, *Technologien zur Einsparung von Energie*, Bundesministerium für Forschung und Technologie, Federal Republic of Germany.
4. O.J. L 295, 77/712/EEC, Council Recommendation of 25 October 1977 on the regulating of space heating, the production of domestic hot water and the metering of heat in new buildings.
   O.J. L 295, 77/713/EEC, Council Recommendation of 25 October 1977 on the rational use of energy in industrial undertakings.
   O.J. L 295, 77/714/EEC, Council Recommendation of 25 October 1977 on the creation in the member states of advisory bodies or committees to promote combined heat and power production and the exploitation of residual heat.
   O.J. L 140, 76/492/EEC, Council Recommendation of 4 May 1976 on the rational use of energy by promoting the thermal insulation of buildings.
   O.J. L 140, 76/493/EEC, Council Recommendation of 4 May 1976 on the rational use of energy in the heating systems of existing buildings.
   O.J. L 140, 76/494/EEC, Council Recommendation of 4 May 1976 on the rational use, through better habits, of energy consumed by road vehicles.
   O.J. L 140, 76/495/EEC, Council Recommendation of 4 May 1976 on the rational use of energy in urban passenger transport.
   O.J. L 140, 76/496/EEC, Council Recommendation of 4 May 1976 on the rational use of energy for electrical household appliances.
   O.J. C 153, Council Resolution of 17 December 1974 on a Community action programme on the rational utilization of energy.
5. O.J. L 52, 78/170/EEC, Council Directive of 13 February 1978 on the performance of heat generators for space heating and the production of hot water in new or existing non-industrial buildings and on the insulation of heat and domestic hot-water distribution in new non-industrial buildings.
6. Commission of the European Communities, 1978, *Proposal for a Council Directive on the Labelling of the Energy Consumption of Domestic Appliances*, Doc. COM(78)358 final + final 2, August.
7. Commission of the European Communities, 1978, *Proposal for a second four year Research and Development Programme*, Doc. COM(78)388 final Volume I and II, August.
8. O.J. L 158, Council Regulation (EEC) No. 1303/78 of 12 June 1978 on the granting of financial support for demonstration projects in the field of energy saving.
9. O.J. L 37, 79/167/ECSC, EEC, EURATOM, Council Recommendation of 5 February 1979 on the reduction of energy requirements for buildings in the Community.

10. O.J. L 231, 2.9.1975, Council Decision of 22 August 1975 adopting an energy research and development programme.
11. Commission of the European Communities, 1979, *Energy Objectives for 1990 and Member States' Programmes*, Doc. SEC(79)27, Technical Annexes, Tables 16 and 17.

Energy Conservation and Thermal Insulation
Edited by R. Derricott and S. S. Chissick
© 1981 John Wiley & Sons Ltd.

CHAPTER 3

Technologies related to new energy sources

R. Derricott*

The National Building Agency, Cardiff

I. TECHNOLOGIES RELATED TO NEW ENERGY SOURCES

. . . a study of the research and development efforts being directed towards making new energy sources available will be conducted with a view to identifying those sources which might be the subject of more detailed examination and of co-operative international research (ECE/SC. TECH./12, Annex).

Introduction

In order to determine the scope of the study of energy technology and its limitations, it would seem appropriate to mention, in a general way, all known natural sources of energy which are now being utilized or which could, at their present stage of development, be considered as possible sources for practical utilization in the near future. All these sources of energy are illustrated in Table 1. All the sources of energy enumerated in the diagram can be conditionally divided into three main groups:

(i) conventional sources of energy (fossil fuels, hydropower or rivers);
(ii) nuclear sources of energy (fission, fusion); and
(iii) other nonconventional sources of energy, including utilization of solar, geothermal, wind, and sea energies.

As mentioned above, the table has been composed from a practical viewpoint of energy sources' utilization. In principle, there are other approaches in systematization of energy sources. For example, Ananichev (1976) divided all sources of energy by origin, into seven systems:

*Edited from restricted papers prepared by the United Nations Economic and Social Council, by permission.

Table 1

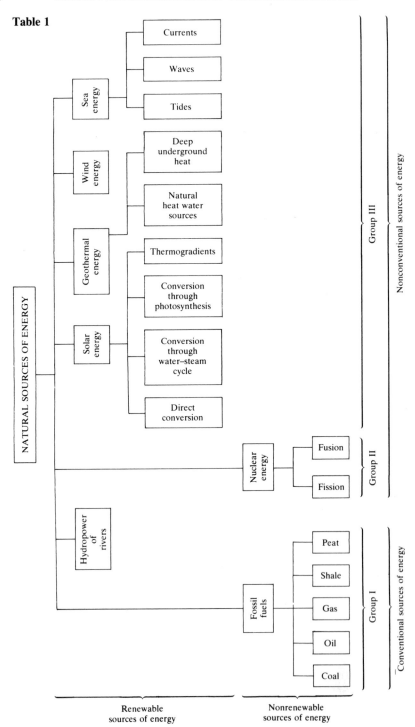

 (i) energy generated by gravitational forces, the Earth's rotation, molecular motion (temperature difference between the atmosphere, lithosphere, and hydrosphere), tides and waves, wind, and geothermal energy;
 (ii) that based on organic life, which enhanced the use of solar energy in plant photosynthesis and the energy of micro-organisms;
 (iii) direct photochemical, photoelectric, and thermoelectric uses of solar energy (i.e. heat or power production by means of electrochemical reactions and optical concentration of sunlight);
 (iv) hydrocarbon fuels such as coal, oil, natural gas, and combustible shales;
 (v) methods of energy production by using intranuclear processes;
 (vi) biochemical energy transformation;
 (vii) hydrogen utilization.

The first three systems covered renewable, and the last four systems nonrenewable energy sources.

Returning to Table 1, it should be made clear that the present paper covers the third group only—other nonconventional sources of energy. Following terminology used recently, all these sources of energy may also be considered as alternative or new energy sources although, of course, they are not new in themselves—only, for example, in their utilization for large-scale production of electricity, being a relatively new scientific and technological as well as an economic problem.

Similarly, this study does not cover possibilities of selected sources of energy utilization such as their use for transportation systems or for local purposes of heating, cooling, and electric power supply of buildings.

As may be seen, the table embraces utilization of natural sources of energy only. Hydrogen and wastes which are very often mentioned in discussions on utilization of alternative energy sources have not been included in the table, and are not covered by the present paper since they are intermediate sources of energy. Hydrogen, the utilization of which is attractive from the environmental point of view (particularly in transport), needs large amounts of electric energy for its production under existing technologies. Consideration of the problem of hydrogen utilization within the ECE is closely related to the work of the Inland Transport Committee and the Senior Advisers to ECE Governments on Environmental Problems. The latter body also deals with the problem of utilization of wastes for the production of heat energy or electricity (since this is mainly an environmental problem) because utilization of wastes cannot influence considerably the energy balance.

The table lists only sources having a practical application at present. In principle, there are other theoretical possibilities which have been omitted. For example, a biochemical process according to which thermal energy can be directly obtained as a product of chemical reactions involving nitrates and phosphates and other related substances is often mentioned. At present, the direct process is known only for explosives—for example, nitrate-based—when a large quantity of energy is released but for an extremely short period of time. Of course, this process cannot be used practically for large-scale production of thermal energy. Unfortunately, the technology for a controllable continuous process, or even a process of controllable continuous series of explosions—for example, the automobile internal combustion engine—has not yet been developed. So it is only one example of possible future technologies of heat and electric energy production which do not have practical applications at the present stage of development and is therefore not discussed in the present paper, particularly as the original source is nonrenewable in character.

Thus, the main aim of this paper is to provide information on scientific and technological, as well as economic, aspects of solar, geothermal, wind, and sea energy utilization on the basis of the review of recent experience of the ECE member countries in this field. Some consideration of new technologies for the utilization of selected sources of energy, as well as of their competitiveness as compared with conventional sources of energy, and of the perspectives of new sources of energy utilization in the near future, is also carried out in the paper.

Before detailed examination of each selected source of energy is made, it would seem advisable to draw attention to some common features of their origin and utilization. From this point of view, it is necessary to stress that all these new energy sources are, as a rule, practically inexhaustible or are renewable. To a certain extent, this explains the great attention paid by ECE member countries to these sources of energy, and in particular taking into account the threat of exhaustion of fossil fuel supplies in the foreseeable future.

At their present stage of development, nonconventional sources of energy are still considered by many east and west European scientists to be the energy of the future. For example, at the 1975 general meeting of the Academy of Sciences of the USSR, which was specially devoted to power engineering, the Soviet Academician Kirillin (1975) reported on the orientation of Soviet science and the economy in perspective to the development of conventional sources of energy—in particular, coal and nuclear energy. He stated that from the technical and economic points of view, it was yet too early to speak seriously about large-scale utilization of solar and geothermal energy for the needs of 'big' power industry.

The same statements may also be found in western sources. For example, a British scientist (Kenward, 1976a) noted that 'the development of new speculative energy resources is an investment for the future, not a means of solving the problem today', and next surveyed the prospects for technology in coal, oil, and nuclear power which, together with natural gas, will in his opinion provide the main contribution to the world's needs. Speaking in particular about the practical utilization of solar energy, another British scientist (Page, 1976) thinks that power from the Sun will not become 'the most important source of energy in the world' until the twenty-first century. He considers that solar energy is now approximately where nuclear power was in 1935. The last assertion is, perhaps, too pessimistic and does not take into account the acceleration in technological progress.

As has already been mentioned, the utilization of most nonconventional sources of energy to meet large-power engineering needs (for large-scale production of heat and electric power) is economically unacceptable at present. However, these sources of energy have considerable potential advantages as compared with fossil fuels, since the latter are, in principle, exhaustible and therefore tend to become more and more expensive. At the same time, large-scale utilization of solar, geothermal, wind, and sea energies will tend to lower the cost price of the final energy intermediate (for example, of most universal electrical energy) created as a result of effective means of mass-producing elements and installations, as well as of new discoveries, inventions, and technological innovations which will result from the process of concentration of research and development efforts on this problem. In any event, if we want to be able to use nonconventional sources of energy in the foreseeable future, we must lay the foundation today. Broad international cooperation in this field, including the organization of cooperative international research, could accelerate this process.

It is clear that the price of electric power production at fossil-fuel-fired power plants and at enterprises converting to nonconventional sources of energy will, finally, be levelled out. Moreover, there are some reasons for concluding that this will happen in the foreseeable future. Two other positive factors in support of efforts to harness new sources of energy are:

(i) their 'cleanliness' from the environmental point of view, and
(ii) their availability to any country (solar and wind energy, the Earth's deep heat) or for most countries (sea energy).

That is why a number of ECE member countries have been paying particular attention to these problems and spending considerable resources on research and development and the construction of experimental and pilot installations, in spite of the fact that new sources of energy are still not yet competitive with conventional sources of energy. This is work for the future. Individual countries striving for energy independence and self-sufficiency are an additional reason for promoting this work.

Solar energy

Among nonconventional sources of energy, the Sun occupies a very important position because it is the most powerful, practically inexhaustible, primary source of energy—and probably the most promising one. Every second, 2×10^{13} kcal of energy from the Sun reaches the Earth's surface, the largest part of this radiation being absorbed by the hydrosphere, lithosphere, and biosphere. Generally speaking, the utilization of solar energy has a long history and there are many examples of solar energy utilization, particularly for local purposes (hothouses, so-called 'solar houses', solar stoves, small steam boilers, solar pumps, etc.). However, as defined below, the most important problem of solar energy utilization at the present stage is its utilization for the large-scale production of heat and electric energy, as well as for production of synthetic liquid and gaseous forms of energy for use in transport and chemical industries instead of exhausting natural resources of this kind (oil, natural gas). Therefore the direct thermoelectric, photoelectric, and photochemical uses of solar energy are mainly discussed in this paper. In practice, this means:

   (i) the thermal conversion of solar energy into electricity,
  (ii) terrestrial plants and perhaps satellites for photovoltaic generation of electricity, and
 (iii) production of methane or alcohol by means of photosynthesis.

(i) Thermal conversion

This method foresees the utilization of reflectors, generally consisting of a great number of mirrors which concentrate sunlight on a steam boiler; or a sun collector consisting of a network of pipes through which water circulates. These boilers are usually installed in high towers surrounded by a system of reflectors which, in principle, depending on the construction, can be immovable or equipped with a special alignment system. Since the boilers produce high-pressure steam at high temperatures up to 600°C, this process allows the use of conventional turbogenerators for further production of electricity. A very important advantage of this method is that it is possible to use existing equipment and, therefore, there is no problem in connecting this system to existing electric power grids.

One of the examples of a pilot installation of this kind is that of Professor Francia, *New Scientist* (1976a), which consists of a flat array of 270 circular mirrors, each 83 cm in diameter, with a total area of 135 m². The mirrors, assisted by a sun-following mechanism, focus solar radiation onto a collector which generates superheated steam at a temperature of up to 600°C. The energy collector in this system is a cone lined with pipes which are triangular in cross section, and any light which is not absorbed by the pipes is reflected into the array onto another pipe. Professor Francia considers that the collector is 95% successful in converting solar energy into thermal energy. The array produces a peak thermal power output of 100 kW. It is envisaged that systems of this type, with an output from 100 kWe, will be put on the market. It has been reported that Professor Francia estimated that it could cost 5 billion lire* to develop and build a 1 MWe system.

Another example of the utilization of a boiler mounted on the top of a 200 ft high tower is the 'solar thermal test facility' capable of collecting 5 MW of thermal power which is being built by Sandia Laboratories for the United States Energy Research and Development Administration (ERDA). The *New Scientist* (1976b) reported that this installation, surrounded by more than 300 large, square mirror modules—so called 'heliostats' (mirrors with slightly concave surfaces)—would be equipped with a follow-up system focusing the Sun on the boiler tower. A boiler will produce high-pressure steam at 555°C. This facility will not generate electricity but will be a testbed for equipment for a solar power station with an output of 10 MWe which, it is reported, should be in operation in 1980. A number of other demonstration installations of this type are now under development in some European countries having an interest in this type of solar energy utilization.

A cooperative research and development project between Switzerland and the International Energy Agency (IEA) is at present under way to create solar energy stations high in the Alps. The *International Herald Tribune* (1977) reported that the plan envisages 40 such stations which could generate enough electric power to cover 10% of Switzerland's energy requirements by the end of the century. The project provides that a station would have mirrors, 7 × 7 m in area, directing the Sun's rays onto a boiler mounted on a 220 m high column. To generate maximum electric power of about 100 MW, the total mirror surface would have to be 500 000 m². A station's annual yield would be about 148 million kWh, with the cost of a kilowatt-hour of solar energy being approximately 17–23 Swiss centimes, which is within the range of cost of energy generated from hydroelectric and thermal stations. Together, the 40 proposed stations would provide a yield of 6.4 billion kWh per year.

---

*Wherever mentioned, currencies have been kept in original sources owing to the complexity of transformation into a single currency (for example, US dollars) because of fluctuations in exchange rates since the time of calculation by the author and uncertainty as to the time when these estimates took place. Since the estimates of effectiveness of new technologies taken from periodicals cannot be considered as complete and fully reliable, no comparisons of this information, except for those made by the authors, are presented.

(ii) Photovoltaic conversion

Utilization of the Sun for photovoltaic generation of electricity is very attractive due to its apparent simplicity. The method of production is represented by concentration of sunlight onto arrays of solar cells by special reflectors kept directed towards the Sun by an automatic alignment mechanism or by an immovable collector having a special form—for example, the Winston collector as described by Chalmers (1976, p 43). The only technical disadvantage of this method is that electricity directly generated is direct current and so has to be changed into an alternating current before transmission to distribution networks or storage facilities. Unfortunately, the production of electricity by photovoltaic cells is still very expensive, mainly owing to the high price of the cells. Thus, it is still not competitive with conventional means or with nuclear power stations and, in addition, it is considerably more expensive than the utilization of thermal conversion of solar energy. That is why examples of solar cell utilization for large-scale production of electricity are not yet available. Therefore, the main technological problem is now to create economically acceptable technologies for solar cell production.

Experimental solar cells, made of silicon grown by a new process known as 'edge-defined film-fed growth' (EFG) at the Mobil Tyco Solar Energy Corporation plant in the United States of America, may serve according to Chalmers (1976, p. 34) as an example of the efforts being made to lower the cost of building photovoltaic power stations to a level comparable with the cost of building fuel-burning stations. In this original process, ribbons of silicon crystal, 2.5 cm wide and 0.15 cm thick, are grown from a pool of molten silicon at the rate of 2.5 cm/min. For the manufacture of solar cells, it is cut into pieces. The final product is a solar cell, 10 cm long and 2.5 cm wide, which can, in full sunlight, generate between one-quarter and one-third of a watt of electricity. The silicon crystal produced by the EFG process is not as chemically pure and crystallographically perfect as those needed for microelectronic devices. It is, however, adequate for making efficient solar cells, and is cheaper.

As previously reported, the United States Energy Research and Development Administration has a programme of utilization of solar cells for large-scale production of electricity. ERDA's goal in this field is to build several 10 MW photovoltaic demonstration systems by the early 1980s. It was stressed that demonstrating the feasibility of generating electricity on a large scale with solar cells will not, however, in itself guarantee their widespread adoption, which depends on the price of the products on the electric-generating system market.

As reported at the Toulouse Congress (1976), both thermal and photovoltaic methods of converting solar energy into electricity face two major difficulties of storage and cost. Since all these systems are dependent on the intensity of solar illumination (affected by seasons, clouds, periods of darkness, etc.), some form of storage system is needed in order to produce a constant supply of energy. At the present stage of research, storage techniques are rather expensive and, in the final analysis, at best double the cost of production and at worst multiply it by as much as ten. Preliminary estimates by the Centre Nationale de la Recherche Scientifique (CNRS) (National Centre for Scientific Research) in France for a prototype 10 MWe solar-powered thermal power station gave the cost per installed kW as being four times more—and for a power station using solar cells eleven times more—compared with nuclear power stations. Taking into account the need for storage facilities, it is considered that, to make this kind of energy competitive, the system cost will have to be reduced by a factor of 100 at least.

The United States of America, which has far-reaching aims in the field of solar energy utilization, pays a great deal of attention to research and development in this field. The Energy Research and Development Administration (ERDA) estimates that the Sun could provide 5–7% of United States' energy needs in the year 2000, rising to 25% by the year 2025. Consequently, research and development financing of solar energy themes has soared from $1.2 million in 1971 to $160 million in 1977. European governments, which are more sceptical about the role of solar energy in the near future, have provided according to *Europe Energy* (1976) only limited government assistance for R&D projects in this field (for example, the Federal Republic of Germany allocated 20 million DM in 1976; France, FFr 20–30 million; and the United Kingdom, £0.5 million).

However, in European countries, societies call for greater attention to the problems of solar energy utilization. For example, the United Kingdom Solar Energy Society, in its last report on that country's solar power prospects (published in 1976), suggested that the United Kingdom might be able to obtain 10–20% of its long-term energy requirements from solar energy sources. The authors of the report called for a long-term research programme in the United Kingdom supported by funds of £2 million in 1976, rising to £20 million. It was also reported at Toulouse that only France, among the European countries, is sceptical about the role of solar energy in the near future since that country is looking increasingly towards nuclear energy to meet its needs, and its main motive for developing solar energy technology is to be able to export it to those countries which are developing from an economic point of view.

At the same time, it was reported that in the USSR, the general energy policy of which according to Kirillin (1975) is oriented towards conventional and nuclear sources of energy, a new plant for the production of solar power installations had been put into operation recently in the city of Bukchara. This plant has started production of water heaters, hothouses, driers, boilers, furnaces, kitchens, and microelectric power plants using solar energy. Construction of the first USSR electric power station using solar energy was also started in the same region according to *Izvestia* (1977a) with the intention of its being in operation by 1980.

Solar satellites

Glaser *et al.* (1970) describe a system of satellites with large panels of solar cells (64 km^2 of solar cell surfaces for 10 000–20 000 MW capacity) in a stationary orbit around the Earth which represents another way of future solar cell utilization for large-scale production of electricity. In this case, energy gathered from the Sun would be transmitted by microwave beam to terrestrial receiving stations, or as suggested by O'Neill (1976) would even be used in space by large-scale industrial satellites of the future. This scheme has some advantages compared with terrestrial solar plants: it would provide energy continuously (if more than one satellite were used), regardless of the time of the day or of weather conditions, and would double the intensity of solar rays at the Earth's surface. Enormous investments, large areas occupied by receiving antennae (100 km^2 for 20 000 MW capacity), and potential effects of the microwave beam on flora and fauna are the main disadvantages of this method which may be mentioned. Although, in principle, there is no doubt that such a power-generating space station could be built, this project seems to be more future-oriented than all the others and cannot at present be discussed seriously as a subject for international cooperation.

(iii) Photosynthesis and vital activity of micro-organisms

There is a variety of emerging energy sources which are essentially based on organic life: plant photosynthesis or activity of microbial organisms. Heat energy produced by burning fuel of vegetable origin, and timber in particular, is the most popular example of the process of utilization of solar energy converted by the photosynthesis process. However, in speaking about the practical utilization of photosynthesis and microbial solar energy conversion, the purposeful activities in plantation and processing of special kinds of plants which are most promising from the point of view of effective production of the most deficient kinds of energy, and for primary production of biomass with respect to microbial energy conversion for hydrogen and methane formation, are meant.

Mention should also be made of a third possibility which occupies a particular position and is the most promising means of solar energy conversion but, unfortunately, the process relates to the remote future: this is artificial reproduction of the photosynthesis process outside a living organism—or, in other words, a highly efficient process of transforming solar energy into chemical energy. At the present time, however, in this respect, only statements by leading scientists exist, with confirmation of the possibility of this mode of energy production in the future. There is little information about any applied research in this field. For this reason, other examples of the utilization of the natural process of photosynthesis—which, of course, are less attractive but more practical—are given below.

Scientists continue to study the possibilities of obtaining fuels and energy from certain kinds of plant grown especially for that purpose (for example, sugar-cane, kelp, trees, and other high-yield plants). It has been reported in *Science* (1977) and in *Socialisticheskaia Industriia* (1977a) that a new source—a shrub which produces a hydrocarbon substance similar to gasoline—has been proposed by Nobel Laureate, Melvin Calvin, of the University of California at Berkeley. The shrub, a member of the genus *Euphorbia*, produces significant quantities of a milk-like sap, called latex, that is actually an emulsion of hydrocarbons in water. A major advantage of two species of this shrub selected by Calvin is that they would grow well in dry regions, on land that is unsuitable for growing food. He estimates that the plants might be capable of producing between 10 and 50 barrels of oil per acre per year. This means, assuming a yield of 40 barrels per acre, that an area the size of Arizona would be necessary to meet current United States requirements for gasoline.

As stated by Calvin, the technology of production in such a case would be as follows. After the plants reach the desired height, they would be cut near the ground and run through a crushing mill in much the same way as for sugar-cane. Hydrocarbons would be obtained from the resultant sap using technology already available for separating emulsions of oil and water. The plants themselves would regrow from the stumps; thus, replanting might be necessary only once every 20 years or so. Calvin optimistically estimated that the cost of crude hydrocarbons obtained in this manner would be somewhere between $3 and $10 per barrel. Furthermore, the oil would be practically free of sulphur and other contaminants.

Thermogradients

There are several ideas for producing electricity from temperature differentials in the ocean. The project of 160 MW semi-submerged power plants, developed recently by the Lockheed Missiles and Space Company (United States) under a National Science Foundation contract, may serve as an example of these efforts. The technique described by McLain (1975) uses relatively warm water near the ocean's surface to heat and vaporize a low-boiling-point fluid, such as liquid ammonia which boils at 29°C. Ammonia vapour would drive a turbine generator to produce electricity for transmission ashore via submerged cables, and after the turbine it would be condensed by heat exchangers cooled by cold water drawn from the ocean depths and then, in a liquid form, would return to the boilers for re-use in this closed circle.

According to estimates made, a 160 MW prototype station would cost $750 million. In mass production, the price would decrease to $1200 per kW installed. The Energy Research and Development Administration is hoping to place an order for a quarter-scale plant for use in 1980. The main advantage of this station would be that its 160 MW output would be produced continuously.

Summary

The main advantages of solar energy are its availability to all European countries; the almost unlimited reserves of this source of energy; the existence of technology for commercial utilization even at its present stage of development; and the relatively acceptable economic indicators, particularly for large-scale thermal conversion and local utilization of the direct conversion method. The organization of large-scale production of currently unavailable elements should contribute to a further decrease in price. The possibility of utilizing already existing equipment for conventional thermal electric power stations (steam turbines, pumps, electric generators, power transformers, etc.) and, therefore, of simplifying interconnection of new installations with existing electric power grids, is an additional advantage of the thermal conversion method. The main disadvantages of both types of solar power station are: daily cyclic character of production, dependence on weather conditions, and considerably high land use. In spite of these disadvantages, the utilization of solar energy seems to be an item of high priority for the organization of joint cooperative research as well as for the utilization of other forms of international cooperation among ECE member countries.

Geothermal energy

The main advantages of geothermal energy are that the reserves of this source of energy are practically unlimited; its utilization does not depend on climatic conditions or the time of the day; and it is available almost everywhere in spite of variations in depth under the Earth's surface. Of course, its commercial utilization is more feasible in volcanic areas where geothermal fields are situated near the Earth's surface. In principle, there are three types of natural geothermal fields:

(i) so-called dry steam fields, or fields with steam under pressure and at relatively high temperatures, which can be used directly in turbines for the production of electric power;

(ii) wet fields, or fields with hot water under pressure and at a temperature above its boiling point at atmospheric pressure (usually 180–370°C)—steam from these fields, consisting of about 10–20% of the discharge by weight, can be used directly for power production; and

(iii) hot water with a temperature of 50–82°C which can be used for electric power production in a system with a low-boiling-point heat absorbent, such as freon or isobutane, as the driving fluid—in this case, the absorbent—moves along the closed circle from heat exchange system, through which geothermal water is pumped, to a system of turbines and condensers, and back to the heat exchanger.

There is one further possibility for the utilization of high-temperature geothermal fields (300–600°C) which do not have any natural water. Here technology envisages, in a simple case, drilling two parallel, deep wells several hundred metres apart, connected with the geothermal field by artificial geological fractures made, for example, by underground nuclear explosions, pumping the water into one of the wells, recovery of superheated steam from the other well and its utilization in turbines for production of electric power. The latter case is, rather, an example of a future technique since deep drilling of geothermal wells is, as a rule, carried out at high temperatures, and creation of the necessary penetrability of the rock is still an extremely complicated technology.

The European region has a great number of geothermal fields of all kinds and a long history of their utilization—but mainly for thermal energy utilization. Mention may be made, for example, of such important dry steam fields as the Larderello field in Italy and the Geysers field in California (USA). Countries such as Hungary, Iceland, Turkey and the USSR are also rich in geothermal fields and have certain experience in their utilization. Recent discoveries in Europe indicate that a potential for exploitation exists in many regions that had not previously been considered. However, very few examples exist at present of the practical utilization of geothermal energy for the production of electricity.

Italian experience (geothermal energy to generate electricity was first used some 70 years ago) might serve as the first example. Kenward (1976b) reported that in Italy there is 400 MW of electric generating capacity linked to geothermal energy sources. The geothermal steam reaches 16 power-generating plants from approximately 200 wells situated in three areas covering 200 km². Those geothermal power stations producing 2.5% of national energy production are operated by Ente Nazionale per l'Energia Elettrica (ENEL), the State-owned electricity utility. Italy is now following three ways of improving its geothermal energy output: development of new resources (Tuscany, Sardinia, Sicily); development of techniques for pumping the waste water (produced when geothermal steam has passed through a power station) back into underground reservoirs; and studies of the possibility of exploiting geothermal 'hot rocks'—underground heat reservoirs that do not have the water needed to bring the geothermal energy to the surface. In this connection, it should be mentioned that ENEL is working with the United States ERDA on a deep-drilling project.

In the USSR (Kamchatka region) the Panjetskaia geothermal power plant has been operating for about eight years. This geothermal power plant, which has a capacity of 5000 kW, produces electricity several times cheaper than that at diesel electric power plants located in various parts of Kamchatka. Preparations were made according to *Pravda* (1974) for reconstruction of this power plant in order to enlarge its capacity.

In the United States of America, the first geothermal power plant (12.5 MW capacity) was commissioned in 1960 on the Geysers dry steam field which is considered to be the largest field discovered so far in the world. Since then, the capacity of the plant—which is operated by the Pacific Gas and Electric Company—has been increased to 192 MW.

The Geothermal Steam Act, adopted by the United States Congress in December 1970, established the development of United States geothermal resources as a national goal, and has provided an impetus in the USA for accelerating progress. in this field. As reported in *Geothermal Energy* (1976), ERDA pressed for Federal loan guarantees for the production of energy from geothermal resources. The Geothermal Loan Guaranty programme is designed to accelerate private-industry development of geothermal resources by increasing the flow of capital into this emergent industry. The latest developments in drilling new geothermal exploration wells in the North Geysers area, in the Utah geothermal area, and in some other places have also been reported.

All these examples show that, in the ECE region, the utilization of geothermal energy has become a problem of practical interest. On the basis of experience gained, some figures about the approximate cost for various applications of geothermal energy may now be compared. It was reported by Barnea (1972) that, in a single-purpose installation producing only electric power at base load, the cost of the power produced was between three and six mills* per kilowatt-hour including full amortization of all the investments over a reasonable period. In addition, in desalination plants, the cost would be about 20–50 cents per 1000 gallons of freshwater yield—far below the cost of other types of system. For heating of houses,

*A thousandth part of a dollar (uncoined).

air conditioning, and similar purposes, the use of geothermal energy enables savings of up to 90% or more. However, the most promising method of geothermal energy utilization would be the development of multi-purpose plants for extraction of all the benefits in crude outflow from the geothermal field. This most reasonable technology of geothermal energy utilization should, among other things, reduce considerably the cost of individual applications.

Summary

The utilization of some sources of geothermal energy is very effective, even at the present stage of development—for example, dry steam fields. However, the availability of such fields is very limited within the ECE region. It is for this reason that the utilization of deep, undergound heat should be of common interest to ECE member countries because of its availability for all countries of the region. This method has some other advantages: continuous cycle of production; relatively acceptable land use; possibility of utilization of certain equipment of conventional power stations and, thus, direct interconnection with electric power grids. Unfortunately, at present, there is no available technology for the practical utilization of this source of energy and the problem of its utilization is at the theoretical-principles development stage, including some special equipment. Nevertheless, this item could be of high priority for future-oriented cooperative international research, owing to the common interest of most ECE countries in the development of this technology.

Wind energy

Wind utilization has a history of several thousand years. In Europe, windmills were used in the Netherlands in the first century, but were more particularly developed there during the sixteenth and seventeenth centuries for various uses. Wind is a very powerful source of energy. The energy of air flows in the Earth's atmosphere is according to Mezentsev (1975) 3000–4000 times higher than the energy of annual coal consumption in the world. In principle, wind energy is available everywhere, although there are regions in Europe which are particularly favoured.

There are many examples of windmill utilization for various local purposes, particularly in agriculture. Here again, large-scale utilization of this kind of energy for electric energy production is intended. There are several conceptions for windmill construction. As is known, the output of rotor installation increases according to the cube of wind velocity and the square of blade length. There are two limitations resulting from this: strength of the materials used in the manufacture of the rotor and blades, and dependence of output—or even installation work—on weather conditions. There are two principal types of wind turbine: with horizontal-axis or with vertical-axis rotor installation. Simons (1975) states that in both types, approximately 35–40% of wind energy is converted into electricity. A number of ECE countries now devote a great deal of attention to this problem.

Described by Hinzichsen and Cawood (1976), the largest installation in the ECE region, with a capacity of 2 MWe, is now under construction in Tvind, Denmark. It has a low investment cost of US $200 per installed kilowatt. In other cases, this increases to over US $2000 per installed kilowatt. The tower of the Tvind windmill is over 50 m high. Each of the two blades will be 27 m long. The following programme for the variable-pitch blades has been determined: wind with a speed of 0–3 m/s, no movement; 3–14 m/s, full pitch; 14–20 m/s, variable pitch; and above 20 m/s, zero pitch, blades locked and inoperable. The total cost of this mill producing 3.6 million kWh of electricity per year (equivalent to 400 tons of oil) was estimated to be approximately US $350 000. One of the advantages of the system is utilization of a silicon control rectifier (SCR diode) which, when installed, will allow Tvind to maintain a standard 50 Hz alternating current. It is hoped that the windmill will be connected to the west Jutland grid. If this is not permitted, it is possible that an 8 ton flywheel, capable of storing energy for one week, will be built. In this connection, it is necessary to stress that Tvind is in a very favourable position for windmill construction. It is situated in Denmark's westerly wind belt where winds blow for 280 days per year at more than 3 m/s at ground level with a maximum lull of less than one week. The Tvind team hopes to perfect six or seven basic windmill designs and to offer them, free of charge, to anyone interested in constructing a similar system.

In the United Kingdon, the Wind Energy Supply Company (WESCO), which produces a small, 5 m diameter rotor, has already started building an 18 m diameter generator, in conjunction with the National Research Development Corporation, for production of heat and electric energy. The company estimated that the grid-feeding generator unit would take two years to develop at a cost of £2 million. Following implementation of this programme, which requires Government support, one more year of testing and monitoring of the rig would be needed. The company also estimated the optimum size of the rotor to be 46 m in diameter. It is foreseen that such a grid-feeding unit would be mounted on a 28 m tower (tower height for the 18 m rotor was specified at 13 m), Sumner (1975) calculates that each installation should generate at least 2 million kWh of electricity per year, with reasonable siting.

A five-year programme for windmill development and utilization has been established in the United States of America. Engineers are working to create a windmill installation of 100 kWe capacity. In the last financial year, the National Science Foundation planned to spend $1 million on research and development in the field of wind energy utilization. The Foundation considers that if certain technical problems could be settled, wind energy would cost no more than other sources of energy, particularly in some regions of the Middle East where strong winds blow continuously. In the Federal Republic of Germany, installations of 70 kWe capacity, which can operate even during weak winds, are being tested.

Summary

In principle, wind energy is available everywhere and can be used successfully at its present stage of development for small-scale, decentralized electric energy production as well as for other local purposes. As far as large-scale commercial production of electric and heat energy is concerned, this type of installation can be competitive with conventional energy sources, in the foreseeable future, but only in particularly favourable regions with a continuous, high-speed wind character. Other disadvantages of the method are its dependence on weather conditions and mechanical limitations on the size of a separate unit. Therefore, at the present stage of development, this technology cannot have high priority for broad international cooperation among ECE countries, particularly for purposes of large-scale production. As far as local utilization is concerned, this problem is perhaps more closely related to the work of the ECE Committee on Agricultural Problems.

Sea energy

Tidal power

Utilization of tidal differences in the level of the sea is, to a certain extent, a problem similar to the utilization of hydro energy of rivers at conventional low-head hydroelectric power plants. In envisaged construction of dams and utilization of hydro-turbogenerators, the main differences are low water head and necessity for construction of long dams, and cyclical change of the water head—and even of the direction of water flow—during the day so that the same water can, in principle, pass through turbines twice (flood and ebb tides).

Two tidal power plants are now in operation in the European region: La Rance (France) and Kislaya Guba (USSR). Others are under consideration, development or construction: in the Bay of Fundy and Ungava Bay (Canada); in the Bristol Channel (United Kingdon); in Cook Inlet, Alaska (USA); and in Mezen and the Sea of Okhotsk (USSR). This technology of electric power construction is considered to be instead of experience gained, of limited practical use mainly because of high investment costs (over US $500 per kW installed) and variable hourly output.

Wave utilization

Wave oscillations could be converted into usable power in various ways, including the possibility of self-contained 'floating factories' for processes such as hydrogen production or uranium separation from sea water. Many devices for extracting energy from waves have been patented. Much attention is paid to this problem in the United Kingdom, for example by Wright (1976), where preliminary studies indicate that it is theoretically possible to provide present-day United Kingdom electrical requirements from a 600 mile stretch of ocean. An evaluation of four methods of extracting energy from waves is to be made in a £1 million research and development programme. These methods are:

(i) Salter duck,
(ii) contouring rafts,
(iii) the air-pressure ring buoy, and
(iv) Russell rectifier.

This list shows the wide differences in approach. According to present plans, the first power station using wave power will be a 10 MW system, to be ready in 1996.

The Salter duck, for example, is a type of oscillating vane devised by S. H. Salter of Edinburgh University. This rocking-float design has shown according to Owen (1976) 90% efficiency under carefully controlled laboratory conditions. The latest developments in Salter's nodding ducks have extended the high efficiency of the units in the area of smaller, high-frequency waves. Salter envisaged strings of ducks mounted on a floating, semiflexible backbone structure, each of which could nod independently of the others. In the North Atlantic, optimum diameter of the ducks is around 15 m and the device would float in water of about 30 m depth. Extraction of the energy from the nodding motion of the ducks would be by means of hydraulic pumps. Each of these ducks would be rated at approximately 200 kW/m in length, and could thus generate up to 2 MW. During one year, average output from a duck would be between 80 and 90 kW/m, or up to 900 kW. The string of ducks in Salter's plan contains between 40 and 50 units.

Others

When reviewing sea energy sources, mention may also be made of two other phenomena reported in *Socialisticheskaia Industriia* (1977b) and in

Izvestia (1977b) which may be used in principle. One of these—sea currents—is well known. A report may be made on the latest discovery, in the south-western part of the Mediterranean, of a constant current with a speed of 1 km/h of a complete mass of water from sea level to sea bed. This current exists even at a depth of more than 3000 m. Some attempts have been made recently to use small pilot installations for local purposes of utilization of river currents. However, there is no indication about practical utilization of sea-current energy. An instance of another phenomenon is the large cleft on the Pacific Ocean seabed to the west of Ecuador at a depth of some 1000 m. Sea water flowing into this cleft is contiguous to scorching magma and gushes out like a geyser at a height of about 45 m. The water temperature under normal conditions is above boiling point. However, it does not boil due to the high pressure. A way of using such 'superheated' water has not yet been found.

Summary

In principle, the various forms of sea energy mentioned above are available to most ECE member countries but, from the economic point of view, tides, waves, and currents can be used successfully only in limited, favourable places in the region. As far as technology is concerned, it is almost available now—but only for tide utilization. Technology for wave and current utilization may become available in the midterm perspective (10–20 years). Inconstancy of the production cycle is the main disadvantage of tide and wave utilization (daily cycle for tides, and annual cycle—dependent on weather conditions—for waves). From the point of view of possible international cooperative research on the problems of utilization of times, waves, and currents, as these are not of common interest for the majority of ECE countries they cannot be considered as an item of high priority.

Coordination and cooperation

The examples presented in this paper indicate the great interest of ECE member countries in utilization of new energy sources. Great efforts have been made at both national and international levels to solve this problem. That is why at the present stage very strong cooperation and coordination is being encouraged and established particularly among the United Nations organs and organizations dealing with the problem of utilization of new energy sources.

Examples of the utilization of new energy sources presented in this paper were taken mainly from recent periodicals and certain other sources, and cannot therefore give a full picture of the situation in the ECE region. However, even individual and, to a certain extent, haphazard examples

Table 2

	Availability of energy source for practical utilization			Availability of technology for practical utilization for large-scale production			Availability of technology for practical utilization for small-scale, decentralized energy production			Method of conversion into usable form (electricity, liquid or gaseous fuel)						
	In most ECE countries	In minority of ECE countries	In limited places	Available now	Available in mid-term perspective (20–25 years)	Available in long-term (more than 25 years)	Available now	Available in mid-term perspective (20–25 years)	Available in long-term (more than 25 years)	Direct conversion into electricity	Conversion into electricity through water–steam cycle	Conversion into electricity through liquid–vapour cycles of low-boiling-point substances	Conversion into electricity through hydro-turbogenerators	Conversion into electricity through hydro-pumps	Conversion into liquid or gaseous fuel through refining cycles	Direct conversion into gaseous fuel
Solar energy	+															
1. Photovoltaic conversion (terrestrial applications)	+				+		+			+						
2. Photovoltaic conversion (satellite system)	+					+	+			+						
3. Thermal conversion	+				+		+				+					
4. Photosynthesis	+				+	+	+				+				+	
5. Microbial conversion (through biomass cycle)	+					+	+									+
6. Thermogradients		+			+			+				+				
Geothermal energy	+															
1. Dry steam fields		+	+	+			+				+					
2. Wet fields		+	+	+			+				+	+				
3. Hot water fields		+	+	+			+				+	+				
4. Hot rocks	+					+			+		+	+				
Wind energy	+				+	+	+			+						
Sea energy	+															
1. Tides		+	+	+			+						+	+		
2. Waves	+				+		+							+		

+ present situation, − not applicable, ○ lack of information, A high, B intermediate, C low.

	Cycle of production				Necessity for storage facilities or interconnection with electric power grids		Means of interconnection with electric power grids		Transportability of most likely form of converted energy			Special limitations on use			State of technical development in ECE countries					Economic acceptability for large-scale production					Economic acceptability for small-scale, decentralized production		Priority for international cooperation
	Continuous	Daily cycle not dependent on weather conditions	Daily cycle dependent on weather conditions	Annual cycle dependent on weather conditions	Unnecessary	Necessary	Direct interconnection	Connection through convertors	Possible with present infrastructure	Possible with suitable infrastructure	New technology required	Land use	Environmental effects	Others	Large-scale commercial production	Local utilization	Pilot installations	Laboratory method	Theoretical principle	Widely acceptable with present technology	Acceptable now in particularly favourable regions	Acceptable now in complex systems (with separation of minerals, etc.)	Likely to be acceptable within the mid-term (20–25 years)	Unacceptable at present stage of development	Widely acceptable with present technology	Likely to be acceptable within the mid-term (20–25 years)	
			+		+		+	+ +		+			+			+	+						+		+		B
	+			+		+			+ +			+	+ +			+			+					+	−	−	C
		+			+ +			+	+ +	+			+		+ +				+		+		+		+		A
	+ +			+		+		−	−			+	O +		+ +				+			+		O	O	C
	+ +			+		+		−		+		+	O +			+	+ +					+		O	O	B
	+			+	+ +		+		+				+		+	+	+					+ +		O	O	C
	+ +			+ +	+		+	+	+			+	+ + +			+ +	+				+		+		+	B
	+ +			+ +	+		+	+	+			+	+ + +		+ + +	+						+ +		+		B
	+			+	+ +		+	+	+				+ + +		+	+	+ +		+		+		+		+	A
	+			+ +	+		+	+					+ +		+	+ +				+		+		+		B
	+ +			+				O +			+ +	+		+	+	+ +						+		O O	C
	+			+		+		+ +	+				+			+	+ +				+		+		+ +	B

Priority for international cooperation

presented in this paper provide food for thought. Some conclusions about the perspective of particular new sources of energy utilization have been given. All the information collected is summarized in Table 2. An analysis of this information allows preliminary general conclusions to be drawn.

Conclusions

(a) The main advantages of the nonconventional sources of energy reviewed are their practical inexhaustibility and their availability to the majority of ECE countries (particularly solar energy). The main disadvantage of most sources is their cyclic character and, in some cases, their dependence on weather conditions (except geothermal energy).

(b) No individual source of energy reviewed is acceptable—from an economic point of view—at its present stage of development for wide, large-scale production of universal intermediate sources of energy (electric energy, liquid and gaseous fuels).

(c) Although solar energy is not yet acceptable for large-scale production of high-temperature heat, it may be used successfully, at present, for large-scale production of low-temperature heat (for example, for heating purposes in houses and buildings instead of conventional installations).

(d) A number of new sources of energy are acceptable at present, under certain conditions, for local utilization (sun, wind). Utilization of some new energy sources is acceptable—or may be acceptable in the near future—in particularly favourable parts of the ECE region, even at the source's present stage of development (geothermal, wind, waves).

(e) No individual new energy source reviewed may be selected as being definitely most favourable in comparison with other new sources or conventional sources of energy, and therefore only a 'pluralistic' energy policy that uses a variety of sources of energy can be accepted.

(f) One of the promising ways of increasing efficiency of new energy sources is the utilization of multipurpose technologies when production of intermediate energy (electricity, for example) is accompanied by subsidiary processes (for example, minerals separation, district heating).

(g) Development of currently unavailable equipment and elements— and organization of their large-scale industrial production (thus reducing their price)—is another important problem, the solution of which could considerably promote practical utilization of new energy sources.

(h) The problem of utilization of new energy sources is closely related to the environmental problem and, when developing new energy sources, it is necessary to take into account, in addition to technological and economic aspects, the aspect of human environment protection, particularly when considering such sources as solar and geothermal energies.

(i) Since most technologies of nonconventional sources of energy utilization are still in the research and development stage, it would be more correct to base economic estimates on the midterm perspective of 10–20 years, which takes into account technological progress as well as price increases and uncertainties in the supply of fossil fuels and uranium.

(j) Since the industrial output from research and development started at present will appear only in the mid-term perspective of 10–20 years, broad, international cooperation in this work should now be initiated in order to accelerate this process in perfect readiness for urgent future energy needs of ECE countries.

(k) From the perspective point of view, and taking into account the availability of nonconventional sources of energy and the interests of the majority of ECE countries, solar energy would seem to be the most promising source at this time for international cooperation, and it should have a high priority.

(l) Utilization of deep underground heat is another promising item of common interest. In contrast with solar energy, it requires more R&D efforts and is therefore more future-oriented; but it also might have a high priority.

II. REVIEW OF RESEARCH AND DEVELOPMENT ACTIVITIES IN SOME ECE MEMBER COUNTRIES IN THE FIELD OF NEW ENERGY SOURCES

Austria

In Austria, priority in the short-term energy research programme was given to research on the *use of solar energy*, and action taken in the past two years has included the establishment of an 'Austrian measurement network for the use of solar energy for heat production'. This network comprises—in addition to dwellings—swimming pools and test benches for researchers in various parts of Austria. The aim is to determine and analyse, according to a uniform programme of measurement, the major values of significance for the economic operation of plants for the utilization of solar energy under given geographical, meteorological, and climatic conditions. The most ambitious plant in this network, from the point of view of its technical equipment, is that of the Institute for Molecular Biology of the

Austrian Academy of Sciences, which is able to cover its energy requirements for space and water heating mainly by means of solar collectors with
a surface area of 150 m², employing an automatic heat pump system.

At present, there are seven large Austrian firms which include the manufacture of solar collectors in their production programmes; in this connection, it has been found that, in practice, such plant can be used efficiently
only if complete solar installations for water or space heating are produced
and offered to consumers.

As a further contribution to the use of alternative sources of energy, an
Austrian university institute has been given the task of planning and
developing a 10 kWe solar power plant (the 'farm' concept), which is
intended for use in developing countries. This small solar power station is
being tested initially in Austria after its completion in 1978.

As the possibility of storing energy is still one of the most difficult problems associated with the use of solar energy, basic research has been
started on the production of hydrogen with the aid of solar energy. In
addition, intensive scientific and industrial research is being undertaken at
present on the direct conversion of solar energy into electricity on a photovoltaic basis.

In order to ensure better coordination of the considerable research and
development work currently in progress or being planned, and in order to
be able to speed up the application of solar energy through the supply of
appropriate information to all concerned, the Austrian Society for Space
Questions, established in 1976 by the Federal Ministry for Science and
Research, has been converted into the Austrian Society for Solar Energy
and Space Questions. Four well known private enterprises also participate
in the work of this Society, a circumstance which demonstrates the particular interest of the Austrian economy in the former's activities.

As regards the *utilization of wind energy*, trials have been carried out to
investigate the possibilities of effectively using such energy under Austrian
climatic conditions. After trials in a wind tunnel, two prototypes of wind
wheel were developed, and in the past two years they have been tested
under natural conditions for efficiency and durability. Another wind-energy
plant with an output of about 8 kW of electricity has been set up to supply
a research station. A wind-power plant with an output of about 20 kW of
electricity is at present under trial: this plant has been built in accordance
with the plans and patents of an Austrian inventor, and is reported to
possess special advantages over plant of the conventional type.

Some experimental research work on *geothermal energy* has already been
carried out in Austria, and further work has been started. Concerning the
economical use of conventional sources of energy, research has been concentrated on 'triple expansion steam processes', 'advanced heat pump systems', and 'materials with superconductivity properties for electrical

machinery'. With regard to *nuclear research*, work has been concentrated mainly on reactor safety and on studies of liquid sodium and potassium cycles. The main research interest in 1978 was possibilities of using the *biomass*.

Belgium

It is recognized that sources of energy other than oil, natural gas, coal, and nuclear power will have to make up an increasing part of world energy supplies, particularly after the year 2000. Apart from studies of nuclear fusion and underground gasification of coal, work on new sources of energy has been undertaken chiefly within the framework of the Belgian national R&D support programme in the energy field. The main emphasis in that programme *at present* is on solar energy.

The following two specific aspects of solar energy are being studied in Belgium.

(a) *Use of solar energy in housing.* The aim of this study is to determine:
    (i) the static and dynamic behaviour of a complete thermal circuit comprising collector, storage tank, and heat exchangers, in the context of the related technical and economic constraints; and
    (ii) the problems raised by the integration of solar units in housing.

The study has entailed:
    (i) installation of equipment and determination of the static and dynamic operational characteristics of a solar heating unit;
    (ii) preparation of a mathematical model of fixed and moving flat collectors, the parameters of which are derived from an analysis of experimental data; and
    (iii) selection and optimization of Belgian solar units from the technical and economic points of view.

Water and air collectors used for the heating of dwellings and of domestic water supplies have therefore been taken into account in this study.

(b) *Conversion of solar energy into electric power.* Studies are being carried out in this field on silicon solar cells and cadmium sulphide solar cells. The aim of these studies is to determine the optimal technical characteristics of solar cells and to bring about a substantial reduction in the cost of cells through the choice and preparation of raw materials, the optimization of units, and automation of production.

Studies of solar energy are part of an *economic study of solar energy*, the aim of which is to take stock of the present situation, to evaluate the economic prospects of using solar energy, and to determine the cost-effectiveness of various methods of using heat collectors.

The national R&D programme provides for the study of the *heat pump* and *low-temperature heating systems*. These new heating techniques are also being studied from the standpoint of energy savings and the use of new energy sources.

A working group on *geothermal energy* has been set up under the national R&D programme for energy with the aims of:

(1) evaluating the different ways of using low-enthalpy geothermal energy; and

(2) making a long-term study of ways of harnessing geothermal energy in Belgium.

R&D activities *contemplated* with regard to new sources of energy may be grouped as follows:

(i) continuation of the study on the use of *solar* energy in housing and the eventual construction of experimental solar plants;

(ii) continuation of studies of new heating techniques especially adapted to the use of solar energy and *geothermal* energy;

(iii) chemical storage of solar energy;

(iv) study of the availability of *geothermal* energy in Belgium itself (exploration, methods, and specific cases), as well as evaluation of possibilities of using low-enthalpy energy; and

(v) study of development of the process using hydrogen and of possible adaptation of *hydrogen* production techniques.

Canada

Introduction

In its 1976 national energy strategy, the Government of Canada adopted the objective of making the country self-reliant within ten years. Self-reliance is interpreted to mean reduced dependence on nonsecure sources of imported oil, with the overall target of holding total oil imports to one-third of total oil demand. In order to achieve self-reliance, five targets were set:

(i) over the next 2–4 years, domestic oil prices are to move towards international levels and domestic natural gas prices are to move to an appropriate competitive relationship with oil.

(ii) over the next ten years, the average rate of growth of energy use is to be brought under 3.5% per annum.

(iii) by 1985, net dependence on imported oil is to be kept to one-third of total oil demand.

(iv) self-reliance natural gas is to be maintained until northern resources can be brought to market under acceptable conditions.

(v) exploration and development in the frontier areas is to be doubled over the next three years.

A number of major legislative and administrative decisions in implementation of these objectives have also been adopted, such as energy pricing, conservation, and development measures, and some other relevant legislative/governmental actions have been taken.

Policy objectives

The Federal Government's Task Force on Energy R&D stated the following overall goal for Canadian energy R&D in 1975: 'To develop the scientific and technical capability to achieve self-reliance in energy with minimal environmental, social or economic costs and maximum industrial and quality-of-life advantages.' In parallel, the Government has adopted a policy of increasing the share of R&D funding going to private industry, and believes that 'R&D is one of the most important instruments [for achieving energy policy objectives], since we are moving towards a future where science and technology will be more and more involved in the decision-making process. Energy R&D must be viewed as one element in the total expenditure on energy policy initiatives dedicated to achieving certain national objectives.'

Legislation and administrative measures for energy R&D

In January 1974, a federal interdepartmental Task Force in Energy R&D was established at the Deputy-Minister level to develop, implement, and coordinate a programme of federal energy R&D. The Office of Energy R&D was then organized to serve as the Secretariat of the Task Force. One outcome of the Task Force's 1975 Report was the creation of a permanent Inter-departmental Panel on Energy R&D.

In 1976–77 this Panel received an allocation of $10 million, supplementary to the regular energy R&D federal budget of $108 million. This was apportioned among six areas as follows: 18% for conservation, 15% for petroleum and gas fuels technology, 25% for coal, 11% for nuclear fuel resource expansion, 17% for transportation of energy commodities and electrical transmission, and 10% for renewable sources. An additional $10 million was approved for the fiscal year 1977–78: 37% for conservation, 15% for fossil fuels, 44% for renewable sources, and 4% for transportation and transmission.

Federal energy research and development new projects to be funded in the financial year 1977–78

The bulk of funding for renewable energy systems will go towards the demonstration of solar heating in new and existing buildings, the development and testing of solar heating components, and the assessment of the energy potential of wind and the development of vertical-axis windmills. Investigations into the production of liquid and gaseous fuels from forestry and agricultural biomass will be accelerated. The prospects for geothermal heat will continue to be assessed. The applicability of these resources depends sharply on local geographical, geological, and climatic circumstances, and also on the cost and availability of other sources of power. Programmes are recommended for the evaluation and demonstration of these resources for specific Canadian conditions:

Assisting Canadian industry to develop solar heating equipment.

Demonstrating and monitoring solar heating systems and components.

Developing capability to retrofit solar heating to existing Canadian buildings.

Building a test facility for seasonal solar heating systems other than collectors.

Developing methodology for researching climatic design for Canadian solar heating systems.

Developing heat pumps and solar heating optimized for Canadian conditions.

Researching basic science of directly converting solar energy to electricity.

Improving measurement of 300–3000 nm solar radiation.

Defining the spectral distribution of solar energy.

Assessing national wind energy potential, particularly for vertical-axis turbines.

Researching the basic biology of anaerobic fermentation.

Maintaining awareness of Canadian prospects for geothermal heat.

Identifying transportation systems for future energy flows from frontiers.

Investigating marine transport of frontier oil and gas.

Designing a liquid natural-gas marine transport system.

Designing a liquid natural-gas pipeline transport system.

Identifying transportation systems for future energy flows in southern Canada.

Evaluating pipeline–ship transfer and sub-sea pipelaying systems for east-coast conditions.

Researching slurry pipeline applications in southern Canada.

Improving electrified rail transport of coal.

Evaluating southern Canada port requirements for energy commodity shipment.

Researching appropriate regulations for transporting hazardous commodities.

Evaluating rail transport of Arctic oil and gas.

Government energy R&D programmes and budgets

Major energy R&D programmes

Federal support of energy R&D falls into five distinct 'tasks', details of which follow below and are given in Tables 3 and 4. In 1976–77, conservation received 7.1%, fossil fuels 11%, nuclear 73.3%, transportation and electricity transmission and distribution 4.8%, and renewable energy sources 3.7% of a total R&D expenditure of $127.8 million. The 1977–78 budget figures are shown in Table 4, from which it will be seen that conservation has increased its share to 9.3%, fossil fuels receive 11.3%, nuclear 68%, transportation, etc., 4.7%, and renewable 6.6%. In terms of current

Table 3 Incremental funding for energy R&D (1976–78), millions of dollars; by subject

	1976–77		1977–78	
	($m)	(%)	($m)	(%)
Renewable energy	1.0	(9.3)	4.4	(44.0)
Energy conservation	1.9	(17.7)	3.7	(37.0)
Fossil fuels	4.8	(44.9)	1.5	(15.0)
Nuclear	1.1	(10.3)	—	—
Transportation and transmission	1.7	(15.9)	0.4	(4.0)
Coordination	0.2	(1.9)	—	—
	10.7	(100)	10.0	(100)

Table 4 Total federal energy R&D budgets (1976–78), millions of dollars; by subject

	1976–77		1977–78	
	($m)	(%)	($m)	(%)
Renewable	4.7	(3.7)	9.1	(6.6)
Conservation	9.1	(7.1)	12.8	(9.3)
Fossil fuels	14.0	(11.0)	15.5	(11.3)
Nuclear	93.7	(73.3)	93.7	(68.0)
Transportation and transmission	6.1	(4.8)	6.5	(4.7)
Coordination	0.2	(0.1)	0.2	(0.1)
	127.8	(100)	137.8	(100)

dollars, the Government expects that its total R&D funding level will more than double by 1985, but states that the distribution cannot yet be accurately predicted.

As far as renewable energy sources are concerned the programmes in this field are as follows:

(a) *River and tidal power*. The three components of the programme are in the early stage of feasibility evaluation:
  (i) a study of Bay of Fundy sites for tidal power, evaluating financial, economic, and environmental characteristics;
  (ii) marine geology concerns; and
  (iii) the environmental impacts of Fundy tidal and James Bay hydro installations.

(b) *Solar energy*. The 1976–77 programme was primarily focused on testing components of solar space heating systems and the building and monitoring of a number of solar demonstration homes covering a range of technologies and local site conditions. Essential to solar energy development are technologies for energy storage, low-grade heat transfer, and heat pump systems.

(c) *Wind energy*. The programme is currently geared to scaling-up demonstrations of vertical-axis turbines in conjunction with conventional generators.

(d) *Agricultural and forest biomass*. Forest biomass is now under study by an interdepartmental taskforce, as well as by several provincial governments.

(e) *Geothermal heat*. The programme is proceeding slowly with very limited resources. It is concerned primarily with Earth physics studies of suitable geological structures.

With regard to the relationship between renewable energy resource utilization and policy, the objectives mentioned above, the resources available, and the situation in other energy fields, the Government of Canada sees the largest potential applications for renewable energy resources, aside from hydro and tidal developments, as being solar energy for space and water heating and forest biomass to produce gas, steam, and electricity. Although estimates suggest that about half the national needs of low-grade heat could be met by solar energy in 2025 (15% of primary energy supply), a more conservative estimate of practical implementation has suggested 10% with 5% by 2000. The contribution of wind turbines to electricity generation will probably be small but significant for remote areas in Canada. A 230 kW vertical-axis wind turbine developed in Canada is now in operation. The economics and practicable scale of deriving liquid fuels from forest biomass are, at the moment, considered to be uncertain. The practical potential of geothermal energy is equally uncertain.

Since these renewable resources require only a minimal level of federal R&D before they could be implemented, the Government has given them top priority for new funding. From the national perspective, the first priority is to determine the practical potential of these energy resources, the impediments to implementation, and the technical goals for R&D which must be reached before large-scale implementation is contemplated.

In another document (*Government of Canada News Release*, 1977), it was indicated that the role of renewable energy in Canada's future received a boost from the Federal Government. Energy, Mines and Resources Minister, Alastair Gillespie, announced that:

> . . . renewable energy has been awarded the largest share of a $10 million increase in Federal energy R&D spending, for the fiscal year 1977–1978;
> a Renewable Energy Resource Branch has been established within the Energy Policy Sector of the Department; and
> a National Advisory Committee on Conservation and Renewable Energy is being created.

In addition to the $4.4 million increase for renewable energy (including solar, wind, biomass, and the use of heat pumps), energy conservation was allotted $3.7 million in new research funds for 1977–78; coal, heavy oils, and oil sands research was increased by $1.5 million; and the area of transportation and transmission of energy received an increase of $460 000. The increased spending will bring the total Federal energy research and development funding up to approximately $138 million in the next fiscal year.

Increases in Federal energy R&D expenditures in the last two years total $20.7 million, with the majority going to developing energy conservation and renewable resource options and extending the fossil fuel resource base (as illustrated in Table 3).

Federal Republic of Germany*

Objectives

The research and technology policies of the Federal Government are to contribute to the maintenance and improvement of the capability and competitiveness of our national economy in order to improve the conditions of working and living of our people and, hence, their quality of life.

---

*Extracts from the document entitled *Energy Research and Energy Technologies Programme 1977–1980*, of the Federal Ministry for Research and Technology, Press and Public Relations Department, Bonn, 1977, transmitted by the Government of the Federal Republic of Germany.

A major precondition of past and future development has been and will be the availability of energy for technical and industrial processes, because all areas of our lives, from industrial and agricultural production through services up to one's individual life, require sufficient supplies of energy. Energy is a prerequisite for the welfare of the citizens of our country.

These are the most important problems which characterize the energy situation in the long run:

(a) The reserves of the fossil sources of energy so far mainly used are limited. Especially the worldwide reserves of oil and natural gas will have been exhausted within a few decades.

(b) Continued economic growth in the industrialized countries and the backlog demand in the developing countries will lead to a worldwide increasing requirement for sources of energy and, hence, to an even faster depletion of useful reserves.

(c) Fuel reserves may become scarce even earlier, if the production of fossil sources of energy falls short of the development of the requirement, either for economic reasons or for technical difficulties.

(d) The conversion of energy is inextricably linked with environmental pollution, which limits the scope of energy utilization in general and the way in which specific sources of energy can be tapped in particular.

In order for research and technology policies to make maximum contributions to the solution of these problems, the measures foreseen under the Energy Research and Energy Technologies Programme, and harmonized with the Energy Programme of the Federal Government, serve the following purposes:

(i) Guaranteeing the continuity of energy supply in the medium and long terms.

(ii) Supplying energy at economically favourable costs in the long term.

(iii) Due and timely consideration of the needs for environmental protection and the protection of the public and the working population from hazards arising from the conversion and application of energy.

(iv) Improving the technological performance capability of our energy technology to maintain its economic competitiveness.

Responsibilities

The Federal Minister for Research and Technology is responsible for the programme. In executing the programme he cooperates with the other Federal Ministries and Agencies responsible for specific project areas.

Co-operation with Federal States, especially with the coal-producing States of North Rhine-Westphalia and the Saarland, exists in many fields.

The Federal Minister for Economics is responsible for funding first-of-its-kind innovation in hard coalmining. Since 1972 the Federal Minister of the Interior has been responsible for reactor safety and radiation protection (except for reactor safety research and technology).

Funding

The programme is based on the medium-term financial planning of the Federal Ministry for Research and Technology and the other Federal Agencies cooperating in the programme. Funds contributed by the Federal States are taken into account in this programme only where these States contribute to the financing of big science research centres, i.e. normally the State hosting a centre provides 10% of the funds. The contributions of the research centres to the objectives outlined in this programme will be referred to below as 'institutional funding'. Table 5 outlines the financial framework of the Energy Research and Energy Technologies Programme, 1977–1980*.

The appropriations for financing the programme are subject to the approval of the annual budgets by Parliament. Changes in updating of the financial budgets may result in changes of the programme.

In addition, some DM 587 million are available in the years 1977 to 1980 from funds earmarked for the Investments for the Future Programme carried out by the Federal Government.

Expenditures for non-nuclear and nuclear energy research (excluding fusion) have shown a pronounced development in favour of non-nuclear research. There will probably be changes in the allocation of funds within each budget which will further increase the fraction of non-nuclear energy research.

Table 5

	1977	1978	1979	1980	Total
Federal Ministry of the Interior	10	12	13	13	48
Federal Ministry of Economics	36	40	41	42	159
Federal States (through 10% contribution by host States)	53	53	55	56	217
Budget of Federal Ministry for Research and Technology	1258	1349	1434	1480	5521
Investments for the Future Programme	79	143	189	176	587
	1436	1597	1732	1767	6532

*Budget appropriations 1977 and financial planning 1978–80 for energy research (in million DM).

New sources of energy

As a result of the recognition that the sources of energy presently dominating, i.e. oil, natural gas, and coal, will no longer suffice to meet the worldwide requirement even in the foreseeable future, the development of new sources of energy has for some time been one of the main goals of applied research. Research has made its first great contribution to a solution of this problem in developing nuclear fission technology. For more than 20 years, nuclear energy research has also tried to exploit the binding energy of atomic nuclei not only by splitting heavy nuclei, such as uranium or plutonium, but also by the controlled fusion of light nuclei, the hydrogen isotopes, deuterium and tritium. Research on controlled nuclear fusion has so far concentrated on studies of the physical conditions and events accompanying the confinement of very hot and dense plasmas necessary for controlled fusion reactions. Major basic physical and technical problems still need to be solved in this field. In addition, also the extremely difficult materials problems will have to be studied in the future.

As far as other possibilities of exploiting natural resources and methods of energy generation by technical processes are concerned, research is now focused on those sources of energy which can be expected to contribute to the supply of energy under acceptable economic and ecological conditions. Despite the short time for which this development work has been going on, the direct use of solar energy is now very much in the foreground of interest. Especially in central Europe, the use of solar energy for heat generation is a promising possibility, first for the preparation of hot water and, in a next step, also for the decentralized supply of space heat to buildings. The conversion into electricity (either by means of photocells successfully used in space technology, or by thermal conversion) hardly stands a chance of economic utilization in central Europe, but there is a considerable potential for specific applications or application in other regions with higher densities of solar radiation incidence.

Work on energy generation from wind, river water, waves, tides or geothermal sources has so far been concentrated on assessing the technical and economic potential and on the development to utilize some of these sources of energy, such as the wind.

(a) Solar energy

Solar energy offers the tremendous long-term advantages of being independent of earthbound resources and not causing additional thermal or other pollution of the Earth, but only a regional change in the energy balance. The amount of energy incident upon the territory of the Federal Republic of Germany corresponds to about 80 times the present use of primary energy. Only a small fraction can be utilized under the technical and economic conditions presently foreseeable.

The following are the most important boundary conditions for technical exploitation:

(a) the low energy density of solar radiation in the Federal Republic of Germany (approx. 1000 W/m^2 as a maximum, approx. 110 W/m^2 as an annual average).

(b) fluctuations in the availability of energy as a function of the time of the day, the weather, and the season. Most of these variations are opposed to the variations in energy requirement.

The second reason links solar energy research with other areas of energy technology, above all the development of high-capacity long-time storage media for heat and electricity.

Solar energy can be used technically by:

(a) photothermal conversion for heat supply;
(b) photovoltaic or photochemical conversion for electricity generation;
(c) biological and chemical conversion for the production of hydro-carbons; and
(d) photolytic conversion for hydrogen generation.

(i) Thermal utilization

Solar radiation energy is most easily converted into heat of a temperature level of up to 100°C. This low-temperature heat can be used to prepare domestic water and for space heating purposes, i.e. in the largest energy consumption area.

The commercialization of low-temperature collector systems is based on the following prerequisites:

(a) component development (collectors, storage media, etc.);
(b) testing systems and specific techniques;
(c) demonstration of feasibility; and
(d) proof of economy.

The R&D measures initiated under the 1974–1977 Basic Energy Research Programme have been able to achieve major progress in all four areas. Demonstration projects are being financed to test the components developed. Special mention should be made of the experimental houses built at Aachen and Essen and of a bigger project (heating of a public swimming pool) at Wiehl.

The commercialization of solar heating systems must be preceded by technical and cost optimization. This includes finding the most suitable materials and insulation measures, also in order to achieve a satisfactory service life of these systems. To increase the capacity of collector systems, additional research must be carried out on selective collector coatings. Also studies of the system's behaviour are necessary.

R&D projects in the storage sector are to make available compact, low-cost and low-maintenance medium- and long-time storage media. The lack of sufficiently mature storage systems is the biggest obstacle to using solar energy for space heating purposes.

Since the storage problem is of less importance in the preparation of domestic water, it can be assumed that solar hot water preparation systems will reach commercial maturity in the next few years. Success of this development would be a major breakthrough, both with respect to energy conservation and environmental protection, especially in view of the low efficiencies of oil-fired heating systems for the preparation of hot water in the summer months.

Most of the R&D concerning solar space heating systems is concentrated on:

(a) technical and economic integration of the storage media into the 'space heating' systems;

(b) development and optimization of hybrid systems in which part of the heat requirement is met by conventional space heating systems; and

(c) development of space heating systems with the lowest possible feed temperatures.

As development goes on, the integration of solar energy systems into existing household technologies and architectural planning is to be demonstrated. Test systems and demonstration projects will also be used to compare different types of collector and storage medium or to qualify a prototype for commercial maturity.

In preparation of commercialization systems, studies are being carried out to assess the prospects of solar hot water preparation and space heating systems and to estimate the economic consequences of such systems, especially with respect to the energy balance.

In the field of concentrating collectors (mirrors, lenses) designed to make available medium- and high-temperature process heat, more basic research will still have to be carried out. In view of the meteorological conditions in this country (only some 1600 h of sunshine per annum), the chances of using these collectors are extremely slight because, unlike plane low-temperature collectors, they practically do not respond to diffuse radiation from the sky. Most of the applied research is therefore devoted to technologies applied in countries with more sunshine, especially the developing countries.

(ii) Use for electricity generation

Solar energy can be converted into electricity either directly, by photovoltaic processes, or indirectly, by thermal processes. In either case, the capital costs are still much too high to warrant economic operation in the Federal Republic of Germany. Because of the low efficiency, the space required by solar power plants operating by either method is also still too large. In the Federal Republic of Germany, a power plant of medium capacity (1000 MWe) would need an area of approximately 100 km^2, given an average density of solar radiation incidence of 110 W/m^2 and a conversion efficiency of 10%. This corresponds to about 25% of the area of West Berlin. In a relatively densely populated country with little sunshine, such as the Federal Republic of Germany, the application of this technology is therefore hardly promising. Compared with the conversion into heat it is less important also because the heat requirement by far outstrips the electricity requirement. For this reason, R&D work in this field is presently aimed mainly at the development of smaller systems which can be used for decentralized energy generation in areas with much sunshine.

However, the construction of large solar power plants in Europe or elsewhere (e.g. in the Mediterranean area) could be envisaged as a long-term possibility to generate a transportable source of secondary energy (e.g. hydrogen) which could satisfy some of Europe's energy requirement.

Solar radiation is directly converted into electricity by photocells which have so far been used only for such specific applications as space technology. Single-crystal silicon cells achieve efficiencies of 10 to 15%, cadmium sulphide cells around 5%. Further intensive research is necessary to develop new types of cell (gallium arsenide, polycrystalline silicon cells, etc.). Since the present market prices of all solar cells known to date are several orders of magnitude higher than those of comparable energy conversion systems, the development of new base materials and low-cost production techniques is of decisive importance. Some success has already been achieved in this approach.

The indirect conversion of solar energy into electricity is possible in solar thermal power plants, both by low-temperature and high-temperature collectors. Although low-temperature collectors do not need concentrators, they must employ low-boiling liquids to drive the turbine. Concentrating collectors with follow-up systems do allow the use of conventional steam processes or the sodium technology known from fast breeder reactor development, but can be used economically only in regions with relatively few days of diffuse radiation from the sky. The components of both conversion systems are being developed in the Federal Republic of Germany. Their combination into systems to be used for pumping and for electricity generation, respectively, will be the next step. These systems must be optimized with respect to fabrication costs, maintenance requirements, lifetime, and the possibilities of fabrication in other countries.

(iii) Biological and chemical uses

In principle, solar energy can be used by means of photosynthesis. Fermentation, pyrolysis, chemical reduction or simple burning of biological material can release some of the solar radiation energy bound by photosynthesis or can make it available as hydrogen or methane. However, the efficiency of photosynthesis is below 1%. It is not yet economically feasible to cultivate plants purely for purposes of energy generation. In addition to aspects of energy, the production of nutrients and other coupling products should be considered and investigated in more detail in the respective research projects.

(iv) Supplementary measures

In order to facilitate the commercialization of solar technologies, a number of supporting measures must be taken. Above all, rules for the installation, choice, and assembly of solar systems, the necessary thermal insulation rules, and guidelines under the law relating to building construction must be found (arrangement of buildings, roof pitch, shadow effect caused by new buildings, etc.).

For the efficient evaluation of these development and demonstration projects, the Federal Government will set up a network of measuring stations with a central acquisition and evaluation capability for meteorological and technical data. The stationary measuring stations will be supplemented by mobile units. The R&D projects will be accompanied by systems analyses.

International cooperation in the exploitation of solar energy has been launched both on a bilateral basis (Australia, Egypt, India, Israel) and on a multilateral level in many ways. Solar energy is part of the Energy Research Programme of the European Communities. For its 1975 to 1979 period, that programme has been allotted funds totalling 17.5 million accounting units (MEUA) which are distributed over six subprojects.

Year	1977	1978	1979	1980
Expenditure (million DM)	26	31	33	37

(b) Other non-fossil, non-nuclear sources of energy

In addition to the fossil and nuclear sources of energy treated above and to solar energy, Nature offers a large number of other possibilities to exploit processes or resources for technical purposes and, at least as far as the theoretical potential is concerned, in this way can contribute to the supply of energy. At the present state of the art, only wind energy and geothermal energy can be utilized in the Federal Republic of Germany on a limited scale. However, in order to promote the development also of other sources of energy with a theoretical potential for application and ensure its being kept abreast of the latest state of the art worldwide, the Federal Republic of Germany participates in international research programmes in almost all areas. The Federal Government will especially contribute to international projects planned within the framework of IEA.

(c) Wind energy

Wind is among the sources of energy which have been utilized by man for transport (sailboats) and to facilitate his work (pumping stations, cornmills) for thousands of years. Wind can be converted into useful mechanical energy directly and relatively simply. However, wind energy is more and more being replaced by low-cost fossil fuels, chiefly because of the major drawback of wind energy, the fluctuations of its availability with place and time. However, in view of the relatively high theoretical potential of wind energy, there is now growing interest in technologies which could exploit it successfully and economically. Yet, wind energy will make major contributions to the supply of energy only if either high-capacity energy storage media are available or integrated wind-power plants can be built and fitted into the existing energy supply structure.

The following power levels can be envisaged for the practical use of wind energy systems:

(a) units of up to 100 kW power for the decentralized supply of energy, preferably in remote areas. This may be an application in which wind energy will be economically most competitive, especially if there is no need for a constant energy flow, and, hence, the use of energy storage media.

(b) larger plants with capacities between 100 kW and the megawatt range to feed supply networks which can balance out variations in time, or to operate pumped storage power plants and compressed-air storage systems.

These are the possibilities of use on which R&D should concentrate in the further development and improvement of known plant concepts and the investigation of new development lines. Since the energy yield of a wind-power plant depends greatly on the windspeed, optimum adaptation to the plant site is a necessary prerequisite for the economic operation of such systems. For this reason, the meteorological data presently available must be processed and properly expanded to furnish precise information on the amount of wind energy available. Current R&D projects pursue these goals:

(a) acquisition and processing of wind energy data for technical applications;
(b) studies of the economic application of small wind-power plants; and
(c) studies of the economic fabrication of wind-power plants of high capacity and their integration in existing energy supply structures.

Future work will help to develop low-cost, simple and low-maintenance small-scale plants and to integrate them into other energy-producing systems. Also larger plants will be investigated (also for operation in integrated systems with other power plants).

The Federal Government intends to participate in international projects within the framework of IEA. IEA has set itself the following tasks in this field:

(a) assessing the magnitude and distribution of the wind energy available.
(b) determining the most important applications.
(c) estimating potential contributions to the supply up to the year 2020.
(d) drafting recommendations for R&D projects.

Year	1977	1978	1979	1980
Expenditure (million DM)	6	5	6	6

(d) Geothermal energy

The possibilities of using geothermal energy are determined by economic and environmental problems inherent in the respective technologies. As a rule, sufficient temperature levels (approx. 200°C) are available only at very great depths (6000 m). The high drilling costs make it uneconomic to exploit these resources. One exception are the so-called geothermal anomalies with relatively high temperature levels at lesser depths. In addition, geothermal energy can be extracted only if water or steam is available underground. In addition, the underground rocks must be sufficiently permeable to allow these liquids to absorb the contained heat while penetrating through the rock and to transport it. Such geological conditions exist only in limited areas in very few places. Geothermal energy is presently used in California, Iceland, Italy, and New Zealand for the generation of electricity (worldwide approx. 1000 MWe) and for the supply of heat (worldwide approx. 5000 MW). Because of the low temperature and the resultant low efficiency in electricity generation, geothermal energy will probably be used only for space heating purposes in this country.

An alternative is offered by the so-called hot dry rock technique presently under development in the United States of America. In this technique, the hot underground rock is artificially broken up by blasting or injecting water, and steam is generated along the fissures created in this way. This steam is extracted through another bore.

Because of the very limited reserves of hot water, geothermal energy can make a major contribution to the energy supply of the Federal Republic of Germany only if the hot dry rock technique is applied. The following initially serve to explore geothermal conditions in the Federal Republic of Germany with a view to the exploitation of geothermal heat:

(a) collection of geothermal data;
(b) geothermal prospecting and exploration; and
(c) testing of new prospecting methods.

Assessments of geothermal energy must also include studies of environmental and safety problems connected with the withdrawal of water and heat from underground and with contamination of the water.

In the development of technologies for extracting the heat from hot rock strata, the Federal Republic of Germany participates in the United States Hot Dry Rock Project and in a programme launched by IEA with the following subprojects:

Assessment of geothermal resources.
Assessment of environmental problems.
Technological development for non-electric applications.
Hot dry rock techniques.
Small electric power plants.

In addition, research and development in the field of geothermal energy is part of the Energy Research Programme of the European Communities. For the period 1975 to 1978, that programme comprised the following project areas:

Compilation of geothermal data.
Improvement of exploration techniques.
Use of hot water reserves.
Technology of steam reserves (high enthalpy).
Training of specialists.

Year	1977	1978	1979	1980
Expenditure (million DM)	2	2	2	2

(e) Other non-fossil, non-nuclear sources of energy

Other possibilities of opening up new sources of energy mainly involve the use of marine phenomena. The low potential inherent in these sources of energy in the Federal Republic of Germany, for reasons of geography, does not warrant the establishment of independent development programmes. In order to be kept abreast of recent developments and, if possible, contribute to the progress in this field by specific efforts, the Federal Republic of Germany participates in several international programmes, above all within the framework of the International Energy Agency. They mainly refer to:

(a) tidal energy,
(b) wave energy,
(c) salinity gradients of the oceans,
(d) temperature gradients of the oceans, and
(e) ocean currents.

The joint projects in all these areas involve the following points:

Assessing the magnitude and location of energy reserves.
Economic and technical assessment of concepts for the extraction of energy from these sources.
Estimating the potential contribution to the energy supply up to the year 2000.
Recommendations about joint R&D projects.

Year	1977	1978	1979	1980
Expenditure (million DM)	1	2	—	—

Hungary

Economic use of solar energy for the production of electricity

The following bodies and organizations are participating in the execution of this programme: State Committee on Technological Development; the Ministry of Metallurgy and Engineering; to some extent, in the solution of individual tasks, the Scientific Research Institute on Electrotechnology (VKI); the State Meteorological Service; and the Scientific Research Institute on Engineering Physics.

The purpose of the programme is to carry out research and development activities in the field of the technology of producing solar elements for the creation of autonomous sources of electricity. The final aim is to create silicon-based solar elements with an efficiency factor of 10–20%, and to carry out climatic tests in operational conditions. Within the framework of the current programme, work is also being done on the development of thermoelectric systems connected with the use of solar radiation energy, thermal energy, thermoelectric conversion, electric energy, and waste heat.

Duration of the programme—five years (1976–80); research team—three persons; service technicians—four persons; financial resources—2 million forints a year. Expected field of application—development of individual autonomous sources of electricity with a power of up to 100 W; the construction of a basic system to develop integrated systems for using solar energy.

International cooperation—research on the subject is at present being conducted on a coordinated basis in cooperation with CMEA member countries. Cooperation would be desirable with member states of the United Nations, especially France, the Federal Republic of Germany, and the United States of America.

Research on fuel elements with alkaline electrolyte (hydrogen–air type)

The following bodies and organizations are participating in the execution of the programme: the Scientific Research Institute on Electrotechnology (VKI); the L. Eötvös University; the Budapest Technical University; and the Veszprém University of Industrial Chemistry.

The purpose of the programme is to create an autonomous system with an energy density of 200 W h/kg and a power density of 100 W/kg. Part of the programme is concerned with research into active electrodes and auxiliary equipment. For economic reasons, the development of an electrode system not using precious metals is aimed at.

Duration of the programme—five years (1976–80); research team—18 persons; service technicians—16 persons; financial resources—expenditure for research and provision of services, 42 million forints over the five-year period. Expected area of application—the introduction of electric transport facilities. If the programme achieves the target parameters of an energy density of 200 W h/kg and a power density of 100 W/kg, the system may be used as an independent power source for road transport facilities.

International cooperation—at present research is being conducted on a coordinated basis between the CMEA member countries. Contacts have been established with Sweden and Italy. Increased cooperation seems desirable and warranted with countries having achieved significant results in this area.

The Netherlands

In February 1976, the Netherlands Government approved the setting-up of a National Research Programme on Wind Energy, the objective of which was to study the technical and economical feasibility of large-scale utilization of wind energy in the Netherlands. The study is restricted to the use of wind power for the generation of electricity to be fed into the existing utility grid.

The usable amount of (mechanical) energy in the windy regions of the Netherlands—the coastlines of North Sea and Ijsselmeer—totals approximately 0.7×10^{10} kWh per annum. This is nearly 15% of the amount of electricity produced in 1974. To utilize the available amount, nearly 5000 wind turbines each with a rotor diameter of 50 m are needed.

The programme is expected to be carried out over a five-year period in phases. The first phase (March 1976–March 1977) was of a preparatory nature. During this period, a literature study was made and the individual projects for research and development were defined more precisely. Two basic wind-turbine designs came under consideration:

(i) the conventional wind turbine (horizontal-axis rotor), and
(ii) the type based on the principle of the Darrieus rotor (vertical-axis rotor).

The latter was patented as far back as 1929, but little is known of its potentialities. Consequently, in order to gain more knowhow, a test facility with a vertical-axis rotor of 5 m diameter was designed and built by the Fokker Aircraft Company.

A computer model was developed of the aerodynamic behaviour of the rigid rotor, which was followed by a dynamic and aeroelastic stability study. Studies on the geartrain and the electric generator were also made.

An important aspect now under investigation is the influence of one wind turbine on the other(s) with respect to positioning. Wake measurements have been performed in a wind tunnel on a 20 cm model of a Darrieus-type wind turbine.

The second phase of the wind-energy research programme covers the period from March 1977 to January 1979 when the siting, economical, and technological aspects were investigated.

The main obstacle to utilization of wind energy on a large scale in a heavily populated country with a high energy consumption is location of areas for the erection of wind-energy conversion systems. Wind turbines cannot be placed in built-up areas or in areas used for industry or by traffic. Nor can they be located in uncultivated areas which are either unsuitable (e.g. woods) or protected. The only possibility, therefore, lies in location in cultivated areas. This would seem to be a satisfactory solution, combining use of land for agricultural purpose and production of wind energy.

However, in the Netherlands a complete change of present-day regional planning policy would seem to be necessary, since at the moment only buildings and structures directly related to agricultural use of the land are permitted to be erected. Although windmills of the old Dutch type are accepted and even appreciated in the rural areas, wind-energy turbines of modern design might well encounter heavy opposition, not only from the general public but also from the authorities concerned as well. In investigating the possibility of siting wind turbines on land, the following factors need to be examined:

(i) the whereabouts of sites qualifying for wind-energy parks;
(ii) characteristics of each site; and
(iii) the amount of power to be harvested from such sites.

Not only the effect on the landscape needs to be studied but also that on telecommunication, navigation, and direction-finding systems, as well as on migration of birds. An alternative solution is the location of wind turbines at sea, but this would inevitably result in a tremendous impact on costs; and location at sea implies many other restrictive factors: navigation, fishing, offshore oil and natural-gas prospecting, and naval defence requirements.

The question as to whether wind energy will be used or not depends mainly on the cost of the energy produced. Therefore, economical studies, including the evaluation of capital and operational costs of wind-energy conversion systems, will be made. (The cost of large rotor blades will contribute substantially to total energy costs.)

Only rotors constructed according to inexpensive fabrication techniques can be expected to generate economically acceptable power. In addition to the wind turbine proper—in principle, consisting of a rotor, a gearbox, a generator, and a tower—attention will be paid to the electrical equipment and cables needed to feed the electric power into the grid. The costs involved could well prove to be of the same magnitude as those of the wind turbine itself.

In studying the technological aspects, parallel approaches of software and hardware investigations are being followed. The second phase of the National Programme comprised the following activities:

(a) initially, an extensive test programme with a vertical-axis test facility of 5 m diameter was carried out. The main objectives were to obtain a thorough insight into the performance and dynamic behaviour of a vertical-axis rotor and, preferably, in such a way that the results might be reliably applied to far larger designs (rotor diameters of up to 50 m and more).

The rotor blades are amply provided with instrumentation. Results of actual experiments will be compared with theoretical results and computed data. Furthermore, the concept of variable geometry will be tested. By varying the effective swept rotor area, the loss of aerodynamic efficiency may be compensated for by an increase of energy supplied at a constant rotor speed.

(b) continuation of fabrication research on all-plastic rotor blades for vertical- and horizontal-axis turbines, whereby use is made of modern bonding techniques as developed for aircraft structures.

(c) the design, fabrication, and assembly of an experimental horizontal-axis wind turbine with a rotor diameter of 25 m and a rated power of 150 kW. The machine commenced operation in the autumn of 1978.

(d) investigation of the tip-vane concept. This project deals with a horizontal-axis wind turbine of which relatively small vanes are attached to the tips of the rotor blades. The vanes deflect the air radially outwards, and this diffuser effect results in an increased mass flow of air through the swept plane, thus leading to a larger energy output per unit of swept area of the rotor blades.

(e) development of an electrical system for the conversion of mechanical to electric energy in which use is made of a modified line-commutated DC–AC converter.

(f) extensive study of all the technological problems related to the erection, running, and maintenance of wind turbines located offshore.

(g) study of wake effects.

(h) development of a double helical epicyclic, hydraulically controlled, system of gears for the conversion of a variable rotor speed to a constant generator speed.

The continuation of the National Programme on Wind Energy into the third phase depends on the assessment of results obtained during the

Table 6   Summary of research and development activities of institutes and industrial enterprises in the Netherlands in the field of wind-energy utilization

Institute or industry	Research and development activities
KNMI (Royal Netherlands Meteorological Institute)	Determination of wind characteristics from long-term measurements at a number of adequately well situated observation stations. Selected short-time wind measurements.

Table 6   (Continued)

Institute or industry	Research and development activities
Fokker Aircraft	Design, building, and operation of a facility for testing components and determining the performance of a vertical-axis rotor wind turbine unit (rotor diam. 5 m). Development of economic fabrication methods for rotor blades.
NLR (National Aerospace Laboratory)	Development of computer programs for parametric investigations of the performance and aeroelastic behaviour of wind turbine rotor systems.
FDO-Engineering Consultants BV	Evaluation of mechanical components, mechanical sub-systems, and control systems for large wind turbines. Assessment of preliminary structural requirements for a conceptual design of the most promising wind turbine unit. Design, fabrication, and assembly of a medium-scale horizontal-axis wind turbine (rotor diam. 25 m).
Eindhoven University of Technology, Department of Electrical Engineering	Development of an electrical system for the conversion of mechanical to electrical energy in which use will be made of a modified line-commutated DC–AC converter.
Smit Slikkerveer BV	Evaluation of electric subsystems for the conversion of mechanical energy into 50 cycle AC power to be fed into the existing utility grid.
Central Technical Institute TNO	Study on wake interactions of wind-turbine clusters.
Delft University of Technology, Department of Aerospace Engineering	Investigation of the tip-vane concept.
Rijn Schelde Verolme	Study of all the technological problems related to the erection, running, and maintenance of wind turbines located offshore.
Rademakers Aandrijvingen BV	Development of a double helical epicyclic system of gears for the conversion of a variable rotor speed to a constant generator speed.
KEMA (Utility Board)	Integration of wind-energy conversion system into the existing electric utility grid. Determination of technical and operational requirements, interfaces and cost goals. Assessment of institutional constraints and requirements. Legal, ecological, and environmental aspects.
ECN (Netherlands Energy Research Foundation)	Project management. Evaluation of programme results. Literature research and documentation.

second phase. During the third phase, the design, fabrication, assembly, and testing of a vertical-axis wind turbine of the same rotor diameter and rated energy output as the horizontal-axis wind turbine will be undertaken. In this way, sufficient information will be assembled to decide upon which type is more suitable.

A list of related research and development activities of institutes and industry in the Netherlands is given in Table 6.

With regard to solar energy, two summaries of recommended general and supporting research projects which are already in operation and a number of new projects are given in Table 7.

Table 8 consists of solar energy utilization projects listed by field of application.

In addition to these programmes, a small-scale programme concerning geothermal energy has been proposed, which concentrates on the factual situation and conditions in the Netherlands. Its most important feature is the improvement of knowledge about structure, composition and temperature pattern of Netherlands soil.

Sweden

Energy research and development programme 1975–78

The National Swedish Board for Energy Source Development (Nämnden för Energiproduktionsforskning, NE) was established on 1 July 1975 with the purpose of evaluating and supporting R&D programmes. The Board's programme 1975–78 included the subprogrammes, budget allocations, and project activities shown in Table 9.

Research and development activities carried out within the various subprogrammes 1975–78 related to new sources of energy can be summarized as follows:

(a) New fuels

Projects within the areas of *new fuels* (*general*) and *fuel cells* supported to date have primarily centred around fuel-cell techniques and studies concerning electrochemical energy techniques.

Within the area of *methanol*, subsidies have been provided to the Swedish Methanol Development Company (Svenzk Metanolutveckling AB) for studies relating to both the Board's planning activities in the area and the technical and economic considerations surrounding large-scale production of methanol.

Research and development concerning *hydrogen* has consisted primarily of participation in the planning stages of an International Energy Agency

Table 7 Summary of the Netherlands' recommended projects for solar energy utilization: general supporting research

Projects	Pre-1980	1980	1981	1982	1983	1984
Meteorological:						
—Processing existing data	■	■	■			
—Identifying data to improve measuring methods	■	■	■			
—Systematic observations			▦	▦	▦	▦
Development of collectors:						
—Continuation research into spectral selective layers to decrease radiation losses		■	■	■		
—Continuation research into reducing convective heat losses		■	■	■		
—Research into reducing random losses:		■	■			
—Studies in increasing intensity of lightbeams on collectors by pre-focusing mirrors or lenses		▦	▦	▦		
—Inventory of knowledge of techniques and low-calorific applications of solar energy				■		
—Production study	■					
—Research into roof-collectors		▦	▦	▦	▦	
—Research into front collectors		▦	▦	▦	▦	
—Research into standardization and quality control					■	■
—Study of problems of construction, corrosion and wear in use				▦	▦	▦
Heat storage:						
—Continuation research into latent heat storage		▦	▦	▦	▦	
—Continuation research into heat storage in chemicals		▦	▦	▦	▦	▦
—Studies in seasonal storage		■	■	■		▦
Social implications:						
—National, economic, legal, financial, and information aspects				▦	▦	
—Dissemination					■	■

■ Research phase ▦ Development phase

Table 8 Summary of the Netherlands' recommended projects on solar energy utilization—already in operation and new—by field of operation

Project	Pre-1980	1980	1981	1982	1983	1984
Warm water production: —Housing, utility buildings, swimming pools, hospitals, government buildings; study of integrated systems	███████████████████					
Room heating: —Experimental projects in house building and other system studies, storage etc.	████████████████████████████████					
Cooling room climate: —Experimentation into room acclimatization, industrial cooling, household cooling (integrated systems)				▪▪▪▪▪▪▪▪▪▪	▪▪▪▪▪▪	▪▪

■ Research phase   ■ ■ ■ Development phase

Table 9

Subprogramme	Expenditure 1975–78 (million Sw. kronor)	Reviewed project proposals	Initiated projects	Completed projects
Nuclear fission	72.6	67	52	19
Conventional electrical and heat production	0.6	18	4	3
Oil and natural gas	0.2	—	—	—
Organic fuels	39.1	61	47	7
New fuel systems	10.4	13	7	2
Waste heat utilization	21.9	28	25	3
Geothermal energy	7.9	17	13	2
Wind energy	20.6	61	44	14
Other ares:				
Solar energy	2.8	12	6	—
Wave energy	2.6	7	4	1
Salt gradients	0.6	3	3	1
Temperature gradients	—	—	—	—
Energy storage	1.2	7	4	—
Nuclear fission	28.3	n.a.	n.a.	n.a.
	208.8	294	209	52

n.a. Not applicable.

(IEA) agreement. According to this agreement, which came into effect in October 1977, Sweden has made a commitment to contribute to an international study concerning the potential market for hydrogen in industry and as an energy carrier.

(b) Geothermal energy

Among the three conceivable alternatives—sedimentary bedrock layers, crystalline rock formations, and crush and crack zones—sedimentary bedrock layers in southern Sweden (Skåne) are the most promising. The programme has been aimed at experimentally appraising this potential from the standpoint of extraction, utilization, and geothermal brine treatment. Studies of other alternatives, possibly through participation in the IEA cooperation, are also being conducted on a limited basis.

The Board is presently analysing data from earlier exploratory oil drilling in southern Sweden. Test drilling for geothermal purposes were started in existing holes in Höllviksnäs in southern Sweden, and test pumping was commenced in 1978. The results of these tests will make it possible to verify during the period 1978–81 the relatively promising prognosis indicated by recently conducted geological investigations.

(c) Wind energy

The overall objective of the wind-energy programme is to make possible a political decision by 1985 concerning large-scale introduction of wind power into the national power grid. The programme is composed of three phases: 'studies and experiment' (1975–77), 'full-scale prototypes' (1977–82), and 'group demonstration' (1983–85).

Results of the 'studies-and-experiment' phase indicate that both wind conditions and capacity of the national power grid in Sweden favour the large-scale introduction and integration of wind energy into the national electrical system. The potential for local, small-scale wind-energy production is, however, considerably less.

The 'full-scale-prototype' phase, now in its initial stages, aims at the development and testing of production-oriented, safe, large wind turbines.

(d) Solar energy

The prerequisites for large-scale production of electricity, heat, and fuels through conversion of solar radiation have been studied and potential development alternatives analysed in an attempt to identify possibilities suited to Swedish conditions. A more detailed study of certain potential alternatives is presently under way.

The Board considers solar energy to be important in the long-term sense, that is, towards the year 2000. To enable Sweden to assume an appropriate role in the global commercial development of solar energy, the Board's programme consists of participating in international cooperative projects and following global developments. Certain chosen Swedish organizations and institutions should be systematically and consistently supported so as to develop and maintain the competence necessary to follow global advances in the field. The Board's activities should complement measures taken by Swedish industry.

(e) Wave power energy

Projects supported by the Board to date have mainly concerned fundamental studies of the prerequisites for utilization of wave energy, including wave measurement. Participation in the IEA cooperative programme has recently been initiated.

(f) Salinity gradient energy

The aim of present projects is to verify the feasibility of the concept and to examine the most crucial technical problems. Results of these studies will enable an appraisal during 1978–81 of the prerequisites for development of salinity gradient energy in Sweden. Limited participation in IEA cooperation has begun.

(g) Temperature gradient energy

Studies indicate that Sweden lacks the prerequisites for utilization of energy from this source. No activities are planned beyond a very limited effort to follow international developments.

(h) Energy storage

Certain projects in the form of feasibility studies have been conducted. Participation in the cooperation has been initiated.

Proposed energy R&D programme 1978–81

The programme organization and economic framework recently proposed by the Board has been reviewed by the governmental Energy Research and Development Commission (Delegationen för Energiforskning), which has recently published its recommendations. The programme structure and budget allocations recommended in the Commission's report are as shown in Table 10.

The relationship between major energy policy objectives and the aims of these programmes is clarified by arranging the programmes in three blocks, with the goals outlined in Table 11.

Table 10

Subprogramme	Million Swedish kronor
1 Biomass	80
2 Peat	33
3 Shale	4
4 Coal	12
5 Oil	—
6 Natural gas	—
7 Fission	60
8 Fusion	59
9 Wind energy	145
10 Geothermal energy	15
11 Aquatic energy	8
12 Solar energy	17
13 Waste heat utilization	40
14 New fuels systems	15
15 Energy production and technology	35
	513

Table 11

	Goal	Programmes
Block A	To attempt to substitute imported oil to a significant extent with other energy sources, primarily those of domestic origin	Biosystems, peat, and shale; Coal, natural gas, and oil; Fission and fusion
Block B	To attempt to utilize renewable forms of energy to a significant extent for electricity, heat, and fuel production	Wind energy; Geothermal energy; Aquatic energy; Solar energy
Block C	To attempt to increase the efficiency, flexibility, availability, and safety of energy production and distribution through system and component development and through increased possibilities for fuel substitution	Waste heat utilization; New fuels; Energy production techniques

The contents of the various programmes on new energy sources recommended by the Energy Research and Development Commission (points 9–12, 14, 15 in Table 10) may be summarized as follows:

Wind energy (pt 9)

Wind energy has the potential to provide a relatively large contribution to the Swedish electricity supply. The recommended programme for wind energy has the largest subprogramme budget for the coming three-year period. The objective is to build three full-scale demonstration prototypes of 2–5 MW size. In a later phase, the programme aims at demonstrating the economic and operational feasibility of wind-power units in groups of ten.

Geothermal energy and aquatic energy (pts 10 & 11)

Limited programmes are recommended in both areas due to the fact that conditions in Sweden are less favourable than those in some other countries. In addition, the prospects for energy potential from these sources are considered to be fewer than for other new energy sources. The programmes aim at following international developments and at clarifying the question of whether continued efforts are justified in these areas.

Solar energy (pt 12)

The recommended programme deals exclusively with solar power. Research concerning solar heating is outlined in a separate programme for solar heating systems and energy storage with a recommended budget of 60 million Swedish kronor for the three-year period 1978–81.

The solar power programme aims at increasing knowledge, primarily by utilizing results of the comprehensive international R&D effort, in order to assess the long-range possibilities for solar power in Sweden.

New fuels systems and energy-production technologies (pts 14 & 15)

The recommended programmes in these areas aim at increasing the efficiency of the existing energy-production system. They also include research into problem areas common to all organic fuel programmes, such as combustion and gasification techniques and methanol production. The recommended programme for energy-production technologies further contains a limited research effort into energy storage, electrochemical and other forms of energy conversion, and systems techniques.

REFERENCES

Ananichev, K. V., 1976, The problem of energy and energy resources, *Environment: International Aspects*, Progress Publishers, Moscow, pp. 68–90.
Barnea, J., 1972, Geothermal power, *Scientific American*, January, p. 76.
Chalmers, B., 1976, The photovoltaic generation of electricity, *Scientific American*, October.
ECE/SC.TECH./12, Annex.
Europe Energy, 1976, no. 16, p. 26, Solar energy: UK trailing on solar research.
Geothermal Energy, 1976, **4**, April, p. 73.
Glaser, P., *et al.*, 1970, *J. Microwave Power*, **5**, no. 4, p. 296 (special issue on satellite solar power stations and microwave transmission to Earth).
Government of Canada, 1977, *Role of Renewable Energy Strengthened*, News Release, Information EMR, Ottawa, 11 February.
Government of the Federal Republic of Germany, 1977, *Energy Research and Energy Technologies Programme, 1977–1980*, Federal Ministry for Research and Technology, Bonn.
Hinzichsen, D. and Cawood, P., 1976, Fresh breeze for Denmark's windmills, *New Scientist*, 10 June, pp. 567–70.
International Congress on Solar Energy, Toulouse, France, 1–5 March 1976.
International Herald Tribune, 1977, 8 March, p. 5, Swiss study solar energy project for Alps.
Izvestia, 1977a, 16 January, Zainutginov, Sovnechnaia energetica, in Russian.
Izvestia, 1977b, 18 March, Superheated water.
Kenward, M., 1976b, Italy: goethermal prospects, *New Scientist*, 16 February, p. 399.

Kenward, M., 1976a, *Potential Energy*, Cambridge University Press, Cambridge.
Kirillin, V. A., 1975, Energetika. Sovremennoe sostoianie i perspectivi, *Vestrik Akademii Nank*, no. 2, p. 5, in Russian.
McLain, L., 1975, Solar energy could come from the sea, *The Engineer*, 17 July.
Mezentsev, V., 1975, Shahty v nebe, *Socialisticheskaia Industriia*, 26 December.
New Scientist, 1976a, 15 February, p. 398, Francias's solar boiler.
New Scientist, 1976b, 15 April, p. 134, Tower power.
O'Neill, G. K., 1976, Space colonies: the high frontiers, *The Futurist*, **10**, no. 1, pp. 25–33.
Owen, K., 1976, Sun, wind and sea may be worth pursuing as an insurance, *The Times*, 26 March.
Page, J., 1976, Scientists will share the sun's secrets, *The Times*, 3 September.
Pravda, 1974, 15 January.
Science, 1977, **194**, p. 46, The petroleum plant: perhaps we can grow gasoline.
Simons, D. M., 1975, Wind power, *Energy Technology Review*, no. 6.
Socialisticheskaia Industriia, 1977a, 2 February, Gorintchee iz . . . dereviev.
Socialisticheskaia Industriia, 1977b, 30 March, Gibraltar 'in profile'.
Sumner, J., 1975, Using wind to generate warmth and business, *The Engineer*, 18/25 December, pp. 24–5.
Wright, P., 1976, £1 m study into harnessing waves for energy, *The Times*, 30 April.

Energy Conservation and Thermal Insulation
Edited by R. Derricott and S. S. Chissick
© 1981 John Wiley & Sons Ltd.

CHAPTER 4

Energy conservation and energy demand and supply in the UK

P. J. Jonas

Department of Energy, London

THE INTERNATIONAL BACKGROUND

The Middle East War of 1973 and the subsequent oil embargo marked a turning point in energy history. It marked the end of the cheap energy period and the beginning of an era of increasingly scarce and expensive fuel. It also confirmed the warnings increasingly sounded even earlier that the unprecedented scramble for growth, higher consumption, and expansion experienced during the 1950s and 1960s could not go on unabated forever. It is widely believed that continuing exponential growth in demand for energy at the historically high growth rate of 5% or 6% internationally and 3% or 4% in the UK is unlikely to recur in the future for any length of time. The consequences of exponential growth are illustrated in Table 1 (Green Paper, 1978) which shows 1975 world energy consumption and recent estimates of proved and ultimately recoverable reserves. There is obviously great doubt about the amount of energy which can ultimately be recovered from the Earth's crust, but it is clear that oil reserves are closest to exhaustion. Although the concensus view about the longer-term prospect for oil as we approach the end of the century is that it will become scarcer and more expensive, views on the future of world oil production differ. There are some who believe that it might plateau during the 1990s and decline around the year 2000 or perhaps later. Others take the view that world oil production may have already peaked. But, as with all forecasts, predicting the exact shape of world oil production is fraught with difficulty. However, we can assume with some certainty that as oil prices rise—and they may well reach twice their present level in real terms by the year 2000—there will be a transition from oil to non-oil substitutes in the non-premium fuel markets, with remaining oil supplies becoming concen-

97

Table 1 Total world fossil fuel and uranium reserves and consumption (Source: Green Paper, 1978)

Fossil fuels	Proved reserves (1000 mtoe)	Estimated ultimately recoverable reserves (1000 mtoe)	Consumption in 1975 (1000 mtoe/year)	Proved reserves ÷ 1975 consumption (years)	Ultimate reserves ÷ 1975 consumption (years)	Duration of ultimate reserves at 4% exponential growth rate (years)
Oil	80.4	233	2.7	30	90	37
Gas	56.5	171	1.1	50	155	51
Coal	329	645[a] 3225[b]	1.9	175	340[a] 1700[b]	71[a] 110[b]

Uranium[c]	Proved reserves (1000 mtoe)	Proved and probable reserves (mtoe)
Up to 15$/lb[d]	19	37
Up to 30$/lb[d]	32	59
Up to 30$/lb in fast reactors	1590	2932

[a,b]10 and 50% recovery rates respectively. There are no published estimates of recoverability factors applicable to ultimate reserves of coal. There is a wide range of possible recovery rates, depending on economic and other factors. A range is therefore shown.
[c]Excluding Communist countries.
[d]Use in current designs of thermal reactors.

trated increasingly on meeting demand in the premium sectors such as transport and petrochemicals.

It is against this background that the major oil-consuming countries are committed, in agreements reached in the International Energy Agency, in the European Economic Community, and at the Venice Summit, to reducing their dependence on oil by working for increased efficiency in the use of all fuels, and through the development of alternatives, most notably nuclear power and coal. At present these are the only options for assured long-term bulk supplies. During the period of transition, risks of further supply interruption in the world oil market will remain very high. The pace of progress in oil substitution will directly affect both the balance between demand and available supply. The 1980s and 1990s will therefore be a very crucial period for the entire world since economic prospects will depend to a large extent on how quickly and effectively the transition can be achieved.

It has to be stressed at the outset that the UK is an integral part of the world economy and that we are therefore unable to cut ourselves off from world trends. We cannot remain a competitive nation if our energy efficiency is lower than that of our competitors. Hence we must ensure that we minimize the cost of supplying and using the energy needed to maintain our economy, also we should not allow the price of energy to fall below the international level because this would stimulate excessive demand which could only be met at a loss compared to the benefits that the nation

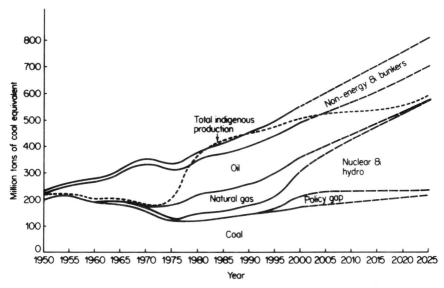

Figure 1   UK primary fuel consumption. (Source: *Energy Commission Paper No. 5*)

could derive from the saving in foreign exchange if the extra demand had not arisen.

The second important fact to stress is the inherent uncertainty of the international scene. With North Sea oil and gas, the UK is temporarily in an exceptionally favourable position; but we would all suffer greatly from a severe world depression caused by further energy crises. In the international sphere, the UK is therefore helping to ensure stability and, in order to enhance its international standing and credibility, the UK has to be able to demonstrate that its own fuel economy is organized along efficient and responsible lines. Our energy policy is therefore important not only from a national but also from an international point of view. This paper sets out broadly some of the requirements for an orderly transition from oil to non-oil substitutes in the UK and how energy conservation can be developed and demand and supply balanced so as to make the maximum contribution to the country's wellbeing.

ENERGY SUPPLY AND DEMAND

The traditional approach to energy supply is to aim to provide adequate and secure energy supplies at the lowest practicable cost to the nation. It will immediately be apparent that there could be a conflict between security of supplies and cheapness, and a great deal of effort is devoted to striking the right kind of balance. It is also necessary to provide a careful definition of what is meant by 'lowest practicable cost to the nation'. In analytical terms, this can be defined as a typical cost minimization problem, the solution to which could be obtained with the aid of linear programming techniques usually applied to resource allocation problems. But energy supply problems cannot be entirely divorced from the nation's wider economic and social considerations. Freedom of consumer choice is one of the cardinal principles of our way of life which can conflict with the technically efficient use of fuel. Cost minimization has, however, to be carried out in terms of real national resources and has to take account of future years including the likelihood of international oil prices doubling or more in real terms by the end of the century.

Since the last war and up to the 1973 energy crisis, successive Governments have concentrated attention on the supply aspects of energy and on energy production. This has involved the rationalization and co-ordination of the nationalized fuel industries to ensure that the cost of supply is minimized. It has now, however, become obvious that this narrow view is inconsistent with the principle of overall optimization of energy supply and demand and that there are many opportunities where the cost of reducing energy demand would be less than the cost of increasing energy supply. Since 1973, therefore, the Government has been undertaking a wide-ranging review of all opportunities for savings in the energy use of the main sectors of the economy. This review is one of the main contributors to the Government's energy conservation policy which is now recognized as an integral part of energy policy.

The extent to which Government can influence the efficiency with which energy consumption is consumed is, however, limited. Consumers take their decisions in a wide variety of circumstances according to a wide range of criteria. Some decisions are taken on strictly commercial criteria whilst others are influenced by often irrational consumer preferences. Government seeks to set the background conditions and prices such that consumers will take decisions which are both in their own and the national interest. A conflict between personal and national interest could arise owing to lack of adequate information and understanding of the situation confronting the individual. It is therefore an important role for Government to provide enough information and advice on the major factors affecting the efficient use of fuel to allow the individual to take its decisions in a rational, well-informed, and objective way. Even so there may be areas such as Building Regulations where mandatory measures are acceptable to consumers. Traditionally Governments have shied away from strong interference with the freedom of choice of individuals. However, as fuel prices increase and expenditure on fuel and power becomes an increasing percentage of national expenditure, there may be a need for Government to provide greater reinforcement for market forces. The extent to which, if at all, this will become necessary in the future will depend on a wide variety of factors, but the actions, in the meantime, of energy consumers, and the resulting energy conservation achievements, will have an important bearing.

Market forces are of course reinforced by placing greater emphasis on realistic energy pricing. Prices should reflect long-run incremental costs of energy supply and present policy is to let prices rise in real terms up to this level. However, even with realistic energy prices, market forces do not always work in a perfect way and one of the obvious areas where market forces are weak is innovation. Hence financial assistance is available in the form of grants to support research, development, and demonstration projects, which involves improvements in energy efficiency and other energy conservation and energy saving features. In the UK over forty novel projects have now been approved for grants of up to 25% and other benefits. And over thirty more projects are in the pipeline as of November 1980.

It is of course also necessary to ensure that, in so far as it is possible, market forces are not distorted. Undistorted market forces should ensure that the price and cost signals which both producers and consumers see and take into account in their decision-taking will properly reflect the true cost in the long term of the additional goods and services involved in the decisions to be taken. Where there is adequate competition, such as, for example, in consumer goods in general, market forces are usually strong enough to ensure this. In the energy sector, however, there are inherent monopolistic forces at work owing to the technical nature of, for example, electric lighting, where there is no ready substitute at present and hence an absence of effective competition. In this situation, Governments have to ensure that monopoly powers are not exercised against the national interest and the financial objectives which the nationalized fuel industries have to meet and the broad principles according to which fuel and energy prices are to be set are therefore subject to Treasury guidelines which are issued from time to time. The main determinant of prices should be the long-run cost of supplying an increment of demand including an adequate return on any additional investment required. The required level of return on net assets in the nationalized industries is also subject to Government guidance which has been set at a real after-risk rate of around 5% in normal circumstances, which was consistent with the average return actually earned in industry as a whole in the UK. This should ensure a reasonable balance between investment in the nationalized and private industries.

FUTURE ENERGY DEMAND AND CONSERVATION

Energy demand projections are usually highly dependent on assumptions about future economic growth and the composition of GDP which it is particularly difficult to estimate. However, in the future economic growth is expected to restart, though perhaps at lower rates than previously. Some other typical forecasting assumptions include the following:

(a) that world oil prices will more than double in real terms by the end of the century and UK fuel prices would roughly follow;

(b) that present policies for coal, nuclear, oil, and gas will broadly continue, subject possibly to variations in our North Sea depletion policy; and

(c) that energy R&D and conservation will be pushed even harder than at present but that, in spite of this, renewable resources such as wind power and solar or wave energy will not, by the end of the century, be able to make a major economic contribution to energy supplies.

Scenarios with assumptions relying on historic trends have been called 'world without conflict' or 'world as we know it today'. It is important to realize that we have no means of knowing if the world will continue to keep on a reasonably stable course. However, if serious conflict were to develop, or if economic growth were to stagnate or even decline, serious problems of readjustment could arise. Assumptions involving very low or no growth in the longer term could therefore lead to a future branded as a 'scenario of conflict and contradiction' and are therefore rather less likely as a permanent and long-term prediction, and would in any case, because of the lower economic growth, pose fewer problems for energy supply than high-growth scenarios. It should not, however, be assumed that there is a unique and absolute coupling between energy growth and GDP growth. A major assumption which affects the GDP/energy elasticity has so far not been emphasized: some energy forecasts assumes that historic relationships between energy and the major independent variables, e.g. industrial output in the industrial sector, will continue into the future, but will be substantially modified by additional energy conservation action. Such action falls into two categories: first, a reduction in energy demand resulting, for example, from the installation of additional insulation and heating controls, and secondly, an increase in the conversion efficiency of primary fuels into useful energy, for example by raising the thermal efficiency of domestic or industrial heating appliances and boilers. In order to allow for differences in the conversion efficiencies of different fuels, historic trends can be evaluated in terms of the demand for useful energy, which is then converted to primary fuel using the appropriate conversion efficiency. Changes over time in the supply pattern and efficiency trends of different fuels in the industrial, domestic and commercial, etc., sector can thus be individually allowed for and it would be unwise to make crude assumptions about changes in the overall national energy/GDP relationships which might occur in the future.

ENERGY CONSERVATION

Even though energy growth assumptions often presuppose the continuation of historic trends of energy relationships and efficiency improvements, it has been thought desirable to make substantial conservation adjustments to the resulting trend level of energy consumption. In recent estimates this amounted in all to a reduction in projected energy demand at the end of the century of over 20%. The extent to which this estimate can be achieved depends on the intensity and perseverance with which all energy users will in the future pursue the goals and objectives of energy conservation. The present pace is not particularly onerous or demanding and does not involve any major change in comfort levels or lifestyle. But there is ample evidence that the adoption of relatively simple and fully cost-effective energy-saving management and investment schemes could, when fully implemented, result in a reduction of perhaps 20–30% in UK energy consumption. The evidence comes from a wide variety of largely independent sources. Detailed energy audit studies by the Energy Technology Support Unit at Harwell and by the Department of Industry of various industrial sectors including iron casting, the brick and heavy clay, and food industries support such estimates, though they do assume that a singificant level of new investment will be undertaken by the end of the century if not long before. Reports by the Department of Industry's Energy Thrift Scheme, also covering most of British industry, suggest similar potential, whilst in the public and commercial sectors there are reports from, for example, Marks and Spencer and from the education, health, and local authorities that savings of this magnitude have actually been achieved. The Property Services Agency and many major industrial firms have also reported astonishing savings whenever a thoroughgoing energy survey has been carried out, and all obvious inefficiencies removed.

The Government's aim is to encourage and promote the universal adoption of all cost-effective energy conservation measures, and the achievement of a uniformly optimal standard of energy efficiency. To this effect the Government is already actively stimulating and assisting energy conservation by, for example, a scheme for financial assistance to tenants and householders, involving grants of up to £65 for insulating lofts and hot-water systems whilst minimum standards for the insulation of new buildings are laid down in Building Regulations. In the industrial, commercial or public sectors the Department of Energy operates the Energy Survey Scheme (ESS) which involves grants towards the services of a consultant for a one-day energy survey, the cost of which is reimbursable up to £75; and there is the extended survey scheme (EESS) with grants of up to 50% towards the cost of a comprehensive survey of all aspects of energy use. There is also a quick technical advice service by telephone (EQAS). There is an accelerated depreciation allowance for insulation of industrial and commercial buildings. Advice, information and help are available from Regional Energy conservation officers situated at the Department of Industry's regional offices, from the Government's research establishments, and of course, from the Department of Energy who issue a comprehensive range of literature, posters, films, and speakers to explain and encourage energy conservation. There are over 4000 Energy Managers, organized in over 40 energy manager groups, assisted by the Department of Energy. Last but not least is the Government's close involvement in the sponsorship and development of research, development, and demonstration projects. Any scheme or project which can show real promise of making a cost-effective and significant contribution to energy conservation and for which there are at present inadequate funds for commercial exploitation can apply for financial assistance. The demonstration scheme for example has as its objective the introduction of known technology into new markets where there has so far been no commercial opportunity to demonstrate financial viability, or it can support novel technology in its first full-scale commercial application. There is also an EEC Demonstration Scheme. Direct assistance to research and development, for example in heat transfer techniques, is also being given by both the UK and the EEC.

In spite of the very real and worthwhile energy savings which have already been achieved, there is still substantial scope for improvement. The basis for the industrial revolution upon which the wealth of the country is founded was built around cheap and abundant energy in the form of coal and oil became cheaper still in the 1960s and 1970s.

Unfortunately old habits die slowly and we are therefore burdened by a historical legacy of substantial but quite unnecessary energy waste. The rooting out of this waste is a priority action area for Government but success depends on the right decisions being taken by millions of individual consumers, both at home and in their place of work. We face a difficult period of transition and adjustment to high-priced and scarce energy, during which Government assistance is provided in the form of information, advice, encouragement, and a limited amount of financial inducement and regulation. It is, however, up to all of us to respond, and much of such response is still lacking.

This book seeks to make a real contribution to the science and technology of energy conservation, in which thermal insulation plays a crucial role. It is perhaps noteworthy that one of the few mandatory measures which successive Governments have been prepared to introduce relates to insulation. I refer of course to the building regulations which lowered the required U-value of the roofs and walls of domestic buildings to correspond to an insulation thickness of at least 50 mm of loft insulation in 1975 but which are again under review and will probably during 1981 stipulate further improvements. Theoretical calculations suggest that at present fuel prices loft insulation thickness of 100 mm could be economic in well-heated new houses using gas central heating, and higher levels are justified with electric heating. New regulations for non-domestic buildings have already been introduced and standards for heater efficiencies are being drawn up. However, there are quite often serious difficulties in predicting the precise savings to be achieved in, for example, intermittently heated homes.

I am therefore delighted to see subjects like the consequences of high insulation and the thermal performance of buildings included in this volume. The case for higher insulation standards in building regulations can only be supported if there is evidence of reasonable cost-effectiveness. A comprehensive effort by the insulation industry, housing authorities, and universities is needed to evaluate the many, and often complicated, interactions which together shape the energy consumption pattern resulting from different levels of insulation. In such calculations, allowance should be made for likely future fuel price increases and for changes in people's heating habits. It is difficult to place a value on the benefit derived from insulating homes where the result is increased temperature rather than fuel saving. One valuation would be based on the fuel cost incurred by increasing the temperature to the same level as that obtained with insulation, but to my knowledge little research has been carried out on people's attitude to bedroom heating or the health, comfort, and social implications. As living standards rise, so one would expect warm bedrooms to become a natural feature of comfort standards. However, there is some evidence that many people are quite happy to do without heating in their bedrooms. From a narrow energy conservation point of view, this may well be the most desirable state of affairs but the energy conservation objective is to ensure that the required heat is generated and used efficiently rather than to restrict energy use. There certainly seems to be scope for further design optimization of the heating system and building envelope to cater for variable requirements of the occupants and their changing needs according to age and occupancy trends over time. It has already been suggested that highly insulated houses designed for low energy use probably may not need bedroom heating, but it remains to be seen if the resulting comfort standards would be generally acceptable.

The evaluation of the cost/benefit of alternative insulation levels must be carried out as part of a comprehensive 'systems approach', involving the type and performance of the building, the heating system, and the occupant. A lifetime calculation is needed, with the objective of maximizing the rate of return on the marginal investment. Successive steps in insulation thickness and alternative combinations of modes of construction and heating appliance types and efficiencies and fuels need to be investigated to ensure an adequate return on the incremental capital involved in the various alternatives. Lifetime optimization involves value 'judgement' about the time preference between expenditures and benefits obtained now and in later years. From time to time the Government carries out a review of the appropriate discount rates to be used in the public sector. The result of the most recent review was published as part of a White Paper entitled 'Financial Objectives for Nationalised Industries'. According to this, the required rate of return (RRR) to be earned in real terms on net assets is set at 5%—which is in general conformity with the actual real returns being achieved in industry at that time.

The test discount rate (TDR) to be applied to individual projects should therefore reflect any uncertainty and other particular circumstances applying to the project, and higher rates could be appropriate where there is long-term risk. In particular, according to a Parliamentary Answer given on 14 April 1978, a rate of 7% may be used for projects whose benefits cannot readily be quantified in financial terms and where there may be significant appraisal optimism. However, if there is a risk that, as frequently happens, the cost of an investment will be underestimated, then an appropriate contingency allowance should be made in the capital and other costs and the TDR should not be increased since this would discriminate against benefits obtained in the longer term, which, especially when considering energy problems, would introduce an undesirable bias.

ENERGY PROSPECTS IN THE UK

The UK is fortunate enough to be able to look forward to a period of energy self-sufficiency in the 1980s which is unique amongst the large industrialized nations of the West. The wealth of the Continental Shelf will not, however, last forever, and the UK is at present still saddled with the legacy of energy inefficiency and waste, as evidenced by the ready achievement of energy savings of 30 to 40% which have frequently been reported or projected after careful study. It is therefore necessary for the UK, whilst there is still time to correct past distortions, to put itself into a position to withstand the rigours of the ultra high-energy cost economy which we will be facing towards the end of the century. Oil and gas reserves both in the UK and worldwide will be insufficient to sustain continuing high and increasing energy demand levels. This requires, first the development of alternative energy supplies which the Government is undertaking as part of its policy of creating and keeping open options which appear to offer adequate long-term scope and competitiveness to meet future demands at lowest cost, but also, and secondly, the achievement of optimum cost-effective levels of efficiency in the use and consumption of fuel and energy which the Government is actively promoting.

For example, consumers both large and small should be investing in all energy-efficiency and energy saving plant which has the certainty of earning and actually achieving a real rate of return of at least 5%. Consumers, particularly in industry, have a long way to go before all such investment is undertaken, and it is inevitable that considerable time will elapse before all major new investments or replacements can be undertaken on a commercial basis. Help from the Department of Energy, is available in the form of assistance towards energy surveys and information, and there is a wide range of technical advice and innovation. Given the resulting improvements in energy efficiency, the long-term future of UK energy supplies, relying on coal, nuclear, and ultimately also renewable energy supplies, should be assured. But there is every prospect that imports of fossil fuels will have to recommence in the 1990s, and there is therefore every incentive to minimize our consumption without, however, interfering with the economic and social well-being of the nation. The short-run advantage we have due to our indigenous energy supplies and any temporary respite in the real rate of increase of international oil prices gives us the required breathing space to prepare for the more difficult long-term energy future. In this connection the contribution which this volume is making to our knowledge and understanding of energy conservation technology and economics is to be welcomed.

To those who expect to be given a magic key to unlock the secrets of the long term or even next year's energy price and availability prospects this paper may appear to be insufficiently explicit. Unfortunately, however, the

Department of Energy does not have the necessary crystal ball and we do not possess the necessary vital knowledge or inside information to be able to predict world energy developments with accuracy and certainty. It is, however, considered likely that 1980 oil prices will at least double by 2000. Now that energy prices are much nearer to their long-run cost than they were in the early 1970s, normal commercial judgements should be adequate to ensure that sound consumer investment decisions are taken. It has to be recognized, however, that there are major national and international risks which can upset even the best laid plans and projections so that it is necessary to aim at flexible policies. Bearing in mind the prospective doubling of fuel prices, sound investment decisions are increasingly likely to involve the achievement of extremely high standards of energy efficiency. These should be based on the choice of energy producing and energy consuming equipment, which has been carefully optimized to minimize, overtime, capital, fuel, and other operating costs.

Thanks are due to the Department of Energy for permission to contribute this paper. The views expressed are those of the author and do not necessarily represent official policy.

REFERENCE

1*Green Paper, 1978, Energy Policy*, Cmnd. 7101, HMSO, London.

CHAPTER 5

US energy needs: the writing on the wall

Ray Dafter

Energy Editor of the *Financial Times* and Fellow of the Center for International Affairs, Harvard University (1978–79)

BACKGROUND

The Middle East—Saudi Arabia in particular—may hold the key to future oil supplies but it is the United States of America that can do most to maintain the supply and demand balance of world energy. It is the thirstiest energy consumer, it still controls a large proportion of the world's fossil fuel resources and, perhaps most important at this time, it has the greatest scope for implementing worthwhile conservation measures.

The figures speak for themselves. The primary energy consumption of the USA in 1977 was the equivalent of 1853 million tonnes, according to an authoritative BP review of the world oil industry. This consumption was no less than 40% of all the energy supplied to non-Communist countries. If the USA had managed to reduce its level of energy demand by just 5%, the saved fuels would have been sufficient to have met the needs of the whole of Australasia, with Portugal thrown in for good measure. A 10% saving—a realistic level of conservation—would have spared enough fuel to supply the whole of Africa or France.

The US society, more than any other, developed at a time when few gave little thought to the availability of energy resources. Everyone assumed that fuels would be available and reasonably cheap. So we find that in the USA, where new generations have become accustomed to big 'gas guzzling' cars, an armoury of domestic labour-saving devices, and overheated (and, in summer, overcooled) homes, the per capita consumption of energy is more than twice that in West Germany or the UK, three times that of Switzerland, and over 50 times that of India.

This paper is an expanded version of the author's article entitled 'Carter in pursuit of an oil policy' which appeared in the *Financial Times* on 2 February 1979. The author would like to express his thanks to the *Financial Times*, the Ford Foundation, and the William Waldorf Astor Foundation for their support of the Harvard Fellowship.

Events in 1973 and 1974, resulting in temporary restrictions on Middle East oil supplies and a fourfold increase in crude oil prices, should have changed society's conception of energy resources. Unfortunately, for most people it was no more than a quickly forgotten shock. It took the civil unrest in Iran—the world's second largest oil exporter—in late 1978 and early 1979 to provide a sharp and timely reminder of the delicate balance that is maintained in the supply and demand of energy. Iran's internal problems prevented some 5.5 million barrels a day reaching the crude oil market, a shortfall amounting to over 10% of non-Communist World oil consumption, and greater than the combined production of the North Sea, Mexico, and Alaska—non-OPEC producing areas which had helped to provide a temporary supply cushion in recent years.

In spite of the warnings implicit in the 1973 and 1974 'oil shocks', the world still relied on crude oil for nearly 55% of its total energy requirements. So OPEC, in the belief that oil is still underpriced in relation to alternative energy sources, took the initiative and implemented further significant crude oil price rises. The spot market, sensitive to temporary supply interruptions, raced ahead of OPEC and, as a result, a number of individual consignments of crude were sold for well over $20 a barrel early in 1979.

The energy shortages that occurred in 1973–74 and again in 1978–79 were a foretaste of what could arise on a more permanent basis in the 1980s or, if we are lucky, the 1990s. For there is a real danger that the development of energy sources will not keep pace with the rising level of demand. When this will be is still unclear. What is important is that such a 'day of reckoning' is recognized. It will be a day when OPEC countries might withhold some of those needed additional supplies for political, diplomatic or economic reasons; it will be a day when the energy follies of the major consuming countries will be counted. The way events are taking shape, the USA—the world's biggest energy consumer—could well find itself out in front, in the most vulnerable position.

Despite volumes of analyses and warnings, frenzied political activity, and desperate presidential pressure, the USA seems almost as far away from forging a meaningful energy policy as ever. Mr Ali Mohammed Jaidah, the Secretary General of OPEC, lamented in August 1978 that the USA had demonstrated a 'psychological inability' to confront its energy problems. The Trilateral Commission in its 1978 report on 'Energy: Managing the Transition' was more temperate in its analysis but concluded that the US energy performance in the 1973–77 period was 'not consistent with the leadership role expected'.

Again, statistics are illuminating. In 1973, when Arab oil producers boycotted supplies to the USA and Holland, imports accounted for about one-third of US oil consumption. Since then, imports have grown to about

45% of US demand. In 1978 these imports were running at a rate of over 8 million barrels a day. US Energy Secretary, James Schlesinger, confirmed in an interview that year that by 1985 the level of imports could be 9 or 10 million barrels a day—an optimistic estimate in the eyes of some in the energy industry and a far cry from the Federal Energy Administration's ambitious 'Project Independence' attempt in 1974.

The dilemma facing the US Administration is that domestic energy demand will continue to be linked to oil supplies for a far longer period and to a far greater extent than it might have wished. The development of alternative energy sources—be they nuclear, coal, solar or some of the more exotic fuels—is taking much longer than once thought. Environmental constraints and the huge costs involved are two of the prime factors. And there is a natural reluctance of consumers to move away from the convenience of petroleum-based energy, i.e. natural gas and crude oil.

But here is the rub. In line with the worldwide trend, Americans are producing—and consuming—petroleum at a faster rate than they are discovering new supplies. For example, the USA produced an estimated average of 8.6 million barrels a day of oil in 1978, a 5.9% increase on 1977. And yet crude oil reserves in the USA fell by about 1 billion* barrels in the same period, ending at 28.5 billion barrels. In 1970 the reserves were nearer 39 billion barrels.

This means that if no more oil is found in the USA—admittedly an extreme scenario—existing reserves could be exhausted in little more than nine years. The production-to-reserve ratio has thus fallen below the 1:10 level generally accepted as the minimum desirable for a healthy oil regime. The scale of the problem was indicated by Dr Herman Franssen of the Congressional Research Service in a November 1977 report prepared for a House of Representatives' Energy Subcommittee. The oil industry, he said, faced the challenge of adding some 4.5 billion barrels of crude oil and natural gas liquids *each year* to reserves between 1977 and 1985 in order to maintain a 10:1 production-to-reserve ratio. Over the past decade, however, the industry has consistently failed to come anywhere near these levels but for one notable exception: in 1970 12.7 billion barrels of Prudhoe Bay oil on the North Slope of Alaska were added to US proven reserves. But even this outstanding discovery failed to redress the country's long-term imbalance of reserves and production.

It would seem then, in this transient period while we are waiting for conservation measures and alternative energy sources to have a major impact on consumption patterns, that the US oil industry needs every encouragement to find and produce more domestic crude. Companies involved in exploration and production have argued, with some justification, that just

*US billion = 10^9.

the opposite has been happening. They have complained that they have been hamstrung by a plethora of regulations, restrictions, and price controls. Not only have these weakened the economic incentives for investment in what will always be a risk business but the rigmarole of regulations have also dampened the enthusiasm of oilmen who have wanted to try something new in exploration or development. In short, bureaucratic red tape and price controls have restricted a potential increase in domestic oil production, the companies have argued.

It is easy to dismiss these concerns as self-pleading by a prosperous industry, particularly in the USA where public scepticism of oil company statements is so rife. On the other hand, while there is no other institution capable of producing the badly needed oil (in spite of the apparent public opposition to big private companies, there is no widespread desire to create a State-controlled oil enterprise in the USA), the warnings ought to be heeded, particularly when the evidence of the production-to-reserves ratio support these warnings.

The target of much of the industry's criticism has been the Energy Policy and Conservation Act of 1975 which included a provision to keep US crude oil prices below world price levels. Early in 1979 it was estimated that US refiners were buying their crude oil at around $12.90 a barrel on average, some $2 less than the cost of imported crude oil. The explanation for this is largely to be found in domestic oil production policies. 'Old oil', defined as that produced from wells drilled before 1972, has been subjected to the tightest price controls on the basis that the development costs were incurred before OPEC transformed the world oil pricing structure in 1973 and 1974. In December 1978, for example, this 'old oil', accounting for some 35 to 40% of US production, was being sold for an average of $5.68 a barrel. In totality, the pricing mechanism is far more complex, involving a number of oil categories, but the effect has been simple to perceive: the ultimate consumer of oil products—the automobile driver or the purchaser of fuel oil—has been shielded to some extent from the full impact of increasing world fuel prices.

While this has some merit in social welfare terms and is of obvious political attraction—particularly at a time of inflation—it is neither conducive to greater energy conservation effort nor an incentive for oil companies to seek out the hard-to-get and costly fuel resources. Taking this second point first, there are a number of known cases where producers have been reticent to spend money on enhancing the recovery from old oil reservoirs because of the prospect of insufficient economic returns. This enforced reticence should not be dismissed lightly for the US oil industry could do much to reduce the need for imports in this way, given a more favourable pricing structure. (The easing of natural-gas price regulations in 1978 saw an immediate increase in production.) The US oil reserves position can be

Table 1 US oil reserves (billions of barrels) (Source: American Petroleum Institute)

Original oil in place (discovered)[a]	442
Produced to date	109
Proved reserves (remaining for production)	33
Remainder	300

[a]Total US conventional crude oil resources, including offshore and Alaska.

summed up as in Table 1. From these American Petroleum Institute figures it is clear that more than two-thirds of the discovered US oil resources could remain in the ground unless companies applied some advanced and costly production technology.

If only a quarter of this remaining 300 billion barrels could be extracted, the effect would be like doubling the current proved reserves of the Western Hemisphere. A 10% improvement in recovery would effectively double the proved reserves of the USA. How much of this hard-to-get oil will be recovered in practice is currently the centre of intense controversy within the oil industry. Recent estimates of the magnitude of this enhanced recovery potential have varied from 7 billion barrels to 110 billion barrels, while estimates of average daily production have ranged from an addition of 500 000 barrels a day to an extra 2.3 million barrels a day on top of the amount expected to be produced by more conventional means.

But that is not all. Apart from its considerable resources of normal crude oil (both discovered and yet to be found and identified), the USA has huge reserves of so-called unconventional oil—oil which may be very thick and difficult to extract but oil, nevertheless, which could be processed to provide a valuable addition to domestic energy requirements. And it should be remembered that every barrel of imported oil saved by the USA is a barrel made available for other energy-consuming nations, less well endowed with their own resources.

Again, reserve estimates are contentious, but British Petroleum calculates that there might be as much as 2200 *billion* barrels of shale oil present in the United States of America (mainly in Colorado, Utah, Wyoming, and the Alaskan Brooks Range). BP feels that, even with much higher oil prices and improved production techniques, it will not be possible to recover more than one-tenth of those resources. But recoverable shale oil reserves of 'just' 220 billion barrels would satisfy current US oil demand for over 33 years. What is more to the point, shale oil and other nonconventional resources like heavy oil and tar sands will supplement the dwindling output of normal crude oil in future years.

Although a number of American energy companies are looking at the possibility of manufacturing substitute natural gas from coal, the US natural-gas resource base is far from being completely depleted. It can be reckoned that some 516 trillion* cubic feet of US gas has been produced so far, and that economically recoverable reserves are around 210 trillion cubic feet. But it is also estimated that a further 200 to 860 trillion cubic feet of potentially recoverable gas remains to be discovered. With domestic gas consumption running at around 20 trillion cubic feet a year, the USA clearly has decades of gas supplies left.

And then, of course, there is coal, the most abundant of tangible energy sources, accounting for 90% of all known US fossil fuel resources, according to Exxon. Total US recoverable reserves of coal are over 200 billion tons, perhaps a third of the world's total. When set against a domestic production level of some 650 million tons a year, it becomes apparent that the USA has sufficient resources to cover the current level of production for over 300 years.

Such figures have to be viewed with caution, however. They reflect only the potential. As with all of the fossil fuels, there are a number of factors—political, economic, and environmental to mention just a few—which could well inhibit their exploitation. Indeed, this is already the case. Environmental restrictions are slowing the development of surface coal-mines; preventing exploration attempts in certain ecologically sensitive areas; and frustrating plans to exploit some reserves of heavy oil and oil shales.

The experiences of three major oil companies—Shell, Amoco, and Northern Michigan Exploration—provides a case in point. In February 1979, the Michigan Supreme Court supported environmental pressure groups and ruled that plans by the companies to explore for oil in Pigeon River Country State Forest must be stopped. The Court held that drilling would be detrimental to Michigan's dwindling elk herd, the only one east of the Mississippi.

As the public become more aware and appreciative of their environment, so such conflicts between Nature and the quest for new energy sources must grow. The USA may well have plenty of fossil fuels left in the ground. Whether all of these potential energy sources are exploited and whether they are developed in the timescale dictated by rising demand is becoming more and more problematical, even at a time when the country is facing shortages.

The Federal and State legislatures have not yet found a successful formula. The US experience with nuclear energy expansion plans is sufficient proof of this fact. With large reserves of uranium, proven nuclear technology, and an electricity demand growing at some 5% per annum, nuclear power would seem the obvious choice for power generation in the USA,

*US trillion = 10^{12}.

particularly as a means of relieving pressure on fossil-fuel supplies. Indeed, in 1977 nuclear power was providing 11% of US electric utility demand. But now nuclear energy has been caught up in confused environmental, safety, and nonproliferation debates. The ordering rate for new nuclear reactors has fallen substantially below the rate of the early 1970s, with the result that even the previously pessimistic growth assumptions for nuclear output seem optimistic.

The Congressional Research Service (1977) report reviews the various constraints on nuclear development and then adds: 'Federal and State legislation to expand the role of State governments in decisions on siting, environmental effects, and operation of nuclear powerplants also could become a substantially new constraint.'

So, it seems, the legislatures have been frustrating their own efforts to provide their country with a secure and growing supply of energy. Perhaps they will be more successful in promoting greater conservation effort.

CONSERVATION

A new report from Harvard Business School 'Energy Future', edited by Professor Robert Stobaugh and Daniel Yergin, rightly concludes that conservation has been undervalued as a benign source of energy. Domestic oil, gas, coal, and nuclear power could not be expected to increase their combined contribution to US energy supply by more than 15% over the next decade, partly because of political constraints, it was found. On the other hand, conservation could effectively make available more additional energy than could be obtained from all conventional sources combined.

James Schlesinger (1979), referring to his difficulties in pushing through a meaningful energy policy, pointed out that in pushing for greater conservation effort and more realistic energy pricing he seemed to be offering only 'pain'. Dr Herman Franssen (1978), of the Congressional Research Service, adopted a similar analogy in his report. An 'aggressive policy' of conservation and interfuel substitution could help to buy time for the development of alternative energy sources, he wrote. 'The government is in a position somewhat similar to that of a physician with a seriously ill patient who feels in fine shape and is in no mood for a major operation. The physician can only inform his patient of his choices and can administer minor treatment until the patient has consented to the operation.'

In short, the US Administration has still to convince the public and thus the Congress (unfortunately many American politicians prefer to follow public opinion rather than to lead it) that the nation is blatantly wasteful in its use of energy. Homes are poorly insulated; buildings have been constructed with little thought of energy consumption—in very many cases they are overheated in winter and overcooled in summer. Americans continue to have a love affair with the automobile, insisting on several per family and preferring to use them, rather than shoe leather, for even the shortest journey. Unfortunately, public transport systems have suffered as a result.

The savings that could be achieved with even 'minor surgery' are spectacular. President Carter's advisers provided some rough estimates in the wake of the Iranian oil cutback early in 1979. Measures under consideration then could have cut US oil consumption by well over 3 million barrels a day. The *Oil and Gas Journal* (1979) provided an insight into these potential savings: by observing the already imposed 55 miles per hour speed limit, motorists could reduce demand by 100 000 b/d; by eliminating unnecessary driving a further 200 000 b/d could be saved; and by resetting home heating thermostats to 65°F another 300 000 b/d could be conserved. More drastic 'emergency' measures were proposed in order to save another 954 000 b/d—they included the weekend closing of service stations (284 000 b/d), downtown parking restrictions (168 000 b/d), mandatory 65°F thermostat settings for commercial buildings (462 000 b/d), and reduced advertizing lighting (40 000 b/d). On top of all this, a further 1 555 000 barrels a day could be conserved by converting industrial and utility boilers from oil to gas, withdrawing oil from the strategic petroleum reserve, and delaying the phasing out of lead in petrol. As the figures show, even these far-from-comprehensive measures could contribute greatly to improving the United States' energy security. Unfortunately, as the *Oil and Gas Journal* (1979) concluded: 'The trouble is America isn't getting serious about conserving.'

This was demonstrated in President Carter's watered-down Energy Act of 1978, a disappointing Congressional response to his battle cry of 'energy—the moral equivalent of war'. Under his original proposals, the USA could have cut oil imports by 4.9 million barrels a day by the mid-1980s. Official Washington estimates indicated that the actual savings arising from the Energy Act could be half of this figure—between 2.4 and 2.9 million barrels a day. The *New York Times* (1978) reported, however, that critics of the Act felt that it might eventually be shown that the real impact would be less than 1 million barrels a day savings on imports in 1985.

It would be wrong to suggest that there has been no real conservation progress in the USA. Industry was particularly quick to react to the higher prices of energy which suddenly became apparent in 1973–74. Oil refiners, for instance, achieved a 16% increase in efficiency of energy use between 1972 and 1977. The average mileage per gallon (mpg) efficiency of US cars has improved from around 13 mpg to nearer 20 mpg over the same period. The Congress has mandated that car manufacturers should achieve an average fuel economy standard of 27.5 mpg by 1985, although the automobile industry has been lobbying Congress for a relaxation of this mandate. A report by the Congressional Office of Technological Assessment, published in March 1979, concluded that fuel economy was still attainable although, perhaps, at the expense of large-car production. A reduction in the weight of cars, through the increasing use of plastics and aluminium, would help. Clearly more milage efficiency is needed, particularly if the Office is correct in its assumption that, in the absence of an oil embargo, the number of cars in the USA could rise from the present number of 66 million to 91 million by the turn of the century.

Taking a wider view of conservation, studies have shown that US energy consumption had been reduced by the equivalent of nearly 5 million barrels a day in the 1973–77 period compared with the demand level that would have been expected if energy prices and consumption patterns were assumed to remain at pre-embargo levels. Exxon, in its 1978 report on America's energy outlook over the period 1978–1990, foresaw the possibility of conservation efforts achieving a demand reduction equivalent to 17 million barrels a day, or a 25% cut, by 1990. Exxon's projections are shown in Table 2.

Figure 1 shows, however, that even with these optimistic conservation assumptions, US energy demand is likely to continue rising, from the equivalent of 38.1 million barrels of oil a day in 1977 to 44.9 million barrels a day of oil equivalent (bdoe) in 1985 and 51 million bdoe in 1990.

This view of a continually growing energy demand is now being increasingly challenged, however. For one thing, higher prices and a move towards greater conservation have halted the traditional pattern which showed energy growth rates moving ahead of economic growth. For instance, during the period 1960 to 1973, US energy consumption grew at an annual rate of 4.1% on average. In the 1973–77 period, the growth was no more than 0.7% on average; a reflection of the much higher energy prices and the resulting economic recession. According to Chevron's energy outlook published in October 1978, US energy consumption might grow at an average of 2.5% over the period 1977–90—a growth which would result in a 1990 demand similar to Exxon's 51 million bdoe projection. The way energy growth could be decoupled from economic growth is shown in the index in Table 3, produced by Chevron. Energy

Table 2 US Energy conservation: projected demand reduction by sector (%)
(Source: Exxon)

	1977	1980	1985	1990
Industrial	14	19	23	26
Transportation	6	12	25	33
Res. Comm.	13	17	21	23
Overall	11	15	21	25

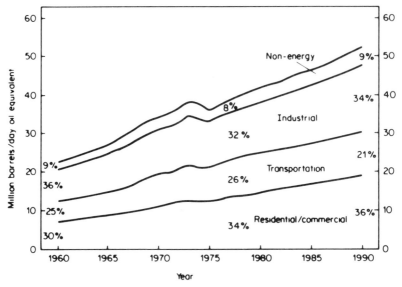

Figure 1 US energy demand by consuming sector, showing percentage shares. (Source: Exxon, 1978)

	Growth rates (% per annum)			
	1960–73	1973–77	1977–80	1980–90
Non-energy	4.1	(1.8)	4.6	2.5
Industrial	3.6	(0.7)	1.7	2.8
Transport	4.1	1.8	1.3	0.6
Res./Comm.	4.6	2.0	2.6	3.0
Overall	4.1	0.7	2.1	2.3

consumption per unit of gross national product was set at 100 in 1972. Table 3 shows the way the index has changed in recent years and the way it could change over the next couple of decades.

But need we assume that energy demand will grow at all in the medium to long term? Amory Lovins, of the Friends of the Earth conservation group, has already told President Carter that by breaking away from the domination of 'hard energies' (fossil fuels like oil, gas, and coal), by matching various energy forms ('soft' or renewable energies wherever possible) with specific demands, and by making a concerted conservation effort, the US, and the world in general, could reduce their total energy consumption.

A similar conclusion resulted from detailed work by the Demand and Conservation Panel of the National Research Council's Committee on Nuc-

Table 3   Energy/GNP index

1972	100
1973	99
1974	98
1975	96
1976	95
1977	94
1985 (est.)	86
2000 (est.)	79

lear and Alternative Energy Systems (CONAES). The panel pointed out (CONAES, 1978) that most people had associated conservation with curtailment. CONAES took the view that, instead, conservation should be seen as a way of enhancing the energy user's perceived welfare, that conservation included technological and procedural changes that allowed us to reduce energy demand without corresponding reductions in the goods and services enjoyed.

CONAES (1978) took quadrillions (quads) as its unit of energy measurement; the US primary energy consumption in 1975 being some 71 quads. With existing policies it was foreseen that US energy demand in the year 2010 might be 136 quads, or almost twice the level of a few years ago. By slowly incorporating measures to increase energy efficiencies, this 2010 level could be cut back to 94 quads, according to the report. Social and economic patterns would be very much like today's.

By adopting a more aggressive approach to conservation and some minor changes in lifestyle, it might be possible to hold primary energy consumption in 2010 down to just 74 quads. Under this scenario, the energy use in buildings would actually fall by 25% from the 1975 levels under the influence of new building and insulation standards. At the same time, the energy used for transport would drop by around 18%, again reflecting new consumer attitudes. Significantly, one-third more energy would be used in industry (itself far more energy-efficient) in order to allow the economy to continue to expand.

In a fourth scenario, CONAES (1978) sees the possibility of the 2010 level of energy consumption falling to an incredible 58 quads. To achieve this level—over 18% below the 1975 level of consumption—some fairly extreme changes in consumption patterns would be necessary. For instance, it was assumed that these would include some shift of the population to warmer climates and a significant proportion of the workforce would have to live close to their factory or office.

The projections are instructive in that they indicate what is possible, in the way of energy conservation, given a positive policy. Unfortunately such policies have been largely lacking. As Chevron's Kenneth Haley (1978) said: '(Energy) Efficiency improvements in the US are comparable to the accomplishments in Japan and the major countries of Western Europe. The progress to date, however, has been primarily a market response to higher prices. The savings in the US have been achieved without the implementation of a grand "Energy Policy".'

At the time of writing, the USA still seems as far away as ever from implementing a meaningful energy policy. The Trilateral Commission put its finger on some of the reasons—deep disagreements between the Executive and Congress, lack of consistent White House leadership, a 'mounting cynicism and sheer lack of understanding of the scope of the problem among the general public', and administrative fragmentation—in particular the inability to co-ordinate Federal, State, and local processes for licensing and siting energy facilities and the myriad of conflicting and cumbersome regulations.

These, then, are some of the obstacles. They form the pessimistic side of America's energy position. On the optimistic side, there is tremendous potential within the USA. The country does have a huge store of its own energy to exploit if it can only break down some of the environmental and regulatory barriers. It also has the ability drastically to reshape its energy consumption patterns through conservation. And, as already stressed, every barrel of imported oil saved is a barrel made available for those countries less well endowed; lesser developed countries which have little opportunity for implementing conservation programmes.

The danger is that these opportunities will be recognized too late. It is still not fully recognized by the Federal Administration and by the public at large that major new energy projects and conservation programmes entail long lead times. A conventional nuclear plant requires some 12 years to move from the planning stage to construction and operation; a major offshore oil or gas field similarly takes five to eight years from the discovery date; an oil shale plant would require a decade of planning and construction.

On the conservation front, it would take five years to replace half the existing US 'gas guzzling' cars, and a decade to replace 90% of them; even if new housing units were built to more energy-efficient standards, it would take 25 to 30 years for such buildings to represent even half of all occupied homes.

Decisions cannot be postponed any longer. The CONAES (1978) report pointed out that Americans had a tendency to want to do things instantly. Adlai Stevenson once remarked: 'We Americans seem never to see the handwriting on the wall until our back is up against it.' Unless the USA changes its energy ways Americans may soon find that their back is against the wall. There they will see what OPEC leaders have been saying and writing for the past few years: that the USA in particular is profligate in its use of energy and laggardly in the development of its own indigenous resources.

REFERENCES

CONAES, 1978, *Science*, **200**, 14 April (Demand and Conservation Panel).

Congressional Research Service, 1977, *Project Interdependence: US and World Energy Outlook Through 1990*, CRS Report.

Dafter, R., 1979, Carter in pursuit of an oil policy, *Financial Times*, 2 February.

Franssen, H., 1978, *World Oil*, March.

Haley, K., 1978, *Fifth Int. Conf. on Energy in the 1980s*, Colorado University, October.

New York Times, 1978, 10 November.

Oil and Gas Journal, 1979, 19 February.

Schlesinger, J., 1979, *Time*, 19 February.

Energy Conservation and Thermal Insulation
Edited by R. Derricott and S. S. Chissick
© 1981 John Wiley & Sons Ltd.

CHAPTER 6

The rational use of gas energy

Robert Jones

Coordinator, Conservation Projects, British Gas Corporation, London

John Rich

Energy Consultant, British Gas Corporation, London

INTRODUCTION

The development of the gas industry since the discovery of the first natural-gas field off the coast of Norfolk in 1965 has been dramatic. Sales have quadrupled and the industry now meets over 30% of the United Kingdom's fuel requirements, excluding transport, compared to just 7% in the early 1960s. With further expansion planned into the 1980s, it is expected that natural gas will meet up to 40% of the UK's heat energy requirements by the middle of the coming decade.

This scale of expansion can find few parallels elsewhere in British industry, but this growth has not diverted the gas industry from its traditional pursuit of utilization efficiency which had enabled it to survive and succeed even when, in 'Towns Gas' days, it had to compete with other fuels at a considerable price disadvantage. Now, when natural gas has brought a degree of pricing advantage to add to controllability, flexibility, and cleanliness, the industry still leads in research and development and customer education in promoting energy conservation. The need to pursue this objective as a major part of national and company energy policies has of course been re-emphasized by the arrival of the energy crisis, with its particular squeeze on oil supplies and prices.

The British gas industry can speak, therefore, from a unique position. Not only has it provided supplies of an indigenous fuel of the highest quality in increasing quantity; but at a time of possible shortage, it has brought a growing awareness of the need to use all fuels wisely.

The importance of making the most of our available energy resources has been understood by British Gas for many years, challenging the supposition that further economic growth and continuing improvements in home comfort necessarily bring a greater appetite for fuel. This view, that much of our future needs can be met by cutting out waste and using fuels efficiently, has now been widely accepted by Government and industry.

This chapter concentrates on describing some of the many initiatives that British Gas is taking to further this end. But, before discussing these, and their impact on the domestic, industrial, and commercial fuel markets, it would be useful to look into the nature of the changing role of gas within the British energy market.

GAS IN THE UK ENERGY MARKET

Today, natural gas is a major force on the UK energy scene. Gas sales totalled nearly 16 000 million therms in the British Gas Corporation's financial year ending March 1979. Some 100 000 people are employed in the many stages involved in the efficient running of the industry, from supply and transmission to marketing and appliance servicing. British Gas meets the daily needs of over 14 million households as well as the complex and often large-scale needs of much of industry and commerce.

But this situation is a far cry from the position in which the industry found itself, only 30 years ago. At the time of nationalization, in 1949, the gas industry was characterized by a large number of small manufacturing companies of mixed viability and in great need of rationalization.

During the 1950s, the gas industry was confronted with a major problem. The availability of suitable gas-making coal, then its principal raw material, became difficult to find, with a consequent effect on price. Prices were rising faster than those for the supply of other coals, particularly for power station use. Whilst this situation remained, the industry had little chance to expand its sales.

Work began to find new feedstocks and to develop the efficient new technology of oil gasification. By 1953 the first new plants were in commission and the industry embarked on a new phase of rationalization and modernization. At the same time, the switch towards these new processes enabled gas transmission pressures to be increased, thereby improving the capacity of the distribution system.

A further landmark in gas industry development was to appear in the late 1950s. By this time, the Gas Council was investigating the possibility of importing natural gas supplies by tanker from Algeria. In the event, an entirely new technology was devised for transporting this gas in liquefied form at low temperatures. A reception terminal with storage facilities was constructed on Canvey Island in the Thames estuary and an initial high-pressure trunkline was established to deliver this gas to eight of the twelve area boards, for use in enriching locally manufactured gas.

Shortly after the first cargoes of liquefied natural gas (LNG) began reaching Canvey on a regular basis, the industry was already preparing itself for a further and very important new phase in its development: the introduction of natural gas from the North Sea, a gas twice as rich in calorific value as existing towns gas and with different burning characteristics.

Even though the use of oil-gasification plants was to cease and the importance of LNG to British Gas was to lessen with the advent of piped natural-gas supplies, the technological achievements made by British Gas during this period remain of considerable importance to our future energy development. The new technology for transporting LNG has opened up an entirely new worldwide trade in energy supply. Increasing quantities are being exported from Africa and many other areas to Europe, America, Japan, and South-East Asia.

Although supplies of natural gas, even without further discoveries, are assured well into the next century, research efforts have continued into gasification techniques—both from oil and coal—in readiness for the eventual time when the industry's activities might include a manufacturing role. Meanwhile, the achievements made at the British Gas research establishment at Westfield in Scotland are already being put into commercial operation by the American gas industry, as it adjusts itself to the prospect of more modest future growth in natural-gas supplies.

The impact of the growing supplies of natural gas on the industry's sales performance from the mid-1960s onwards is clearly shown in Figure 1. As

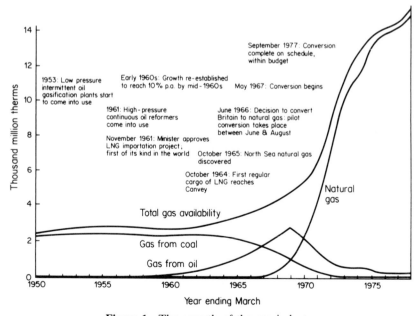

Figure 1　The growth of the gas industry

previously mentioned, sales virtually quadrupled over the ten-year period from 1966 and, even in the last few years, have continued to expand in the face of static overall demand for energy and the decline in sales of some fuels.

To achieve this growth, the gas industry has been faced with a series of daunting challenges. It has had to construct an entirely new national transmission system to distribute supplies throughout England, Wales, and Scotland. This system is shown in Figure 2. Five natural-gas terminals were built and these will shortly be joined by a sixth near Barrow-in-Furness to receive gas from the Morecambe Bay field. Over 3000 miles of high-pressure pipeline have been constructed, plus a series of gas compressor stations to boost the gas through the system. Added to this, the development of underground storage facilities and the extension by several thousand miles of the regional distribution network have brought the total investment to over £2000 million.

To enable natural gas to be burned in existing appliances, a national conversion programme had to be mounted to accommodate the different burning characteristics of the new fuel. By the time this exercise had been completed, successfully both to budget and time constraints, over 13.5 million domestic customers had been visited, representing seven in every ten households in Britain, and 35 million appliances were converted. In addition, some 400 000 commercial premises and 75 000 industrial establishments made this programme the largest single gas conversion exercise undertaken in the world.

Whilst undertaking these massive physical exercises, British Gas continued to place a premium on the high quality of its product and its obligation, within a national energy framework, to ensure the most efficient and rational use of natural gas. Sales for bulk-heat purposes, such as the generation of electricity, have been consistently discouraged. In the early days some agreements were entered into but only on an interruptible basis and to meet contract obligations to the producers. Today, as Figure 2 has shown, this is no longer the case, and, whilst some interruptible agreements remain in force with British industry, British Gas now enjoys a greater flexibility in supply and has much less of a need to place gas in this way. Future gas marketing will remain firmly committed to meeting the needs of that part of the market where the additional values of the fuel can be more fully realized. Attention to these 'premium' markets, British Gas envisages, will allow the industry to continue to provide a growing share of the heat energy supplied.

Current reserve figures provide for the further expansion of sales along these lines well into the 1980s with perhaps an additional 30–40% increase on current levels. Thereafter, British Gas is confident that it can hold such an additional load well into the next century. Whether this can constitute

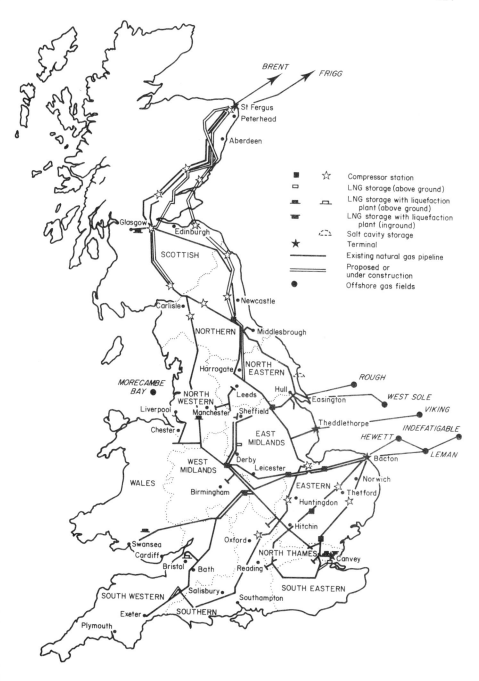

Figure 2 The natural gas transmission system

the platform for the meeting of an even greater share of the country's energy needs, in an increasingly oil-short world, will depend on the degree to which consumers themselves take steps to minimize waste and replace inefficient plant and appliances.

The Government's present estimate of UK continental shelf gas reserves stands at 55 trillion* cubic feet. Of this total, 32 trillion is to be found in the relatively newly-found fields in the northern sector of the North Sea and in the Morecambe Bay field. But these estimates do not include the known reserves from the Norwegian sector of the Frigg field, which are already contracted to British Gas. Other Norwegian fields are thought to contain an additional 8–10 trillion cubic feet of gas, although it is not clear to what extent these will be made available to the UK or whether they will be supplied to West Germany for use in continental Europe.

Plans are also being drawn up to supplement known British sector supplies by the construction of a gas-gathering pipeline system linking smaller gas fields and the many oil fields in the North Sea which have associated gas supplies. At present, much of the gas associated with oil fields is either being flared wastefully or reinjected into the oil reservoirs to await collection later on. The latest increases in international oil prices are reported to have convinced the Department of Energy that, even without links to the Norwegian sector, these reserves are on a scale, and the future supply of additional natural gas so important, to warrant the early start to the construction of such a system.

In this way, British Gas appears well placed to continue as a major indigenous energy provider. But, at the same time, it has been continuing its efforts to ensure that its supplies are used in the most efficient and appropriate ways possible.

GAS INDUSTRY INITIATIVES FOR ENERGY SAVING

The promotion of fuel efficiency is fundamental to the British Gas marketing strategy, in the domestic market and in its approach to industry and commerce.

A major part of the research and development work at three of the four British Gas Research Stations is dedicated to the improvement and refinement of gas-burning appliances, plant and equipment. These are discussed more specifically in the following two sections of this chapter. For the moment, discussion is concerned with the many new initiatives that British Gas is taking within its overall business strategy, as follows.

---

*UK trillion = 10^{18}.

1. Marketing policies

As previously noted, British Gas has pursued a restrictive policy towards providing gas supplies for purposes which can be met adequately by lower-grade fuels, such as coal and the heavier fuel oils. An example of this, and a contrasting picture to the European energy scene, is the restriction on supplies to power stations for electricity generation. In Britain, natural gas provides less than 4% of the fuel used for firing power-station boilers, compared to around a 10% contribution in France and Italy, and as much as 30% in the Netherlands and West Germany.

In contrast to this restrictive policy, in the UK sales to the domestic market have been rising steadily. Here the intrinsic or 'premium' values of gas, such as its controllability, cleanliness, on-tap availability, and high burning efficiency, can best be realized. Sales to industry and commerce have also been rising. The substitution of new gas-fired equipment, displacing older less efficient fuels, can often bring dramatic improvements in energy saving and the working environment.

British Gas actively promotes the use of home insulation with the installation of central heating systems and research has shown that gas-fired homes have a higher proportion with insulation that those using other fuels. The availability of a range of servicing agreements, to suit all types of appliances and conditions, also ensures that appliances are maintained to the highest standards of efficiency and safety.

The controllability of gas is also an important feature, as witnessed by the high level of use of time clocks and thermostats with gas-fired heating systems, bringing lower fuel bills for the consumer and less energy consumption for the country.

2. Appliance design and technical developments

The active marketing of new domestic appliances brings associated savings in energy through the displacement of older, less efficient equipment. Work on domestic boilers and heating systems has brought considerable gains, whilst the replacement of pilots by spark ignition on cookers has also saved worthwhile quantities of gas.

The employment of advanced engineering design in the industrial area, often to individual equipment specification, has brought very considerable savings, up to 40 or even 50% in many cases. Specific examples are given in the section dealing with industry and commerce, and it will be seen there that, where large-scale steam systems are in use, very substantial energy savings can be achieved by their partial replacement by individual gas-fired appliances.

3. Information and advice

Recognizing that the bulk of gas used in the UK is used in buildings, British Gas is providing an ongoing programme of assistance to architects. A computerized program, THERM, is available, on a commercial basis, to assist in the design of energy-efficient buildings. Recognizing, too, the key position of young architects in shaping an energy-efficient future for the UK, British Gas has introduced its 'Design for Energy Management' competition for architects in training. In addition to cash prizes, part of the award is a course in energy efficiency at the Watson House laboratories. Along with the other major energy suppliers, the Corporation also gives financial support to the RIBA Energy Initiative, chiefly concerned with retraining midcareer architects in energy conservation in the built environment.

Advice to industry and commerce is provided on a local level by teams of qualified engineers attached to the Technical Consultancy Service (TCS) of the twelve gas regions. The activities of the TCS are highlighted every year by the Gas Energy Management (GEM) competition, with awards made to those companies which, with the support of the local region's TCS, have achieved significant savings in energy use. Supported throughout energy and commerce, and with the backing of the Department of Energy, the GEM awards have become a major event in the energy-saving calendar.

A notable British Gas 'first' in the provision of information was the establishment, shortly after the energy crisis of 1973–74, of the British Gas School of Fuel Management. Located at the site of the industry's Midlands Research Station, the School has already trained more than a thousand managers and operatives in the more efficient use of fuel in industry and commerce. It was also used to start the Department of Energy's National Energy Managers' Courses, providing first a venue and then ongoing administration from the autumn of 1979.

The year 1979 also saw a new and important initiative in the field of information supply with the opening of the first British Gas Energy Advice Centre. Located in a prime shopping position in the centre of Birmingham, the Centre provides information and advice to customers in all markets on how to use gas and energy in the most efficient ways. Lecture rooms, video facilities and a library are amongst the facilities available. The centre is an experiment in consumer education and is providing valuable experience to back the Corporation's view that energy efficiency has to be marketed, like any other worthwhile product, if it is to have a significant impact and make a worthwhile contribution to the nation's energy strategy.

4. Longer-term research

The continuing work at the British Gas Research Station at Westfield in Scotland into coal- and oil-gasification techniques has already been referred to. As the world leader in the production of substitute natural gas (SNG), British Gas is accumulating knowledge of how to use these feedstocks to make gas again when Britain eventually needs to return to gas-making processes. High production efficiencies are already available and further work is continuing to extend the range of feedstocks for which the processes may be used.

5. Energy saving within the gas industry

As a major employer and fuel user in its own right, British Gas has taken significant steps to improve its own energy efficiency. Improvements have been made in good housekeeping practices in offices and showrooms, and campaigns have been introduced to improve the running of the transport fleet. Results are constantly monitored and the programme added to as appropriate opportunities arise. Gas regions have featured as prizewinners in local fuel-efficiency competitions alongside other industries.

In addition to these specific measures, British Gas is regularly developing new ideas in the field of education for both domestic and business consumers. The industry has long held the view that improvements in home comfort and future economic growth can be achieved without substantially increasing demand for fuel.

GAS IN THE DOMESTIC ENERGY MARKET

From source to user, gas is the most efficient fuel available. Figures derived from official statistics show that conversion and distribution losses are highest with electricity (72%), followed by oil (11%), solid fuel (9%) and, finally, gas (7%) (Department of Energy, 1979a).

Losses are also associated with the use of the different fuels within the home. Electric appliances can be operated at very high levels of efficiency but, even here, losses are associated with such problems as the continuing operation of the equipment at times when it is not required. The controllability of gas in home heating applications eliminates such waste.

Table 1 sets out estimated efficiencies for the various fuels in different applications. The figures are considered accurate for the results typically achieved when modern appliances are correctly installed and are maintained in good condition.

One of the main problems associated with using energy in the home is to

Table 1 Estimated fuel efficiencies in the home

Process	Fuel	Efficiency (%)
Fires	Electricity	100^a
	Gas	55
	Solid fuel	40^b
Water heaters	Electricity	85^a
	Gas	50
Central heating	Electricity	90^a
	Gas	70
	Oil	70
	Solid fuel	65

[a]Electricity is converted into heat at 100% efficiency but, as storage heaters produce heat throughout 24 hours and at times when it is not required, overall efficiency rates are lower. Heat losses from pipes and the tank have been taken into account for water heating.
[b]Efficiency improves up to 65% in open fires with high-output back boilers.

minimize losses due to part-loading. Work at the Watson House Research Station on domestic boilers and domestic heating systems has shown how it is possible to design systems so that losses due to such part-load operations can be virtually eliminated. Modern gas boilers can now be operated throughout the year at around 70% efficiency; that is to say, at a primary energy efficiency of 65%, compared with $26\frac{1}{2}$% for electricity.

It has also been demonstrated that natural gas can provide heat and hot water, even for 'low-energy' housing, at lower fuel use and consequently less cost, than other fuels. The promotion of insulation with central-heating sales remains a high priority. By the end of 1975, 73% of homes with gas central heating had some kind of roof insulation, while equivalent figures for electrically heated and solid-fuel heated homes were 49% and 58% respectively.

The consistently high rating of gas in efficiency terms in Table 1 is based partly on its premium characteristics of controllability and cleanliness and partly on the advances made in instrumentation and controls available for the correct programming of gas-fired heating systems. Among the equipment now available to the householder, and essential in his fight to keep down fuel bills, are room and cylinder thermostats, time controls, and individual radiator thermostats. Attention to the careful setting of these brings substantial rewards. Turning down the room air thermostate by 3°F, for example, is normally expected to bring a 10% saving in gas used for heating.

The direct use of gas in individually owned or controlled appliances has been shown to use less energy than the same customer might consume in a scheme employing communal heating. Studies have shown that, on average, communal systems use as much as 60% more fuel than schemes with individual central-heating appliances. This analysis has made a strong case for the continuing supply of fuel to households on an individually metered basis wherever possible. In this way, energy savings of a substantial scale will be achieved whilst preserving the right of the individual to select his own heating system and decide when he requires to use it.

In the final section of this chapter, it is shown that sales of gas to households have more than doubled in the last ten years, taking the share of the domestic heat market supplied by gas to 44%, compared to around 10% in the early 1960s. This increased adoption of gas, mainly at the expense of less efficient solid-fuel appliances and systems, has resulted in the amount of fuel supplied to the domestic sector having hardly increased over this period, despite the increase in new house construction and the general improvement in home heating comfort. The increased penetration of gas in the domestic market has been central to achieving these important objectives without the use of more fuel. Many opportunities also exist for similar moves in the industrial and commercial sectors.

GAS IN INDUSTRY AND COMMERCE

The expansion of gas sales in the industrial and commercial markets has been even greater than growth in the domestic sector. In 1969, sales to these markets accounted for 36% of total gas send-out, whilst, in 1978, this proportion had risen to 54%.

The implications of this growth, in terms of the market shares of gas and the other fuels in these two markets, are shown in the charts in Figure 3. Again, gas has been able to increase its sales at the expense of less energy-efficient fuels. These are now left to cover applications that would be inappropriate for gas, and continued gas sales increases are likely to be confined to replacement of the lighter oil products.

Gas sales to the smaller consumer, those taking less than 25 000 therms a year, are maintained on a tariff basis in a similar manner to the domestic market. Loads larger than this are supplied on contract at market-related prices.

It has already been noted that the need to meet initial supply contracts and the limited storage facilities available in the early days of natural gas gave rise to the need to sell quantities of gas on an interruptible basis in the face of fluctuating seasonal demands. These loads could be discontinued during times of peak heating demand in periods of severe winter weather. Sales for nonpremium uses, made on this interruptible basis, are today proportionately less necessary, as gas storage facilities have been expanded and the early contracts for supplies from the southern fields have

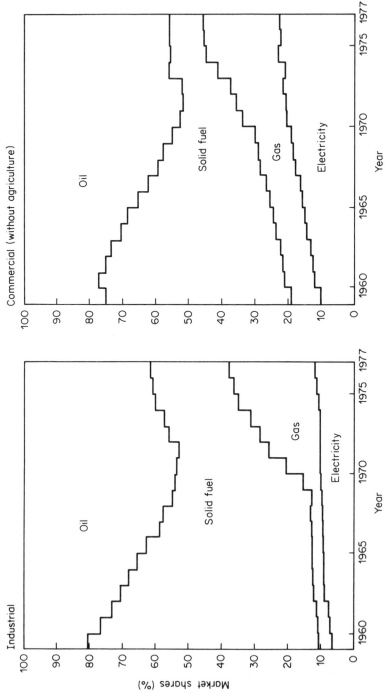

Figure 3 Market shares in the industrial and commercial energy markets: heat-supplied basis

been reconsidered in the light of increasing availability of Frigg gas. The Morecambe Bay field will also assist, not only because of its north-westerly location, but also as its full ownership by British Gas will allow it to be used, to some extent, as a seasonal supply facility.

Supplies of gas for nonpremium uses are unlikely to increase in future years. Instead, efforts will be concentrated on deriving the maximum benefit from the fuel in industrial and commercial operations. In many areas, this does not only mean the replacement of old, less efficient systems and appliances which are already using a premium fuel, such as the lighter-grade oil products. It also means the replacement of complete systems that currently require the input of bulk heat. Steam systems, fired by large-output boilers, are increasingly being investigated to see whether it would not be more economical to use systems based on smaller boilers and individual gas-fired equipment. In this area, the British Gas Technical Consultancy Service is active in cooperation with engineers at the Midlands Research Station (MRS).

Work at the MRS is geared primarily towards improving fuel use efficiency in industry and commerce. A selection of recently developed equipment follows.

Self-recuperative burners

Heat recovery by preheating burner combustion air is a well established practice but, hitherto, generally limited to large, continuous, purpose-built furnaces. British Gas engineers have now designed a range of self-contained, high-efficiency burners using the recuperative principle. These are now available for use in a wide variety of batch furnaces. The replacement of traditional burners typically brings a raising of efficiency—by as much as 30% on existing equipment. Up to 50% is obtainable on new furnace designs. Winners of Gas Energy Management (GEM) awards in recent years who have installed recuperative burners have confirmed the achievement of payback periods for investments as a low 18–24 months. These burners are now becoming increasingly available on the open market.

Rapid-heating furnaces

Traditional designs of metal reheating furnaces have changed little over the years. Situated apart from the production-line, furnaces of low thermal efficiency are in widespread use. With long heating-up times, they are frequently wasteful in material losses. Designs are now available using purpose-built, rapid-heating units as an integral part of the production line. These furnaces offer rapid heat-up time, lighter weight, high thermal efficiency, less material wastage, and improved environmental conditions, to add to significant monetary savings. Payback periods normally fall in the 18–24 month range.

Vat and tank heating

The heating of industrial vats and tanks constitutes a major part of the industrial heat energy load. It has been estimated that some 2500 million therms of heat energy are consumed annually on this process, equivalent to around 10% of the total energy consumption in industry.

Most of this load is met by central steam boilers, with lengthy distribution systems and, consequently, of low overall efficiency. By using specially developed immersion tube burners, incorporating the recuperation technique, overall efficiencies are greatly improved with substantial fuel and cost savings.

Other new developments

Following extensive fieldwork in British industry, a recent report from the Department of Energy (1979b) has concluded that as much as 30% of the fuel used by industrial companies could be saved. The report quantified not only the general improvement available from better housekeeping practices but also that arising from the replacement of inefficient processes and equipment by more advanced techniques.

Keeping itself at the forefront in the development of these new techniques and in the supply of the necessary advice and information, British Gas R&D continues to devote much of its work to the examination of current systems requiring bulk heat input to a central source and where flexibility of output cannot be readily matched to variations in seasonal and daily heat demand. This has produced pioneer work in two areas to limit this large-scale and wasteful use of fuel. These are:

(a) the design of modular boiler systems, whereby a number of boilers are linked, all being in operation during periods of peak demand but some being shut down entirely in off-peak periods, allowing the remaining ones to operate at full efficiencies; and

(b) the development of free-standing appliances and equipment located throughout the factory or offices, such as radiant tube space heaters, which avoid the need for a central boiler system.

With many of these central boiler systems now operating on high-cost oil products, further developments in this area and the increased adoption of gas options in industry and commerce are now expected.

LONGER-TERM CONSIDERATIONS

Earlier in this chapter it was shown that, from a supply stance, British Gas is well placed to meet further demand for its fuel into the next century. As the price of oil continues to rise, the incentives for further offshore exploration will increase and additional reserves will most probably be found. The likelihood of the commissioning of a gas-gathering pipeline system is also of major importance and this should go a long way towards avoiding the wasteful flaring of gas.

Future supply prospects for coal are also bright, with the country's known reserves able to meet demands for the next 300 years. Developments in new coal technology have been dealt with elsewhere in this volume and it is evident that coal and natural gas provide this country with its long-term energy insurance: coal for the supply of the necessary bulk heat for electricity generation and essential crude heat purposes in industry; gas to meet the growing needs for premium fuel applications in the home, in industry, and in commerce.

For the UK, other future energy sources, such as nuclear and hydro, tend to involve social, financial, and environmental difficulties that some consider to be out of proportion to the likely contribution. For most countries, the decision to build the controversial fast breeder reactor in quantity is probably one of the hardest facing their governments. As far as North Sea oil is concerned, we cannot depend on these reserves for more than a temporary respite to our energy problems.

In this context, the developments and initiatives by British Gas take on added significance. Much of the work has contributed to keeping the total

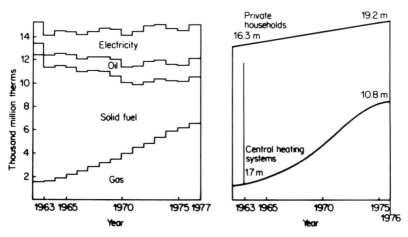

Figure 4   Heat supplied to the UK domestic market compared with growth in comfort demand

level of increase of fuel consumption down to a marginal annual rate of less than 1%. This has been achieved whilst still maintaining economic growth and improvements in home comfort on the scale to which we have become accustomed. That this is attainable is demonstrated in Figure 4, which charts the level of fuel supplied to the domestic market each year for the last fifteen years. It can be seen from this figure that the total amount of fuel supplied has remained remarkably static, despite the growth in the number of private households by nearly three million and the increase in the number of central heating systems installed by over nine million.

Similar improvements in efficiency are quite possible in industry and commerce. The natural-gas industry is leading the way in many areas and, if others followed, there could be a significant reduction in national energy demand. With techniques to produce SNG at an advanced stage of readiness, British Gas is looking forward to its future position as the energy leader in Britain, not only in the fields of research and development, to take us into a new era in energy efficiency, but also in the supply of advice and information to the benefit of its customers on the rational use of fuel.

REFERENCES

Department of Energy, 1979a, Digest of U.K. Energy Statistics, HMSO.
Department of Energy, 1979b, Energy Conservation: Scope for New Measures and Long Term Strategy. Energy Paper No. 33, HMSO.

Energy Conservation and Thermal Insulation
Edited by R. Derricott and S. S. Chissick
© 1981 John Wiley & Sons Ltd.

CHAPTER 7

Coal utilization

R. C. Payne and W. Lawson

National Coal Board, Cheltenham

A. COAL AS A BASIC ENERGY SOURCE

1. What coal is

Coal is a sedimentary rock of vegetable origin. It was formed over a period of about 80 million years—the Carboniferous Period—which itself ended over 200 million years ago. Vast deposits of plant remains accumulated in semitropical swampy areas, were subjected to bacterial action, became submerged due to gradual land subsidence, and were overlain by mud or sandy deposits often borne along by huge rivers forming deltas in the swampy region. This sequence of biochemical decomposition, submergence, and sedimentation was repeated many times. Deposits consolidated due to the depth of overburden; pressure, heat, and earth movements acting over geological time gradually produced a series of coals:

wood and plant debris,
peat,
lignite,
brown coal,
bituminous coal, and
anthracite.

The banded appearance of many coals is due to differences in the original plants and in the conditions under which they were turned to coal.

The position of a coal in this 'coalification series' is described as the *rank* of the coal, the measure of its maturity; so geologically younger coals are low rank and the older coals are high rank. Change of rank is manifested by progressive changes in coal properties and behaviour. Thus with increasing rank, carbon content and calorific value increase, oxygen, hydrogen, inherent moisture, and volatile content decrease, and caking power rises from zero to a maximum in the higher-rank bituminous coals and disappears again as anthracites are approached.

139

In general, the deeper the seam the higher the rank of the coal, but the differences in rank at different depths in the same colliery are usually less than the differences in rank between different parts of the same seam across a coalfield.

2. Coal classifications

There have been many classifications of coal, both scientific and practical. Thus scientific classifications may be based on the proportions of carbon and hydrogen in coals, subclassified according to caking power, calorific value, volatile matter, and so on, in the coal substance calculated to a moisture- and mineral-matter-free basis.

In British coals, as rank increases, carbon content increases from about 75 to 93 wt.%, hydrogen decreases from about 6 to 3%, and oxygen content from nearly 20 to less than 3%. These data relate to the vitrinite of coals, the main and characteristically bright 'maceral' of coal; the other macerals are inertinite (dull ground mass) and exinite (often rich in plant remains and spores).

The UK National Coal Board classifies coals on their volatile matter and caking power (Table 1), characteristics that are related to their commercial applications, though in evaluating coals commercially their size grading, moisture and ash characteristics, and their variability are taken into account. An international (ECE) classification has a broadly similar structure and is based on dry ash-free volatile content, calorific value on a moist ash-free basis, and swelling and caking indices of coals.

3. Coal, oil, and natural-gas reserves

Although coal accounts for less than one ten-millionth part of the Earth's mass, estimated world reserves are still enormous at some 6000 000 million tons, sufficient for 250–300 years at the present rate of consumption. They are far larger than those of oil or gas, which appear to be sufficient for only about 40 years (see Table 2). The reserves are those considered to be economically recoverable at present prices and using current technology.

Rising energy prices and novel extraction methods could enhance the amount of recoverable reserves, especially for coal, for which new exploration lagged prior to the 1973 oil crisis.

Much research and development effort is being given to improving methods of extracting all the fossil fuels to maximize the finite available reserves. In the USA, some $75 million was officially allocated for oil and gas extraction R&D alone in 1978, while the NCB spent £15½ million on

Table 1 The coal classification system (revision of 1964)[a] (Source: National Coal Board)

Coal rank code[d,e]			Volatile matter dry mineral matter free (dmmf) (%)	Gray-King coke type[b]	General description
Main class(es)	Class	Subclass			
100			under 9.1		*Anthracites*
	101[c]		under 6.1	A	
	102[c]		6.1–9.0	A	
200			9.1–19.5	A–G8	*Low-volatile steam coals*
	201		9.1–13.5	A–C	
		201a	9.1–11.5	A–B	Dry steam coals
		201b	11.6–13.5	B–C	
	202		13.6–15.0	B–G	
	203		15.1–17.0	E–G4	Coking steam coals
	204		17.1–19.5	G1–G8	
300			19.6–32.0		*Medium-volatile coals*
	301		19.6–32.0	A–G9 and over	
		301a	19.6–27.5	G4 and over	} Prime coking coals
		301b	27.6–32.0	G4 and over	
	302		19.6–32.0	G–G3	Medium-volatile, medium-caking or weakly caking coals
	303		19.6–32.0	A–F	Medium-volatile, weakly caking to non-caking coals
400 to 900			over 32.0		*High-volatile coals*
400			over 32.0	A–G9 and over	
	401		32.1–36.0	G9 and over	} High-volatile, very strongly caking coals
	402		over 36.0	G9 and over	
500			over 32.0		
	501		32.1–36.0	G5–G8	} High-volatile, strongly caking coals
	502		over 36.0	G5–G8	

Table 1 (Continued)

Coal rank code[d,e]			Volatile matter dry mineral matter free (dmmf) (%)	Gray–King coke type[b]	General description
Main class(es)	Class	Subclass			
600					
	601		over 32.0	G1–G4	
	602		32.1–36.0 over 36.0	G1–G4	High-volatile, medium caking coals
700			over 32.0	E–G	
	701		32.1–36.0 over 36.0	E–G	High-volatile, weakly caking coals
	702		over 32.0	C–D	
800			32.1–36.0 over 36.0	C–D	High-volatile, very weakly caking coals
	801		over 32.0	A–B	
	802				
900					
	901		32.1–36.0 over 36.0	A–B	High-volatile, noncaking coals
	902				

[a]Coals with ash of over 10% must be cleaned before analysis for classification to give a maximum yield of coal with ash of 10% or less.

[b]Coals with volatile matter of under 19.6% are classified by using the parameter of volatile matter alone: the Gray–King coke types quoted for these coals indicate the general ranges found in practice, and are not criteria for classification.

[c]In order to divide anthracites into two classes, it is sometimes convenient to use a hydrogen content of 3.35% (dmmf) instead of a volatile matter of 6.0% as the limiting criterion. In the original Coal Survey rank coding system, the anthracites were divided into four classes then designated 101, 102, 103, and 104. Although the present division into two classes satisfies most requirements, it may sometimes be necessary to recognize more than two classes.

[d]Coals that have been affected by igneous intrusions ('heat-altered' coals) occur mainly in classes 100, 200, and 300, and when recognized should be distinguished by adding the suffix H to the coal rank, e.g. 102H, 201bH.

[e]Coals that have been oxidized by weathering may occur in any class, and when recognized should be distinguished by adding the suffix W to the coal rank code, e.g. 801W.

Table 2   Fossil-fuel reserves and consumption

Fuel	World reserves (10^3 mtce)	(%)	World consumption 1976 (10^3 mtce)	(%)	Estimated lifetime of reserves (years)
Coal	607	72	2.8	30	250
Oil	152	18	4.4	46	35
Gas	84	10	2.3	24	40
Total	843	100	9.5	100	

mining R&D compared with £8½ million on all aspects of coal utilization (Kaye, 1979).

There is a great imbalance between the occurrence of oil and population. Saudi Arabia, for example, with a population of about seven million, had 23.2% of the world's estimated oil reserves (1978), compared with India's 548 million population having only 0.5% of world oil reserves.

Coal distribution, production, and population, on the other hand, are more evenly matched, though coal is more abundant in the northern hemisphere. Thus in 1976 the distribution of coal reserves was estimated to be: USA 28%, USSR 17%, China 16%, UK 7%, Germany 6%, India 5%, Australia 4%, South Africa 4%, rest of world 13%.

In any event, there seems to be little doubt that, by the year 2000, the demand for oil and gas will outstrip the supply. As reserves of these fuels are depleted, a change in the usage pattern of fossil fuels will occur. Coal will gradually return to its former dominant position, initially in those fields where it can most easily replace oil and gas because the technology is known: these are power generation, steam raising, and heating. Indeed where possible it should prove more economic to burn coal instead of oil or gas than to manufacture oil or gas from coal, and to restrict the fluid fuels to premium uses such as transport and petrochemicals. A recent International Energy Agency investigation into coal's potential as a replacement fuel (IEA, 1979) indicates that if OECD governments encouraged coal production and trade, the OECD's net oil imports might be reduced by as much as 7 million barrels/day by the year 2000 (the present OECD consumption is 40 million barrels a day).

4. UK coal reserves

Fortunately Britain has ample reserves of coal to allow us to take such measures; our coal reserves are far more abundant than those of oil or natural gas, as Table 3 shows.

Table 3   UK fossil-fuel reserves

Fuel	Reserves (10^3mtce)
Natural gas[a]	1.2
Oil[a]	2.4
Coal[b]	45.0

[a]Proven reserves. (Source: Department of Energy, 1978b.)
[b]Technically and economically recoverable reserves. (Source: IPC, 1978.)

The British coalfields appear to have been deposited in four large areas:

(i) the Midlands, Lancashire, and North Wales;
(ii) Northumberland, Durham, and Cumberland;
(iii) Scotland; and
(iv) South Wales, Forest of Dean, Bristol area, and Kent.

Geological and other forces worked on these deposits so that, as well as increasing in rank, some areas were removed to leave detached exposed coalfields, while others were concealed by newer noncarboniferous rocks.

The final distribution of the various types of coal is shown in Table 4, from which the particular importance of the English central coalfields is evident. The table also hints at the uses to which the various types of coal are put, though the term 'gas coal' no longer has its original connotation since natural gas now supplies that market.

5. Coal's share of the energy market

Although the UK has far greater reserves of coal than the other primary fuels, its recent (1976) pattern of energy consumption is similar to that of the developed countries of the world as a whole, as may be seen from Table 5. Figure 1 shows how coal lost ground to oil and natural gas in supplying the UK's primary energy requirements between 1960 and 1977. In recent years, however, the respective contributions of coal and oil appear to have stabilized, while gas is increasing at the rate of about 6% per annum.

In the UK, the major use of coal is for electricity generation, as Table 6 shows. Essential factors which influence the selection of a fuel for a given application from the economic (not necessarily conservation) viewpoint are:

(i) availability of the fuel at the point of use;

Table 4 UK coal reserves by type and location

NCB coal rank code	Description of coal	Main reserves	Subsidiary reserves
100	Anthracite	South Wales	Kent
201/202	Low-volatile dry steam	South Wales	Kent
203/204	Low-volatile coking steam	South Wales	Kent, Lancashire
301	Prime coking	South Wales, Durham	South Wales, Lancashire
400	Coking/gas—very strongly caking	Durham, Yorkshire	
500	Coking/gas—strongly caking	Yorkshire, Durham	Northumberland, Lancashire, Nottinghamshire and North Derbyshire, Cumberland, North Staffordshire, South Wales, Bristol and Somerset
600	Coking/gas—medium caking	Yorkshire, Nottinghamshire and North Derbyshire, Lancashire, Northumberland	Central Scotland, North Staffordshire, Durham, North Wales
700	General purposes—weakly caking	Yorkshire, Nottinghamshire and North Derbyshire, Northumberland, Central Scotland	Lancashire, Cannock, Ayrshire and Dumfriesshire, Forest of Dean
800/900	High-volatile steam and house—very weakly caking and noncaking	Nottinghamshire and North Derbyshire, Yorkshire, South Derbyshire, and Leicestershire, Warwickshire, Fife Durham, Central Scotland	Central Scotland, Lothians, Ayrshire, South Staffordshire, Lancashire, Cannock and Northumberland
100H/200H/300H	Heat-altered		Ayrshire, Fife. South Wales, Bristol and Somerset

Table 5   World energy consumption in 1976 (Source: British Petroleum, 1977)

Region	Consumption (mtce)					
	Oil	Solid fuel	Natural gas	Water power	Nuclear	Total
North America	1544.1	621.9	952.9	218.3	90.6	3427.8
South America	316.9	88.2	68.9	61.0	1.2	536.2
Western Europe	1200.9	442.3	284.4	158.1	49.5	2135.2
Rest of World	1832.1	2146.9	637.3	187.9	31.8	4836.0
Total	4894.0	3299.3	1943.5	625.3	173.1	10 935.2
World percentage	44.7	30.2	17.8	5.7	1.6	100.0
UK percentage	44.3	35.0	16.6	0.5	3.6	100.0

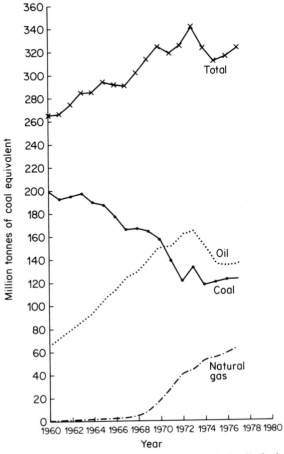

Figure 1   UK annual consumption of fossil fuels. (Source: Department of Energy, 1978b)

Table 6   UK use of fossil fuels in 1977 (Source: Department of Energy, monthly, *Energy Trends*)

Final users	Consumption (mtce)			
	Coal	Oil	Natural gas	Total
Electricity generation	78	18	2	98
Iron and steel	14	5	2	21
Other industries	11	30	22	63
Transport	—	52	—	52
Domestic	14	6	26	46
Other	3	15	6	24
Total	120	126	58	304

(ii) cost of the fuel per useful therm of heat produced, involving combustion plant efficiency and effective heat utilization;

(iii) capital cost of combustion plant and ancillaries; and

(iv) labour costs and maintenance charges.

Introducing a conservation factor should favour the use of coal, because of its vastly greater reserves. In general, solid fuel costs less per bought therm than oil and gas, but capital, labour, and maintenance costs of the plant are higher, so that utilization research and development are directed towards their reduction. This is considered more fully later.

6. Marketing for conservation

It is self-evident that the most effective utilization of coals, and, indirectly, their conservation, demands careful matching of the coal to the equipment and conditions in which it will be utilized. Among other things, this has required a knowledge of the coal rank code, size grading, calorific value, ash content and ash characteristics (e.g. propensity for clinkering), moisture content, and any special requirements such as sulphur, chlorine or phosphorus contents. Consistency also rates high in marketing coals, and for this reason the National Coal Board go to considerable effort and expense in operating an effective Scientific Control service.

From the conservation viewpoint, it is desirable not only to be selective about using coals that are required to have specific properties, but also to divert to less demanding markets those coals that can be used in more tolerant applications, such as steam raising. Nearly all coals may be used for the latter, providing the plant is suitably designed and operated. It is often economic, for instance, to raise steam with high-ash coals which have had minimal preparation between the seam and the furnace. The new technique of fluidized bed combustion is particularly adaptable to this approach.

In pricing industrial coals, account is taken of the nominal size, rank, calorific value, the volatile, moisture, sulphur and ash contents, and ash fusion temperatures. A full 'marketing analysis' can be supplied when technical considerations relating to industrial plant warrant it, but, since coal is a natural product, the analysis is typical only and not necessarily precise in respect of each individual consignment.

Most coals are cleaned and some are graded for the market in preparation plants incorporating washing and screening processes. All coal-washing plants depend on coal's lower specific gravity compared with inert mineral matter. The larger graded coals are mainly used in the domestic market or are crushed and rescreened for industrial use. 'Graded' coals are those sized by separation between two screens, to distinguish them from 'smalls' which have only an upper size, and 'large' coals for which only a lower size is stated. Bituminous coals are graded into large cobbles (75–150 mm typically), cobbles (75–125 mm), trebles/large nuts (50–75 mm), doubles/nuts (25–50 mm), and singles (12½–25 mm). Welsh anthracites are graded to still smaller sizes, to suit the gravity-feed boiler market, for example.

B. MINING THE COAL

1. Ensuring future capacity

At first sight, conservation and coal extraction seem incompatible, for coalmining is an extractive industry: once mined, prepared, and used, there is no resource to conserve. But further consideration will show that efficient and economic mining and preparation followed by wise utilization in efficient equipment is indeed conservation, for it is making the best use of a valuable natural resource.

The necessity to expand the industry as we approach the year 2000 and beyond has already been noted: coal production will need to rise as oil and natural-gas reserves decline. This responsibility to expand the industry and to develop processes for manufacturing oil and gas from coal is linked to the need to modernize the industry (the average age of British collieries is about 80 years) and to keep it competitive with the other energy industries.

Competition from cheap imported oil in the 1960s meant that many smaller older collieries were then no longer viable and many closures were necessary; the proportion of coal reserves regarded as economic fell by three-quarters (NCB, 1974) and total output declined from 184 million tonnes in 1960 to about 120 million tonnes in 1974. The 1973 oil crisis highlighted the need to arrest these trends, for losses through exhaustion of capacity amounted by then to some 3 million tonnes every year and even the highly productive collieries in the central coalfields shared this experience, putting their longer-term future at risk.

A £1400 million investment plan to 1985 was therefore agreed between Government, NCB, and unions, to provide 42 million tonnes of new capacity, comprising:

(a) 9 million tonnes by extending the lives of existing pits that would other-wise exhaust, by opening up new reserves;

(b) 13 million tonnes from major improvement schemes to increase outputs from existing pits; and

(c) 20 million tonnes by opening new mines, notably near Selby, in the Vale of Belvoir, and in Staffordshire.

Increased opencast output (then 10 million tonnes per annum (mtpa)) was planned to take advantage of its lower mining costs. (An important feature of deep-mined coal winning is that labour costs account for more than half the production costs.) The 1973 plan was later extended by the Department of Energy (1978c), to continue increasing new capacity by 4 mtpa to the end of the century. These proposals were essential elements in the Government Green Paper on energy policy (Department of Energy, 1978a).

2. Winning the coal

Since nationalization in 1947 and particularly in the 1960s, the productiv-ity of coalfaces has increased several times over, from about 700 tonnes per day (tpd) to as much as 7000 tpd in the best conditions, thanks to the introduction of new machines like the Anderton shearer loader (in 1952), self-advancing roof supports (1956), and the flexible armoured face con-veyor. This combination made the longwall prop-free-front face possible. The shearer, mounted on the conveyor, travels along the face like a bacon slicer, without obstruction, there being about 2 m clearance between the row of hydraulic roof supports and the coalface. Coal falls into the con-veyor and is assisted along it by scraper plates attached to moving chains within the conveyor to the end of the face, where it is transferred to another conveyor in the roadway, thence into tubs for haulage to the pit bottom and out of the mine.

Average performance in the industry has not increased to the extent anticipated in 1974, however, for reasons identified by the NCB's Director of Mining Research (Tregelles, 1978) as these:

(i) coal production at the face has outstripped the capacity to wind it to the surface—hence the need for reconstruction and concentration schemes for existing collieries and for new mines;

(ii) problems of work payment—hence the introduction of incentive schemes, based on ten years of intensively mechanized production; and

(iii) too great a proportion of time spent when the machines were not cutting coal.

Improved methods of mining, machine control, and achieving reliability are being applied to increase productive running time, the whole being known as MINOS (Mine Operating System), utilizing automatic monitoring and computer control where it is possible and cost-effective.

This concept of systems engineering is being extended from coal winning to ventilation of the mine, coal preparation, and other ancillary plant, and is leading to real improvements in performance. Further application of this principle is envisaged in the next decade or so, to raise the productivity of men and capital employed. Possible goals beyond 2000 include virtually automatic mining controlled from the surface and underground treatment or utilization of coal by solvent extraction, chemical and even microbiological treatments.

C. EFFICIENT COMBUSTION AS A MEANS OF CONSERVATION

1. General

Reference to Table 6 shows that no less than 78 million tonnes of coal are burned for electricity generation, equal to 65% of NCB output, and this apparently offers the greatest scope for energy conservation through increased thermal efficiency. In the majority of coal-fired, power-generating boilers, the coal is supplied as a finely ground powder, known as pulverized fuel, which is blown into a large combustion chamber with a blast of air such that the particles burn in suspension. This technique allows a wide range of coals to be burned.

It is salutary to learn that the national overall efficiency of generating and transmitting electrical energy averages only 28%, and that the best fossil fuel-fired power station operates at only 36% thermal efficiency. This low efficiency arises from practical limitations of the 'heat to mechanical energy' conversion cycle, resulting in the wasteful rejection of low-grade heat at the pass-out end of the converter. Useful gains in primary fuel efficiently can be made by linking power stations to local district heating schemes that can utilize this waste heat by mains hot-water circulation networks, though these are long-term, capital-intensive operations difficult to execute in established urban areas. District or group heating schemes have been in operation in the UK for several decades and the advantages they offer, particularly for coal firing, include:

(i) high standard of heating at high efficiency, giving an economic cost;
(ii) a high level of amenity equivalent to that of piped fuels; and
(iii) heat provided to the dwelling in a manner which provides maximum safety, especially for elderly people.

2. Domestic Heating

Introduction

During the early 1950s, increasing environmental concern in the UK over air pollution resulting from the combustion of fossil fuels, particularly coal for domestic and industrial purposes, led to the Clean Air Acts of 1956 and 1968. Legislation and recommendations derived from these acts have progressively tightened control over the emission of smoke and particulate matter and the proper disposal of sulphur dioxide from fossil-fuel-burning equipment, and have been very successful in cleaning the air over our towns and cities.

The effect of this legislation on the domestic market for coal has had two results. First, the development of a range of smokeless fuels and of appliances for their efficient combustion, to cater for the increasing number of smoke control areas. Secondly, an alternative programme to develop domestic appliance designs capable of burning commercial grades of highly volatile, bituminous coal directly and in such a manner as to be efficient and convenient in use and giving acceptable levels of smoke reduction to satisfy clean air requirements. This alternative is important for conservation, since highly volatile coals occur widely and plentifully in the UK, whereas natural smokeless fuels like anthracite do not. UK domestic solid-fuel consumption for 1977 is given in following Table 7 (Department of Energy, 1978b).

A primary aim of coal research for the domestic sector is its concern to ensure that coal utilization is optimized to provide the most efficient, convenient, and inexpensive means of home heating. In terms of primary energy, the simple open fire burning bituminous coal is as efficient (about 30% including useful heat gained from the chimney) as the average power station. If the coal is converted to a smokeless fuel by carbonization (itself

Table 7 UK domestic solid-fuel consumption, 1977 (Source: Department of Energy, 1978b)

Fuel	Consumption (10^6 tonnes)
Bituminous house coal	9.3
Anthracite and dry steam coal	1.8
Manufactured smokeless fuels	3.7

an expensive process), the primary fuel efficiency remains about 30%, since the carbonizing efficiency is perhaps 80% and the thermal efficiency of a coke-burning open fire without boiler is 38–40%. Replacing the fireback by a high-output back boiler increases the thermal efficiency to about 41% when burning coal or to some 60% (about 50% primary efficiency) when burning a carbonized fuel. Replacing an open fire by a room heater makes a useful fuel saving, since its thermal efficiency when equipped with a high-output back boiler is around 80% (about 65% in primary energy terms for a carbonized fuel, but as much as 75–80% if a natural smokeless fuel like anthracite is burned).

To help maintain good standards of efficiency, design, and construction, and approval scheme (the Domestic Solid Fuel Appliance Approval Scheme, DSFAAS) is operated by the NCB in association with fuel producers and the Department of Energy participation. This involves type testing of new appliances and works inspection visits. (Similar schemes are operated by the other energy industries.) Table 8 lists current levels of efficiency requirements and also indicates typical appliance efficiencies under testbench and field conditions.

Field tests on smokeless-fuel-burning appliances

The majority of dwellings which are fully heated by solid fuel are served by individual systems of up to about 12 kW output. It is very important to know how efficiently such heating appliances perform in domestic practice compared with laboratory test standards.

Information has been obtained on the overall house heating performance of a range of modern room heaters and open fires with high-output boiler for typical trial-site installations. These appliances, which are designed to heat directly the room in which they are situated and supply hot water for additional heating and domestic purposes, are in popular demand and provide full heating for an estimated 1.5 million dwellings (7.5% of total domestic dwellings). Values of appliance heating efficiency have been determined for typical patterns of domestic usage. The useful heat released from the chimney structure to the living space of the house (known as chimney heat gain) has also been taken into account.

The results have demonstrated that the heating efficiency of solid-fuel appliances with high-output boiler when operated under domestic conditions remains significantly above the approval limits set by the DSFAAS for laboratory high-output tests. Measured efficiency values (including chimney heat gain) were above 75% for room heaters and above 60% for open fires, both with high-output boiler, when burning approved smokeless fuels. An important result is that appliance efficiency increases with reducing output, giving higher values at normal operating loads than for rated

Table 8 Solid fuel appliance efficiencies (Source: Solid Fuel Advisory Service, 1975)

	Efficiency (%)	Chimney gain (%)	Total (%)
1. *Open fire without boiler*[a]			
Burning coal	26	4	30
Burning smokeless fuel	37	3	40
DSFAAS requirement (smokeless fuel)			37 min.
2. *Open fire with domestic boiler*			
Burning coal	34	3	37
Burning smokeless fuel	47	3	50
DSFAAS requirement (smokeless fuel)			45 min.
3. *Open fire with high-output boiler*[b]			
Burning coal	41	3	44
Burning smokeless fuel	57	3	60
DSFAAS requirement (smokeless fuel)			50 min. (soon to be raised to 60)
4. *Exempted openable room heater (Rayburn Prince 76)*[c]			
Burning coal (smokelessly)			
open	50	5	55
closed	55	5	60
DSFAAS requirement (pending)			45 min.
5. *Room heater without boiler*			
Burning smokeless fuel	60	5	65
DSFAAS requirement			60 min.
6. *Room heater with boiler*			
Burning coal (smoke reducing)	65	5	70
Burning smokeless fuel	70	5	75
DSFAAS requirement			65 min.
7. *Independent boiler*[d]			
Burning smokeless fuel	65		65
DSFAAS requirement			65 min.
Burning small anthracite	75		75
DSFAAS requirement			70 min.

(Source: Solid Fuel Advisory Service, 1975).
[a]The installation of a throat restrictor will increase heating gains from the system by some 2 to 3%.
[b]There are many appliances in this category, the best of which exceeds 70% efficiency, but figures shown are typical average operating efficiencies and assuming some throat restriction.
[c]This appliance gives a higher flue gas temperature than open fires with a consequent increase in chimney gain.
[d]No credit is given to independent boilers for casing or chimney heat gain, although in fact a typical installation with a cast-iron flue pipe connection may provide up to about 1 kW of extra warmth to the kitchen.

output. This is illustrated in Figures 2 and 3 which give typical curves of efficiency versus load and indicate that a high value of thermal efficiency is maintained from full output to slumbering conditions. For the room heater example, the chimney heat gain from a lined, centrally located structural chimney makes a significant contribution of about 8% towards the overall house heating efficiency.

These findings represent a major advantage of solid-fuel usage in the home and allow realistic heating costs for an important range of solid-fuel appliances to be estimated.

To take full advantage of these high thermal efficiencies for appliances which supply both heat to room by means of radiation and convection and heat to water for central heating use, the proper design of controls is an important factor in efficient utilization of the heat supplied. The use of a room heater with water thermostat control of burning rate and room-

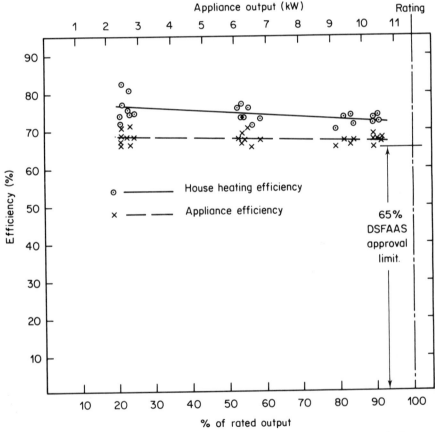

Figure 2   Efficiency versus load for a room heater with high-output boiler

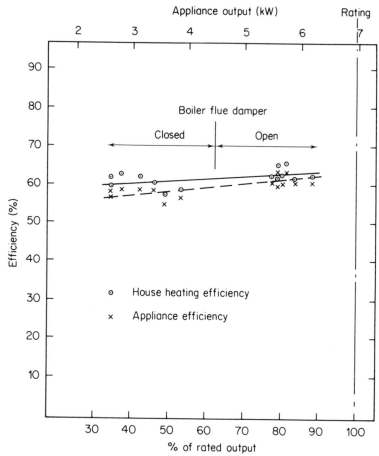

Figure 3 Efficiency versus load for an open fire with high-output boiler

temperature-sensitive thermostatic radiator valves has been shown to give high heat savings and to ensure that each room of the house is at the temperature required regardless of changing patterns of use and errors in heat estimates.

Smoke-reducing coal-burning appliances

As mentioned earlier, as an alternative to appliances which burn processed or naturally occurring smokeless fuel, appliances have been developed specifically to burn bituminous coal with acceptable smoke reduction.

The majority of British bituminous coals contain between 30 and 40% volatile matter, about one-third by weight, which is evolved when coal is heated to temperatures above about 400°C. When such coal is fired on a conventional domestic grate, the released volatiles pass upwards through the firebed along with the combustion air to the cooler zones of the firebox. Here, the volatiles are chilled quickly, to below their ignition temperature, and are only partly burned giving inefficient combustion and causing the emission of yellowish-brown smoke from the chimney.

Initial studies indicated that the essential requirements for burning volatile matter were that it should be well mixed with air (oxygen) and maintained at temperatures above 600°C for a time (about 0.5 s) sufficient for all the combustion reactions to be completed. These requirements of turbulence, temperature, and time are well known to combustion engineers as the 'three Ts'.

The objective, therefore, has been to satisfy these criteria and complete the burning of coal volatiles within the limited space of an appliance design. In most cases, the design has been required to fit into a standard fireplace opening. The means employed to meet these stringent conditions is the principle of downdraught combustion. This is an idea which dates originally from 1680 when it formed the basis of Dalesme's Heating Machine which is shown schematically in Figure 4. The principle was later rediscovered and patented by James Watt in 1785.

In a downdraught system, air enters the combustion chamber, or the bowl in Figure 4, at the top and combustion products leave from the bottom of the firebed. The important technical point is that the burning fuel front travels upwards within the fuel bed, in the opposite direction to the downward primary combustion air. Any volatiles evolved are entrained by the downward airstream and pass into the incandescent zone of the firebed

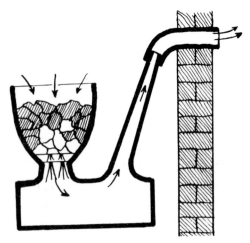

Figure 4   Dalesme's heating machine, 1680

to be burned. That is, the fire consumes its own smoke, in contrast to the updraught combustion system described previously for the conventional grate. For low burning rates, a second combustion chamber and air supply may be needed to provide a sufficient level of smoke reduction.

The first generation of fully approved commercial appliances, developed jointly by the NCB and major appliance manufacturers, was launched in the early 1970s. These models were continuously burning room heaters with high-output boilers capable of providing full heating for the two/three-bedroomed house. They were of sophisticated design with fan-blown combustion air and burned prepacked singles, a commercial-grade of coal in the size range 12.5 to 25 mm. The appliances were somewhat selective on fuel, preferring coals of non- or weakly caking quality.

Development has continued of a range of 'second-generation' units of simpler design which operate solely on natural chimney draught, are not sensitive to coal type, and are able to burn household grades of doubles and trebles coal of 25 to 100 mm size with high thermal efficiency. This enables them to burn locally available house-coal grades wherever they are used within the UK. The first commercial appliance, an open room heater (Agaheat Prince 76) was marketed in 1975, and several other manufacturer's designs are expected to be available commercially in the near future. The Prince 76 is shown in Figure 5.

Figure 5 The Agaheat Prince 76 room heater

The attractions of the coal-burning development programme from the viewpoint of the householder, the manufacturer, and of energy conservation respectively are that:

(i) bituminous coal is a readily available fuel, of relatively low cost, and with guaranteed long-term supply—when burned in an efficient manner, heating costs can be maintained extremely competitive compared with alternative fuels;

(ii) manufacturers are able to market coal-burning appliances to the increasing number of smoke control areas, in addition to units which may burn only smokeless fuel—at present, smoke control orders cover about 30% of the 20 million domestic dwellings in the UK; and

(iii) commercial grades of bituminous coal can be burned in a domestic heating appliance as effectively as a smokeless fuel, without the need for processing to provide smoke control, thus utilizing primary energy supplies more directly and efficiently.

To date, sales of bituminous-coal burning appliances are in excess of 60 000 units and account for over a quarter of a million tonnes of coal burned annually.

3. Energy conservation measures for domestic dwellings

Thermal insulation

United Kingdom building regulations for new domestic dwellings were first introduced in 1965 to rationalize existing national and local authority requirements and to improve standards of construction. A complete revision of the regulations was made in 1972 and an amendment introduced in 1975 to set improved standards of thermal insulation for the perimeter walls, floors, and roof of a dwelling.

As a result of the increasing cost of fuels, a Government campaign was launched in 1976 to encourage householders to conserve energy, particularly heat and fuel saving, by means of improved thermal insulation in the home.

These measures have led to local authorities and private builders providing good standards of insulation in new housing and to improvements in insulation for older dwellings. However, some of the benefits of additional insulation may be taken in improving thermal comfort conditions without any direct energy saving. This could be true of older homes built to poor thermal standards and with part-heating systems installed, served, for example, by open fires with medium- or low-output boilers. About 12 million of the 20 million UK dwellings are of solid-wall construction and could fall within this category. The Building Research Establishment (1975) study estimated that a 5 to 6% saving in total national primary energy consumption was possible from improved insulation without affecting environmental standards.

Improved house insulation, in addition to reducing the heat loss rates and thereby the length of the heating season, gives other useful benefits. Householders' assessment of thermal comfort is not only determined by the air temperature of the environment, but of almost equal importance in this respect is the mean radiant temperature.

When the insulation level of a dwelling is increased (e.g. by cavity wall insulation, double glazing or loft insulation) and the heat transfer rate to the outside air is thus reduced, this inevitably produces an increase in the inside surface temperature of the insulated areas. This gives a higher value of mean radiant temperature which will increase the occupants' subjective assessment of 'warmth' and could if desired be compensated by the reduction in the inside air temperature, and thus, by increasing house insulation levels, further energy savings can be achieved.

The NCB has developed a computer program which can predict comfort levels in a typical living room. The computed results clearly illustrate a further beneficial consequence of improved insulation in that the area of comfort is extended closer to the outside walls and windows.

Low-energy-consuming dwellings

As potential heat savings become realized by the pressure of increased energy costs, etc., they lead to reduced house heating requirements and the need to provide solid-fuel appliance designs to match the reduced demand. Several schemes are now in progress to build low-energy consuming houses with ultra-high standards of insulation and controlled ventilation, which require only 3 to 4 kW for full heating. Warm-air heating systems fired by bituminous-coal-burning appliances with special forms of control are being developed for these applications. The standard of insulation is reflected in an estimated coal consumption for the three-bedroomed house of about 1 tonne per year to meet the heat load.

In these low-energy house designs, the opportunity has been taken to provide a very high level of amenity with regard to the convenience of refuelling and de-ashing the appliance and the siting arrangements for fuel and ash storage.

Heat-saving techniques (appliance and system)

Investigations have continued within the National Coal Board to assess heat-saving techniques and their possible application towards improvements in solid-fuel appliance and system design.

(i) Appliance

In the past, open fires have been associated with excess ventilation and inefficient operation. However, recent developments of high-output boiler models have included design features which limit the amount of air drawn up the chimney and have shown that savings in whole house heating of up to 10% can be achieved by this means. Measurements have also shown the importance of similar devices called 'throat restrictors' in simple dry-back open fires. In the case of the various closed solid-fuel room heaters now available, proper design ensures that only sufficient air is drawn in for the efficient combustion of the fuel. This quantity of air has been measured to be about one air change per hour for the room in which the appliance is situated. The chimney serves as a controlled way of providing ventilation without the use of any additional equipment.

(ii) System

In centrally heated dwellings in the UK, the most common method of distributing the heat throughout the house is by the use of water-filled radiator panels fitted to walls. When radiator panels are fitted to external walls, up to 15% of the heat transferred from the water can be lost directly to the outside atmosphere through the wall. It has been demonstrated how this loss can be simply and cheaply reduced. Figure 6 shows the savings which can be achieved by fitting aluminium foil reflectors onto either or both of the back radiator panel and facing inside wall surfaces. The energy saving is clearly dependent on the thermal resistance of the external wall. Since 12 million of the 20 million dwellings in the UK have solid walls, and in these cases the cost of fitting reflectors is more than recovered during a heating season. The heat loss can be decreased further by fitting insulating panels between the radiator and the wall, and Figure 6 shows the effects of fitting 12 mm thick sheets of expanded polystyrene. The greatest potential benefits will be seen to be obtained in dwellings with solid walls.

Combined solid-fuel and alternative energy systems

Practical assessments have been made of schemes to combine solid-fuel heating with alternative energy systems such as solar heating and heat pumps. A solid-fuel appliance with boiler, in addition to providing water heating, provides direct space heating which is not required during warm summer months. It is difficult to separate the functions of water and direct space heating in a solid-fuel unit, and the common practice is not to use the appliance during the summer and use electric immersion heaters for water heating. However, increased running costs for electrical water heat-

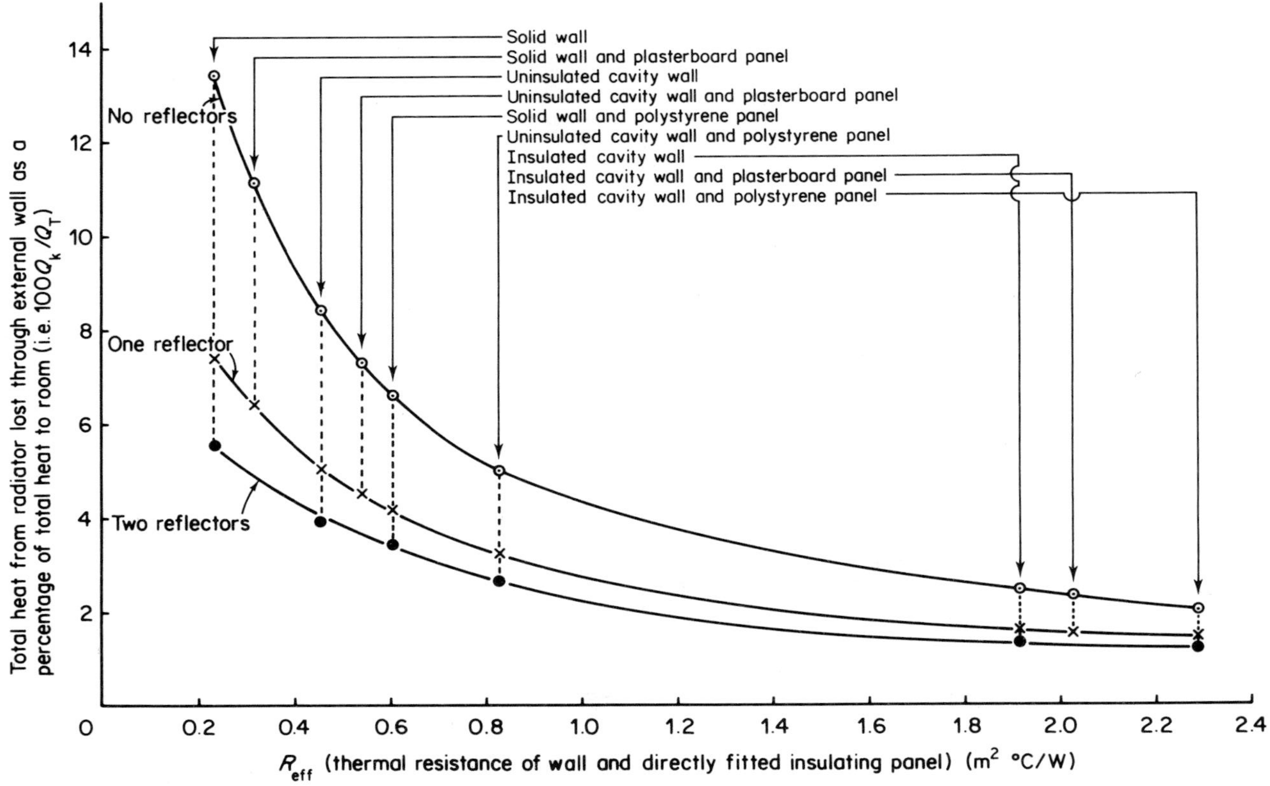
Solid wall
Solid wall and plasterboard panel
Uninsulated cavity wall
Uninsulated cavity wall and plasterboard panel
Solid wall and polystyrene panel
Uninsulated cavity wall and polystyrene panel
Insulated cavity wall
Insulated cavity wall and plasterboard panel
Insulated cavity wall and polystyrene panel
No reflectors
One reflector
Two reflectors
Total heat from radiator lost through external wall as a percentage of total heat to room (i.e. $100Q_k/Q_T$)
R_{eff} (thermal resistance of wall and directly fitted insulating panel) (m^2 °C/W)
0
0.2
0.4
0.6
0.8
1.0
1.2
1.4
1.6
1.8
2.0
2.2
2.4
2
4
6
8
10
12
14

Figure 6 Dependence of thermal loss on the effective thermal resistance of the wall

ing may tend to dissuade potential customers from having solid-fuel systems installed.

Therefore, two alternative methods of heating water during the summer period were investigated. In the first, a solar panel collector was used to supply part of the heat energy normally supplied by an electric immersion heater; in the second, a heat pump converted low-grade heat in the ambient air into a form suitable for providing domestic hot water.

A realistic assessment of the savings which can be achieved by the use of solar collectors has been obtained from direct measurements when such a system was used in a typical domestic setting. The annual saving was found to be less than 4% of the capital cost of the commercial solar system (£700 at 1977 prices). It is therefore concluded that at present solar heating cannot be justified on economic grounds in the UK for this application.

Tests using a heat pump showed that the heat transferred from ambient air (at about 10°C) to maintain hot water at 60°C was about three times the total electrical energy supplied to the unit. It is estimated in this case that such a saving represents a 6% annual return on the capital cost of the unit which is comparable with that of a solar heating system.

4. Industrial combustion of coal

Introduction

The present industrial market for coal (1978–79) amounts to about 11 million tonnes burned annually. Some 8 million tonnes is used in boilers for hot water or steam raising, ranging from units of about 300 kW in schools to boilers of 30 MW and above for industrial power generation. Almost all of the remaining 3 million tonnes is used in the manufacture of cement. Here, coal is used exclusively as the fuel and this market is expected to continue.

The size of the total industrial market, excluding power generation, is estimated at about 70 million tonnes of coal equivalent (mtce) and therefore represents a considerable potential for increased coal consumption, mainly at the expense of oil.

During the 1960s when oil was relatively cheap and plentiful, packaged boiler units were developed to provide very compact designs capable of giving oil firing rates about twice those available from conventional coal-fired equipment. This means that the size and capital cost of a conventional coal-burning installation can be typically two or three times that for an oil-fired installation of similar output and have reduced amenity because of coal- and ash-handling considerations.

Since 1974, coal has had a price advantage over oil and this has resulted in an increased demand from industry for coal-fired equipment to use in substitution for oil firing in new and existing boilers and furnaces. However, a fuel price advantage alone is not sufficient to ensure that coal will be used rather than oil for industrial purposes. Therefore, the requirement was to burn more coal in less space in order to reduce the physical size of combustion chamber required and so give reduced capital costs.

In 1972, in order to expand the industrial market for coal and in anticipation of future market changes, the NCB embarked on a research programme to identify and develop new coal-burning systems. The requirements were that such combustion systems should be able to provide similar heat release rates and operational amenity to oil and gas firing and be of comparable capital cost. The application of fluidized combustion has emerged as a convenient method for burning coal for both the power generation and industrial markets. The technique is attractive because of its basic simplicity, its ability to burn a wide range of coals, and its potential for improved pollution control.

Fluidized combustion

Coal-fired fluidized combustion was developed initially during the latter part of the 1960s to investigate means of steam raising for power generation as an alternative to pulverized-fuel firing. Preliminary work was carried out by the CEGB but, because of the involvement in development programmes for pulverized-fuel firing and nuclear power at that time, the development of fluidized combustion was taken over by the NCB and allied interests. Development proceeded through several small experimental rigs culminating in a unit of 0.5 MW thermal output built to represent a section of water tube boiler for power generation.

By 1971 this work had led to designs being commissioned for a 20 MW (thermal) demonstration steam boiler. However, Government funds were not made available to support the venture and the particular project on atmospheric pressure combustion was discontinued.

The NCB had also sponsored, at the British Coal Utilisation Research Association (BCURA) laboratories, the construction and operation of a 2 MW combustor, operated at pressures up to 6 bar, to obtain data for combined-cycle power generation.

American government interests, particularly on the use of fluidized combustion to reduce sulphur emission, had led to joint sponsorship with the NCB of the BCURA developments. Subsequent to 1971, these US interests took over complete support for the work at elevated pressures at BCURA and this has continued.

Because of the interest shown in fluidized combustion by other countries, particularly the USA, a joint company, Combustion Systems Ltd (CSL), was formed by the National Research and Development Corporation, British Petroleum (to cover oil-burning interests in fluidized combustion), and the National Coal Board to exploit the expertise abroad. In association with CSL, Babcock & Wilcox converted a 13.5 MW (thermal) water tube boiler to fluidized combustion and this has been operated successfully. Subsequently, in 1975, a new company, Babcock-CSL, was formed to allow British equipment as well as expertise to be sold abroad.

The basic concept of fluidized combustion is that a fuel such as coal is fed to a hot, highly turbulent bed of finely divided, inert particles, which may be formed from coal ash. Turbulence of the bed is achieved by passing through it an upward, evenly distributed flow of air. The airflow is controlled to give the bed the appearance of a boiling liquid with considerable splashing of particles occurring above the surface. For this condition, rapid mixing of particles occurs within the bed and it behaves like a fluid, finding its own level and possessing hydrostatic head.

When burning coal it is necessary to keep the bed at temperatures below that at which coal ash begins to fuse or sinter. Therefore, normal bed operating temperatures are maintained within the range 750 to 950°C and coal combustion results in the production of a soft fine ash. Because of the high degree of turbulence, coal is distributed rapidly within the bed and the temperature can be maintained relatively uniform.

Two variants of coal-burning, fluidized-bed systems are being investigated. The initial method developed for power raising application involves the combustion of crushed coal (below 3 mm size) injected pneumatically at several feedpoints into a bed up to 1 m deep consisting of ash from the coal. The combustion chamber height above the bed needs to be about 3 m to prevent excess loss of particles by splashing. With fine coal there is inevitably some carry-over of particles by the flue gas and it may be necessary to collect and refire unburned fines to give acceptable combustion efficiency.

A second method has been developed specifically for industrial boiler and furnace applications where available headroom and space limitations are important considerations. This alternative allows direct burning of uncrushed commercially available grades of coal up to 50 mm top size, with acceptable combustion efficiency. The coal is fed mechanically to the top of a relatively shallow bed of inert particles such as silica sand, in which it floats. The advantage of using uncrushed coal is that a reduction is given in carry-over of unburned fines, bed and combustion chamber heights are reduced, and the coal does not require special preparation at the customer's site (an important cost consideration).

Other important features given by this bed system are:

(i) a simple coal feed arrangement whereby coal can be dropped by gravity directly to the surface of the bed—rapid mixing distributes the coal across the complete bed area, eliminating the need for mechanical sprinklers or spreaders;

(ii) no mechanical moving parts within the combustion area, which reduces maintenance costs;

(iii) an even temperature within the bed, allowing the use of a simple automatic output control system; and

(iv) the production of a soft coal ash which is blown from the bed and can be collected by conventional equipment, e.g. standard grit arrestors.

Plant operating experience has identified the following major advantages offered by the fluidized combustion of coal, particularly for industrial boiler applications.

(i) The ability to burn a wide range of coals regardless of caking property, ash and moisture content. This allows the use of low-grade coals not previously considered suitable for combustion and effectively increases the quantity of coal reserves which can be economically used.

(ii) The ability of the fluidized bed to transfer heat to immersed cooling surfaces at very high rates compared with conventional boiler designs. It has been demonstrated that about half of the heat generated within the bed from combustion can be extracted directly by immersed cooling tubes. This permits increased combustion intensities and more compact designs of coal-fired plant.

(iii) The low combustion temperature of about 950°C, which reduces the vaporization of alkali salts from the coal ash. This minimizes fouling of the smoke tubes in a boiler caused by deposition of these salts and reduces their emission to the atmosphere.

(iv) The ability to reduce pollutants such as sulphur dioxide from high-sulphur-containing fuels by the addition of limestone to the bed. As yet, sulphur retention is not a requirement in the UK.

The industrial development programme

In order to expand the industrial market for coal, the present NCB development programme is aimed at demonstrating the technology of fluidized-bed combustion systems and arriving at commercial designs of boiler and furnace. The programme is being carried out in collaboration with boiler manufacturers. The main features of the programme are:

(i) The development of retrofit systems for existing shell boiler installations previously fired by oil or gas, to convert them to coal or dual firing without loss of output. Work is concentrated on systems for horizontal shell boilers up to about 20 MW output as these represent a large proportion of the UK industrial boiler market. Units under test include a compact 3 MW boiler designed commercially for oil firing.

(ii) The development of novel boilers of similar size to oil- and gas-fired boilers designed to take full advantage of the properties of a fluidized bed. Identified as particularly suitable for this application are:

(a) vertical shell boilers for both steam and hot water output up to about 5 MW—two prototype boilers of 2.8 MW output (one steam, one hot water) have operated satisfactorily at commercial sites since mid-1977, and commercial designs based on these units are being developed (a cross-sectional view of the prototype 2.8 MW steam-raising boiler is shown in Figure 7);

(b) a 10 MW prototype boiler of double-ended locomotive design which has been operated at a commercial site since 1978 providing steam at medium pressure;

(c) a 30 MW coil boiler of a type used with oil firing to produce high-pressure, superheated steam—this is being used as the basis for coal-fired units and is expected to operate at a commercial site during 1979; and

(d) purpose-designed horizontal shell boilers ranging in output from 5 to 45 MW—such designs are being progressed in collaboration with manufacturers.

(iii) The conversion of industrial water tube boilers with outputs of up to 50 MW. Development here is towards the use of commercial rather than crushed coal sizes.

(iv) Development of furnaces to provide hot gas for direct heating in agricultural and industrial drying. Prototype units have been operated to provide hot gas for grass drying. Based on operating experience gained with these prototypes, commercial designs are now available for the many applications where a small amount of ash is acceptable in the product. Development is continuing of units to produce 'clean' hot gas.

(v) The production of low-calorific-value fuel gas from coal using a fluidized bed, for electricity production at high efficiency in combined-cycle systems. Similar types of gasifier are likely to find applications in industry.

(vi) Improvements in amenity of coal-burning systems with full automatic control to ensure that combustion systems are operated at optimum efficiency and by the development of improved ancillary equipment, probably for coal- and ash-handling.

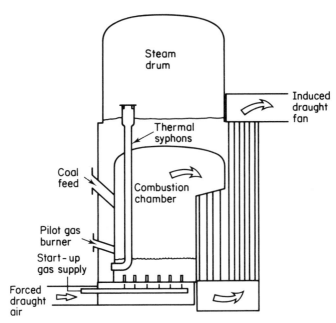

Figure 7 A 2.8 MW vertical steam boiler

Based on this development work and that carried out by manufacturers independently of the NCB, commercial designs of fluidized-bed boiler are expected to be available to the UK industrial market from 1980. Resulting from increased coal-fired equipment sales, the objective is to attain an industrial market size of approaching 20 million tonnes of coal per annum after 1985. It has been predicted by the Department of Energy (1978c) that this market could rise to between 30 and 50 million tonnes per annum by the end of the century.

5. Incineration

The inherent characteristics of fluidized combustion make it ideal for incineration of waste materials, with or without heat recovery depending on the calorific value. A method of incinerating the tailings slurry from coal preparation plants has been developed by the NCB, and described by Randell *et al.* (1978). Support for the work has been given by the European Coal and Steel Community.

The slurry is concentrated to about 50% water content (w/w) and sprayed onto a fluidized bed operating at a temperature of 850°C. There are sufficient combustibles present to evaporate the water and produce ash which is safe for tipping, but fluidized combustion is the only method by which the combustibles can be burnt.

The expertise gained in this work is being applied to other industrial slurries, sludges, municipal garbage, and wastes. A commercial plant has been built at Caernarvon to incinerate sewage sludge, and dries and treats sewage from a population of 25 000. Efficient combustion in the fluidized bed ensures that there is no emission of odorous compounds.

NCB experience has confirmed that many materials which would not be classified as fuels in normal circumstances can be usefully burnt in fluidized beds. The usual requirement is the production of a dry inert solid for safe disposal; high throughput rather than combustion efficiency is the main consideration.

6. Power generation—elevated pressure

The use of gas turbines in power generation has been restricted by their need for a clean fuel. Previous attempts with coal or fuel-oil firing have failed after rapid corrosion or erosion of the turbine blades. Fluidized combustion at elevated pressure now offers a real prospect that turbines might be fired by hitherto unacceptable fuels, including coal, primarily because of the reduced emission of alkali salts and vanadium at the low combustion temperature. Furthermore, the ash is nonabrasive and the ultrafine particles which pass through the gas cleaning system do not erode the turbine blades. Operation of the BCURA pressurized combustor has indicated that it is possible to use coal to produce a hot gas which is suitable for a gas turbine.

With pressurized fluidized combustion, gas turbines may now be considered for applications where they have been economically unacceptable because of the high cost of clean fuels. In particular, there is the prospect of increasing the efficiency of electrical power generation by combining a gas turbine cycle with a conventional steam cycle (Figure 8). Pressurized fluidized combustion retains all of the advantages described previously, plus a further reduction in boiler size. Because of the lower capital cost and the higher power generation efficiency, electricity produced is expected to be about 13% cheaper than for pulverized-fuel power generation, due to an increase in conversion efficiency from 36% (for the best conventional fossil-fuel-fired station) to 45%.

The potential importance of pressurized fluidized combustion has been recognized by the International Energy Agency Working Group on Coal Technology. A research facility for design optimization studies for fluidized combustion systems is nearing completion at an NCB site at Grimethorpe. The plant will be capable of operating with a thermal input of 85 MW under a wide range of conditions including atmospheric operation and includes provision for later installation of a gas turbine for combined-cycle studies.

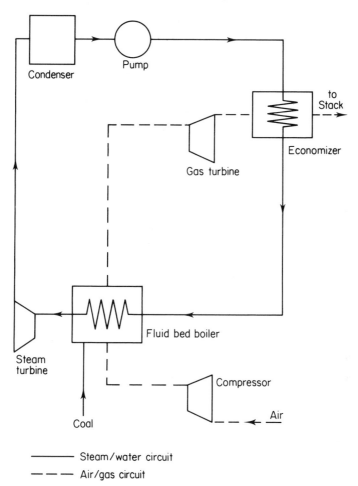

Figure 8 Pressurized fluidized-bed combustion combined cycle

D. COAL IN THE CONTEXT OF ENERGY MANAGEMENT

1. UK coal utilization projects (national and international)

Energy in its various forms is so essential to our way of life that it should be the duty of all governments to ensure that adequate and secure supplies are available to their peoples. So energy supplies and utilization become matters of national and international concern, especially since the 1973 oil crisis and the prospect of ever-escalating oil prices. To demonstrate this, we may examine how these views have influenced the role of coal in the management of UK energy requirements.

Reference should be made to the work of the Coal Industry Tripartite Steering Committee or Group (1974, 1977) and to the Government Green Paper on energy policy (Department of Energy 1978). Among other matters these deal with the need to ensure that the coal industry will have the capacity for increasing production for the foreseeable future, and also for research and development programmes to improve coal conversion and utilization technology.

The NCB (1974) plan stated that the NCB intended to expand its mining R&D programme without calling for Government financial support. As to coal conversion, however, although the NCB had researched several coal conversion projects to the stage of being ready for scaling up to pilot-plant size, the cost of pilot-plant work (about £40 million) 'was more than the industry could bear by itself given the timescales before the prospect of financial return.'

The 1974 Tripartite Working Group concluded that the combination of NCB expertise, relevance to the UK economy, and commercial prospects justified Government support in (a) fluidized-bed combustion, (b) coal liquefaction by solvent extraction, and (c) pyrolysis. The NCB sought international participation and joint funding, successfully for (a) and (c), through the International Energy Agency (see below). In late 1975 the UK, USA, and the Federal Republic of Germany agreed to share equally the £17 million cost of a project to design, build, and operate an 85 MW thermal/28 MW electrical pressurized fluidized-bed combustion facility at Grimethorpe near Barnsely. A pyrolysis project was also agreed, jointly financed by West Germany, the UK, and Sweden, to investigate the fundamental nature of pyrolysis at high pressures and in various atmospheres.

In addition the NCB and BP Ltd agreed to conduct detailed studies of the potential of NCB processes for manufacturing synthetic liquid fuels and feedstocks through the liquefaction of coal by liquid and gas solvents. Another example of international cooperation is to be found in the joint study and demonstration programme of these NCB processes by the NCB and the EEC.

A substantial market for substitute natural gas (SNG) is envisaged beyond the year 2000, estimated to represent a demand of several tens of millions of tonnes of coal. Consequently, the NCB has a joint interest with the British Gas Corporation to develop a 'composite' gasifier concept.

Recognizing the importance of coal in the world energy scene, the International Energy Agency was set up in 1975, to coordinate international collaborative activities in coal resources, mining, and utilization research. It involved twelve countries, namely Austria, Belgium, Canada, West Germany, Italy, Japan, the Netherlands, New Zealand, Spain, Sweden, the UK, and the USA. Although the Agency is managed by NCB (IEA Services) Ltd on behalf of member countries, it is funded entirely by IEA members participating in projects under five headings:

Technical information service.
Economic assessment service.
World coal resources and reserves databank service.
Mining technology clearing house.
Fluidized-bed combustion.

The Grimethorpe fluidized-bed combustor mentioned earlier is the major demonstration project under the last heading; it is an excellent example of how cooperation between countries can fund costly projects of common concern.

Consideration is also given to overseas developments relevant to NCB R&D programmes. Through the medium of international exchange agreements and membership of multinational organizations, the NCB keeps in close touch with developments in coal utilization. A consultancy service is also offered through Coal Processing Consultants, a subsidiary set up by the NCB with Woodall-Duckham.

2. Coal burning and public opinion

Until the 1939–45 war, Britain enjoyed cheap and abundant energy from her ample coal resources, overseas oilfields, and well developed technology in power generation and gas engineering. Conservation and efficient utilization in a cheap fuel economy were less important than they are now.

But war and its shortages underlined the wisdom of better energy management. Bodies like the Fuel Efficiency Branch of the Ministry of Fuel and Power, and the Boiler Availability Committee were set up, and did much to reduce energy waste in the industrial boiler and power-generation fields.

A start was made, too, on encouraging the more efficient use of domestic fuels, for example by Government insistence that local authorities install only those appliances which authorized laboratories had shown reached acceptable levels of efficiency and constructional standards. Also, the Egerton (1946) and, later, the Parker Morris (1961) reports set requirements for the heating of domestic dwellings that demanded improved insulation standards, subsequently written into Building Regulations. If these standards did not always lead to fuel savings, they did ensure enhanced comfort levels for the occupants, a pattern which is being repeated to the present day as people insist upon more comfortable environments at both work and home.

As fossil-fuel resources gradually become exhausted, however, increasing reliance will have to be put on alternatives—nuclear, solar, wind, waves, tidal, geothermal, etc. So far, most interest and investment has been in nuclear energy, though its rapid expansion has been hedged by political constraints. Even the UK with 300 years supply of coal, has invested heavily in nuclear generating capacity, and at the present time (1978–79) some 3.6% of UK energy requirements are produced by nuclear stations. This is inevitably a controversial development for its consequences cannot yet be fully assessed. A recent incident at Harrisburg, Pennsylvania, USA, caused much public alarm with the possibility that radioactive contamination might have occurred over a wide area if steam and hydrogen had escaped from an overheated nuclear pile.

In one aspect of environmental control, that of reducing smoke pollution, the UK has been both active and successful. We estimate that between 1952 (the year of the disastrous five-day London smog) and 1977, domestic smoke emissions decreased from some 1.1 million tonnes per annum to 0.30 mtpa, due mainly to the reduction in the quantity of house coal burned in appliances not designed to reduce smoke. Up to the present time more than 60 000 domestic smoke-reducing appliances have been installed, and since these burn highly volatile coal with virtually no smoke emission and at about twice the efficiency of the appliances that they replace, the twin objectives of better utilization and an improved environment are both attained.

3. Coal conversion processes

In order to meet the longer-term needs of the energy market, new coal conversion processes are being developed to make coal an attractive alternative to fossil-derived supplies of oil and gas, as these fuels become relatively scarce at the turn of the century.

(i) Oil from coal

Converting coal to oil involves two basic steps. These steps are the separation of the coal substance from the ash and impurities associated with the coal, and breaking down the complex coal molecules into simpler molecules and increasing the hydrogen-to-carbon ratio.

Three projects are in progress concerned with the production of liquid fuels and petrochemical feedstocks from coal. These projects are liquid solvent extraction, supercritical gas solvent extraction, and pyrolysis. The first two methods of coal liquefaction are being developed at the small pilot-plant scale and are expected to proceed shortly to fairly large scale pilot-plant (Figures 9 and 10) to allow process evaluation and provide essential design data.

In the liquid solvent extraction process, coal is treated with a hot liquid solvent (itself derived from coal) and the viscous tar-like solution obtained is

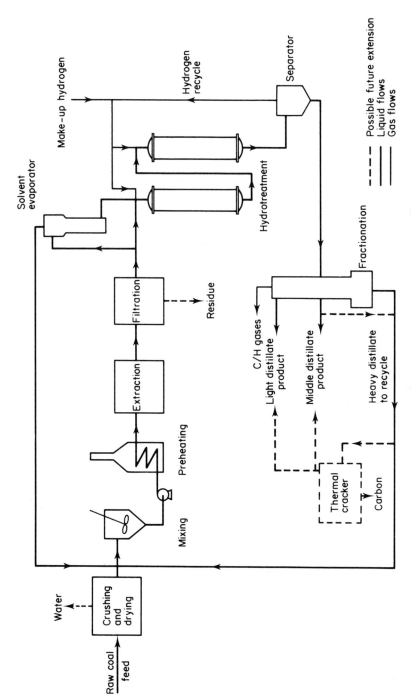

Figure 9 Proposed pilot plant for coal liquefaction using liquid solvents

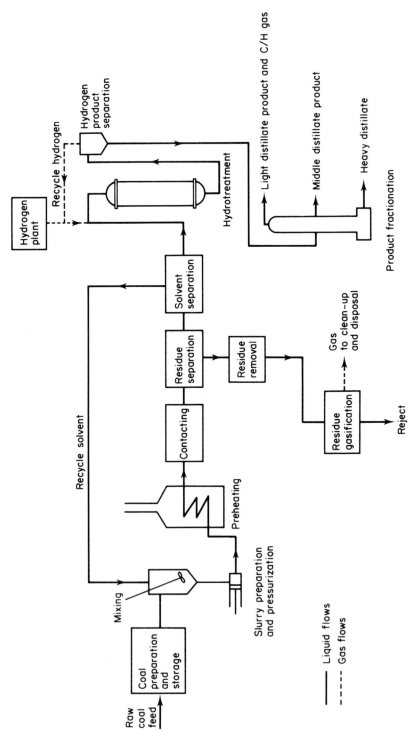

Figure 10 Proposed pilot plant for coal liquefaction using supercritical gas extraction

separated from insoluble coal components and mineral matter by filtration. The solution is then reacted with hydrogen, and a synthetic crude oil, and recycle solvent are the main products.

The extraction process is presently being operated on the scale of about 0.2 tonnes of synthetic crude oil per day. It has been demonstrated that this material is suitable for refining to transport fuels and petrochemicals feedstock.

The supercritical gas solvent extraction process takes advantage of the capability of gases at high pressure to dissolve considerable proportions of complex organic substances including components of coals.

Solutions of the coal-derived material in the gas are readily separated from coal residues. The material in solution is recovered by reducing the pressure, and the solvent gas, being virtually unaffected by the process, can be recycled. The extracted material shows considerable promise for the production of aromatic chemicals feedstock. So far, this technology has only operated on the small scale of 5 kg/h.

Both of the above processes preserve some of the aromatic structures in the coal, together with their partial conversion to naphthenes. Consequently, liquids suitable for inclusion in high-quality motor fuels are produced directly. The middle distillates are less immediately suitable for diesel fuels, but could produce fuels suitable for gas turbines.

By direct pyrolysis (chemical decomposition by action of heat, usually in the absence of air), it is possible to separate the tars and gases from the coal. These are more hydrogen-rich than the main body of the coal and, being relatively ash- and mineral-free, form a natural precursor for the production of liquids from coal. Laboratory experiments under conditions of rapid heating have shown that this could provide the most effective method for maximizing the yields of the more important components, especially as chemical feedstocks have so far been rather low.

The NCB small-scale work on high-intensity pyrolysis has identified areas of uncertainty in the basic process mechanisms. The NCB is participating in a cooperative International Energy Agency project designed in the first instance to investigate the basic principles of pyrolysis processes.

(ii) Low-calorific-value gas for power generation

This project is concerned primarily with developing techniques for generating electricity at high efficiency, using high-temperature gas turbine systems (now under development) in conjunction with boilers and steam turbines.

Coal is gasified with steam and air in a fluidized bed operating at relatively low temperatures so as to minimize the vaporization of alkali salts in the coal, which would damage the gas turbine. The fluidized-bed gasifier includes a gas-cleaning system, a gas combustor, and a coal residue combustor. Present NCB small-scale work is being supported partly by funds from the European Coal and Steel Community.

(iii) Substitute natural gas (SNG)

Following experience gained from their operation of standard Lurgi gasifiers and their own synthesis technology, the British Gas Corporation have subsequently developed a high-throughput, high-temperature slagging gasifier. This forms the basis for a design of a $1.7 \times 10^6 \, \text{m}^3/\text{day}$ SNG demonstration plant which is being considered for construction soon in the USA.

An objective of this design is to investigate and develop a system capable of converting a wide range of coal types and sizes to substitute natural gas.

E. NCB THERMAL INSULATING PRODUCTS

1. Introduction

Thomas Ness Ltd, a National Coal Board Company, have developed and produced a variety of materials of value to the building industry and energy conservation generally, either for waterproofing absorbent surfaces or for use as insulating sheets or panels. They include:

Synthasil—a colourless silicone water-repellent.
Synthaprufe—a waterproofer and adhesive.
The Ness silicone injection dampcourse system.
Presomet—a black bituminous paint.
Hevikote—a pitch epoxy coating.
Hyload—a pitch/polymer dampcourse material.
Ness-Board—a phenolic foam roof insulation.
Heviprufe—a surface damp-proof membrane.
Multi-Plas—a multilayer bitumen/polymer roofing sheet.

We may consider two of these more fully, namely Synthasil and Ness-Board.

2. Synthasil

An external wall coating of Synthasil, for example, can justify its cost, in terms of fuel saved, in a very few years. To see why this is so, it is well known that the more porous building materials are poorer heat conductors than denser materials, essentially because air is a much worse conductor than solids and porous materials enclose entrapped air. If some of this entrapped air is replaced by water, as when the material is exposed to rain, the material becomes much more conductive and the building loses heat more rapidly. The Building Research Station found that for various types of wet brick the average value of thermal conductivity was about 2.1 to 2.2 times that of dry brick; concrete was sligly less susceptible than brick, but on wetting lightweight concretes increased their conductivity to 2.7 to 3.0 times their dry values.

Consider three common types of external wall referred to in Table 9 all having a half-inch (13 mm) thickness of plaster on the inside wall surface. Let us assume that the nominally dry wall contains 1% moisture, and that the same wall in a damp condition contains 5% moisture. Taking typical areas of wall for the three types of house referred to, also assuming their lofts are lagged with a 2 inch (50 mm) thick layer of fibreglass between the joists and the rafters are boarded, Table 9 shows the percentage heat losses through the external walls.

It also indicates the reduced percentages of total heat loss, as savings, effected by painting the external wall (Synthasil is colourless and transparent) with the product, assuming the application allows the wall to dry from an ambient 5% moisture down to 1%. Maximum heat savings are achieved, of course, for the detached house with its maximum external wall area of solid construction (3.8%). BRE have estimated that there are 12 million dwellings in the UK with solid external walls and 8 million with cavity wall construction. A saving of 2.4% can be expected after coating the external walls of a typical Victorian terrace house with Synthasil, if it has 9 inch (225 mm) thick solid brick walls. The savings are reduced to 1.6% for cav-

Table 9 Percentage of total heat loss through exterior walls

Wall construction	'Victorian' terraced		Semidetached		Detached	
	Loss (%)	Saving (%)	Loss (%)	Saving (%)	Loss (%)	Saving (%)
9 in (225 mm) solid brick						
average moisture 1%	18.8	2.4	29.5	3.3	32.0	3.5
average moisture 5%	21.2		32.8		35.5	
13½ in (340 mm) solid brick						
average moisture 1%	17.3	2.6	26.7	3.3	30.4	3.8
average moisture 5%	19.9		30.0		34.2	
11 in (280 mm) cavity wall						
moisture in outer leaf 1%	15.7	1.6	25.0	1.6	27.6	1.6
moisture in outer leaf 5%	17.3		26.6		29.2	

ity wall constructions, but if the cavity has been insulated by foam infilling no significant saving can be expected.

3. Insulating board

There are now many case histories of impressive savings to be made by insulating buildings. For example, in lining a warehouse roof with 50 mm of insulation material, one company were saving £5000 a year—more as fuel costs rise—for an initial outlay of only £11 000.

Factors affecting selection include low thermalconductivity, low flammability, low smoke emission, adequate strength for handling on site and

Figure 11   Properties of Ness Board demonstrated by heating

walking on for maintenance, dimensional stability over a wide temperature range, chemical inertness, and moisture resistance. Ness-Board has these properties to a substantial degree. It is a closed-wall, fire-resistant, dimensionally stable, laminated foam roof insulation board, consisting of a core of rigid phenolic insulating foam, laminated on each face with glass fibre tissue. Figure 11 demonstrates these properties: after 10 minutes heating a slab of Ness-Board is still supporting ice cubes, which have not melted. It can be used on concrete, timber or metal roof decks before waterproofing directly with built-up felt, asphalt or single-layer roofing materials; however, it must have a vapour barrier between the roof deck and the insulation board to protect the board against the ingress of moisture from within the building and to prevent interstitial condensation. If the roof is to be subjected to continuous foot traffic, the waterproofing should be protected by concrete or similar composite tiles or by a screed.

4. Other Ness products

Thomas Ness Ltd also operate, jointly with Texsa of Spain, a new plant at Caerphilly, South Wales, for producing Multi-Plas, a bituminous roofing sheet that can be laid directly on top of Ness-Board. Further, a useful building maintenance aid which saves energy and surface finishes, by controlling rising damp in older buildings that lack damp courses, is the silicone fluid injection scheme mentioned above. Hyload, on the other hand, is a coal-derived pitch/polymer dampcourse material particularly suitable for difficult and stepped cavity trays in new buildings. It will withstand higher loadings than normal bitumen dampcourse and is stable over a wide temperature range.

REFERENCES

British Petroleum, 1977 *Our Industry, Petroleum.*
Building Research Establishment, 1975, *Energy Conservation: A Study of Energy Consumption in Buildings and Possible Means of Saving Energy in Housing*, BRE Working Party Report CP 56/75, HMSO, London, June.
Coal Industry Tripartite Group Steering Committee, 1974, Coal Industry Examination, Department of Energy.
Coal Industry Tripartite Group R&D Working Party, 1977, Coal Technology—Future developments in the conversion utilisation and conventional mining in the UK, Department of Energy.
Department of Energy, monthly, *Energy Trends.*
Department of Energy 1978a, *Energy Policy* (Green Paper), Cmnd. 7101, HMSO, London.
Department of Energy, 1978b, *Digest of UK Energy Statistics*, HMSO, London.
Department of Energy, 1978c, *Coal for the Future; Progress with 'Plan for Coal' and Prospects to the Year 2000*, D. of E., London.

Egerton, A. (1946) Heating and Ventilating of Dwellings. Ministry of Works, HMSO, London.

IEA, 1979, *Steam Coal Prospects to 2000*, International Energy Agency, Paris.

I.P.C., 1978, *World Energy Resources, 1985—2020*, IPC Science and Technology Press, Guildford.

Kaye, W. G., 1979, Combustion—achievements and tasks, *Conf. on Future Energy Concepts*, London, February. Conf. Pub. 171, Inst. Elec. Eng.

NCB, 1974, *NCB Plan for Coal*, National Coal Board, London.

Parker Morris (1961) Homes for today and tomorrow. Ministry Housing and Local Government, HMSO, London.

Randell, A. A., Gauld, D. W., Dando, R. and LaNauze, R. D., 1978, *Proc. 2nd Engineering Foundation Conf.*, April. Ed. Davidson, J. F., and Keairns, D. L., Cambridge University Press.

Solid Fuel Advisory Service, 1975, *Centrepiece*, 15. N.C.B., London.

Tregelles, P., 1978, *Coal Research 2000*, Robens Coal Science Lecture, London, October; *Mining Engineer*, Feb. 1979.

Energy Conservation and Thermal Insulation
Edited by R. Derricott and S. S. Chissick
© 1981 John Wiley & Sons Ltd.

CHAPTER 8

Research on energy conservation in houses carried out by the Electricity Council Research Centre

P. Basnett

Electricity Council Research Centre, Capenhurst, Chester

INTRODUCTION

The Electricity Council Research Centre was set up in 1965 to service the needs of the electricity supply industry in England, Scotland and Wales for research relating to the distribution and utilization of electricity. As far as the utilization of electricity is concerned, our brief is to promote the wise use of energy. In pursuit of this aim, we carry out research directed at developing and improving the application of electrical processes in four main areas:

Metallurgical industries.
Chemical and allied industries.
Other manufacturing industries.
Human environment.

Within the 'human environment' area we work on ventilation, lighting, odour control, and the development of equipment, as well as on thermal insulation and 'the low-energy house', a concept which involves matching the performance of equipment with building needs and hence providing a combination which requires a minimum amount of heating energy to satisfy comfort criteria. It will be seen from this brief outline and from Figure 1 that work directly associated with energy conservation in buildings forms only a very small part, about 2%, of our total research effort.

The work on energy conservation in buildings can itself be subdivided into three classes:

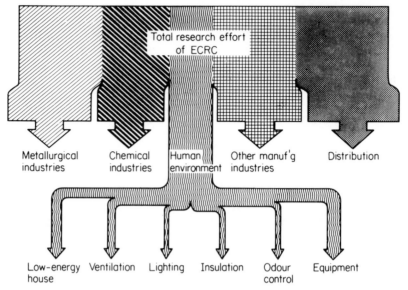

Figure 1 Division of ECRC research effort into five main areas, with 'Human environment' subdivided

(i) design work, collaborating with insulation manufacturers to develop insulation systems with the properties we believe to be necessary;

(ii) computation work, studying the effects of various conservation measures on comfort and energy consumptions in buildings, using a computer program which simulates the transient thermal behaviour of a building and its heating system; and

(iii) experimental work, carried out in collaboration with our colleagues from the Electricity Council's Marketing Department, studying the effects of conservation methods applied to actual buildings to ensure that they do in fact work in the manner that we expect.

The remainder of this paper will deal with the first two facets, design and computation, of our work on energy conservation in buildings.

DESIGN WORK

Nearly half of the 19 million houses in the UK have solid external walls. They were built before about 1935, so that many of them lack modern amenities, and are the subject of political and social attention. Many people condemn the demolition of aging houses, particularly in city centres. Radical upgrading and rehabilitation of such dwellings need cost no more than half as much as demolition and rebuilding.

One problem making rehabilitation difficult has been that there was little that could be done to improve the familiar 9 inch solid brick walls, which, as well as having poor U values, are liable to suffer from rain penetration and condensation. Mould (1976, 1977) has shown that if insulation is applied internally there is a considerable danger of interstitial condensation occurring, both in the insulation itself and in the brickwork of the wall, unless a vapour barrier can be provided and maintained on the room side of the insulation. This would mean, for example, that no fixings could be allowed into the wall. If, however, insulation is applied to the outside of the wall, there is no likelihood of condensation occurring. Figure 2 shows how the danger of interstitial condensation arises. Furthermore, external insulation will protect the brickwork from rain penetration, while this can still occur with internal insulation.

A 225 mm (9 inch) solid brick wall has a U value of 2.1 W/m^2 K. The addition of 25 mm of mineral fibre or foamed plastic insulation will bring this value slightly below the current Building Regulations standard of 1.0 W/m^2 K, 50 mm of insulation will make the wall comparable with a foam-filled cavity wall with a U value of approximately 0.6 W/m^2 K, and 100 mm of insulation would reduce this figure still further to about 0.33 W/m^2 K.

As a rough guide, an unimproved semidetached house with solid brick walls would have an annual space heating energy consumption of approxi-

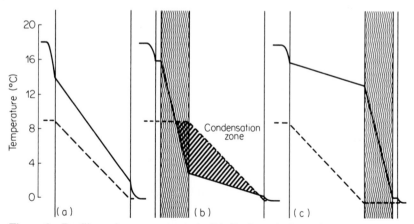

Figure 2 Profiles of temperature (solid line) and dewpoint temperature (dashed line) in uninsulated and insulated walls: internal conditions 18°C, 50% relative humidity; external conditions 0°C, 90% relative humidity. (a) Uninsulated solid brick wall; (b) solid brick wall with internal insulation—condensation occurs where dewpoint is above all temperature; (c) externally insulated solid brick wall—no condensation. (Reproduced by permission of The Electricity Council Research Centre, Capenhurst, England)

mately 50 GJ, or 14 000 kWh. With minimal thermal insulation, i.e. 75 mm of mineral fibre in the loft space, this would be reduced to about 42 GJ/yr (12 000 kWh/yr), a saving of approximately 15%, while with a package of insulation as recommended by Mould (1977), i.e. 100 mm of insulation in the roof and walls, plus double-glazed windows, the space heating energy requirement could be reduced to 15 GJ/yr (4200 kWh/yr), a saving of 70%.

The essential features of an external insulation system are the insulation itself, a protective covering, and a method of fixing it to the wall. One system which has been used for a number of years in West Germany, and has also been used on an extremely exposed block of maisonettes in Scotland, consists of 40 mm boards of expanded polystyrene stuck to the face of the wall with strips of high adhesion mortar and covered with a laminate of mortar and glass fibre matting built up *in situ* and finally coated with a sand-based decorative finish.

A rather similar British system is under development, and this has a continuous bed of mortar adhesive together with mechanical fixing by metal pins. This system should provide improved impact resistance. Another system, also under development, uses a mineral fibre mat held against the wall by a galvanized wire mesh fixed by polypropylene 'Hilti' pins. The wire mesh is then used as to support a cement–mortar rendering.

A different approach, which is already being test marketed, is to pin vertical timber battens to the wall over a sheet of polyvinyl chloride. The battens are then clad with oil-bound hardboard and the space so formed is progressively filled with a urea-formaldehyde foam, similar to that used for cavity fill insulation. The completed wall is then finished with a decorative coat of resin-bound stone chippings.

None of these systems has yet been in use for long enough in British conditions to be able to guarantee their weathering and impact resistance qualities.

Another approach, which is suitable for 'do-it-yourself' application, was devised with help from ECRC for a house renovation demonstrated on the Granada TV programme 'A House for the Future'. An old coach-house was converted into a dwelling house with several energy-saving features. These were a solar collector on the roof, mechanical ventilation with heat recovery, and a wind generator, as well as external thermal insulation. The design work and thermal performance predictions were reported by Siviour et al. (1977).

In this conversion, the details of the external insulation were as shown in Figure 3. 50 mm battens and counter-battens were nailed to the wall and the spaces between them filled with 100 mm of insulation, a mixture of mineral fibre, used next to the wall to take up irregularities and reduce convection and expanded polystyrene board to help retain the fibrous material.

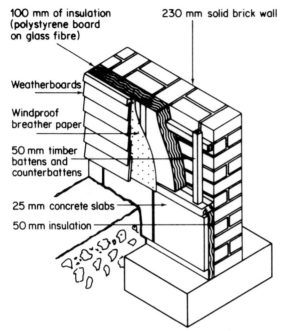

100 mm of insulation
(polystyrene board
on glass fibre)

230 mm solid brick wall

Weatherboards

Windproof
breather paper

50 mm timber
battens and
counterbattens

25 mm concrete slabs

50 mm insulation

Figure 3 External insulation as applied to the Granada TV 'house for the future'. (Reproduced by permission of The Electricity Council Research Centre, Capenhurst, England)

This was then covered with windproof breather paper and weather-boarding. Below damp course level the insulation was by 50 mm closed-cell polystyrene foam boards protected by 25 mm concrete paving slabs stood on edge. The overall U value of the wall is 0.36 W/m^2 K.

Whatever form of external insulation is used, great care is needed in detailing the edges and corners. The thickness of the insulation will normally mean that rainwater pipes and so on have to be removed and replaced, as there is not enough space behind them for the insulation. If there is sufficient overhang at the eaves, no additional protection will be needed, while if the insulation is thicker than the projection of the eaves, flashing will be required to prevent rain water penetrating into the insulation. This is shown in Figure 4. In order to prevent thermal bridging and rain penetration, the insulation should be carried up to the top of a gable wall rather than being stopped at the level of the first-floor ceiling.

If the building renovation involves replacing window frames, the new ones should be planted on the outer face of the brickwork and specially designed to protect the edges of the insulation, as shown in Figure 5. On the other hand, if existing window frames are used, the insulation and

Figure 4 Eaves details for external thermal insulation. Left: insufficient overhang—top of insulation needs protection by flashing. Right: eaves have sufficient overhang to protect the insulation. (Reproduced by permission of The Electricity Council Research Centre, Capenhurst, England)

cladding will need to be returned to meet the window frame and an extension cill will also be required.

COMPUTATIONAL WORK

A computer program 'HOUSE' (Basnett, 1975) has been developed at ECRC to simulate the thermal behaviour of a building and its interactions with its occupants, its heating system, and the weather. This program has been used (Basnett, 1978) to study the effect of thermal mass on the behaviour of an occupied, well insulated house.

Two different constructions were considered for semidetached houses of identical layout and design day heat loss. The first was a house which was of 'lightweight' construction, with as little thermal inertia as can conveniently be built. In this house the front and back walls were of timber frame construction with 13 mm of foil-backed, dry-lining plasterboard, 75 mm of mineral-wool fibrous insulation, 9 mm of plywood sheathing, and finished on the outside with softwood boards on battens for the ground floor, or concrete tiles hung on softwood battens for the first floor. The gable wall consisted of a cavity brick and block wall, with 50 mm of foam insulation in the cavity and 13 mm dry-lining plasterboard on plaster dabs. The ground floor was a carpeted suspended timber floor over a crawl space with 180 mm of mineral-wool insulation below the floorboards. Internal partitions were made of two layers of 13 mm plasterboard supported on 75 mm × 50 mm timber framing.

The second construction contained as much thermal inertia as is practicable. The external walls were of 150 mm hollow dense concrete blockwork with a 19 mm plastered finish on the inside and externally insulated with

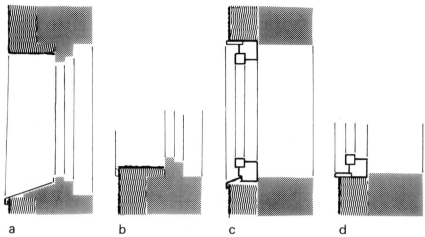

a   b   c   d

Figure 5 Window details for external thermal insulation. Vertical (a) and horizontal (b) sections through insulation applied with existing window frames. Insulation and cladding have to be returned at top and sides to meet window frame. Metal extension sill to protect the insulation below the window. Vertical (c) and horizontal (d) sections through insulation applied with replacement window frames planted on the face of the brickwork and designed to protect the insulation

mineral wool under 19 mm of cement rendering. The thickness of the insulation was adjusted to give the same overall U value as the corresponding walls in the lightweight house.

The ground floor consisted of a carpeted concrete floor slab 100 mm thick with 150 mm of rockwool insulation between it and the ground. Edge insulation was arranged to give the same steady-state heat loss as for the suspended floor. The internal partitions were of 190 mm cellular concrete blockwork with 19 mm plaster finish on both sides.

Both houses had double glazing, 200 mm of mineral-wool insulation above the first floor ceiling, and intermediate ceiling and floor consisting of 13 mm plaster, 180 mm × 50 mm joists, and a 19 mm suspended carpeted plywood floor, and identical furnishings. The houses had a floor area of approximately 84 m², and the design day heat loss was 2.92 kW ($\Delta T = 20$K) assuming a ventilation rate of one air change per hour.

Two different occupancy patterns were used, one to represent a two-person family with the house empty all day, and the other a four-person family, with the house occupied all day.

At first sight, one would expect the lightweight house to require less energy than the heavy one to maintain the same comfort conditions, since it should respond more quickly to heat inputs. This is indeed the case in cold weather where it will be seen from Table 1 that the lightweight build-

Table 1 Predicted average daily heat requirements for light and heavy houses

| Occupancy | Two-person | | Four-person | |
Construction	Light	Heavy	Light	Heavy
12–16 Jan. 1977—Average external temperature 1.7°C				
Internal gains (MJ/day)	44.1	44.1	67.5	67.5
Solar gains (MJ/day)	12.3	12.3	12.3	12.3
Heating system (MJ/day)	63.9	71.2	63.9	73.1
Total heat required (MJ/day)	120.3	127.7	143.7	152.9
Ventilation heat loss (MJ/day)	18.6	19.2	23.7	18.8
17–21 Nov. 1976—Average external temperature 8.3°C				
Internal gains (MJ/day)	44.1	44.1	67.5	67.5
Solar gains (MJ/day)	7.0	7.0	7.0	7.0
Heating system (MJ/day)	25.1	27.4	24.4	23.2
Total heat requirement (MJ/day)	76.2	78.5	99.0	97.7
Ventilation loss (MJ/day)	8.9	8.1	15.3	11.4
17–21 April 1976—Average external temperature 10.4°C				
Internal gains (MJ/day)	39.3	39.3	63.7	63.7
Solar gains (MJ/day)	54.3	54.3	54.3	54.3
Heating system (MJ/day)	4.3	8.9	6.0	8.6
Total heat requirement (MJ/day)	98.0	102.5	124.0	126.6
Ventilation loss (MJ/day)	18.3	9.5	39.6	23.4
Estimated 180 day heating season— Average external temperature 5.9°C				
Internal gains (MJ/day)	43.3	43.3	66.9	66.9
Solar gains (MJ/day)	17.1	17.1	17.1	17.1
Heating system (MJ/day)	37.8	42.6	37.8	41.6
Total heat requirement (MJ/day)	98.2	103.0	121.5	125.5
Ventilation (MJ/day)	14.5	13.0	22.8	16.5

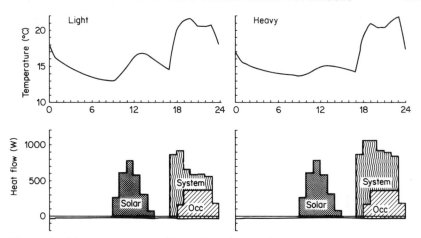

Figure 6 Air temperature and heat flows for the (south-east facing) lounge of light and heavy buildings with two-person occupancy on a sunny January day. The baselines of the heat flow diagrams show the heat loss due to ventilation. (Solar—solar heat gain through the windows; system—controlled heat gain from the heating system; occ—uncontrolled heat gain due to occupancy)

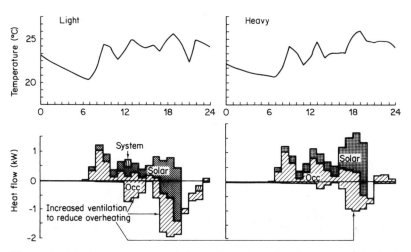

Figure 7 Air temperature and heat flows for the (north-west facing) kitchen–dining room of light and heavy buildings with four-person occupancy on a sunny April day. The baselines of the heat flow diagrams show the heat loss due to ventilation. (Solar—solar heat gain through the windows; system—controlled heat gain from the heating system; occ—uncontrolled heat gain due to occupancy.) Note increased ventilation to reduce overheating at midday and in the evening, especially in the lightweight building

ing only requires about 90% of the energy needed by the heavy one for two-person occupancy, or 94% for four-person occupancy. However, in mild winter weather the differences are only very small and near the ends of the heating season the heavy house can use less energy than the light one, especially on sunny days and with four-person occupancy. This is because the heavy building damps out temperature swings due to peaky internal and solar heat gains better than the light one, and hence makes better use of these gains. These effects are demonstrated in Figures 6, 7, and 8, where the temperatures and energy inputs to the kitchen and lounge of the two buildings are contrasted for January and April weather.

Over the heating season as a whole, it is estimated that the lightweight building would use, for either occupancy pattern, approximately 700 MJ less heating energy than the lightweight one, which is approximately 10% of the energy used by the heating system alone, or 5% of the total energy billed to the building (including cooking, lighting, water, heating, etc.). If all of the heating energy was on-peak electricity, the difference in cost would be approximately £6 per year at 1978 prices.

Calculations were also made for heatwave weather, such as occurred during the summer of 1976 and, as was expected, the extent of overheating which occurred in the heavy building was much less than in the lightweight one. This can be seen, for example, from the curves presented in Figure 9.

We believe that the small additional energy cost of using a heavy build-

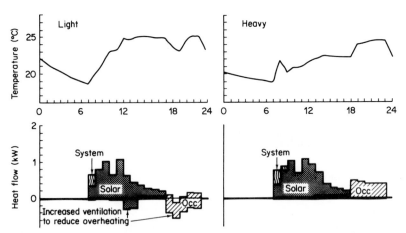

Figure 8   Air temperature and heat flows for the (south-east facing) lounge of light and heavy buildings with four-person occupancy on a sunny April day. The baselines of the heat flow diagrams show the heat loss due to ventilation. (Solar—solar heat gain through the windows; system—controlled heat gain from the heating system; occ—uncontrolled heat gain due to occupancy.) Note that no increased ventilation is needed in the heavy building

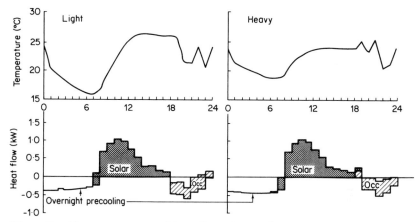

Figure 9   Air temperature and heat flows for the (south-east facing) lounge of light and heavy buildings with four-person occupancy and heatwave conditions. The baselines of the heat flow diagrams show the heat loss due to ventilation. (Solar—solar heat gain through the windows; system—controlled heat gain from the heating system; occ—uncontrolled heat gain due to occupancy)

ing construction is acceptable considering the greatly increased comfort of a heavy building in comparison with a light one.

ACKNOWLEDGMENTS

The work described in this paper is part of the Electricity Council's Research Program and thanks are due to the Director of the Electricity Council Research Centre for permission to publish it. The design work on external thermal insulation was carried out by A. E. Mould.

REFERENCES

Basnett, P., 1975, Modelling the effects of weather, heating and occupancy on the thermal environment inside houses, *Mathematical Methods for Environmental Problems*, ed. C. A. Brebbia, Pentech Press, London.

Basnett, P., 1978, *A Theoretical Study of the Thermal Behaviour of Occupied, Well Insulated Houses*, ECRC/M1175, Electricity Council Research Centre, Capenhurst.

Mould, A. E., 1976, House improvement: the solid external wall, *Building*, 7 May.

Mould, A. E., 1977, *External Insulation—A Practical Way to Improve the Thermal Performance of Existing Solid External Walls*, ECRC/M1033, Electricity Council Research Centre, Capenhurst.

Siviour, J. B., Stephen, F. R., and Mould, A. E., 1977, *Thermal Design and Performance Predictions of the Granada house*, ECRC/M1109, Electricity Council Research Centre, Capenhurst.

Energy Conservation and Thermal Insulation
Edited by R. Derricott and S. S. Chissick
© 1981 John Wiley & Sons Ltd.

CHAPTER 9

Energy recovery from refuse and pollution control

G. R. Winch

University of Manchester

TRENDS IN REFUSE

During the past decade in Europe and North America, increasing attention has been given to the disposal of refuse by incineration, due to a combination of several factors.

Most refuse is disposed of by dumping in open ground and the decreasing availability of convenient sites near large centres of population, coupled with the increasing volumes of refuse produced, has made the cost of open tipping rise considerably due to the transport involved. The decreasing density of refuse with an increase in calorific value makes it more suitable for burning and less suitable for dumping. Further, the substantial increase in the costs of energy, following the oil crisis of 1974, has caused attention to be given to the contribution which may be made by production of heat from refuse.

The quantities of refuse produced are impressive. Currently in the UK the annual quantity collected by local authorities is estimated as being 17 million tonnes and the total production 40 million tonnes per annum, including industrial and trade refuse. In the USA it is thought that the current level of refuse collection is 150 million tonnes per annum and the total refuse produced 250 million tonnes per annum. In terms of fuel equivalent, the refuse produced in the UK is approximately equal to 5 million tonnes of fuel oil per annum and in the USA 30 million tonnes of oil per annum. In Europe as a whole, the current level of refuse collected is 0.7 kg per person per day and the figure for the UK is slightly higher, at 0.8 kg per person per day. In comparison, in the USA the current level of refuse collected is thought to be close to 1.9 kg per person per day. Refuse produced is generally rising at a rate of 1% per annum by weight but at 4% per annum by volume, which broadly speaking means that refuse volumes will double every twenty years. As an indication of the decline in refuse

Table 1   Trends in refuse combustible content (%) for the UK

	1935	1968	1980
Dust & cinders	57	22	12
Paper	14	37	43
Textiles	2	2.5	3
Plastics	—	1	4
Putrescibles, etc.	14	18	18

density and increase in volume, in the UK in 1935 the typical density of refuse collected by local authorities was of the order of 290 kg/m³. By 1980 it is thought that the density will fall to 120 kg/m³.

The refuse produced by society is an indication to some extent of a way of life. Study of Table 1 will show the major trends in combustible refuse content from 1935 to a projected figure for 1980. The table indicates two major trends: the decline in open-fire solid-fuel heating, suggested by the steady decline in the content of dust and cinders in refuse collected, and the rise in paper and plastics content, which indicates a substantial increase in the purchase of packaged foods and products. As far as can be foreseen, these trends will continue into the 1980s.

REFUSE AS A FUEL

Refuse disposal by incineration is not a novel concept and, as far as is known, municipal refuse has been disposed of by central incinerators in both Europe and the USA since about 1870. Before and since, the cottage grate throughout the world was used to dispose of refuse produced in the house or dwelling.

As long ago as 1896 the city of Hamburg constructed an incinerator plant as a defence against epidemics. The plant utilized the heat produced for generating electric power and steam for industrial use, the power being used to drive a sewage pump and battery charging station for electrical refuse vehicles, and the steam was supplied to a disinfection institute and to a cement factory. Subsequently several other cities in Germany followed this example. In 1927 a similar incineration/power plant was installed in Glasgow.

Many of the early municipal refuse incinerator plants or refuse destructors attracted an unenviable reputation. The refuse tended to be batch fed onto fixed grates with limited provision for the control of smoke and dust emission from the chimney. The arrangements for dust removal from flue gases were usually of the form of a gravity settlement chamber or contrived by passing the flue gases over a water trough or through water sprays, giving the inevitable plume of steam from the chimney. As recently as 1965 in the UK there were no municipal incinerators provided with continuously fed moving grates and modern dust removal systems.

In the UK in 1967 approximately 8% of waste collected by local authorities was treated by incineration. By 1980 it is estimated (Department of the Environment, 1971) that approximately 20% of collected refuse in the UK will be disposed of by incineration using over 30 modern units, but probably less than five of these units will take advantage of heat recovery from refuse in the provision of steam for power generation or hot water for district heating.

This may be compared with the situation in the Federal Republic of Germany where in 1975 there were 31 large municipal incinerator plants of which 21 produced both heat and electric power and seven produced heat for district heating or industrial use. By 1980 it is estimated that over 40 large incinerator plants will be in operation in German cities, the great majority of which will provide heat and electric power for the municipality they serve, dealing with 25% of refuse collected. It has been said (Thomen, 1976) that in West Germany the limit for the proportion of the population which can be served by large municipal incinerators with energy recovery is 33%, which will require the installation of an additional 20 new incinerator plants, each for 250 000–280 000 customers.

Whilst West Germany may be said to possess the most well developed structure for municipal refuse disposal with energy recovery in the world, a number of European countries have installed such units during the past decade and these include Sweden, Denmark, Austria, Italy, France, and Holland. In North America, municipal incineration for the disposal of refuse is widely practised but most plants are not provided with facilities for energy recovery. During the past decade an increasing number of townships have installed modern large incineration plants with facilities for heat recovery with provision for pollution control in compliance with modern standards.

The incineration plant is not always located in buildings specifically designed for the purpose. In West Germany a city power plant often consists of oil- and coal-fired boilers, raising steam for use in turbines generating electrical power, and alongside is a refuse-burning plant of similar size for the same purpose. In some cases refuse and coal are fired together. In the UK and in the USA trials have been undertaken to burn refuse together with solid fuel in normal power station boilers.

In addition to central municipal incineration plants, a number of buildings throughout the world possess their own incinerator plant for the disposal of refuse generated within the building or group of buildings. Hospitals and large factories are typical instances of this. Few of these on-site or local incineration plants are provided with heat recovery, although in certain industries this is practised. In the woodworking industry, a difficult refuse disposal problem is solved by the incineration of wood refuse on site, and usually the heat recovered from the combustion of the refuse is used to supplement the energy requirements of the factory.

Refuse as a fuel is unlike most fuels in that it is extremely heterogeneous in nature. On assessing data available on municipal refuse in the UK, it is thought that a typical average figure for the gross calorific value of municipal refuse would be 10 000 kJ/kg, but that this could vary between the limits of 6000 kJ/kg to 12 000 kJ/kg. In comparison, the gross calorific value of coal is close to 30 000 kJ/kg. Figure 1 shows trends in refuse calorific value.

The ash and moisture content of typical municipal refuse is usually of the order of 60% in the UK whereas that for coal is usually in the region of 15%. As an indication of trends on the Continent, published figures for West Germany show that in 1960 the net calorific value of municipal refuse was an average of 4600 kJ/kg, rising to 8000 kJ/kg in 1975. The projected figure for 1980 is 8300 kJ/kg. The average incombustible content of refuse in West Germany is 65%, of which 35% is water. The moisture content of refuse in the UK is close to 25% on average (see Table 2).

Two other points of comparison between coal and refuse as a fuel are of interest. One is the sulphur content, which on combustion gives rise to SO_2, which is a serious pollutant of the atmosphere. Typical sulphur contents for British refuse are between 0.2 and 0.5%, and from West Germany the figure indicated is between 0.2 and 0.7 wt.%. In comparison the

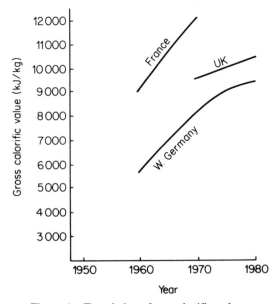

Figure 1   Trends in refuse calorific value

Table 2 Probable contribution (%) to gross calorific value of refuse in 1980 in the UK

Dust & cinders	7
Paper	60
Textiles	5
Plastics	16
Putrescibles, etc.	12

sulphur content of typical coal is between 1.0 and 2%. The other factor is the content of total volatile matter, which for refuse is usually between 60 and 80 wt.% compared with coal which is usually close to 35 wt%. In the case of refuse, this means that much of the combustion will take place above the grate rather than on it, as in the case of solid fuels.

The refuse from commercial and public buildings is generally similar to that produced from domestic buildings with the exception that, unless catering or food sales are involved, the vegetable and putrescible content is lower than domestic refuse and the paper and cardboard content higher. The factory waste from industrial buildings is of a widely varying nature and is indicative of the nature of the process or operation carried on within the works. Industrial waste of a specialized nature of either a solid, liquid or gaseous form cannot be considered as suitable material for general-purpose generators and, if incineration is practicable, special purpose units are required. However, the general refuse from factory buildings is on the whole similar to that produced by commercial buildings with a somewhat lower density and slightly higher calorific value than domestic refuse and usually with a lower putrescible content.

The refuse from hospitals has special characteristics, which are important when considering disposal by incineration. In addition to the usual materials found in domestic refuse, hospital waste contains quantities of infected material and an increasing quantity of disposable plastics. The density and calorific value can vary widely, with moisture contents up to 80% and gross calorific values as low as 2500 kJ/kg. When considering the incineration of plastics, it should be borne in mind that PVC (polyvinylchloride) on burning produces traces of gaseous hydrochloric acid which can result in the corrosion of the incinerator itself and local pollution of the atmosphere. Another particular problem arising in dealing with hospital waste is the presence of quantities of stainless-steel needles from disposable hypodermic syringes.

One of the main difficulties in dealing with the receipt of municipal refuse at a central incineration plant is the mechanical handling of the refuse from the delivery vehicle to the furnace grate. Refuse is an extremely variable material and one would expect to receive refrigerators, bedsteads, and bicycles, through the range of assorted small materials, to rolls of carpet and linoleum. In one instance, the undetected receipt of a pair of magnesium-alloy motor car wheels caused a hole to be burnt in the grate of a municipal incinerator, causing the complete shutdown of the boiler unit in question until repairs had been effected.

Clearly where uncontrolled reception of refuse is in operation, some form of initial screening is desireable, particularly if the magnetic separation of ferrous material is of economic advantage. Large metal objects can be passed to the grate without causing hold-ups and blockages in the grate feed system, although it is probable that difficulties will be experienced at the ash conveyor system of the incinerator.

Ideally, in order to provide good control of combustion and chimney emission, municipal incinerators should operate continuously, although many plants operate sucessfully on a two-shift basis. However, most collection and delivery organizations operate on a single-shift system for five days per week, and therefore considerable storage is required at the incinerator plant to enable refuse to be stored, particularly during weekends and holiday periods.

TRENDS IN POLLUTION CODES AND EFFECT ON DESIGN

It is inappropriate to consider incinerator design without close reference to current thought in environmental protection, atmospheric pollution control legislation and clean air codes, and also future trends in these matters. From study of atmospheric pollution codes in Europe and the USA, it is clear that level of engineering design in incinerators corresponds closely with the standards of control applied.

As long ago as 1964, West Germany gave a world lead in the control of effluent from incinerators in specifying a smoke limit of Ringelmann scale 2 with dust limits of 150–200 mg/m^3 of flue gas at NTP (normal temperature and pressure) and 7% CO_2. The lower figure applied to incinerators which burnt more than 20 tonnes of refuse a day (Federal Ministry of Health, 1964).

Even as late as 1971, there was no national code in France, and the levels of operation were determined by the local prefect at his discretion, under the powers given by the General Anti-pollution Law of August 1961.

In the USA the only national law in operation at that time was the National Air Quality Standards Act of 1970 covering the operation of the incinerators in Federal installations only. Otherwise each state and city enacted specific legislation of pollution control over a wide range from the maximum of 230 mg/m^3 flue gas in Chicago to 1200 mg/m^3 in New York State. The maximum level of smoke emission was Ringelmann 1 for Federal installations and generally Ringelmann 2 elsewhere. The chimney emission from Federal incinerators was limited to a dust emission of 460 mg/m^3 at 12% CO_2, except for incinerators burning less than 90 kg/h where the limit was 700 mg/m^3.

In the United Kingdom at that time, control existed under the Clean Air Acts of 1956 and 1968 under which the chimney emissions from incinerators were not allowed to exceed Ringelmann 2 density for smoke, with the exception of permitted periods, and where the accepted level for dust emissions was 460 mg/m^3 of dust in flue gases. In certain urban areas, smoke control orders prohibited the emission of visible smoke, and an exemption certificate was required before incinerators could be installed and operated.

By 1973 control of solid particles in flue gases from incinerators in the USA had come into line with the 1964 West German level of 150 mg/m^3 and the figure for Sweden was 180 mg/m^3. The figure for Holland, Norway, and Denmark was 215 mg/m^3.

In the UK the Department of the Environment (1974) published their second report containing recommendations for the maximum allowable emission of grit and dust from incinerator chimneys. For municipal incinerators having a capacity exceeding 4.5 t/h, the maximum recommended emission is 230 mg/m^3 at 10% CO_2 content in the flue gases. For smaller incinerators, the maximum recommended dust content in the flue gases is progressively increased such that for incinerators having a capacity not exceeding 270 kg/h, it is recommended that the flue-gas dust content should not exceed 920 mg/m^3. These recommendations as yet have no legislative authority.

The long-standing excellence of the West German environmental protection legislation on incinerator emissions has encouraged the development of municipal and on-site incinerators designed to operate at a higher performance level than comparable units elsewhere. A further development in West Germany is the Federal pollution control law prescribing a limitation for inorganic gaseous chlorides, with a limit for maximum HCl emissions of 100 mg/m^3 of wet flue gas. It is thought that this is the first time that an attempt has been made in a legislative manner to control hydrochloric gas emission from incinerator chimneys. The source is mainly PVC material contained in the refuse, and this implies that limits must be imposed on the PVC content of the refuse, or that some form of wet scrubber system designed to remove HCl from flue gases is installed, should the emission limit exceed the figure quoted.

The trends in the properties of municipal refuse already mentioned, such as increasing volume and heat content and reducing density, do not impose undue technical difficulties in design nor affect the operation of modern incinerator plants, once these changing properties and their trends have been recognized. Similarly, developments and trends in legislation on pollution control must be studied at the outset of any design project, as trends in pollution control measures are just as important as trends in refuse properties as determining factors in technical design levels, consequent costs, and level of operational performance required. Technological improvement in any field of engineering usually incurs a substantial additional cost which must be paid in order to secure improvement.

Public awareness of the value of high performance incineration has been limited in the past, but with rising consciousness of environmental protection, an incinerator plant which may last for 20–30 years must anticipate future trends rather than recognize the levels current at the time of its inception. Current thought in environmental protection is that the pollutor must pay, and this fact, together with the dominance of legislation as a design determinant, is indicated by a study of conditions existing in West Germany and its neighbour France in the mid 1970s. In West Germany at that time, the legislation controlling emission from incinerator chimneys was, and probably still is, the most stringent in the world, whereas in France the comparable legislation was minimal. A comparison of the standards of design of German on-site incinerators as compared to French on-site incinerators will illustrate this point. Another point of interest is that the largest municipal incinerators in France were based on German technology responding to German antipollution legislation.

As an indication of the general design requirements of an incinerator, it is clear that it must be capable of receiving and storing the refuse from the area or building it serves without visual or olfactory offence. It should then be capable of burning the range of refuse for which it is designed without the emission of smoke and with the minimum emission of dust in relation to existing environmental protection codes and without the emissions of odour, and so reducing the main combustible components in the refuse, carbon and hydrogen into carbon dioxide and water vapour. The efficiency of combustion should be such that the proportion of carbon monoxide in flue gases should not exceed 0.1 vol.% in large incinerators and 0.25% in small incinerators with a low proportion of unburnt hydrocarbon gases. In addition, the proportion of the putrescible material or unconverted unorganic matter in the ashes, should not exceed 0.25 wt.%.

Under these operating conditions, the refuse is reduced to a sterile ash of approximately 10% of the original volume which may be used as an aggregate for landfill or for other construction purposes. This may be compared with environmental effect of dumping untreated raw refuse on open ground. If during combustion process the energy released by the combustion of the refuse is recovered in the form of hot water, steam or a combustible gas, then a factor of social significance emerges in the form of resource re-use, in that material rejected by society can be converted into material valued by society, particularly when it is in the form of energy, the natural reserves of which are tending to decline at an increasing rate.

Whilst this chapter is concerned with energy recovery from incineration and the resulting requirements for pollution control, it must be borne in mind that the primary purpose of incineration is to destroy refuse, and therefore some thoughts on design parameters for certain types of incinerator are relevant at this stage, particularly in relation to the two major factors, the properties of the refuse to be burnt and the antipollution codes controlling emissions from the incinerator chimney.

The study of on-site incineration practice provides a clear indication of the relationship between technology and legislation. On-site incinerators fall into three main classifications. The first is the single-chamber incinerator, consisting of a combustion chamber with a flat grate and a flue for the passage of effluent gases, together with an ashpit under the grate for the receipt of combusted material (see Figure 2). The second form is the double-chamber incinerator, which is a single-chamber incinerator with an additional chamber providing for extended combustion to burn out combustibles in suspension and a gravitational settlement chamber for the deposition of the denser suspended particles in the flue gases (see Figure 3). The third type is the advanced double-chamber incinerator, which is a double-chamber unit with an inclined grate and an additional high-efficiency dust arrestor through which the flue gases pass before exit to the chimney. It is also provided with supplementary burners and sealed ash-bins. In the early 1970s the single-chamber unit was widely available in France. The double-chamber unit was in general use in the USA and UK and the advanced double-chamber units were in use in West Germany.

Modern on-site incinerators must be convenient to operate and should be capable of being charged mechanically by means of a chute or conveyor, or be provided with an easily operated inclined charging door. Combustion should take place automatically without the necessity for manual work involved in fire trimming. The conventional flat grate does not allow this, unlike the inclined stepped grate used in the double-

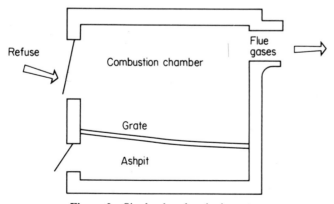

Figure 2   Single-chamber incinerator

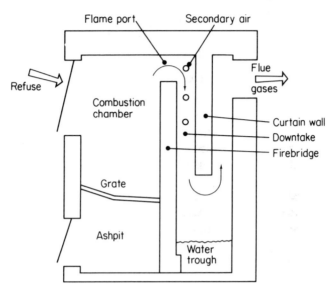

Figure 3   Double-chamber incinerator

chamber advanced units, in which the refuse moves down and compacts the firebed under its own weight. A sealed ashpit to avoid dust emission and a transportable container into which ashes can be easily deposited from a tilting or dumping grate is essential for this concept, so that the light labour normally found in public buildings may undertake the duties required. In order to meet the low smoke emission and minimum odour requirements, it is essential to maintain a temperature of between 800 and 900°C in a secondary combustion chamber. Due to the large proportion of volatile matter in refuse, it is also necessary to inject substantial quantities of secondary and tertiary air at high velocity in order to provide the turbulence required for a quick and complete burn-out. Preheated secondary air has a considerable advantage over cold air in stimulating good combustion, and it is felt this should be an inherent part of any design. Adequate quantities of heat are of course available for preheating air.

Control of dust emission is no longer satisfactory when conventional inertial separation chambers are used. The usual solution is to use a dry arrestor system using high-efficiency cyclones with induced draught fan in which the dust is automatically deposited into sealed bins. The efficiency of the cyclone system is proportioned to the resistance to airflow through the unit and inversely proportioned to its diameter. In order to achieve the high efficiencies required by modern antipollution codes, multiple units are necessary and their resistance requires the use of a fan. As the operating temperature of steel cyclones is normally limited to 300°C, cooling is necessary before the cyclone to reduce flue-gas temperatures to within this limit. This is often accomplished by the admission of dilution air, which has the disadvantage that the capacity and cost of the cyclone and induced draught system is increased due to the extra air weight handled, and the admission of dilution air usually increases the weight of the gases to be handled by a factor of 3 or so. This encourages the use of waste heat recuperators, as, although the heat recovery device has to be paid for, the size of the dust arrestor system and fan is substantially reduced and so is the energy consumption at the fan motor.

The dust burden in flue gases leaving the combustion chamber of on-site units with fixed grates is lower than that arising in municipal units. Dust burden depends upon the ash content of refuse fired, the under-grate air volume per unit grate area, and the degree of disturbance of the refuse on the grate. With typical commercial and domestic refuse and primary air volumes of $180-270 \text{ m}^3/\text{m}^2$ of grate per hour, dust burdens of up to 1100 mg/m^3 of flue gas at NTP may be expected; thus high-efficiency dry cyclones at 80% efficiency will enable particulate emissions of less than 230 mg/m^3 to be maintained.

Due to the necessarily intermittent nature of the operation of on-site units, supplementary firing under automatic control using gas or oil is required to maintain smoke and odour control within acceptable limits. A secondary or afterburner in the downtake, together with a forced secondary air supply, will maintain the conditions of 800°C minimum temperature and high turbulence which are required for smoke and odour elimination. If refuse having a moisture content of much over 25% is to be burnt, a primary burner over the grate is also required for preheating purposes and a burner capacity of about $11\,000 \text{ kJ/kg}$ of moisture in the refuse is usually found to be adequate. The grate combustion rating is an important design parameter. It depends upon the moisture content and the nature of the refuse together with incinerator capacity. Data are given in Figure 4. Combustion chamber heat release based on the volume of primary and secondary chambers varies from a minimum of $550 \text{ MJ/m}^3 \text{ h}$ to $900 \text{ MJ/m}^3 \text{ h}$ and is inversely proportional to incinerator capacity. Flue-gas temperatures of the order of 1000°C are reached at these ratings. This is a lower temperature than attained in large municipal units due to the higher excess air figures used in on-site units and the higher heat losses from the combustion chamber itself. Design excess air figures of 200% or so with a distribution of total combustion air of 10–20% under-grate, 60–70% secondary air, and 10–30% tertiary air provide a good basis on which to meet high levels of incinerator performance. Ideally, secondary and tertiary air should be preheated to not less than 200°C.

Other important design parameters are the gas velocities in the flame port between the primary and secondary chambers and the gas velocity in the

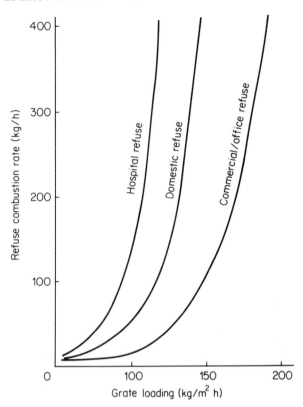

Figure 4 Variation of grate loading with capacity for double-chamber incinerator

downtake, which should not be less than 15 m/s and 7.5 m/s respectively, in order to provide good mixing and turbulence to achieve good burn-out. The uptake velocity should be 2–3 m/s to provide not less than 0.5 s residence time, thus assisting in settlement of heavier particles and completion of combustion. A minimum draught of 25 Pa is necessary in the primary combustion chamber to induce primary air through the grate and to prevent gas leakage and flashback through the charging door. The primary and secondary combustion chambers should be fitted with refractory lining capable of a maximum service temperature of 1300–1500°C, depending upon location, and adequate insulation should be provided between the refractory and the steel or cast-iron casing of the incinerator. The casing itself should be well finished and of a standard comparable with that applied to boilers and air-conditioning plant in modern buildings.

In the case of incinerators designed for municipal use, substantial developments have been made in the past 10 years in the design of such plant. Traditionally, municipal incinerators were of the mechanically or manually fed batch type, in which the refuse was dropped onto the firebed from above, one charge at a time. The units were generally equipped with a fixed grate and usually the only attempt to control dust emission was by means of a comparatively large gravitational separation chamber for gases before the chimney, which provided enough draught to enable operation to proceed. Batch feeding caused the introduction of large quantities of excess air into the furnace and also volumes of relatively wet material often cause a reduction in firebed temperature, so that heavy smoke emission followed the deposition of each charge. Ash removal was also on a batch basis, so that ash cleaning operations also caused large volumes of cold air to enter the combustion chamber, with detriment to the combustion process, and the disturbance of the firebed tended to give rise to substantial ash emission from the chimney. Thus the typical prewar incinerator of this type did not enjoy an enviable reputation as far as the public was concerned.

Changing attitudes after World War II lead to the development of wetted baffles in separator chambers equipped with water sprays but with natural draught, the allowable pressure drop through such chambers was small and hence the efficiency of dust collection was limited. With good operation and with fixed grates, the dust content of gases entering the chimney could be maintained at about $920 \, mg/m^3$ flue gas. Substantial quantities of water required, however, of the order of $7 \, m^3$ per tonne of refuse burnt, of which approximately 50% was evaporated after coming into contact with the hot flue gases. Hence the smoke plume from an incinerator fitted with a wet dust control system tends to be saturated and is visible throughout much of the heat. This in itself may give rise to objections from the public.

Where higher dust collection efficiencies were required, the dry mechanical cyclone systems were used and with conventional units of this type, with moderate initial dust burdens, the dust content in the chimney could be maintained at approximately $460 \, mg/m^3$ of flue gas. This improvement in dust removal performance involved a cost penalty, in that dilution or cooling air admission was necessary prior to the cyclones. Cooling by the use of water sprays avoids this problem but complications can arise with corrosion and possible malfunction. If the water spray system fails, as has been known to occur, then severe damage to the cyclones is probable. One advantage of wet systems over dry dust removal methods is that a proportion of the soluble gaseous pollutants can be removed by absorption in the circulated spray water.

The most efficient dust control device is the electrostatic precipitator and this is coming into increasing use, due to its comparatively high reliability and its capability for meeting the most stringent clean air codes. Again, this unit is not operated with gas temperatures above 300°C or so due to problems associated with corrosion from zinc chloride or HCl contained in flue gases, and in addition the use of steel separator plates requires some form of flue-gas cooling. The most satisfactory gas cooling solution in current practice is usually found to be the use of incinerators with heat recovery surface installed. These surfaces are usually in the form of water wall boiler units, with substantial radiant and convection cooling surfaces in the gas stream, in the manner of modern solid-fuel-fired power station boilers. Thus current and future pollution control requirements encourage the installation of heat-recovery incinerator units and, when a purchaser for the heat produced is available, the economics of incineration are much improved.

In order to exercise consistent control of incinerator performance, continuous operation is necessary, and this implies mechanical feeding of refuse onto a mechanical moving grate. Modern municipal incineration plants in Europe and the USA are tending towards the type of plant one would expect to find producing steam for power generation with solid fuel, and these plants are on the whole based on practice and experience developed in West Germany.

The major objective in grate design is to open out the incoming refuse to form as homogeneous a fuel bed as possible and to keep this in a compacted form moving steadily towards the rear of the grate and the ashpit, so that a complete burn-out is achieved without the formation of holes in the grate which, if allowed to occur, result in the reduction in firebed temperature, ash carry-over, and smoke emission. The grate loading or refuse combustion rate is much higher on large continuous grates than on the small fixed grates used with on-site incinerators, and is normally in the range 240–340 kg/m^2 of grate surface per hour. It has been established that a well designed and operated continuous modern incinerator plant is capable of disposing of town's refuse and of supplying heat or power to a community whilst meeting the most stringent requirements stipulated in clean air codes. Whilst continuous supplementary firing is not required in the case of well designed continuous municipal incinerators, adequate supplies of preheated secondary air injected in a correctly predetermined manner vital in maintaining smokeless operation. Many of these plants use air preheaters set in the flue-gas stream after the main convection tubes, but difficulties have been encountered when large quantities of wet refuse must be burnt, resulting in a fall in secondary air temperature when a high temperature is most necessary. Operators of such plants have only a limited degree of control of refuse quality and therefore the use of steam-heated air preheaters is an improvement over the conventional flue-gas air heater, in that a sudden fall in flue-gas temperature does not result in a critical drop in secondary air temperature.

Another design element which is increasing in use is some form of mechanical shredder, often steam turbine driven, to reduce the extremely heterogeneous refuse as delivered to more homogeneous form, permitting consistent combustion. Clearly some form of metal separation is required to avoid undue wear or damage to the shredder units, but the recovery of ferrous materials from the refuse is often of economic advantage and contributes to the reduction in cost of incineration.

As in the case of large power plants, the siting and height of the main incinerator chimney are critical factors in design in meeting clear air requirements. Whilst the dust concentration in the flue gases leaving a plant equipped with an electrostatic precipitator may well be below 100 mg/m^3 of flue gas, the pollution potential from a chimney is a product of gas weight per unit time and the dust concentration of that gas. In view of the substantial volumes of flue gases discharged from a large plant, a tall chimney with a minimum height of some 60–80 m is necessary.

In connection with compliance with clean air codes, a factor which must be considered within the next decade is the increasing proportion of plastics in refuse. With suitable plant the combustion of plastics does not pose a problem, although in the case of PVC the emission of HCl gas does result in a corrosion problem within the incinerator itself and a pollution problem in the environs of the plant. In this connection, the absorption characteristics of wet scrubber systems are an advantage. The efficiency range has been said (Little, 1970) to be between 50 and 90% of hydrogen chloride in flue gases removed. In the case of dry systems, such as the electrostatic precipitator or the mechanical cyclone, no HCl is removed.

When dealing with solid particulates, a dust burden of the gases leaving the incinerator furnace may be 4000 to 7000 mg/m^3 and this factor has an important bearing on the choice of gas cleaning equipment. High-efficiency multicyclones operate in the range of 70–80% efficiency, Venturi scrubbers in the range of 85–93% efficiency, and electrostatic precipitators (ESP) at 95–99% efficiency. The space required by ESP units is considerable and is three to four times that required by cyclones or scrubbers. This is due to the necessity to limit gas velocities to 1 m/s through the unit. The design of the boiler unit for municipal incinerators follows conventional practice for stoker-fired solid-fuel boilers, with a radiant section of tangent tube or membrane wall construction designed to cool the gases from the peak combustion temperature of about 1100°C to a temperature below the softening point of suspended ash, to avoid heavy build-up of slag on the superheater tubes. The ash fusion temperature depends upon the constituents of the refuse but good practice today indicates that the flue-gas temperature entering the superheater bank should ideally be below 700°C. Generous tube spacing is necessary if an availability of 80% or above is to be maintained (equivalent to 7000 h operation per year) and a minimum of 100 mm between superheater tubes and 75 mm between convection tubes is required. Even so, it is necessary to clean the tubes whilst boiler is on-load at frequent intervals, and steam or compressed air soot blowers are a vital part of the boiler equipment so as to achieve a satisfactory operating period before shutdown for cleaning.

Because of the chemical and physical properties of the ash suspended in the flue-gas stream, the possibility of corrosion and erosion of the steel boiler surfaces is always present and must be guarded against by preventive design. Erosion by impinging abrasive ash particles is reduced to acceptable levels by controlling the flue-gas velocities to 5 m/s or less when passing through the superheater and convection tube banks. There are two main forms of corrosion which are experienced in incinerator heat-recovery equipment, low-temperature corrosion and high-temperature corrosion. As with all fuel-fired appliances, if the flue gases are cooled to a temperature below the dew point of the gases, then precipitation of moisture will occur and, dependent upon the sulphur content of the flue gas, low-temperature corrosion due to the acid content of the flue gases will develop. The sulphur content of municipal refuse is lower than many fuels and low-temperature corrosion does not appear to be a serious problem in incinerators provided that the flue-gas temperatures are not allowed to fall below 200°C. The main cause of high-temperature corrosion is the presence of chlorine in the combustion gases, and corrosion from this source can be controlled by maintaining the metal surface temperatures below 300°C. This does not pose a problem when the heat recovered from the incinerator is to be used for space or process heating, but, when it is desired to generate electrical power, then for efficiency and economy it is desirable to generate steam at a high temperature and pressure. Boiler tubes in the furnace zone are often protected by the application of silicon carbide refractory material and superheater tubes which are prone to high-temperature corrosion are protected by a combination of a limit on steam superheat temperature of 400°C or so and by the use of alloy corrosion-resistant steel. Provision should be made for the easy replacement of superheater tubes.

In addition to reducing the flue-gas temperature at the entry to the gas cleaning equipment to a safe operating temperature of less than 300°C, the use of water wall incinerator units has another major advantage. Satisfactory combustion can be undertaken with between 50 and 100% of excess air as compared with the older style of incinerator with solid uncooled refractory walls which require about 200% excess air to reduce the peak combustion temperature to within limits that the refractory material can tolerate. The flue-gas volume from water wall incinerator units is less than the solid wall units and this makes possible a reduction in the size and cost of gas cleaning equipment.

As is the case with the smaller on-site generator units, combustion air must be supplied over the fire to burn out the volatile matter and, in order to achieve adequate turbulence for good burn-out over the large grates, high-velocity preheated secondary air jets are required with nozzle pressures of 7.5 kPa or so in order to achieve good penetration in the larger furnaces. It is also necessary to have the facility for extensive control of under-grate air distribution in view of the variable nature of the refuse which has to be burnt.

NEW METHODS OF INCINERATION

Whilst the conventional form of incineration developed in West Germany has influenced practice in both Europe and the USA, new systems of combustion of refuse have been developed to a satisfactory level. Two systems are of particular interest: the pyrolysis process and the fluidized-bed process. The fluidized-bed process is capable of burning a much wider range of refuse than the conventional grate method and virtually any kind of liquid or solid fuel or refuse can be used. The fluidized-bed principle consists of a large reactor at the base of which is a distributor plate covered with a thick bed of inert particles. Preheated air is blown up through this distributor plate causing the particles to whirl in a violent turbulent motion. The particles become in effect fluidized. The bed is initially preheated by an ignition burner and when a temperature of approximately 900°C is reached, the refuse is introduced by an injection pump or screw conveyor. Most of the combustion takes place in the turbulent bed and if necessary secondary combustion air is injected over the bed to ensure complete burn-out. Solid refuse is shredded to approximately 100 mm size to enable convenient feeding and consistent combustion. The reactor is lined with refractory material and may be water-cooled.

An increasing problem in many cities is the disposal of sewage sludge. Typical quantities of raw sewage are 0.2 m³ per person per day, having a water content of over 90%. Sewage sludge can be burnt in an incinerator provided that the water content is reduced to about 70% by mechanical dewatering or drying. The fluidized-bed incinerator is particularly suitable for disposing of sewage sludge and other wet fuels as the bed contains a large mass of particles of great thermal capacity, so that the temperature within the bed varies little over the entire volume and, with a long residence time and good mixing, good combustion is ensured even with a fuel of a variable nature. A further advantage of disposing of sewage sludge by incineration is that the digestion process at the sewage treatment works may be omitted and the sludge from the settling tank may be taken direct to the incinerator.

Intensive research in the USA has developed the pyrolysis method of incineration to an advanced level. Pyrolysis, which is generally unknown in

Europe as a means of refuse disposal, consists of a starved air combustion process in a sealed reactor, the output from which is a liquid slag and a combustible fuel gas. In comparison, the conventional incinerator produces a dry friable ash of approximately 10% of the volume of the original refuse. The liquid slag from the pyrolyser is in the range of 3–5% of the original volume and is quenched so as to form a dense black granular material which may be used in construction work as an aggregate or roadway base. Thus the process is an excellent example of resource re-use, in that material rejected by society can be converted into energy and a useful construction material. Special tests have indicated that the process can dispose of undigested sewage sludge with approximately 80% water content when mixed with normal municipal refuse.

Energy recovery from the fluidized-bed and pyrolysis processes is obtained by means of separate waste heat boilers producing hot water for district heating or steam for power generation. Clean air codes are complied with by conventional flue-gas cleaning equipment such as the elec-·trostatic precipitator.

EXAMPLES OF HEAT-RECOVERY UNITS

An interesting range of on-site incinerators from West Germany offering facilities for heat recovery and compliance with stringent clean air codes is the Incitherm range with capacities from 50 to 1250 kg of refuse per hour and in which hot water up to 180°C can be produced or steam up to 20 atm be generated, with an alternative of a provision of warm air for drying and heating purposes. Integral tubular recuperators are used for heat recovery. The Incitherm units are of the advanced double-chamber design consisting of a welded steel combustion chamber, fully water-cooled. A primary burner is installed in the combustion chamber with a secondary burner in the downtake, with facilities for the provision of preheated secondary air which is admitted through a hollow firebridge, with the object of cooling the refractory brickwork in the hottest zone of the furnace. The units are fitted with mechanically operated grates for continuous cleaning and ash removal and part of the grate is water-cooled. The ash from the grate and from the multiple-cyclone dust cleaners is deposited directly into sealed bins. Incombustible materials such as metal cans have to be removed from the grate manually. Refuse is fed into the combustion chamber manually in the smaller and medium-sized units, but can be undertaken automatically by special feeders in the large units.

In the case of on-site incinerators, the case for heat recovery is in a sense greater than with the larger municipal units because of the greater need for supplementary firing with oil or gas, to ensure compliance with clean air codes. This is particularly so where refuse of a variable and difficult nature is to be burnt, such as in a hospital where refuse quantities of 2–4 kg per bed per day can arise, having calorific values on occasions as low as 2500 kJ/kg but probably averaging 8500 kJ/kg. In order to ensure satisfactory operation, supplementary oil or gas firing at the rate of 3000–5000 kJ/kg or refuse may be required and, without heat recovery, the heat in this supplementary fuel and in the refuse itself is lost to the atmosphere. Data from West Germany indicate that the Incitherm unit can operate at an efficiency in the region of 66%, so that only one-third of the heat value of the refuse and supplementary fuel is lost instead of the total value. Where the Incitherm unit is an integral part of the heat generation system of a building, when refuse quantities are low the supplementary burners can be brought into operation and the heat output of the unit may be maintained on supplementary fuel alone, and the efficiency with this mode of operation is similar to that achieved by a conventional hot water or steam boiler.

As an example of the economic value of the heat recovery in the case of a 500-bed hospital producing 3 kg of refuse per bed per day, with a heat recovery efficiency of 66%, the daily energy recovered would be 8.4 GJ or approximately 3000 GJ per anum. Taking the cost of energy produced in the form of hot water or steam at £1.50/GJ, the annual value of heat recovered would be £4500. According to information received, over 70 Incitherm heat recovery incinerators are in operation in West Germany and neighbouring countries.

The Incitherm units and others available can operate with flue-gas dust levels of less than 200 mg/m^3 at NTP and 7% CO_2 in the flue gases. With special agreement, operating levels of less than 150 mg/m^3 can be obtained. The units are also designed to operate at less than Ringelmann 2 smoke density at the chimney.

In the case of municipal incinerators, it is clear that their primary function is to dispose of refuse hygienically and in compliance with the clean air codes. For the reasons given, heat recovery has technical and economic advantages in disposing of town's refuse, However, the Department of the Environment (1971) stated that, on a national basis, heat utilization from incineration of refuse is not of importance. Whether the working party concerned will come to revise their attitude in view of the subsequent energy crisis remains to be seen. As in the case of on-site incinerators, the West German style of technology leads the world in municipal incineration with heat recovery, and some results from this application of technology are of interest.

From consideration of information available, it is apparent that, with adequate planning and resourceful incineration of all refuse from large cities, approximately 5–10% of the entire electrical energy demand of the city can be generated by refuse burning, and quantities of waste heat are available by condensing the steam from the turbines for district heating or other purposes.

In Amsterdam, the steam produced by the incineration of all of the city's refuse produces 5% of the electricity used in the city. In Munich, 9% of all consumed electricity is said to be produced by refuse incineration, together with some heat for district heating schemes from the same source. The incineration plant in Rennes produces approximately 33% of the entire requirements of the district heating system from combustion of refuse.

In 1971 the city of Nottingham commenced construction of a large refuse-incineration plant with supplementary coal firing connected to a large district heating scheme which by 1980 will serve the equivalent of 17 000 dwellings. The scheme was financed jointly by Nottingham Corporation and the National Coal Board and will be operated by the NCB. The realization that there was no longer any suitable areas of land for open dumping of refuse, a desire to improve the environment, and the realization that the annual output of refuse from the city had the same heat value as 40 000 tonnes of coal gave encouragement to a rational approach to refuse disposal.

For similar reasons the large disposal plant at Edmonton in London came into being, but in this instance the energy recovered is generated as high-pressure, high-temperature steam and is used for power generation, and the intended average figure for output was 30 MW with an average throughput of refuse of 1330 tonnes/day. The refuse in this plant is burnt in the crude as received form, without any sorting or pretreatment, and hence the hoppers and furnaces are capable of handling bulky items such as bedsteads and pianos. This is an interesting contrast to practice in the USA where there is a tendency to remove metal and shred or pulverize the refuse before combustion, so as to obtain more consistent operating conditions and the elimination of difficulties associated with metal objects in moving mechanical grates and ash conveyor systems.

The steam pressure and temperature at Edmonton are 44 atm and 450°C respectively, which is in accordance with normal West German practice. An interesting comparison is the plant at Rotterdam, also based on West German practice, but specifically designed to increase reliability and reduce boiler fouling and corrosion by operating at a somewhat lower pressure and temperature of 27 atm and 360°C. The capacity of the Rotterdam plant is 1150 tonnes/day and the electrical power output is 14 MW. This is less in proportion to the refuse burnt than the Edmonton plant and is no doubt partly accounted for by the modest steam operating conditions. Whether the best policy is to aim for a high generation efficiency with a possibility of boiler operating problems or to aim for maximum reliability and a modest power output is difficult to determine in general terms.

In the first five years of operation of the Rotterdam plant there was no indication of high-temperature corrosion due to chlorine and only a slight local corrosion caused by sulphur in the fly ash.

Two of the largest installations in the world are in Paris. The unit at Ivry has a capacity of 2640 tonnes/day and the other at Issy les Moulineaux has a capacity of 1800 tonnes/day. The Issy plant commenced operation in 1965 and provides both electric power and heat for district heating. The availability of these Paris incinerator plants is reported to be over 80%.

Whilst it is clear that large cities benefit from modern techniques of refuse disposal and energy recovery, there are many small towns with refuse disposal problems and an interesting installation is at Hinwil near Zurich (Sulzer, 1976) where the incinerator has a capacity of only 120 tonnes/day and produces 2.3 MW of electricity and 1.9 tonnes/h of steam for heating. This unit has been in use since 1971 and in the first four years of its operation achieved an availability of over 80%, with an operating period of 2500 h between shutdowns. The steam pressure and temperature is 42 bar and 400°C respectively and the steam output averages 2.68 tonnes per tonne of refuse. Increasing quantities of refuse necessitated the installation of a second incinerator unit in 1976 and an interesting development is that since 1973 the calorific value of the refuse has been falling, as a result of efforts to recycle some of the refuse. The second plant was designed to use refuse with a calorific value as low as 5000 kJ/kg.

The second incinerator was designed to achieve 4000 h of operation between shutdowns and, in view of some corrosion problems in the superheaters of the first incinerator, the steam temperature was reduced from 400 to 385°C and the spacing between tubes was increased from 120 mm to 140 mm. In addition, the silicon carbide refractory lining in the combustion chamber was taken up to a greater height than in the first incinerator.

In many cases in Europe the incineration of refuse is combined with normal power generation for a city, and in Stuttgart, Munich, and Vienna, for instance, coal or oil are used in conjunction with refuse to provide heat and power for the city.

An interesting experiment commenced in 1971 in St Louis, Missourri, USA, on the basic theory that if municipal refuse were prepared before firing and if the percentage of refuse to the total fuel were kept to a practical minimum, operational problems in large boiler installations fired with both refuse and coal as a fuel would be little, if no, different from when the fuel is entirely coal. It was thought that this concept could become the primary means of refuse disposal for many large US metropolitan areas, as not only is the process thought to be less expensive than conventional incineration but large efficient furnaces are already in existence for the application of the process. The initial provisions for the preparation of refuse included particle size reduction to approximately 40 mm and the removal of magnetic material. The boiler in quesion is a 125 MW corner-fired unit and the proportion of refuse is 10–20% of total fuel. The project is of interest in that, whilst there are a number of combined refuse- and coal-fired units in Europe, these are nearly all stoker-fired and the use of prepared refuse in a pulverized-fuel-fired boiler is of interest due to its wider application in large modern power stations. The shredded refuse is stored in large bins and then transported in a compacted form by road vehicle to the power plant 12 miles away where it is again stored in bins. The refuse is discharged from the storage bins onto a belt conveyor and then conveyed pneumatically to the boiler. The pneumatic system consists of a maximum run of 365 m and includes a 30 m rise to burner level. The initial rate of refuse disposal was 300 tonnes/day.

In another interesting experiment in England, hand-sorted municipal refuse was successfully burnt in a conventional rotary cement kiln, together with pulverized fuel, without any reduction in cement quality or through-put.

The recent installation of three 200 tonnes/day incinerators in Luxembourg is of great interest. Two of the units are conventional grate-type incinerators, but the third is an Andco-Torrax Pyrolysis unit fitted with a waste heat boiler steaming at a pressure of 35 atm and temperature of 385°C. The steam production is designed to be 2.8 kg per kilogramme of refuse, with a refuse calorific value of 10 500 kJ/kg on the net basis. The steam from all three refuse units passes to a 7 MW turbine generator with provision for the installation of a second generator when refuse volumes increase. Two other pyrolysis units are being installed in France and West Germany.

This is a reversal of the trend in that German incinerator technology has tended to dominate practice in the US, and now it appears that the application of research on solid waste disposal in the US has initially been applied in Europe.

The economic advantage of heat recovery from refuse incineration depends largely upon the availability and value of heat and/or power in the area around the incinerator. There are indications that generation of power only, where there is no useful heat, will reduce the cost of incineration by approximately 25%. The predicted saving for the Edmonton plant is a 25% reduction in overall costs due to the sale of electricity. Figures for the Hinwil plant in Switzerland showed that over a four-year period there was a saving of 22% in incineration costs due to the sale of electricity. This is encouraging as the Hinwil plant is only one-tenth of the size of the Edmonton plant. In the larger European cities where both heat and power are generated from refuse, the economic situation appears to be improved, as might be expected. In one West German city, the sale of steam and electricity is said to reduce incineration costs by 40%, and published figures for Rotterdam show a cost reduction of 30% by energy recovery. There is of course a concurrent cost penalty in that the heat recovery equipment and its distribution network has to be purchased in the first place. The additional cost of the heat recovery and power generation plant at Edmonton was given as 35% higher than incineration without recovery, with an annual return on the additional capital of 11%.

The costs of energy recovery equipment for straight heat utilization are much less than when power generation is involved, although the heat distribution pipe network would be much more expensive than an electric cable system.

A modern municipal incinerator plant, whether of the well developed and conventional steam boiler type, or the newly developed fluidized-bed or pyrolysis types, may be regarded as a factory or workshop which converts the garbage rejected by society into materials valued by society in the form of energy, heat and/or power, and a sterile landfill material in the case of ash or a construction material in the case of slag from pyrolysis. A modern incinerator is therefore a recycling plant which must be economic in order to be successful. There is little point in recovery of energy if it has limited value. However, it may be expected that energy prices will double every five to ten years for the next few decades and under these circumstances the value of energy recovery is high. The modern incinerator should not only anticipate and meet trends in clean air codes but also trends in refuse properties which reflect society at large. Much of the combustible content in refuse today indicates wasteful use of materials in packaging and should this trend be reversed, then there will be a substantial fall in the calorific value of refuse, which will have a bearing on incinerator design and possibly its economic viability.

REFERENCES

Department of the Enviroment, 1971, *Refuse Disposal*, HMSO, London.
Department of the Environment, 1974, *Report of the Second Working Party on Grit and Dust Emissions*, HMSO, London.
Federal Ministry of Health, 1964, *Technical Directive for Maintaining the Purity of the Air*, West Germany.
Little, A. D., 1970, *Systems Study of Air Pollution from Municipal Incineration*, National Air Pollution Control Administration, U.S. Department of Health, Education and Welfare.
Sulzer, 1976, *Sulzer Technical Review*, March.
Thomen, K. H., 1976, Energy recovery by the incineration of solid waste, *Proc. Inst. Mech. Eng.* **190**, 64/76.

BIBLIOGRAPHY

ASME, 1974, Resource recovery through incineration, *Proc. ASME Conf.* Miami.
Institute of Fuel, 1969, The incineration of municipal and industrial waste, *Proc. Institute of Fuel Conf.*, Brighton.
Winch, G. R., 1972, Design of incinerators to comply with clean air codes, *Proc. Clean Air Conf.*, Melbourne.

The Edmonton Refuse Incinerator Plant, referred to in this paper on pages 212 and 214, is discussed more fully in Occupational Health and Safety Management, edited by S. S. Chissick and R. Derricott, Wiley, 1981.

Energy Conservation and Thermal Insulation
Edited by R. Derricott and S. S. Chissick
© 1981 John Wiley & Sons Ltd.

CHAPTER 10

Air-to-air heat recovery

George Applegate

Curwen & Newbery Ltd, Westbury

1. INTRODUCTION

We have to ask ourselves a simple question: 'How much energy can we afford to waste?' Fuel costs have risen at such an alarming rate during the last few years that energy costs just cannot be ignored by anyone who is motivated by profit nor by Governments who have to import a large proportion of their consumption of fuel, and with world shortages it becomes a worldwide problem.

A great deal has been done since 1973 in this respect, and industry and government throughout the world take energy conservation very seriously, but there are still large savings to be made. One must ask the questions: 'What proportion is your energy cost of

(1) turnover,
(2) manufacturing costs, and
(3) profit,

and has the proportion changed during the last five years? Were any energy-saving schemes introduced and were they successful?'

Having regard to the dramatic increase in the cost of fuel and to the further increases that must be expected in addition to the effect of continual inflation, this apart from the need for conservation, an energy-saving scheme becomes essential for all industrial consumers. This should be designed to save a minimum of 10% of consumption, but in the case of major consumers a saving of 50% of consumption is not unknown.

Energy fact finding backed with management action suitably monitored will produce good results. Target savings will help and they should be set in conjunction with the foreman or supervisor who will have the responsibility for achieving them. You will have to obtain:

(a) total energy costs for a given period;
(b) your energy cost per unit of heat;
(c) fuel consumption for each process;

(d) energy used for each separate manufacturing process; and
(e) energy used for heating, lighting, etc.

The Electricity, Gas, and Coal Boards all have energy-saving schemes and some excellent literature has been made available. The Department of Energy produces booklets and other technical information and also holds seminars for those actively involved in the responsibility for saving energy; the following list indicates areas where heat recovery is both practicable and economic:

Offices
Laboratory exhausts
Hotels
Hospitals and nursing homes
Cinemas
Theatres and auditoriums
Public buildings
Swimming pools
Sports centres & indoor rinks
Public houses
Restaurants & commercial kitchens
Schools and universities
Banks
Flats and high-rise buildings
Supermarkets
Computer rooms
Research laboratories
Industrial buildings & work areas
Agricultural buildings
Laundries

The methods of recovery can be listed as:

(1) The use of furnace, kiln or boiler flue heat to provide heat for preheated combustion air.
(2) Recycling recovered heat for drying or space heating.
(3) Boiler feed water preheating or other processes to use heat recovered by blowdown, waste water, flue heat, etc.
(4) Production of hot water or steam by the use of waste heat boilers.
(5) Greater use of the heat content in condensate, the use of flash steam, etc., and other widely accepted methods.
(6) Air-to-air heat recovery from ventilation systems, discharges from process plant, driers, etc.

Always assuming that the building and plant are well insulated, the majority of commercial and industrial buildings produce a lot of heat. All electricity used ends up as heat; occupants themselves produce heat and, of course, there can be solar heat gains. During the winter time, all these sources of heat are valuable and a well designed heat recovery system will recover up to 80% of this excess and redistribute it for re-use.

Buildings operated under a negative air pressure are more wasteful than those operated under a positive air pressure, and air-to-air recovery makes this latter feature possible without adding to running costs. When air is exhausted from a building, it is always replaced, through design or accident, by outside air and there is always a potential saving to be made when the temperature of the air discharged is higher than the outside air temperature.

If one's efforts are to improve the living or working environment without wasting energy and increasing costs, a method of introducing fresh air without raising heating costs will be necessary, and there are several proven and acceptable methods of achieving this.

The volume of air exhausted from a manufacturing or space heating process which is to be kept at a given temperature and/or humidity imposes a necessary demand on the primary mechanical equipment designed to provide and maintain these conditions, since the fresh air must be brought in from the outside and raised to the temperature required. The air that is being mechanically exhausted must be contaminated or good engineering design practice should indicate that it should be re-used in some way.

My objective is to provide a simplified formula and method for evaluating the feasibility of incorporating air-to-air heat recovery equipment. The principles of air-to-air heat regenerators will be outlined and their basic thermodynamic principles reviewed. Simplified formulae to establish approximate recovery rates will be developed and applications and case histories given.

2. PRINCIPLES OF OPERATION

Air-to-air regenerators can be divided into two basic types, recuperative and regenerative.

In the recuperative type, the contaminated gases are passed through one series of passages while fresh air is passed through alternate passages. Thus the two streams of air are kept apart while heat is transferred from the hotter stream to the colder stream by conduction.

In the regenerative type, a single heat transfer medium is exposed alternately to the contaminated airstream and to the fresh airstream and—acting as a reservoir—gives up its stored heat from the hotter to the colder stream. With proper design, mixing of the airstreams can be kept well below the limits established by the most exacting standards of the US and Canadian public health services as well as individual county and city

Figure 1 Standard CN Single-Wheel Heat
Regenerator

requirements. It is evident that many more square metres of heat transfer area can be incorporated in a given volume of the regenerative type. As a result, properly designed regenerative air-to-air rotary heat exchangers are, perhaps, half the size and weight of the recuperative type of the same capacity and are also less costly.

The usual type of regenerative air-to-air heat exchanger in use is constructed on a rotary principle. In the rotary exchanger (Figure 1), a number of sector-shaped elements forming a cylindrical mass of heat transfer medium contained in a rotor, which is mounted in a suitable framework, continually rotates alternately through the contaminated exhaust airstream and the fresh airstream.

Assuming a hot exhaust, Figures 2 and 3 show the heat transfer which takes place in this cycle. The hot contaminated exhaust passes over one-half of the heat transfer medium, giving up some of its heat to the medium, and the cool contaminated exhaust is ready for disposal in the usual manner. Cold air is delivered to the other half of the heat transfer medium in a counterflow manner. The heated medium travels through the

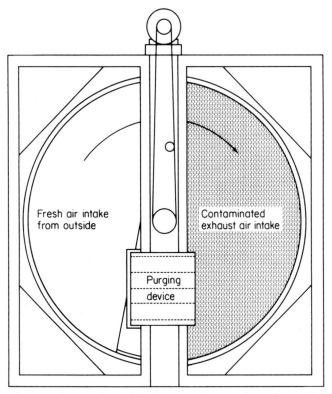

Figure 2 Illustration of fresh air and contaminated air sides
of regenerator with purging device

fresh cold airstream and gives up some of its stored heat, thereby providing fresh hot air for use in the process.

On a cooling cycle, a similar heat transfer takes place but in the opposite direction, and, in addition, on many cooling cycles, condensation may take place.

3. BASIC THERMODYNAMICS

To assure that the formulae which will be developed stem from a solid foundation, a brief outline of the basic thermodynamic principles involved is at this stage appropriate. The logical starting point is the fundamental equation for an energy balance in a steady flow system. The terms comprising the components of the total energy at each point are

$$E_p/J + E_k/J + E_r + u + Pv/J + q + w$$

in which all terms are in Btu/lb mass and where E_p/J = potential energy,

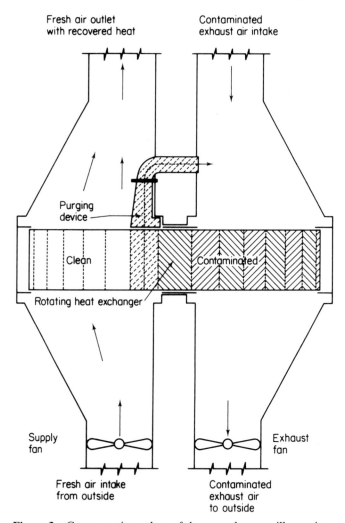

Figure 3 Cross section plan of heat exchanger illustrating clean and contaminated sections of the rotor and correct fan positions

E_k/J = kinetic energy, E_r = reactive (chemical) energy, u = internal energy, Pv/J = flow energy, q = heat flow, w = work, and J = mechanical equivalent of heat. Fortunately, in the areas of thermodynamics under consideration, a number of terms become negligible or unnecessary even for relatively precise calculations and certainly can be eliminated for this purpose. Thus, for practical applications, the terms can be reduced to

$$u + Pv/J + q$$

Since the terms u and Pv/J appear in most flow problems, a further simplification can be made. The sum of internal energy and flow energy, or $u + Pv/J$, is by definition called 'enthalpy' (sometimes referred to as 'total heat') and is designated by the symbol h and has units of Btu/lb mass. Thus the equation can be further reduced to $h + q$. For a steady flow process between points a and r; we can write

$$h_a + q_a = h_r + q_r$$

or

$$q_a - q_r = h_r - h_a$$

The major component of the gaseous mixture with which we will be dealing is dry air. The balance of the mixture will be made up of water vapour and contaminated vapours. Although the mass of vapours may increase or decrease during a flow process, the mass of dry air itself will remain constant, providing the temperature is also constant. The advantages of the use of properties related to an unchanging parameter is apparent so that our equation for a particular mass flow of dry air becomes

$$Q = {}_aQ_r = m_{da}(h_r - h_a) \tag{1}$$

in which ${}_aQ_r$ = total net flow of heat from points a to r in Btu, m_{da} = mass flow of dry air in the mixture in lb mass, h_r = enthalpy of the mixture at point r in Btu/lb mass dry air in the mixture, and h_a = enthalpy of the mixture at point a in Btu/lb mass dry air in the mixture.

Although it may readily be apparent from the process under consideration whether heat is to be added or rejected, it is as well to note here that the sign of the right-hand term of equation (1) will determine this. If the value is positive, heat is added from points a to r, and, if negative, heat is rejected.

Also note that if m_{da}, the mass flow of dry air, is taken for one hour, the units of ${}_aQ_r$ will be Btu/h. which are familiar units.

For many processes in which the main constituent other than dry air is water vapour, the values of the enthalpies may be read directly from a psychrometric chart with sufficient accuracy. These processes include normal ventilation applications in the comfort range and some drying applications. The volume of contaminated mixture under consideration, in units of cubic feet per minute (cfm), is known either by field tests or as part of the basic design of the project. The amounts of moisture and other contaminated vapours present in the mixture are also known, and therefore the specific volume of the mixture in units of cubic feet per lb mass of dry air may be determined from the psychrometric chart or found by use of tables. By dividing one of these two quantities by the other, a reasonably accurate value of m_{da} may be estimated and equation (1) may be solved.

However, many industrial processes do not fall within the range of a psychrometric chart for a number of reasons, such as:

(1) elevated temperature,
(2) high moisture content,
(3) excess variations from the barometric pressure from which the psychrometric chart was constructed, or
(4) sufficient contaminant, other than water vapour, seriously to influence the properties of the mixture so that a psychrometric chart does not give a valid representation of the mixture properties.

In these cases, it is necessary to calculate the various properties by reference to appropriate tables, but, although at times tedious, the calculations are not particularly difficult. Fortunately, one common type of process which does not fall within the range of the usual psychrometric charts available lends itself to the use of a simplified formula. This process takes place at elevated temperatures with little water vapour present.

In such cases, it can be assumed with reasonable accuracy that the mixture is dry air. A valid assumption can also be made that dry air acts as a perfect gas, having variable specific heats. For a perfect gas, the specific heat at constant pressure C_p, is either a constant or a function of temperature only. In addition, since the variation in C_p of dry air is modest over the temperature range normally considered, a mean or average value may be used satisfactorily. Referring to the definitions of enthalpy, which is

$$h = u + Pv/J$$

and since we are dealing with a perfect gas for which

$$Pv = RT$$

enthalpy may be expressed as

$$h = u + RT/J$$

or in differential form

$$dh = du + R\, dT/J$$

For a perfect gas the specific heat C_v at constant volume is

$$du = C_v\, dT$$

so that

$$dh = C_v\, dT + R\, dT/J = (C_v + R/J)\, dT$$

But since

$$C_p = C_v + R/J$$

it follows that

$$dh = C_p \, dT$$

Thus integrating the expression between the limits of T_a and T_r, we obtain

$$\Delta h = h_r - h_a = C_p(T_r - T_a)$$

Substituting an average value of $C_p(\text{air}) = 0.24$, converting the temperatures to degrees Fahrenheit, rather than degrees R (Ranking), since we are only concerned with the temperature difference, and introducing a mass flow of dry air m_{da} to both sides of the equation, we obtain:

$$\Delta H = m_{da}(h_r - h_a) = m_{da} \times 0.24 \times (t_r - t_a)$$

Finally from equation (1) substituting $_aQ_r$ for $m_{da}(h_r - h_a)$ we arrive at the familiar equation:

$$_aQ_r = m_{da} \times 0.24 \times (t_r - t_a) \tag{2}$$

It is as well to point out here that equation (2) is also valid at any temperature where experience indicates that the quality of the vapour present in the mixture does not seriously influence the results. This situation may occur when winter ventilation from a comfort process is under consideration. Unfortunately, many of the processes encountered in practice have appreciable amounts of vapour present. Therefore, a method of calculating the enthalpy of such a mixture is a necessity. For any mixture of gases, the total enthalpy is merely the total of the products of the masses as specific enthalpies of the various constituents. The total enthalpy of any mixture of gases, which includes dry air and water vapour and contaminated vapour, may be expressed as

$$H = m_{da} \times C_p(\text{air}) \times (t_d - t_O) + (m_{da} \times W_v \times h_v) + (m_{da} \times W_{vc} \times h_{vc})$$

in which t_d = dry-bulb temperature of the mixture in degrees Fahrenheit, t_O = datum and temperature above which the enthalpy of dry air is measured, W_v = a ratio of lb mass of water vapour/lb mass of dry air, h_v = specific enthalpy of water vapour, Btu/lb mass water vapour, W_{vc} = a ratio of lb mass contaminated vapour/lb mass dry air, and h_{vc} = specific enthalpy of contaminating vapour/lb mass contaminating water.

By dividing both sides of the equation by m_{da}, we arrive at the enthalpy of the mixture of 1 lb mass dry air in the mixture. (Not the specific enthalpy of the mixture.)

$$h = C_p(\text{air}) \times (t_d - t_O) + (W_v h_v) + (W_{vc} h_{vc})$$

The datum and temperature above which the enthalpy of dry air is measured is taken as $0°F$. Substituting an average value of $C_p(\text{air}) = 0.24$ and using the very close approximation that for water vapour $h_v = h_{gd}$, where h_{gd} is the specific enthalpy of dry saturated steam at the corresponding dry-bulb temperature (since in the superheat region the enthalpy lines are practically coincident with the constant-temperature lines), we arrive at our third basic equation:

$$h = 0.24t_d + (W_v h_{gd}) + (W_{vc} h_{vc})$$

It is as well to note here that in some cases the properties of the contaminated vapour are not readily available, so that experience must dictate whether its influence on the mixture is sufficient to make it an important factor. In summary then, we have developed a general equation for evaluating the total net heat flow from points a to r:

$$_a Q_r = m_{da} (h_r - h_a) \tag{1}$$

a special equation which can be used with reasonable accuracy for essentially dry air:

$$_a Q_r = m_{da} \times 0.24 \times (t_r - t_a) \tag{2}$$

and the third general equation for evaluating the enthalpy of any mixture of gases:

$$h = 0.24t_d + (W_v h_{gd}) + (W_{vc} h_{vc}) \tag{3}$$

4. APPROXIMATE RECOVERY RATES

Up to this point we have been discussing regenerative air-to-air heat exchangers in general, and particularly those based on rotary principles. Now we must be even more specific and state that all further data regarding performance and physical characteristics are based on certain laboratory and field tests.

For a specific application, the facts concerning the condition of the contaminated exhaust mixture are known, as well as conditions of the fresh air to which the recovery will be made. Thus it is most practical to develop formulae relating the conditions of these two mixtures to the final condition of the fresh air after recovery has taken place. In order to accomplish this, certain empirical factors designated as thermal efficiencies have been established by evaluating the result of extensive field tests. The outlines referred to in the particular sections of the two airstreams are:

c contaminated exhaust entering regenerator from which recovery will be made;

b contaminated exhaust leaving the regenerator after recovery has been made—for disposal;

a fresh air entering the regenerator to which recovery will be made; and r fresh air leaving the regenerator after recovery has been made for introduction into the process.

Other symbols used are:

d dry-bulb temperature (°F);
dp dewpoint temperature (°F);
w wet-bulb temperature (°F);
s sensible heat (Btu); and
e enthalpy (total heat, Btu).

Thermal efficiencies are defined as follows:

$$\eta_s = \text{sensible heat efficiency} = \frac{t_{dr} - t_{da}}{t_{dc} - t_{da}}$$

$$\eta_e = \text{enthalpy (total heat) efficiency} = \frac{h_r - h_a}{h_c - h_a}$$

Since we have defined our thermal efficiency, we can now empirically determine certain properties of the recovered airstream and by substitution in the fundamental equations previously developed we can arrive at approximate heat recovery rates.

From the definition of sensible heat thermal efficiency, we can estimate the dry-bulb temperature of the fresh air subsequent to the recovery process:

$$t_{dr} = t_{da} + \eta s \ (t_{dc} - t_{da}) \tag{4}$$

and substituting this value in equation (2), the whole net quantity of sensible heat transferred becomes:

$$_aQs_r = m_{da} \times 0.24 \times (t_{dr} - t_{da}) \tag{5}$$

We can also estimate dry-bulb temperature of the contaminated exhaust air subsequent to the recovery process:

$$t_{dr} = t_{da} + (1 - \eta_s) \ (t_{dc} - t_{da}) \tag{6}$$

From the definition of the enthalpy efficiency factor, we can estimate the enthalpy of the fresh air subsequent to the recovery process:

$$h_r = h_a + \eta_e \ (h_c - h_a) \tag{7}$$

and substituting this value in equation (1), the total net change in the enthalpy becomes

$$_aQe_r = m_{da} \times (h_r - h_a) \tag{8}$$

An additional formula that will prove useful in actual practice concerns the additional mass of fresh air that can be blended with the hot recovered airstream to reduce its dry-bulb temperature to a suitable level for introduction into the process. The additional mass of fresh air to be blended m_{dab} to produce a mixture dry-bulb temperature in the recovered airstream t_{dm} which has a mass flow of m_{dar} and is given by the relationship:

$$m_{dab} = \frac{m_{dar} \times (t_{dr} - t_{dm})}{t_{dm} - t_{da}} \tag{9}$$

and the total mass of fresh air at a dry-bulb temperature of t_{dm} available for delivery to the process is

$$m_{dam} = m_{dar} + m_{dab} \tag{10}$$

Since the thermal efficiencies vary depending on the type of process, we must now establish certain general categories and indicate the value of the thermal efficiencies to be used for estimating purposes.

The processes may be broken down into two main classes for the basis of this thesis. The first might be termed a 'heating cycle' process in which the dry-bulb temperature of the contaminated exhaust is higher than the dry-bulb temperature of the fresh air to which recovery will be made, i.e.

$$t_{dc} > t_{da}$$

The second might be termed a 'cooling cycle' process in which the dry-bulb temperature of contaminated exhaust is lower than the dry-bulb temperature of the fresh air to which recovery is made, i.e.

$$t_{dc} < t_{da}$$

These two classes of process may be further broken down into two additional categories; the first might be termed 'dry', in which no significant condensation occurs, and the second 'wet', in which appreciable condensation occurs. A heating cycle can either be dry or wet, whereas a cooling process cycle is normally always wet.

A dry heating cycle process is defined as a process in which no condensation takes place during the recovery cycle. For this cycle in which any sensible heat is transferred, the value of sensible heat thermal efficiency η_s which can be used for estimating is 80%. Also, heat recovery calculations must be based on sensible heat only (using equation (5)) and the result will be an accurate approximation of the enthalpy (total heat) recovery. This is valid since, with no condensation, the quantity of vapour fresh airstream does not change appreciably during the recovery cycle and thus the vapour component of the enthalpy of fresh air remains essentially constant.

In addition, we can include here those processes that have insignificant condensation, which in normal applications can include those with a vapour content of less than 120 grains per lb mass of dry air.

It is apparent that condensation is not to be expected when the dry-bulb temperature of the cool contaminated exhaust after the recovery cycle is higher than the dewpoint of the contaminated mixture, i.e.

$$t_{dl} >> t_{dpc}$$

or when the dry-bulb temperature of the fresh air to which recovery will be made is higher than the dewpoint temperature of the contaminated mixture, i.e.

$$t_{da} >> t_{dpc}$$

A wet heating cycle process is defined as a process in which condensation takes place during the recovery cycle. For the cycle where both sensible and latent heats are transferred, the value of the sensible thermal heat efficiency to be used for estimating is $\eta_s = 60\%$, and the value enthalpy normal efficiency to be used in estimating is $\eta_e = 50\%$. Sensible heat recovery calculations can be based upon equation (5) and the enthalpy calculations based upon equation (8).

It is apparent that condensation will occur when the dry-bulb temperature of fresh air to which recovery will be made is less than the dewpoint temperature of the contaminated mixture, i.e.

$$t_{da} < t_{dpc}$$

A cooling cycle process will normally have condensation taking place during the recovery cycle. In this cycle where both sensible and latent heat are transferred, the value of the sensible heat thermal efficiency to be used for estimating is $\eta_s = 80\%$, and the value of the enthalpy thermal efficiency to be used for estimating is $\eta_e = 60\%$. Sensible heat recovery calculations can be based on equation (5) and enthalpy calculations are based upon equation (8).

In a wet heating cycle process and a cooling cycle process where enthalpy calculations are required after estimating the enthalpy of the recovered airstream h_r from equation (8) and the recovered dry-bulb temperature t_{dr} from equation (5), the dewpoint temperature and wet-bulb temperature of the recovered airstream may be read directly from a suitable psychrometric chart if available. If reference to appropriate tables is required by substituting the h_r and t_{dr} values in equation (3), the value of the ratio W may be approximated. With the value of W, the barometric pressure, the partial pressure of the vapour may be estimated directly and the value of the dewpoint temperature obtained. Also with the value of W, the barometric pressure, and the value of h_r, the adiabatic saturation temperature, which the practising engineer uses as the observed wet-bulb temperature, may be evaluated by a number of successive trials.

It is wise to note here that, based on experience to date, the values of sensible heat and enthalpy thermal efficiencies indicated above are conservative and actual performance will probably exceed these values. However, the values shown provide a very acceptable basis for an economic analysis. Since we can now evaluate the quantity of heat saved by the use of a regenerator, we can now calculate the savings in mechanical equipment. Further, by introducing factors of 'hours of operation' and 'cost of fuel' and other energy, we can calculate the savings in operating costs.

5. CONSTRUCTION CONSIDERATIONS

It is not the purpose of this article to go into the exact details of construction of a regenerator, but to outline how some of the factors encountered in the field influence design would seem to be in order.

The materials from which the heat transfer medium rotor and parts of the frame are constructed is to a very large extent influenced by the contaminants in the air or gas stream, dewpoint of the mixture, and the final temperature of the mixture. The heat exchange medium can be aluminium, stainless steel or other suitable material. In the case of aluminium or stainless steel, the medium is carefully folded in layers and compressed to a predetermined and uniform density to fit the individual rotor sectors. The drive is mounted outside of the ductwork for easy access and cooling.

Figure 4   Small regenerator illustrating drive and duct connections

Figure. 4 indicates typical duct connections for a small Curwen & Newbery (CN) Regenerator.

It is apparent that if the final temperature of the mixture is below dewpoint, condensation will occur and corrosive acids may be formed in combination with the water present in the condensate. Thus applications involving contaminants such as sulphur, nitric acid, etc., should not normally be considered.

Applications where fouling of the medium may occur must be given special consideration with regard to medium construction and type of cleaning attachment used.

Total airflow and space limitations are strong factors due to pressure drop considerations. A reduction in the size of the regenerator to allow first capital cost or to meet a specific space requirement can be accomplished, but the penalty of higher velocity through the medium must be paid. Since the pressure drop varies as the square of velocity, the increment of total static pressure allocated to the unit may increase out of all proportion. If the drop across the unit is in excess of 1.3 inch water gauge, it is conceivable that less expensive fans may be required, to say nothing of increased fan horsepower and possible reinforced ductwork. Thus for the usual applications, a conservative design utilizing low velocities and low pressure drops seems more appropriate. The material and construction of all components is influenced by the entering mixture temperature. Where mixture temperatures exceed this limit, they can be tempered by mixing cool exhaust gases or outside air with the primary exhaust.

The construction is also dependent on the pressure differential between the two airstreams. Acceptable cross-leakage control can be maintained in this standard unit with difference in pressures up to 25 inch water gauge.

In addition to being built for specific job conditions, all units are designed for almost free maintenance operation to include drive motor outside both airstreams, provisions for each adjustment, readily available sector-shaped segments of medium, easily replaced medium segments, and purging devices, etc.

6. SELECTION DATA REQUIRED

For selecting a proper size and type of regenerator, the following information should be available:

(1) exhaust quantities;
(2) exhaust analysis;
(3) exhaust temperatures;
(4) temperatures and conditions of air in respect of which recovery will be made;

(5) end use of air after recovery has been made; and

(6) permissible pressure drop.

In addition, for economic analysis, the following information should be available:

(7) cost of energy delivered to the area (gas, oil, electricity);

(8) hours of operation;

(9) estimated installed cost of conventional heating equipment per million Btu; and

(10) estimated installation cost of refrigeration, per ton of refrigeration, when used in conjunction with air conditioning.

Given the above information, an appropriate selection of a regenerator can be made and its installation cost estimated. Against the average estimated installation cost, some credits are available for:

(1) savings in capital expenditure for conventional heating equipment; and

(2) operating costs savings as compared to an installation without a regenerator.

7. INSTALLATION EXAMPLES

The following case histories include a number of specific installation examples complete with their proven economies. They are listed here and discussed individually:

1 Brickworks.

2 Ceramic industry.

3 Kettle manufacturer.

4 Rubber goods factory.

5 Plastics manufacturer.

6 Textile factory.

7 Food manufacturer.

8 Electrical goods factory.

9 Packaging manufacturer.

10 Paper manufacturer.

11 Local authority civic hall.

They are adapted here from information sheets compiled by Curwen & Newbery Ltd under the heading 'Thermal Engineering: Energy and Money Saving Case History' Nos. 1 to 11.

Energy and money saving case history no. 1

Type of industry

Brickworks

Description of installation

A model SW288B Heat Regenerator was installed to recover waste heat discharged from a natural-gas-fired brick tunnel kiln. The recovered heat was used to heat brick-drying corridors which were originally gas-fired.

Technical data

Exhaust air volume	3.07 m³/s(6500 cfm)
Exhaust air temperature	210°C (410°F)
Supply air volume	1.9 m³ (4000 cfm)
Supply air temperature	35°C (95°F)
Supply air temperature after heat regenerator	176°C (349°F)
Amount of heat recovered	328 KW (1.1 × 10⁶ Btu/h).

Annual savings

The amount of heat recovered was almost identical to the heat requirement of the gas-fired fittings corridors. As a result, the existing gas

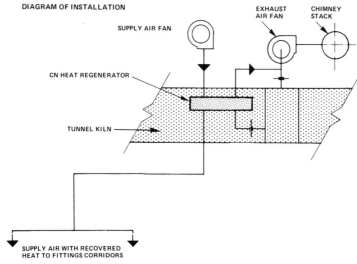

Case history no. 1

burners have not been used since the completion of the installation. The total cost of the installation in 1976 was £19 000 and the payback period was 18 months.

Energy and money saving case history no. 2

Application

Tunnel kiln—ceramic industry

Description of installation

Site measurements showed that the exhaust discharge from a tunnel kiln was a major source of waste heat. The client requested that the kiln balance was to be maintained, and to achieve this the recovered heat was used to heat batch dryers. A model SW10A Heat Regenerator was installed directly over the kiln.

Technical data

Exhaust air volume	1.78 m^3/s (3763 acfm)
Exhaust air temperature	115°C (239°F)
Supply air volume	1.4 m^3/s (3000 cfm)
Supply air temperature	7.2°C (45°F) average
Supply air temperature after heat regenerator	93.3°C (100°F)
Amount of heat recovered	140 KW (5 therm/h)

DIAGRAM OF INSTALLATION

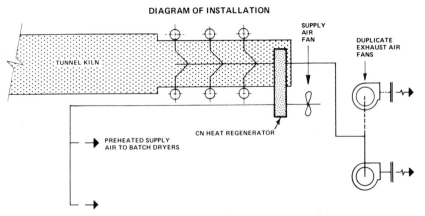

Case history no. 2

Case history no. 2

Annual savings

As the kiln operates 24 hours a day, the capital outlay of about £3500 for the complete scheme was recovered in about nine months (1974 figures).

Energy and money saving case history no. 3

Application

Aluminium furnaces—kettle manufacturer

Description of installation

A model SW4A Heat Regenerator was installed to recover waste heat from the exhaust gases discharged from eight aluminium bale-out furnaces. The recovered heat was used for space heating of an adjoining factory workshop.

DIAGRAM OF INSTALLATION

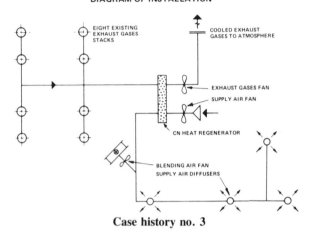

EIGHT EXISTING
EXHAUST GASES
STACKS

COOLED EXHAUST
GASES TO ATMOSPHERE

EXHAUST GASES FAN

SUPPLY AIR FAN

CN HEAT REGENERATOR

BLENDING AIR FAN
SUPPLY AIR DIFFUSERS

Case history no. 3

Technical data

Exhaust air volume	0.71 m³/s (1500 cfm)
Exhaust air temperature	166°C (330°F)
Supply air volume	0.5 m³/s (1000 cfm)
Supply air temperature	15°C (60°F)

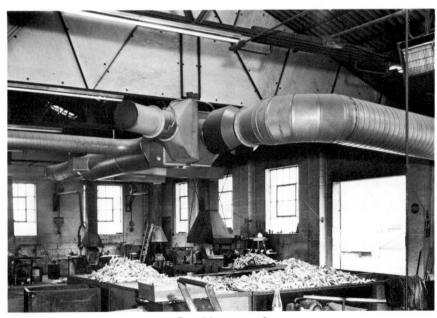

Case history no. 3

Supply air temperature after heat regenerator
(before dilution) 136°C (276°F)
Amount of heat recovered 70 KW (237 635 Btu/h)

Annual savings

As the heat regenerator is only used for space heating, the fuel cost savings occur over 30-week heating seasons. A traditional heating scheme would have had to have been installed in the factory workshop, and the extra invested capital covering heat recovery was recovered in under two years.

Energy and money saving case history no. 4

Application

Steam boiler installation—rubber goods factory

Description of installation

Waste heat discharged from the flues of two dual-fired G.W.B. Powermaster Boilers, each of 1800 kg/h (4000 lb/h) steam capacity, was recovered

Case history no. 4

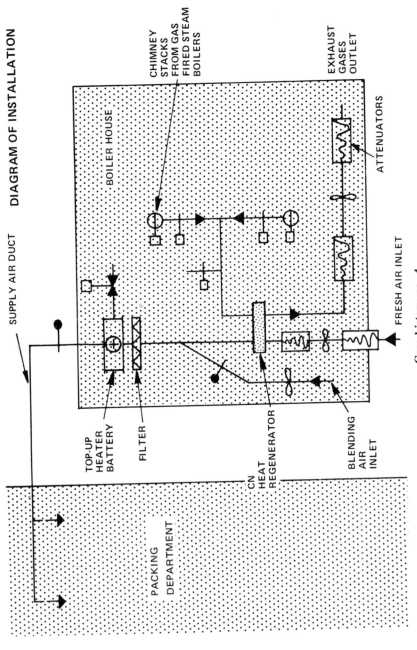

DIAGRAM OF INSTALLATION

SUPPLY AIR DUCT

CHIMNEY STACKS FROM GAS FIRED STEAM BOILERS

BOILER HOUSE

EXHAUST GASES OUTLET

ATTENUATORS

FRESH AIR INLET

TOP-UP HEATER BATTERY

FILTER

CN HEAT REGENERATOR

BLENDING AIR INLET

PACKING DEPARTMENT

Case history no. 4

by a model SW10B CN Heat Regenerator. The recovered heat was used to space heat a packing department some 30 m (100 ft) away. The packing department was previously heated by two ducted oil-fired air heaters. The heat recovery project was a finalist in the 1977 GEM (Gas Energy Management) Award Scheme.

Technical data

Exhaust air volume	1.70 m³/s (3596 acfm)
Exhaust air temperature	283°C (541°F)
Supply air volume	0.91 m³/s (1920 cfm)
Supply air temperature	7.2°C (45°F)
Supply air temperature after heat regenerator	230°C (446°F)
Amount of heat saved	249 KW (847 880 Btu/h)

Annual savings

The amount of heat recovered represents 8.5% of the gross boiler fuel input. The total installation cost was £13 500 and the annual fuel saving was £7200.

Energy and money saving case history no. 5

Application

Film dryer—plastics manufacturer

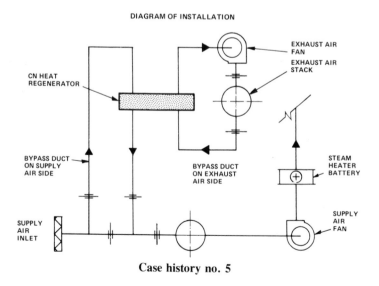

DIAGRAM OF INSTALLATION

Case history no. 5

Description of installation

A model SW4C Heat Regenerator was fitted on a bypass to the main exhaust air duct from a plastic film dryer. The supply air, preheated by the heat regenerator, is further heated by an existing steam heater battery before entry to the drier. A stainless-steel unit was used as the exhaust air contained small amounts of terephthallic acid.

Technical data

Exhaust air volume	0.47 m³/s (1000 acfm)
Exhaust air temperature	180°C (356°F)
Supply air volume	0.31 m³/s (650 cfm)
Supply air temperature	10°C (50°F)
Supply air temperature after heat regenerator	146°C (295°F)
Amount of heat saved	51 KW (175 200 Btu/h)

Annual savings

The cost of the installation and the savings are confidential to the client, but it is known that the drier consumes 30% less steam. A further eight units have been supplied to this client.

Energy and money saving case history no. 6

Application:

Stenter—textile factory

Description of installation

Heat contained in the exhaust air from a Proctor Swartz textile stenter was recovered by a model SW20A CN Heat Regenerator. The supply air was further heated by steam heaters in the first, second, and sixth chambers and by direct gas firing in the third, fourth, and fifth chambers.

Technical data

Exhaust air volume	3.4 m³/s (7200 acfm)
Exhaust air temperature	130°C (266°F)
Supply air volume	2.6 m³/s (5500 acfm)
Supply air temperature	34°C (93°F)
Supply air temperature after heat regenerator	118°C (244°F)
Amount of heat recovered	251 KW (857 200 Btu/h)

Annual savings

The total cost of the installation was £5000, which included £1200 for ducting. As the fuel cost savings were £5000, the capital was recovered in 14 months. The steam consumption to the drier was reduced by about 30%.

Energy and money saving case history no. 7

Application

Process dryer—food manufacturer

Description of installation

To recover heat discharged from an existing Proven Air/Heenan & Froude Rusk Dryer, a model SW20A CN Heat Regenerator was installed on a bypass duct to the exhaust air ductwork. The recovered heat is used to heat clean fresh air which is further heated by a steam heater battery and then fed back to the dryer. Both existing fansets were re-used but upgraded to meet the additional pressure requirements.

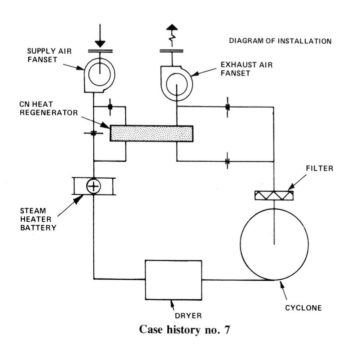

Case history no. 7

Technical data

Exhaust air volume	2.93 m³/s (6200 acfm)
Exhaust air temperature	43°C (110°F)
Supply air volume	2.7 m³/s (5700 cfm)
Supply air temperature	10°C (50°F)
Supply air temperature after heat regenerator	37°C (98°F)
Amount of heat recovered	89 KW (304 000 Btu/h)

Annual savings

The total installed cost was £4500 (1975 prices) and the payback period was 19 months. The client now has a total of three units in operation.

Energy and money saving case history no. 8

Application

Paint plant—electrical goods factory

Description of installation

Sheet-metal components are passed through a Tycoate eight-colour powder coating plant which incorporates a pretreatment section and dry-off oven. Heat was recovered from the exhaust stacks of three natural-gas-fired immersion tubes on the pretreatment section using a model SW4B CN

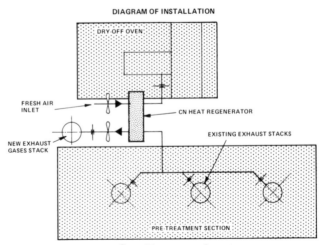

Case history no. 8

Heat Regenerator. The recovered heat was used to preheat supply air to the dry-off oven which was further heated by a direct gas heater.

Technical data

Exhaust air volume	0.41 m³/s (874 acfm)
Exhaust air temperature	226°C (439°F)
Supply air volume	0.24 m³/s (508 cfm)
Supply air temperature	18°C (65°F)
Supply air temperature after heat regenerator	184°C (363°F)
Amount of heat recovered	103 KW (350 000 Btu/h)

Annual savings

The total installed cost was about £5000 and the savings were £2400 per annum, giving a payback period of two years. These are 1975 prices and the savings based on a gas tariff of 9.5p/therm.

Energy and money saving case history no. 9

Application

Colour printing machine—packaging manufacturer

Description of installation

Three model SW50A CN Heat Regenerators were fitted to a new Tecmo seven-colour printing machine. To dry ink, fresh air is drawn over a heater. This heated air impinges on the wet ink, driving out the fast-drying solvents and leaving the ink dry on the paper. The air after drying the ink is exhausted to the atmosphere. The heat regenerators are fitted to recover the

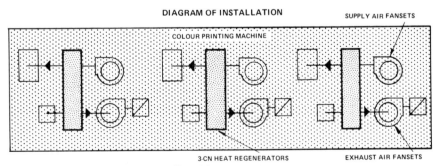

Case history no. 9

Case history no. 9

waste heat in the exhaust air, transfer this heat to fresh air, and direct it back to the machine. Extensive tests showed an insignificant solvent transfer to the supply air side of 1.25%, confirming the suitability of the heat regenerator.

Technical data (per unit)

Exhaust air volume	8.26 m³/s (17 000 cfm)
Exhaust air temperature	100°C (212°F)
Supply air volume	6.49 m³/s (13 750 cfm)
Supply air temperature	23°C (73°F)
Supply air temperature after heat regenerator	85°C (185°F)
Amount of heat saved	616 KW (2.1 × 10⁶Btu/h)

Annual savings

The fuel consumption of the dryer was reduced by 55% and the total cost of the heat recovery project, including ductwork and fans, of £35 000 (1975 prices) was recovered in under 2½ years.

Energy and money saving case history no. 10

Application

Board-making machine—paper manufacturer

Description of installation

A Model SW80A CN Heat Regenerator was fitted to recover heat contained in the exhaust air from the canopy covering the last nine drying cylinders of a board-making machine. The recovered heat is used to heat clean fresh air which is then further heated and then fed back to the paper machine at a temperature of 128°C, where drying of the board is carried out within the hood.

Technical data

Exhaust air volume	14 m³/s (30 000 cfm)
Exhaust air temperature	72°C (162°F)
Supply air volume	12 m³/s (25 000 cfm)
Supply air temperature	29°C (84°F)
Supply air temperature after heat regenerator	63°C (146°F)
Amount of heat recovered	502 KW (1.7 × 10 Btu/h)

Annual savings

The last nine cylinders covered by the hood were equipped with a steam meter, and tests showed that the amount of steam used was reduced from 1180 kg/h (2600 lb/h) to 500 kg/h (1100 lb/h). This represents a saving of 680 kg/h (1500 lb/h) which was achieved within an increase in production. This represented a reduction in oil consumption of 50 l/h (11 gallon/h) and gave a fuel cost saving of £10 000 per annum. The installed cost was £18 000, giving a payback period of 22 months.

Energy and money saving case history no. 11

Application

Heating and ventilating system—local authority civic hall

Description of installation

The main auditorium was designed to be multipurpose and to hold a total of 500 people. At an early stage, it was realized that a considerable amount of lighting and body heat would be available for recovery. As a

Case history no. 11

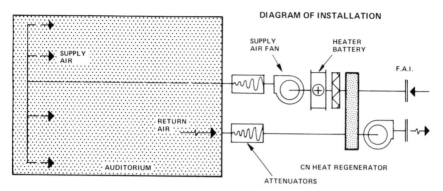

Case history no. 11

high air change rate was demanded, a model SW50 Heat Regenerator was installed to ensure that the system was economic to run.

Technical data

Exhaust air volume	8.25 m³/s (17 500 cfm)
Exhaust air temperature	21 °C (70 °F)
Supply air volume	8.25 m³/s (17 500 cfm)
Supply air temperature	−1 °C (30 °F)
Supply air temperature after heat regenerator	16.6 °C (62 °F)
Amount of heat recovered (at design conditions)	181 KW (616 000 Btu/h)

Annual savings

Because of internal heat gains and varying operating hours, the annual savings were difficult to quantify. As the heat regenerator reduced the size of the boilers and ancillary equipment, the payback period was estimated to be about 3 to 4 years.

8. ESTIMATED PERFORMANCE

Example 1

Fresh air 95 °FDB*, 75 °FWB*
Exhaust air 75 °FDB, 62 °FWB (48% RH*)
Sensible heat efficiency = η_s = 80%
Enthalpy (total heat) efficiency = η_e = 60%

Fresh air, after
recovery (R)

Fresh air,
entering (A)

(1) 79.0°FDB
(2) 32.15 = h_R
    67.8°FWB
    84 grains
    62.3°FDP*

95.0°FDB
75.0°FWB
99 grains
66.8°FDP
h_A = 38.6 Btu/lb

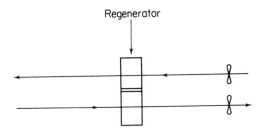

Regenerator

Contaminated exhaust,
from space (C)

75.0°FDB
62.0°FWB
62 grains
54.0°FDP
h_C = 27.85 Btu/lb

---

(1) Recovered DP at η_s = 80%

(2) Recovered enthalpy at η_e = 65%

t_A = 95.0°FDB
$-t_C$ = 75.0°FDB
    20.0°FDB Possible recovery

h_A = 38.60 Btu/lb
$-h_C$ = 27.85 Btu/lb
    10.75 Btu/lb Possible recovery

20.0 × 0.80 = 16.0°F
                  Design recovery
95.0 − 16.0 = 79.0°FDB = t_R

10.75 × 0.60 = 6.45 Btu/lb
                 Design recovery
38.60 − 6.45 = 32.15 Btu/lb = h_R

---

* °FDB = degrees Fahrenheit, dry bulb.
  °FWB = degrees Fahrenheit, wet bulb.
  °FDP = degrees Fahrenheit, dewpoint.
  RH = relative humidity.

Required *conventional* tonnage per 1000 scfm (75 lb/min) of fresh air:

75 lb/min × 60 min/h × (38.60 − 27.85) Btu/lb = 48 375 Btu/h (4.0 tons)

Required tonnage *with regenerator* per 1000 scfm (75 lb/min) of fresh air:

75 lb/min × 60 min/h × (32.15 − 27.85) Btu/lb = 19 350 Btu/h (1.61 tons)

SAVINGS = 2.39 tons of refrigeration per 1000 scfm of fresh air.

Example 2

Fresh air   95°FDB, 78°FWB
Exhaust air 75°FDB, 62°FWB (48% RH)
Sensible heat efficiency = η_s = 80%
Enthalpy (total heat) efficiency = η_e = 60%

Fresh air, after
recovery (R)

Fresh air,
entering (A)

(1) 79.0°FDB
(2) 33.35 = h_R
      69.1°FWB
      91 grains
      64.5°FDP

95.0°FDB
78.0°FWB
117 grains
71.6°FDP
h_A = 41.6 Btu/lb

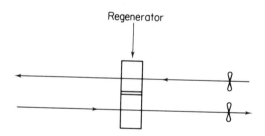

Regenerator

Contaminated exhaust,
from space (C)

75.0°FDP
62.0°FWB
62 grains
54.0°FDP
h_C = 27.85 Btu/lb

(1) Recovered DB at η_s = 80%	(2) Recovered enthalpy at η_e = 65%

t_A = 95.0°FDB	h_A = 41.60 Btu/lb
$-t_C$ = 75.0°FDB	$-h_C$ = 27.85 Btu/lb
$\overline{20.0°\text{FDB}}$ Possible recovery	$\overline{13.75 \text{ Btu/lb}}$ Possible recovery
20.0 × 0.80 = 16.0°F	13.75 × 0.60 = 8.25 Btu/lb
Design recovery	Design recovery
95.0 − 16.0 = 79.0°FDB = t_R	41.60 − 8.25 = 33.35 Btu/lb = h_R

Required *conventional* tonnage per 1000 scfm (75 lb/min) of fresh air:

75 lb/min × 60 min/hr × (41.60 − 27.85) Btu/lb = 61 875 Btu/h (5.12 tons)

Required tonnage *with regenerator* per 1000 scfm (75 lb/min) of fresh air:

75 lb/min × 50 min/h × (33.35 − 27.85) Btu/lb = 24 750 Btu/h (2.06 tons)

SAVINGS = 3.06 tons of refrigeration per 1000 scfm of fresh air.

Air conditioning cycle

Fresh air, after
recovery (R)

Fresh air,
entering (A)

76.5°FDB
66.5°FWB
81.5 grains
61.3°FDP
h_R = 31.25 Btu/lb

92.5°FDB
74.0°FWB
97 grains
66.3°FDP
h_A = 37.65 Btu/lb

Regenerator

Contaminated exhaust,
from space (C)

Contaminated exhaust,
after regeneration (L)

72.0°FDB
63.5°FWB

86.0°FDB
70.0°FWB

74 grains
58.7°FDP
$h_C = 28.95$ Btu/lb

84.5 grains
62.3°FDP
$h_L = 43.1$ Btu/lb

Sensible heat efficiency = η_s Total heat (enthalpy) efficiency = η_e

---

(Dry-bulb temperatures only)   (Includes moisture removal)

$t_A = 92.5°$FDB $t_A = 92.5°$FDB $h_A = 37.65$ Btu/lb $h_A = 37.65$ Btu/lb
$\underline{-t_C = 72.0°\text{FDB}}$ $\underline{-t_R = 76.5°\text{FDB}}$ $\underline{-h_C = 28.95\,\text{Btu/lb}}$ $\underline{-h_R = 31.25\,\text{Btu/lb}}$
20.5°FDB   16.0°FDB   8.70 Btu/lb   6.40 Btu/lb
Possible recovery   Actual recovery   Possible recovery   Actual recovery

---

$$\eta_s = \frac{16.0}{20.5} \times 100 = 78.0\% \qquad\qquad \eta_e = \frac{6.40}{8.70} \times 100 = 73.6\%$$

Cooling cycle recovery

	Outside fresh air temperature	
Exhaust air temperature[a]	35°CDB, 24°CWB	35°CDB, 26°CWB
24°CDB, 17°CWB	26/20	26/21
26°CDB, 20°CWB	27/22	27/22

[a]Degrees Centigrade, dry bulb and wet bulb.

Typical recovery conditions and savings per 5 m³/s

Exhaust air temperature	Outside air temperature				
	−18°C	4°C	24°C	35°CDB, 26°CWB	
24°CDB, 17°CWB	16°CDB 201 kW	20°CDB 94 kW	— —	26°CDB, 21°CWB 116 kW	Supply air temp. Savings in refrigeration
49°C	116°CDB 805 kW	120°CDB 697 kW	124°CDB 604 kW	126°CDB 550 kW	Supply air temp. Savings

Estimated CN Regenerator performance for design purposes

Air conditioning cycles, balanced weight of airstreams

Exhaust condition	Fresh air condition	Fresh air condition after regenerator	Refrigeration		Refrigeration saved (kW per m³/s)
			Conventional design (kW per m³/s)	With regenerator (kW per m³/s)	
24°CDB, 17°CWB (48% RH)	35°CDB, 26°CWB	26°CDB, 21°CWB	38	15	23
24°CDB, 17°CWB	35°CDB, 24°CWB	26°CDB, 20°CWB	30	12	18
27°CDB, 19°CWB	35°CDB, 26°CWB	26°CDB, 22°CWB	28.5	11.5	17
27°CDB, 19°CWB	35°CDB, 24°CWB	28°CDB, 22°CWB	20	12	8

Tempering hot regenerated air

When dealing with hot exhaust, the recovered air temperatures after the CN Regenerator will usually be too high for use directly for space heating. Two simple techniques are available to overcome this problem are now discussed.

(1) Three-fan system

The essential elements of the three-fan system are a standard CN Regenerator installation, based on balanced weights of exhaust and outside air, plus an additional blending fan. The blending fan mixes sufficient outside air with the regenerated air to reduce the mixture temperature to acceptable levels.

The three-fan system is the most efficient since it recovers the maximum amount of heat at all times. Since the quantity of make-up air varies widely due to fluctuations of outside temperature, it is particularly suited to applications where there is a severe negative pressure, and the air can be discharged into an area without a sophisticated ducting system.

Three-fan system example

Given

Exhaust   16 238 cfm dry air at 400°F
Outside dry air 0°F at design temperature, 40°F average temperature
End use   Plant heat to relieve severe negative pressure
             Deliver at 70°F

Design conditions, 0°F

Mixture 70°F	*Blending air 0°F*	*0°F*
3429 lb/min	2679 lb/min	750 lb/min
v = 13.34 cu ft/lb	v = 11.58 cu ft/lb	v = 11.58 cu ft/lb
45 743 cfm	31 023 cfm	8685 cfm (10 000 standard cfm)

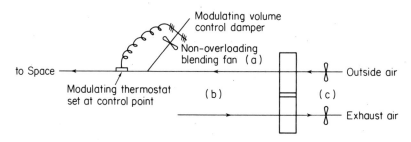

(a) 320°F, 750 lb/min
(b) 400°F, 750 lb/min, v = 21.65 cu ft/lb, 16.238 cfm (10 000 scfm)
(c) 80°F, 750 lb/min, v = 13.59 cu ft/lb, 10.193 cfm

Average conditions, 40°F dry

Fans are constant-volume devices. For illustrative purposes, fan volumes from the previous diagram are used with no allowance made for changes in previous loss or density.

Mixture 70°F	*Blending air 40°F*	*40°F*
6624 lb/min	5934 lb/min	690 lb/min
v = 13.34 cu ft/lb	v = 12.59 cu ft/min	v = 12.59 cu ft/min
88 364 cfm	74 709 cfm	8685 cfm

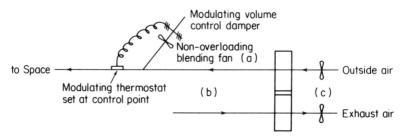

(a) 328°F, 690 lb/min
(b) 400°F, 708 lb/min, v = 21.65 cu ft/lb, 15 328 cfm
(c) 118°F, 708 lb/min, v = 14.40 cu fit/min, 10 193 cfm
Overbalanced exhaust; assume balanced weights at lower level of 690 lb/min

(2) Fixed bypass system

In this system, in addition to the standard balanced-weight CN Regenerator installation, a fixed outside air bypass is added. The CN Regenerator operates normally with balanced weights of air and sufficient outside air is bypassed and then blended to obtain an acceptable mixed air temperature. Using this technique, maximum heat regeneration is accomplished at external design conditions. As the outside temperature rises, however, capacity control should be employed to maintain the desired mixture temperatures. Since the air volumes remain relatively constant, this system lends itself to applications where a duct system must be used to distribute the make-up air.

Fixed bypass system example

Given

Exhaust   16 238 cfm of dry air at 400°F
Outside dry air 0°F at design temperature
              40°F at average temperature
End use   .   Plant heat deliver at 70°F

Design conditions, 0°F

Mixture 70°F	*Bypass 0°F*	*0°F*
3429 lb/min	2679 lb/min	3429 lb/min
v = 13.34 cu ft/lb		v = 11.58 cu ft/min
45 743 cfm		39 208 cfm

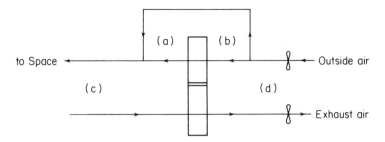

(a) 320°F, 750 lb/min
(b) 0°F, 750 lb/min, 10 000 scfm
(c) 400°F, 750 lb/min, v = 21.65 cu ft/lb, 16 238 cfm (10 000 scfm)
(d) 80°F, 750 lb/min, v = 13.59 cu ft/lb, 10 193 cfm

Average conditions, 40°F

Fans are constant-volume devices. For illustrative purposes, fan volumes from the previous diagram are used with no allowance made for changes in pressure loss or density. Assume same percentage through bypass in winter design.

Mixture 103°F	*Bypass 40°F*	*40°F*
3154 lb/min	2464 lb/min	3154 lb/min
v = 14.17 cu ft/lb		v = 12.59 cu ft/lb
44 692 cfm		39 707 cfm

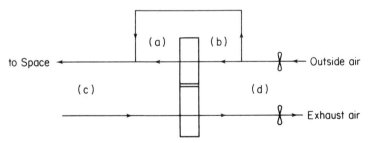

(a) 328°F, 690 lb/min
(b) 40°F, 690 lb/min
(c) 400°F, 708 lb/min, v = 21.65 cu ft/lb, 15 328 cfm
(d) 112°F, 708 lb/min, v = 14.40 cu ft/lb, 10 193 cfm
   Mixture temperature well above the required 70°F; should have capacity control
   Overbalanced exhaust; assume balanced weights at lower level of 690 lb/min

Paper mill application

Heat requirements

Without regenerator	With regenerator

Total heat required is

$$Q = \text{CFM} \times \rho \times 60 \times \Delta T$$

where Q = quantity of heat (Btu/h), CFM = volume fluid flow rate (cubic feet of air per minute), ρ = air density, C_p = specific heat, 60 = constant (minutes in an hour), and ΔT = temperature difference (°F). So

$$Q = 17\ 000 \times 0.076 \times 0.24$$
$$\times 60 \times (180 - 80)$$
$$Q = 1860\ 480\ \text{Btu/h}$$
$$M = 1918\ \text{lb/h saturated steam}$$

By using the regenerator, 80% of the heat exhausted from the machine may be recycled to the process after conditions have stabilized. Therefore the heat available is:

$$Q = 1860\ 480 \times 0.80$$
$$Q = 1488\ 384\ \text{Btu/h recovered}$$

When the regenerator is operational, the steam load will be equivalent to:

$$Q = 1860\ 480 - 1488\ 384$$
$$Q = 372\ 096\ \text{Btu/h}$$
$$M = 384\ \text{lb/h saturated steam}$$

Annual savings

Assumptions: Boiler efficiency 70%; 5760 operating hours per annum;

Heat recovered, $Q = 1488\ 384$ Btu/h

$$\text{Annual savings} = \frac{1488\ 384 \times 17.5 \times 5760}{100\ 000 \times 0.7 \times 100}$$

Annual savings = £21 433

Application to the paper and board industry

This is an industry exhausting vast amounts of heat energy by way of moisture-laden air from drying hoods and roof spaces over the machines. Whilst temperatures may be considerably lower than actual drying cylinder temperatures, the exhaust volumes are large, and hence the total heat energy exhausted is without doubt high-grade and very worthy of recovery. The Curwen & Newbery Heat Regnerator is now available, and able to contribute towards this end.

The usual sources of waste heat are:

(1) drying hoods;
(2) conditioner hoods;

(3) roof spaces over machine;
(4) M.G. cylinder hoods;
(5) air caps;
(6) gas-fired steam boiler stacks;
(7) turbo alternator set roof spaces, including heat from transformers;
(8) stock preparation plant roofs, especially when steam is used for stock heating; and
(9) chemical plant roof spaces, such as over starch cooking.

Points of recovered heat utilization

Savings can be made around the year by feeding hot air, instead of air at ambient temperatures, to:

(a) felt dryer heater batteries;
(b) high-velocity air systems;
(c) wet end roof circulation;
(d) machine house-intake air;
(e) inlet air to dryer cylinder M.G. hoods and air caps; and
(f) dryer pit.

Savings can be made for part of the year by feeding recovered air for heating and ventilating at controlled temperatures and humidity to salles:

(a) quality control testing laboratories;
(b) engineering workshops and stores;
(c) warehouses and paper stockrooms;
(d) offices and canteens;
(e) air curtains; and
(f) stock preparation floors.

It would appear that, in the past, machine house ventilation in mills has not been carried out effectively for the following reasons. Plant has been costly in terms of capital investment and operating costs, and of comparatively low efficiency, offering low returns on investment. Secondly, large fan and motor sets have been necessary to move the volumes of air really needed to ventilate properly without brining about discomfort for operatives and condensation from roof members both in winter and summer. The higher the temperature of air, the greater will be the weight of water absorbed. In paper mill case histories, there has been a tendency to introduce smaller volumes at higher temperatures to achieve low capital and operating costs, whereas greater volumes at lower temperatures would have achieved a better extraction of moisture and resulted in generally cooler machine houses. The heat regenerator achieves an 80% saving in exhausted heat, and output flow volumes can be increased beyond exhaust volumes to pressurize machine houses with greater volumes of air at lower temperatures. This has the effect of obviating draughts into machine houses and creates improved working conditions. The more air in movement, the less likelihood there is of stagnant pockets. The reductions in stagnant moisture-laden pockets, even their complete elimination, can lead to increases in machine speed, where drying has previously been the barrier.

A further advantage for the older mills in moving more air about and avoiding stagnant pockets will be in a reduction of decay in roof linings, and the extension of life expectancies in consequence.

Example

(This is an initial preliminary assessment from an actual case.) Paper machine hood discharging 120 000 CFM at 150°F dry-bulb with ambient temperature standing at 60°F. Operational year 5760 hours. Boiler fired by natural gas at 17.5p per therm. Boiler efficiency 82%. Cost of grid electricity 2.5p per unit. Three supply air fansets each driven by 60 horsepower motors. Existing extract fans serving contaminated side. Regenerator efficiency 80%. Add the aspects of possible machine speed increase and reduction in roof replacements to this attractive return on capital, then the total savings and increase in profitability are certainly worthy of careful consideration.

Savings

Heat units recovered per hour by regenerator

= CFM × 60 × weight of air/cu ft × specific heat of air
 × rise in temperature × regenerator efficiency
= 120 000 × 60 × 0.0765 × 0.24 × 90 × 0.8
= 9500 000 Btu

Extent of saving per annum reflected in fuel cost reduction

$$\frac{\text{Btu saved/h} \times \text{h/year} \times \text{cost per therm}}{\text{Btu/therm} \times \text{boiler efficiency}}$$

$$= \frac{9500\,000 \times 5760 \times 17.5}{100\,000 \times 0.82 \times 100}$$

= £116 781

Capital costs

Three model 100 regenerators at £13 000 each	£39 000
Three 60 HP fan & motor sets at £2000 each	6 000
Probable cost for ducting	£15 000
Total	£60 000

Operating costs

Three 60 HP motors operating for 5760 h with unit cost 2.5p	£19 336
Probable maintenance costs	£1 000
Total	£20 336

Return on capital

Assuming a life expectancy of 10 years, although in practice 15–20 years could be expected, the return is as follows:

Gross savings over 10 years	£1167 810
Capital cost + 10 years operating costs	£60 000
	+ £203 360
	= £263 360
Nett savings over 10 years	£1167 810
	− £263 360
	= £904 450
Average annual savings	£90 445
Return on capital each year for 10 years	£90 445
	£60 000
	= 151%

Comments

The above typical examples outline certain commercial and industrial applications where the use of the regenerator has proved beneficial in many respects. Its uses, however, are very wide, and when used constantly 24 hours per day can prove most beneficial. Units operating at variable speeds can be modulated to suit varying conditions; they are suitable for recovering heat from kitchen extract plant, cafeterias, gymnasiums, swimming pools, classrooms, hotels and office buildings. In the case of high-rise blocks of flats with mechanically operated ventilation, this often accounts for 30–40% of the total building heat loss. While the use of a rotary heat regenerator has not a direct effect on the heat losses due to transmission or infiltration, it will reduce the mechanical ventilation loss by 80% or more. If electric heat is used, the regenerator will often pay for itself in one heating season. With the trend towards the greater use of electricity for heating buildings, there is a strong case for using a regenerator. This could have the effect of reducing the running cost of an electrically heated building below that of one heated by gas, oil or solid fuel without heat recovery.

Electricity has the advantage of providing a more reliable form of heat, lower capital cost, low maintenance costs, low flue-gas losses or atmospheric pollution, elimination of boilerhouse, pumping plant, etc., and avoids the duplication of other services, as electricity has, of course, to be provided for lighting purposes. One can envisage a wider use of electrical energy for heating purposes, provided that heat which would normally be lost by ventilation is recovered wherever possible. If a wide use of electrical heating systems in schools and public buildings occurred, such systems could be designed around packaged installation consisting of heating elements, filtration equipment, and heat regeneration.

9. TWIN-WHEEL UNITS

Figure 5 is a photograph of a CN Twin-Wheel Regenerator. The rotors are driven from a double-shaft worm wheel reducer with a 50 to 0 ratio, the required performance for this particular assembly being to handle 50 000 scfm heated from $-10°F$ to $+58°F$ when exhausting 56 000 scfm at 75°F and with a pressure drop of 0.94 inch water gauge. The twin unit requires a one horsepower electric motor to drive the wheels and three $1\frac{1}{2}$ inch drains are required to take off condensation.

10. NONREGENERATIVE HEAT-RECOVERY EQUIPMENT (CN RECUPERATOR)

The previous sections have described in detail the CN Heat Regenerator. This Company also handles the nonregenerative recuperator which is an air-to-air heat-recovery unit of the fixed-plate type. This unit is nonregenerative; a superconducting heat exchanger which has no moving parts is arranged so that one airstream flows through one side of the unit and the other airstream counterflows through the opposite side. Heat is then transferred from one airstream to the other (Figure 6).

Depending upon the volume of air handled, the relative humidity of the exhaust air, and the type of unit selected, efficiencies in excess of 70% can be obtained.

Essentially, the CN Recuperator consists of a closely spaced series of

Figure 5 CN Twin-Wheel Heat Regenerator

thin aluminium plates set on edge within a boxlike enclosure. The plates are mechanically secured to the enclosing structure at the edge and, in addition, are firmly sealed in place by a chemical-resistant polyester resin of the isophthallic type. At each end the aluminium plates are brought together top and bottom and securely sealed. Construction is completed with the addition of the casing, which is mild steel, and the divided end construction and sleeve. The CN Recuperator is a perfect counterflow heat

Figure 6   CN Recuperator

exchanger through which exhaust air, fully isolated into multiple thin streams, is continually exposed to the make-up air in each alternate slot. There is no chance of intermixing or contamination of the make-up air and, since there are no moving parts in the CN Recuperator, there is nothing to wear out or maintain.

Installation

The CN Recuperator can be installed in any position with respect to heat transfer. However, where the design condition indicates that the exhaust air will be cooled to a temperature below dewpoint, it becomes necessary to arrange for adequate drainage. In this case the unit should be installed level or slightly graded towards the drain. Care should be taken to ensure that the drain opening is used for the exhaust air side of the system. In freezing conditions, special precautions will be required to prevent a build-up of frost and ice on the exhaust end of the exchanger. Where the climate is severe enough to create such a problem, the economy of the CN Recuperator is especially attractive, and it may be desirable to install a duplex system, permitting one system to be shut down for defrosting while the other continues to operate. Several alternatives are available to solve this problem and a number are illustrated in the following examples.

Example No. 1: Electric heater battery in the external air duct

Fresh air quantity constant. Exhaust air quantity constant. The fresh air is heated by means of the electric heater battery to the minimum inlet temperature. No danger of frost to the pre-heater.

Example No. 2: Separate fresh air intake with hot water heater battery and damper

Fresh air quantity constant. Exhaust air quantity constant. The minimum temperature of the fresh air is maintained by the progressive control of the dampers. In winter, the air heater battery is continuously in operation (pay attention to freezing). Take care to see that the two airflows are well mixed. As the damper in most cases is not tight, there is always some air heated.

Example No. 3: Hot water air heater battery in the fresh air duct

Fresh air quantity constant. Exhaust air quantity constant. The fresh air is heated to the minimum intake temperature by means of a preheater. The battery has to be protected against frost damage.

Example No. 4: Bypass with incorporated heater battery

Fresh air quantity constant. Exhaust air quantity constant. The minimum fresh air temperature is maintained by the automatic control of the heater battery. The bypass air quantity is constant.

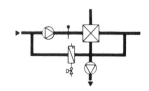

Example No. 5: Bypass with incorporated heater battery

Fresh air quantity variable. Exhaust air quantity constant. The minimum fresh air temperature is maintained by the progressive control of the damper. In winter, the air heater battery is continuously in operation (no danger of freezing). The bypass air quantity depends on its temperature.

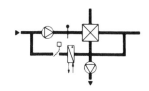

Example No. 6: Bypass with incorporated damper

Fresh air quantity variable. Exhaust air quantity constant. The minimum fresh air temperature is maintained by the progressive control of the damper. The bypass air quantity depends on the intake air temperature. This arrangement is only suitable for a high intake air temperature.

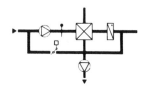

Example No. 7: Air mixing before the heat exchanger

The air flows through the heat exchanger are constant. The fresh air and exhaust air quantities depend on the external air temperature. This arrangement is specially suitable for swimming pool ventilation systems.

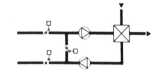

Filtration and cleaning

As the CN Recuperator is a heat-exchange device, its efficiency will decrease when dirt builds up on an aluminium heat-exchange surface. This makes it advisable to use a filter on both the exhaust and make-up air ducts leading to the unit. In critical applications where heavy fouling cannot be dissipated, as in kitchen exhaust systems, the use of water washing equipment becomes desirable. Where there is a dirt build-up over a period of years which seriously affects the operation of the CN Recuperator, it can be cleaned by the use of steam and/or hot water detergent spray. In unusual situations, such as paint spraybooths or woodworking exhaust systems, special care must be taken to remove particles in suspension which could shorten the life of the unit.

Construction

The Standard CN Recuperator construction includes aluminium plates for heating exchange surfaces and a resin to seal the edges of the metal. For special applications outside the normal temperature range, advice from the manufacturers should be taken.

Corrosion

In considering the potential of the CN Recuperator under corrosive conditions, the normal characteristics of both the aluminium plates and the polyester resin should be considered. The resin will stand up to most applications and aluminium is well known to resist normal ventilating corrosion problems. For saline applications, the aluminium can be anodized. Certain detergent compounds will attack aluminium, and great care should be taken in the selection of materials for use with water wash units.

Tests

The CN Recuperator has been extensively tested and checked for performance for a period of more than two years. Operating tests include repeated on/off cycles with temperature differences in excess of 100°F. In the units tested, there has been no evidence of any failure of the seal or of metal fatigue. The performance data related to air volumes, efficiency, and pressure losses are the result of multiple tests using duplicate instrumentation to avoid error.

Special precautions

To make certain that both sides of the CN Recuperator are subject to the same type of pressure, either positive or negative, make sure that (a) the drain connection is adequately trapped, (b) the drain is installed on the

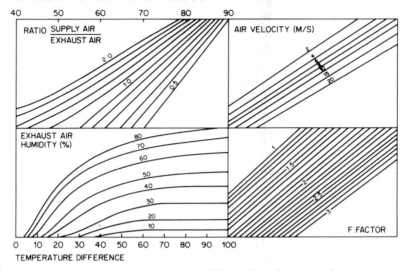

SENSIBLE HEAT EFFICIENCY SUPPLY AIRSIDE (%)

Figure 7   Heat recuperator sizing and performance data

exhaust air side of the unit, (c) the CN Recuperator is not distorted when marrying up to ductwork. Figures 7 and 8 give performance data.

Use of tables

Since variations in the temperature difference between exhaust and make-up air do not significantly affect the performance of the CN Recuperator expressed as an efficiency ratio, it has been possible to present selection data in a very simple and easily used form. This is one advantage of the fixed plate nonregenerative type of recovery unit as against the rotary heat regenerator. The CN Recuperator is available in standard models.

11. NONREGENERATIVE HEAT-RECOVERY UNITS (Heat Pipe)

Experiments in recent years have resulted in the production of thermal recovery units made up of heat pipes. A typical heat-recovery unit of this type (Figure 9) consists of a bank of many heat pipes. For example, one rated at 2000 cfm is made up of six rows of 16 heat pipes in each row. The unit is separated in the middle with a diaphragm to ensure that no cross-contamination of outgoing and incoming air can take place. Such a complete unit is inserted between the incoming and outgoing ducts of an air system similar to the CN Recuperator. As with the CN Recuperator, the heat-pipe exchanger is primarily an exchanger of sensible heat. Under certain conditions, however, one airstream or the other may approach saturation sufficiently to cause condensation on the finned surfaces. If this

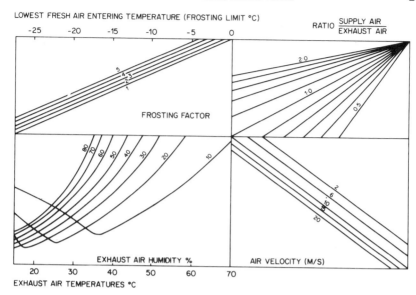

Figure 8 Heat recuperator frost limits

Figure 9 Standard CN Heat-Pipe Recovery Unit

CN HEAT PIPE RECOVERY UNIT — Flat plate fin

CN HEAT PIPE RECOVERY UNITS:

The CN Heat Pipe Recovery Unit is designed to recover energy normally exhausted to atmosphere in air or gaseous discharges. The recovered energy is transferred to a counterflowing supply airstream.

Working with dry airstreams, the CN Heat Pipe Recovery Unit is a sensible heat exchanger, but when handling humid exhaust airstreams a degree of latent heat recovery is achieved with the advantage of not transferring moisture. Generally speaking the transfer of moisture in the vapour phase is only required for air conditioning systems and in such cases the CN Heat Pipe Recovery Unit should not be used.

As both the supply and exhaust airstreams are separated by a division plate the two airstreams cannot intermix.

The CN Heat Pipe Recovery Unit ideally should be used where:

- Air/gas streams are between − 10°C to + 300°C.
- High static pressure differences exist between the two airstreams.
- There is limited headroom.
- Fan positions are fixed.
- No intermixing of air/gas streams can be permitted.
- There are low to medium levels of contaminants in the air/gas stream.
- Moisture is likely to condense out on the exhaust air/gas side.
- There is a considerable imbalance of masses or volumes between the two airstreams.

CONSTRUCTION:

There are two types of CN Heat Pipe Recovery Units available:

    Flat Plate Fin

    Spiral Fin

Flat plate fin:

For low temperatures up to 200°C (400°F) and certain industrial applications flat plate fin units may be used. The flat plate fins can be either of aluminium, pre-painted aluminium, copper or pre-tinned copper. The heat pipes would be of seamless copper tubes pre-tinned if required.

The heat pipes are desiged on a staggered pattern 40 × 40 mm allowing for maximum fin efficiency. The heat pipes are mechanically expanded and thoroughly fixed to the plate fins to ensure maximum heat transmission from the heat pipes to the fins. The framework for the flat plate fin units is manufactured from galvanised sheet steel. The framework can also be made from aluminium or stainless steel.

Spiral fin:

For higher temperature applications up to 315°C (600°F) the secondary heating surface would be spirally wound onto the heat pipes. The heat pipes and secondary heating surface are usually of copper and can be electrotinned if required. The framework for spiral fin units is manufactured from black mild steel.

Figure 10   Heat Pipe Recovery Unit

condensation does occur, the unit will have a greater recovery capacity because the air temperature on the wet side of the unit does not change as much for each unit of energy transferred, and the air side transfer rate of the wettest surface is substantially decreased.

Figure 10 shows the Heat Pipe Recovery Unit and related discussion, and Figure 11 shows the application data.

Figures 12–14 show factors which affect efficiency.

12. PREHEATING COMBUSTION AIR WITH RECOVERED HEAT

The original Lungstrom Regenerator was generally applied to preheating combustion air, but the increased efficiency of modern boiler plant has led to this process being sadly neglected in recent years.

With the increasing use of natural gas, however, the author visualizes the

IVISION PLATE:

)th types of CN Heat Pipe Recovery Units are supplied with division ates which separate the two air/gas streams. The division plate is nstructed of the same material as the framework. To accommodate balanced airstreams the division plate can be located between ¼ d ¾ of unit length from one end.

ND COVERS:

e ends of the heat pipes are protected by airtight end covers nstructed of the same material as the framework. The side panels of e frame are designed to allow free expansion of the heat pipes into e depth of the end covers.

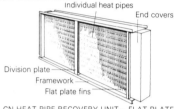

Individual heat pipes
End covers
Division plate
Framework
Flat plate fins

CN HEAT PIPE RECOVERY UNIT – FLAT PLATE FIN

APILLARY WICKS:

hen fitted horizontally, or near to horizontal, all heat pipes will be ed with capillary wicks. All vertical heat pipes are designed to erate as thermosyphons. In the latter case the evaporator must ways be below the condenser.

PERATION:

e CN Heat Pipe Recovery Unit consists of between 40 and 240 dividual heat pipes. Each heat pipe is a super conductor capable of gh thermal conductance. The basic heat pipe consists of an external ell, two end plugs, a capillary wick and a small amount of working id. When one end of the heat pipe is heated, the evaporator, the orking fluid vaporises and moves to the cold end of the heat pipe, here it condenses, the condenser. The condensate then returns back the evaporator via the capillary wick to complete the cycle. The nount of heat transported depends on the latent heat of the working id. Large amounts of heat can be transferred with small working id mass flow rates and small temperature differences.

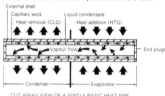

External shell
Capillary wick   Liquid condensate
Heat removal (CLG)   Heat addition (HTG)
Vapour flow   End plugs
Condenser ——— Evaporator
CUT AWAY VIEW OF A SINGLE BASIC HEAT PIPE

FFICIENCY:

he efficiency of the CN Heat Pipe Recovery Unit is a function of the ²/gas velocities, the number of heat pipe rows, and the number of flat ate or spiral fins. The optimum face velocity is 2.5 m/s (500 f.p.m.) d ideally units should be selected between 2 and 3.5 m/s (400 to 700).m.). At these velocity limits units may be selected with efficiencies nging between approximately 70% and 57%.

RESSURE DROP:

he pressure drop through the heat exchanger depends on the number fins, the air gas velocity and the number of rows of heat pipes. For ormal use it is recommended that units are selected in the pressure op range of approximately 75 Pa to 250 Pa (0.3″ to 1.0″ w.g.). When e exhaust air is saturated and condensation forms on the heating urface, the pressure drop will increase and the appropriate correction ctors must be used.

EMPERATURE RANGE:

N Heat Pipe Recovery Units may be used where air/gas streams are stween – 10°C and + 300°C. For higher temperature discharges, nbient air may be blended into the exhaust airstream to bring the final temperature within the above limits.

APPLICATION:

Limited Space:

The CN Heat Pipe Recovery Unit is well suited for applications where there is a limited amount of space of low headroom. Units may be supplied with finned heights of 400, 600, 800, 1,000 and 1,200 mm. For the optimum selection the height should be 50% of the length.

Fan Positions:

Both the supply and exhaust air fansets can be fitted in any position relative to the CN Heat Pipe Recovery Unit. Therefore if fan positions are fixed and cannot be moved the heat pipe is a good choice of heat exchanger.

AIRSTREAM INTERMIXING:

As the two airstreams are separated by the division plate there is no way that air or gas can flow from one side to the other.

CONTAMINATED EXHAUST AIRSTREAMS:

When the exhaust airstream contains low to medium quantities of dust/fibre the exhaust airside of the CN Heat Pipe Recovery Unit should be selected with a low number of fins, one fin per 2.5 mm, 3.2 mm (10, 8 f.p.i.). In these cases the supply airside can have a high number of fins, one fin per 1.8 mm, 2.1 mm (14, 12 f.p.i.). For very dirty applications filters must be used and several CN Heat Pipe Recovery Units are preferable to one large unit. For example a single 8 row unit may be supplied as 2 × 4 row units. Ductwork connections between the two units should be provided with access doors.

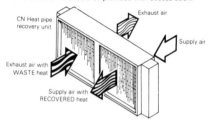

CN Heat pipe recovery unit
Exhaust air
Supply air
Exhaust air with WASTE heat
Supply air with RECOVERED heat

MOIST EXHAUST AIRSTREAMS:

When exhaust airstreams are very humid, moisture is likely to condense out, wetting the primary and secondary heat transfer surfaces. With process drying applications the transfer of moisture is not desired or required and the CN Heat Pipe Recovery Unit is a good choice of heat exchanger. A condensing situation on the exhaust airside increases the efficiency of the Heat Pipe Recovery Unit gaining some of the latent heat released.

HIGH STATIC PRESSURES:

In some cases where fan positions are fixed the CN Heat Pipe Recovery Unit may be subjected to high static pressures. The division plate is designed to accept high static pressures. A typical example is pre-heating boiler combustion air where the division plate is subjected to high static pressure from a forced draught combustion air fan.

IMBALANCED AIR MASSES:

For balanced air masses or air volumes at stp conditions the division plate separating the two airstreams will be central. If the masses or volumes are not equal the division plate can be supplied off-centre to ensure that optimum recovery is achieved. This feature is very useful when recovering heat from steam discharges.

Figure 11   Application data

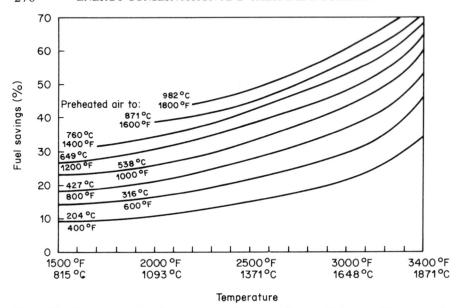

Figure 12 Graph showing how speed of rotor affects efficiency: Curwen and Newbery regenerator

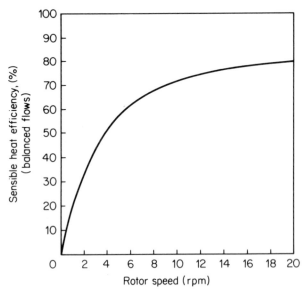

Figure 13 Graph showing how unequal airflow affects efficiency

standard of boiler efficiency being raised to as much as 90% by using the CN Heat Regenerator to preheat combustion air, a process which increases the theoretical flame temperature by approximately half the extent of the preheat.

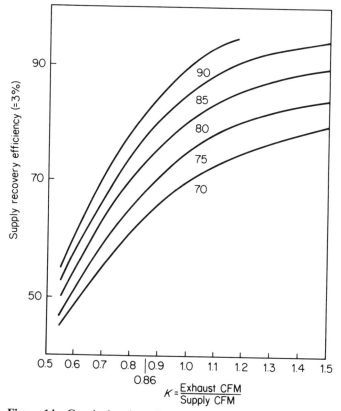

Figure 14 Graph showing affect of counterflow versus parallel airflow and loss of efficiency for Curwen and Newbery regenerator. Note: If the airstreams are arranged to enter from the same side (parallel airflow), the efficiency is reduced by almost half (30–40%) compared with counterflow. A counterflow airstream helps to keep the media clean, which is not the case for parallel flow

It has previously been stated that for solid and liquid fuels with approximately 25% excess air, the fuel saving is some $2\frac{1}{2}$% for each 100°F preheat given to the combustion air, which is more than the equivalent of the heat supplied as preheat owing to the reduction in the flue-gas quantity. With heat recovered from flue gases, the fuel saving is

$$100\% \left(\frac{\text{Drop in stack gas temperature}}{\text{New stack temperature minus ambient}} \right) \% \text{ of the original consumption}$$

It is a fairly simple matter by using a regenerator to reduce the flue-gas heat losses from a gas-fired boiler by transferring the heat from the hot flue gases to the cold combustion air. If the combustion air is preheated in this manner, less fuel is used in heating the air within the furnace and

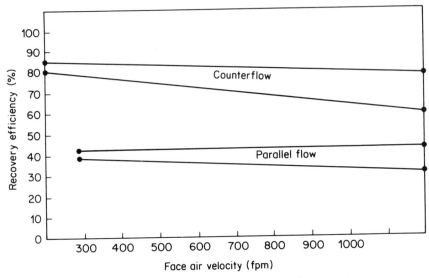

Figure 15   Fuel savings by preheated combustion air

more becomes available for useful heating—a process which, of course, constitutes a considerable gain in thermal efficiency. The various devices used to achieve this gain are called heat exchangers, air preheaters, recuperators or regenerators. We are dealing with the method of rotary heat regeneration and Figure 15 shows the percentage of savings in fuel resulting from using preheated air for an appliance with the same temperature throughout. Precautions which must accompany the use of air preheaters are, of course, larger burner/air openings to accommodate larger volumes of heated air and refractory life may be affected by the change in temperature levels. It is unfortunate that in this country there are a very small number of burners which will accept air above ambient temperatures. Figure 20 shows the CN Heat Regenerator fixed in a horizontal position, taking heat from a boiler plant where the recovered heat can be used for preheating combustion air or space heating purposes. Where stack temperatures are reduced to dewpoint levels, stacks should be manufactured from stainless steel or from modern glassfibre materials such as Elkalite. The regenerator will, of course, be made from stainless steel with stainless-steel heat exchange medium and the usual precautions for draining the regenerator should be made in accordance with the maker's recommendations.

The advantages claimed for combustion air preheating can be outlined as follows:

(a) The recovery of heat from the flue gases reduces the heat losses from this source. It has been proved that every 30–35°F reduction in the flue-gas temperature results in an approximate fuel saving of $1\frac{1}{4}-1\frac{1}{2}\%$.

(b) The flame temperature in the combustion space is raised, with a consequent greater rate of heat transfer by radiator.

(c) With the higher flame temperature, the fuel can be burnt with less excess air.

A heat regenerator is continuous as long as the burner plant is working and the control of the regenerator should be tied in electrically with the boiler/burner control system.

13. POSSIBLE SOURCES OF USE FOR HEAT-RECOVERY EQUIPMENT

Typical heating, ventilating and air conditioning applications

These apply to mechanically ventilated buildings of all types, including:

Offices Hotels
Laboratory exhausts Hospitals and nursing homes

Figure 16   Domestic CN Regenerator

Cinemas
Theatres and auditoriums
Public buildings
Swimming pools
Sports centres and indoor rinks
Public houses
Restaurants and commercial kitchens
Schools and universities

Banks
Flats and high-rise buildings
Supermarkets
Computer rooms
Research laboratories
Industrial buildings and work
  areas
Agricultural buildings

Typical Industrial Applications

Paint-drying ovens
Food dryers and process plant
Grain and grass dryers
Plasticizer curing ovens
Foundry furnaces
Paint dryers
Film dryers
Laundry dryers

Dehumidifiers
Food processing ovens
Pharmaceuticals
Rubber manufacture
Cosmetic manufacture
Oil industries
Paint manufacture
Boiler burner preheaters

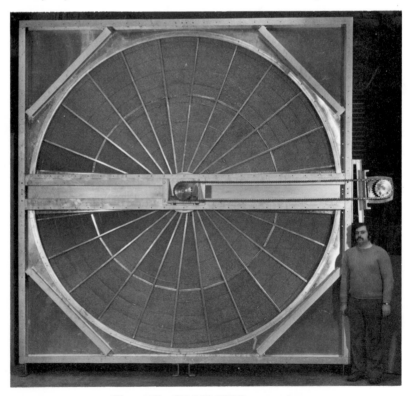

Figure 17   CN 100 TH Regenerator

Heat-treating ovens	Breweries, meat products, etc.
Paper dryers, pulp mills	Glass manufacture
Textile ovens and stenters	Car manufacture
Chemical dryers	Aircraft
Paint spray booths	Plating tanks
Ceramic kilns	Pollution control
Ceramic dryers	Mineral processing

Figures 16–21 show various regenerators and applications.

14. CN HYDROFIN WASTE HEAT BOILER

The CN Hydrofin Waste Heat Boiler is designed and manufactured for the recovery of waste heat from furnaces, kilns, engine exhaust systems, ovens, incinerators, and all processes discharging high-grade heat. It is suitable for a wide range of temperatures and working conditions.

The economy and efficiency of a wide range of industrial processes is greatly improved by the recovery of waste heat and energy savings of up to 50% are possible. The CN Hydrofin Waste Heat Boiler is suitable for steam or hot water and operates with inlet temperatures up to 700°C, standard range 250–500°C.

Figure 18   CN regenerator fitted with fan-assisted purge and spray cleansing nozzles for use with contaminated gas flows

Figure 19   Section of rotor of CN regenerator and automatic cleaning nozzles

The modular concept can be easily applied to the standard range of CN Hydrofin Boilers with the same advantages of all modular boiler plant.

When used in conjunction with internal combustion engines, the CN Hydrofin Waste Heat Boiler will increase overall thermal efficiency by 15–20%. They are suitable for horizontal or vertical inlet and outlet connections (see Figures 22 and 23).

Rated output is dependent upon volume and temperature of the exhaust gas and whether steam or hot water is generated. It is possible to calculate the amount of energy saved and the appropriate fuel cost reduction for a given application as follows:

Figure 20   CN regenerator when applied to two gas boiler flues for heating factory from boiler flue gases

$$\frac{\text{Btu/lb steam} \times \text{lb/h steam capacity}}{\% \text{ efficiency of fuel-fired boiler} \times \text{Btu/gall or cu ft of gas used}}$$

$\times$ cost per gall or cu ft of gas

$=$ fuel cost per hour

Net heat recovery gains are high with the CN Waste Heat Boiler range because they have very low pressure drops, their light weight reduces installation and time costs, and they are suitable for a wide range of gas types, volumes, and temperatures. They can be preassembled for quick assembly of heat exchanger.

Typical applications are:

Internal combustion engines (where they also make good silencers)
Gas turbines   Glass manufacturers
Foundry furnaces   Heat Treatment ovens
Kilns

It is almost impossible to find a single area in a factory where there is more scope, increased profit potential, and personal satisfaction than in the recovery of waste heat from exhaust gases, but the waste heat recovery

Figure 21 Standard CN Type TH Regenerator for air-conditioning applications

equipment must be correctly sized and engineered. Our 25 years of specialized heat-recovery experience is available for you to use. Rapid heat transfer, strength, ease of cleaning of all heating surfaces, and at lower initial cost, all put the CN Hydrofin Waste Heat Recovery Boiler in a very special class.

The transfer of heat must be regarded as a science; the same rules apply to everyone. There are very many technical books on the subject of heat transfer, but the universal formula is simple, and we like to keep it that way. It is:

$$Q = h \times A \times \Delta T$$

where Q = heat transferred, h = heat transfer rate, A = surface area, and

Figure 22 Typical waste heat boiler for horizontal or vertical mounting

ΔT = temperature difference. This will apply to flue gas transferring heat to a boiler tube and the transfer of heat from tube to the water. The heat passing from the tube to the water must obviously equal the heat input to that tube from the combustion products. That is just a little too simple, as the abilities of fluids to transfer heat are not equal.

Boiling water has a clean heat transfer coefficient of 6000 Btu/h ft^2 °F, but the flue-gas heat transfer rate is limited from 3000 to 30 000 Btu/h ft^2 °F. The high heat transfer efficiency of the CN Hydrofin Waste Heat Boiler is assured by the use of purpose-made finned tubing with extended heating surfaces which increase the total rate of heat transfer by

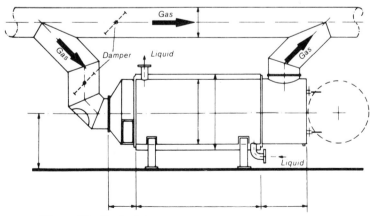

Figure 23 Typical duct connections for waste heat boiler

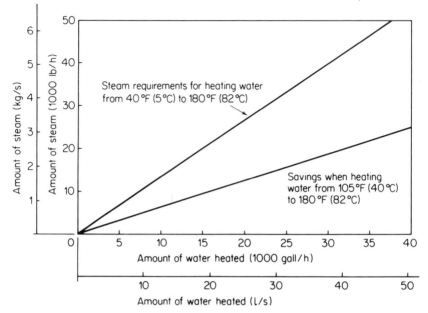

Figure 24   Steam saving (100 psig, 6.9 bar) chart when recovering heat from warm waste water

almost $3\frac{1}{2}$ times and at the same time allows for easy cleaning due to the wide spacing of the fins.

15. CN TUBEX WATER TO WATER HEAT RECOVERY UNIT

One of the major costs in running a laundry or dye house is heating the water used for all the processes. Your continuing payments for water, fuel, and the running costs of maintaining water heating equipment when geared to inflation is a never-ending headache.

As much as 20–40% of the money used for these purposes is wasted by the loss of the heat contained in the dirty water going down the drain, and on top of that you have to pay high sewerage disposal charges. In the majority of cases it is not economical to re-use the dirty water, but a considerable amount of heat contained in it can be used.

To make a quick assessment of what savings can be made from reclaiming the heat energy contained in the waste water, multiply the total cost of producing the hot water by three-tenths and the answer is the approximate amount that this form of heat recovery will save you because the installation of the CN Tubex Water to Water Heat Recovery Unit will save approximately 20–40% of the water heating costs, and for the average laundry, this will equal 30% savings on the overall fuel costs.

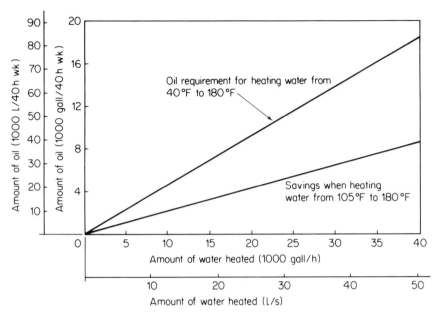

Figure 25   Oil saving chart when recovering heat from warm waste water

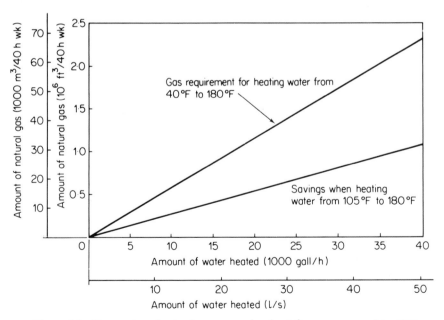

Figure 26   Gas saving chart when recovering heat from warm waste water

The graphs in figures 24, 25, and 26 show the savings that can be expected.

The capacity of a heat exchanger depends upon the quantity of the temperature rise it has to handle. When cold incoming water is preheated by heat recovery, your existing heater will handle your requirements with ease, and you will probably get 20–40% more hot water from the same heater with the same heat input. If your existing plant (including boiler) is overloaded, you will soon recognize the importance of heat recovery.

The payback period will be from 1 to 3 years. An opportunity to make substantial savings exists and should be investigated, and a careful analysis of estimated savings should be balanced against the investment required to obtain these savings. There are a number of engineering decisions that must be considered before the scheme is regarded as being viable. Any waste water discharged at a higher temperature than the incoming cold water will, of course, represent some waste, but to be worthy of reclaiming, this must be available for collection at a sensible cost relative to the volume which should be adequate. It is not usually sensible to try to recover heat from a source less than 90°F, but the incoming volume can be as small as 10 gall/min of fresh water. There is a further bonus that must not be overlooked. Many Local Authorities are starting to enforce the legislation that exists which prevents the discharge of hot water above 140°F into public sewers. A heat-recovery system can reduce the final discharge temperature so that it is acceptable by the Local Authority.

It cannot make sense to anyone to discharge heat down the drain anymore than it does to allow it to escape to the atmosphere via an extract ventilating duct.

Energy Conservation and Thermal Insulation
Edited by R. Derricott and S. S. Chissick
© 1981 John Wiley & Sons Ltd.

CHAPTER 11

Energy conservation: its influence on future building design

A. C. Hardy

University of Newcastle upon Tyne

The United Kingdon has a temperate maritime climate, so that externally it is rare for the outside temperature to be high enough to cause thermal discomfort. Often, in summer, the internal thermal conditions within build-ings are unacceptable due to high temperatures brought about by either internal heat gains or the admission of solar radiation. In other words, while buildings may be capable of modifying the external climate in winter they do not necessarily adequately modify the summer conditions, with the result that energy is required to suppress the peak summer temperatures. To design buildings for low energy consumption involves not only con-sideration of the heat loss situation but also the heat gain.

From a thermal point of view, there are three basic building types which are determined by their total internal heat gains. The first type is a build-ing with the minimum of internal heat gains, where its major energy requirement is for space heating and the thermal design of the building is such that it does not require any energy for summer cooling. The second is a building with moderate internal heat gains, where again the major energy consumption is for space heating but it requires a proportion of summer cooling by the use of mechanical ventilation. The third type is where the internal heat gains are such that the building needs air conditioning and therefore the major energy consumption is for cooling rather than heating.

The factors which determine the thermal building type are as follows. The function of the building will determine its internal heat gains in rela-tion to density of occupation, artificial lighting during daylight hours, and installed equipment. The building type will also decide to some degree its plan form, which may be modified by site constraints. Basically there are three types of plan form. Where a building is required to be daylit, unless

283

it is single-storey, room depths are limited to 5.5 m, if the vertical glazing is not to be an excessive area of the external facade. This gives a plan width of about 14 m. Where the functions of the building or site constraints require room depths in excess of 5.5 m, unless it is a single-storey building where rooflighting is acceptable, the building will require supplementary lighting and mechanical ventilation. In this situation thermal balance calculations have to be made to determine if mechanical ventilation is adequate to suppress peak summer temperature. With supplementary lighting, rooms can be up to 10 m deep. Where the building function requires large open-plan areas deeper than 10 m, it is most likely to need air conditioning, but this is dependent on the internal heat gains. Plan areas up to 30–40 m are possible, but the maximum plan dimensions are usually determined by the need for the occupants to be provided with an adequate view out.

Plan forms are therefore related to the environmental systems to be used —daylighting and space heating, supplementary lighting and space heating with mechanical ventilation, or permanent artificial lighting with vision windows and air conditioning.

Where the major energy requirement is for space heating, heat loss is the most important design parameter. High thermal insulation values, an airtight form of construction to minimize infiltration heat loss, and minimum fresh-air ventilation rate are the main design factors. As walls always have a lower thermal insulation value than roofs, because they always contain a proportion of glazing for a given plan area, calculations are necessary to determine the most thermally efficient building form. As the thermal insulation values of the opaque elements are increased, so the glazed area becomes of greater importance to the conductive heat loss calculations. With high insulation values in walls, the difference between the internal surface temperature of the wall and the interior of the glazing increases, with the result that the risk of thermal discomfort in winter due to downdraughts and cold radiation is much greater. With single glazing this requires that the heat input into the space is under the window. Research into the thermal discomfort caused by windows suggests that when walls have high thermal insulation values this problem can be overcome by the use of double glazing. This avoids the need to have the heat input under the window. The new Building Regulations Part FF require a wall insulation value of 0.6 W/m^2 °C for most building types and they also state maximum permissible areas of single glazing for specific building types. In all situations the design of glazing should be carefully considered so that it is the practical minimum in relation to the window function. In general, rooflighting should be avoided, whenever possible, as it has a detrimental effect on thermal comfort in winter, problems associated with solar heat gain in summer, and high capital cost.

Field research has shown that different types of building construction vary widely in their infiltration rate. All forms of prefabricated construction tend to have a high infiltration rate. This is due mainly to the great lengths of exterior jointing which, although weatherproof, are far from airtight. Most wet forms of construction—brickwork, blockwork or cast *in situ* concrete—have the lowest infiltration rate. It is important to detail the joints between all external building elements to ensure that these are airtight.

The use of windows as a means of ventilation is a very inaccurate means of controlling ventilation rate. While windows may be openable for ventilation outside of the space heating season, other means must be found to control the fresh-air ventilation rate during the space heating season. This may be by adjustable ventilators to provide either natural cross or stack ventilation. They must be designed and located so that they do not cause draughts. All openable windows must be gasket-sealed as well as all external doors. The main entrance doors to the building must be double with an adequate vestibule.

The thermal performance of the external walls, ground floor, and roof must have detailed consideration. The location of the thermal insulation material has to be related to the time pattern of occupation of the building as this will affect the thermal speed of response of the building interior. Predictions have to be made regarding the risk of interstitial condensation within building elements, which requires that their moisture vapour porosity is known. Attention should also be given to ensuring that improved insulation values do not increase the thermal stress in external building materials. This particularly applies to flat roof construction.

Thermal time lag of the external building envelope is also an important design requirement, in particular as a means of suppressing peak summer temperature. As thermal time lag is determined more by the thickness of the construction than by its density, thin forms of construction should be avoided, otherwise peak solar heat gain will be coincident with peak external air temperature and the effectiveness of mechanical ventilation as a means of cooling will be lessened. A thermal time lag of five hours can be attained with a thickness of 150 mm of material.

The thermal capacity of the materials within the building will determine its thermal speed of response and therefore also its preheating time. It would appear from predictions made on nonsteady-state thermal performance that this factor has little effect on space heating energy consumption. Studies of the energy consumption of both lightweight and heavyweight houses show that, provided the installed space heating load is twice the design heat loss for steady-state conditions, the preheating times are similar, although they start from different base temperatures due to the varying rates of overnight cooling.

The internal thermal capacity can have an influence on the utilization of incidental heat gains, in particular the solar heat gain through windows. As the conductive and ventilation heat loss of a building is reduced, so the incidental heat gains become an increasing proportion of the internal heat demand, and the risks of overheating through solar gains are enhanced. If an interior is heavyweight, incidental heat gains can be absorbed and reradiated later as useful heat, so lessening the risks of overheating. With a lightweight, quick-response interior the risks of overheating due to solar gains are increased. This can result in the opening of windows during the space heating season.

It is vital in all low-energy buildings that the heat input into each room is accurately controlled, otherwise incidental heat gains will not reduce space heating energy consumption but only increase internal temperatures.

All the factors discussed so far relate to all building types. In the daylit building, because it has the greatest glazed area, orientation of the glazing is an important design factor with respect to both the utilization of solar heat and the avoidance of unacceptable peak summer temperatures. This is particularly important when the glazing is unobstructed.

Where predictions show that it is necessary to use mechanical ventilation because of either the plan depth or the suppression of peak summer temperature, attention must be paid to the design of the system to minimize its energy consumption. Field research has indicated that the factors determining mechanical ventilation energy consumption are duct sizing, fan power, and the percentage of fresh air. Ducts should be as large as possible to reduce fan power. Distribution systems should be as straight as possible to reduce frictional losses in bends. The system should have adequate controls to be able to minimize fresh air during the space heating season. Where a system is designed to operate on two speeds, low speed for space heating and high speed for summer cooling, the fan power required for summer cooling should be checked as it has been found that some systems consume more power in the fans than would be used if the low speeds were maintained in summer with the addition of a cooling coil. In a number of smaller buildings with sufficient thermal capacity, summer cooling can be attained by running fans on a thermo-time control. This senses the internal and external air temperatures and runs the fans in the early morning to drain out the heat stored in the building structure from the previous day.

Should it be found that peak summer temperature control demands high fan power, consideration should be given to improving the thermal performance of the building before investigating the need to provide cooling. Situations where a cooling plant is needed for only a small proportion of the year should be avoided.

The decision on designing a building for air conditioning should be made from the outset. Heat-recovery air conditioning, which is usually the most efficient in energy utilization, requires that the recovered heat can be used. It appears to be uneconomic to store recovered low-grade heat.

It is possible to improve the thermal insulation values of an air-conditioned building to the extent that it can increase the cooling load. Investigations suggest that for the United Kingdom thermal insulation values should not be increased beyond 0.6 W/m^2 $°C$ for walls and roofs. The heat recovered from the compressor usually has a lower cost than that provided from a boiler, and therefore great care has to be given to the control design of a heat-recovery air-conditioned system. Design studies have indicated that a basic rule for the design of the energy distribution system is that air distribution systems should be horizontal and fluid distribution systems vertical. Situations have been found where free cooling has not been as efficient in energy utilization as actual cooling. This was due to the need to provide heat from a boiler to temper the incoming fresh air, whereas less heat energy and fan power was required to recover heat from exhaust air and the compressor to provide the heat. In some instances the maximum cooling energy was between 9.00 and 9.30 a.m. This was due to the air-conditioning plant heating the building to design temperature before occupation and then the incidental heat gains required cooling immediately after occupation.

A building designed for air conditioning must have adequate plant room space to ensure ease of access for maintenance and plant replacement and adequate ceiling voids to ensure a low-fan-power low-velocity distribution system. No plant should be placed above suspended ceilings, even if access trap doors are provided. It has been found that frequently there is inadequate maintenance on such plant and that it can malfunction for long periods. Access to such plant by maintenance engineers often causes damage to ceiling decorations. It is important for reasons of energy conservation that there is accurate control of each temperature zone and that the plant as a whole has a dependable optimum start thermo-time control. Attention should be paid to ways in which recovered heat can be used to the maximum to minimize the necessity of using energy to throw away heat at the cooling tower.

In building types which need energy for the suppression of peak summer temperatures, it is important that any artificial lighting should be designed to a high efficiency, otherwise this adds to the cooling energy required.

In all building types, it has been found that it is very inefficient to provide domestic hot water from the main space heating system, outside the space heating season, due to low efficiency of the boiler at low and intermittent loads and the high distribution losses from central plant. Domestic hot water should be provided by local plant whenever possible. Domestic hot water in air-conditioned buildings may be supplied from recovered heat, depending on the distribution losses involved.

Studies of the wide variations in building energy consumption for similar building types have shown that the major factor influencing building energy consumption is not the thermal performance of the building but the design and control of the environmental services.

In the design of buildings today, which will have an economic life of at least 60 years, during which time the world fuel situation will change a great deal with regard to availability and cost, the design of plant rooms should take into account the space requirements for major changes in fuel systems. Unless plant rooms have sufficient space for changes, for example, from gas to oil or gas to solid fuel or possibly the use of electrode boilers on off-peak tariff, with hot water thermal storage, the building may be restricted in its fuel choice to a high-cost fuel during its economic life.

Sufficient information is available for low-energy buildings to be designed today, but it will usually require a higher capital cost, possibly in the region of $+5\%$, and larger spaces for the provision of high-efficiency thermal plant.

Energy Conservation and Thermal Insulation
Edited by R. Derricott and S. S. Chissick
© 1981 John Wiley & Sons Ltd.

CHAPTER 12

Designing for energy conservation: the way forward

Alan M. Berman

Greater London Council

One answer to the question, 'How can energy be conserved?' is 'Turn off the heating.'

The approach to energy conservation above reveals immediately the need to define very carefully the objectives we are seeking to achieve and the constraints we wish to impose before looking at the fine detail of the methods to be used (Greater London Council, 1974). Considerable fruitless debate about these methods has occurred where different objectives or constraints are involved leading to diametrically opposed but equally valid approaches. Such a difference occurs between UK legislation and EEC requirements, where in the UK the objectives are generally to take such measures to save energy as are cost-effective but in the wider European community saving energy is an objective in itself, whether the methods employed are cost-effective or not.

It is a worthwhile preliminary, therefore, to look at a few possible objectives for energy conservation. To start with, the EEC approach to conserve energy develops from a catalogue of the existing resources of fossil fuels, postulates a rate of use, and hence extrapolates a time by which they will be exhausted either worldwide or nationally. The argument may be further developed to indicate that conserving energy will buy time for the sciences and technologies to develop their fast breeder reactors, their thorium reactors or eventually hydrogen fusion. The view is overlaid sometimes with the consideration that we really ought to leave some fossil fuel for succeeding generations or, more emotively, for our children, and this argument can be summed up as the posterity consideration. This does, however, have a strong counter-argument which briefly says, 'What has posterity ever done for me?'

An extension of the dwindling-energy argument is what has come to be called the 'Club of Rome' approach. It indicates that all raw material resources are being exploited at an ever-increasing rate and proposes a no-growth society to extend the availability of raw materials. This view to some extent has its roots in Malthus (1797) who considered that exponential population growth combined with a linear growth in the supply of food was a formula for disaster. This logically leads to a further modified objective which is to conserve nonrenewable resources and to husband these with forestation schemes, etc.

For some time the UK has been beset with balance-of-payments problems; in simple terms we import more than we export, and oil and oil products have been a significant part of our import bill. This can lead to a further objective which is to improve our balance-of-payments position and can be achieved by reducing energy imports from particular parts of the world and substituting them with indigenous resources or with energy from alternative areas where balance of payments do not present a problem. The result of such an objective could lead to a policy of the rapid exploitation of United Kingdom coal, oil, and gas reserves which would be quite the reverse of a conservation policy.

Some parts of the world, and unfortunately this includes several where oil is found, are politically unstable and the threat of a disruption in the supply of oil is of considerable economic and strategic concern. This concern leads to an objective of requiring a stable supply of energy independent of foreign political considerations.

Coalmines and power stations in the United Kingdom are themselves not exempt from interruption through industrial disputes and strikes, so the same consideration applied above can be slightly modified to emerge as yet another objective, to have a supply of energy independent of local political considerations. This objective leads to arguments against large centralized power supply sources and towards local community resources.

There is also an ecological argument, that by using energy and indeed other resources profligately we are adversely affecting the environment in one way or another. The worries concern sulphur dioxide, carbon dioxide and the greenhouse effect, nuclear power and radiation hazards, and the destruction of forest and despoliation of land etc.

Another modification of the ecological view considers the urban environment to be an intrusion and deplores centralized facilities. This view seeks to have each unit independent of central energy resources and looks to autonomous housing where the unit puts back into the environment what it takes out, utilizes solar heat, processes its own sewage, etc., and has the broad objective, therefore, of leaving the ecology undisturbed in addition to having constructions which are independent of central energy resources.

For our purposes I have adopted here the very mundane objective that because we are paying too much on our energy bills and since we want to avoid the wasteful use of energy we can define as our objective 'to take whatever measures are cost-effective to reduce the use of energy'.

It should be noted that at the time of writing and for the foreseeable future the use of solar plate collectors would not comply with this objective since they are not, in the present state of technology, sufficiently cost-effective.

Not only do all these differing objectives lead to differing approaches to the practical problem of the construction and insulation of buildings but the Government's energy policy at any particular time can also greatly affect the situation by applying subsidies in one area rather than another or by imposing regulations with respect to the energy consumption in buildings.

THE CONSTRAINTS

It is not difficult to achieve our objectives in energy saving if we cast aside all constraints and decide to accept standards vastly differing from those that are current. To make any sense of design requirements, we are therefore required to impose some constraints on the energy-saving programme, and the following constraints seem to be reasonable although in other circumstances they may need to be modified.

We will probably decide that the first constraint is that we do not wish to reduce our standard of living significantly and this implies that our gross national product must be maintained or increased. Since in industrialized countries the gross national product is closely correlated with industrial fuel consumption, this will clearly affect the level of savings to be made in the industrial sector. A second constraint could be that we do not wish to reduce our comfort standards, that is our water temperatures, room temperatures or perhaps ventilation rates and illumination levels in our buildings. A third constraint could well be that we do not wish materially to alter our habits, that is to wear more clothes or to revert to Victorian long woollen underpants or to switch off the heating and go to bed earlier. In other words, we want to conserve not only fuel but all those other evolved living conditions that we now enjoy.

There is also the need to consider whether we require rapid short-term solutions or are concerned with twenty years hence.

New construction gives us the widest scope for improving energy usage, but at current construction rates it would be some decades before any changes in new construction would have a significant effect on the total building stock. If we are therefore considering a short timescale of a decade or two, then energy use in existing buildings will need to be the prime target for improvement, and in addition to any legislation applying to new construction there would need to be retrospective action aided by subsidy.

THE MODEL

The situation of coping with a harsh climate with severely restricted fuel supplies is certainly not new and other societies have attempted to do this in various ways with more or less success. In the situation we are faced with, it would be no bad thing to look at the methods employed by those societies which have succeeded in this adaptation and to look more closely at the evolved successful conclusions.

Arctic conditions are amongst the most onerous in winter (for example, Fort Yukon in Alaska has recorded temperatures of −71 °F in winter and 100°F in summer) and the availability of fuel, food, and construction materials in winter is extremely limited. North American Indian societies using skins and tents were unable to adapt to this climate and suffered ill health and dwindling populations, but the Eskimo societies flourished and a look at their methods of construction with the application of some of the

Figure 1   Igloo under construction. (Courtesy of the Canadian High Commission)

lessons learned could be rewarding. It is not widely known what conditions are like inside an igloo in an Arctic winter and the following description by Steffanson (1921) might come as a surprise.

When an Eskimo comes into such a house as the one in which I lived in 1906 to 1907, he strips off all clothing immediately upon entering, except his knee breeches and sits naked from the waist up and from the knees down. Cooking is continually going on during the day and the house is so hot that great streams of perspiration run down the face and body of every inhabitant, and are being continually mopped up with handfuls of moss or of excelsior, or, according to later custom with bath towels; and there is drinking of cup after cup of ice water. At night the temperature of the house will be only 10 or 15 degrees lower; or if it drops more, people will cover up with fur robes instead of sleeping nearly uncovered.

So where have we gone wrong? The igloo (illustrated in Figure 1) is an example of an ideal in energy conservation and is a useful model with which to compare modern constructions.

SHAPE

The shape of the igloo is hemispherical and we do not have to reach too far back into our school mathematics to remember that the sphere possesses the minimum surface area for a given volume and that the section of the sphere, that is a circle, possesses the minimum perimeter for a given area.

Figure 2 illustrates a few shapes of the same area and indicates the advantage of circles, and rectangles approximating to a square configuration, over the 'H' like shapes and flat slabs.

Where volume is the consideration, the construction 40 units high on each of these shapes would all have volumes of 40 × 40 cubic units and surface areas in the same ratio as their perimeters, varying from an area of 880 units in the case of the cylinder to 1760 units in the case of the 'H' shape, plus of course a roof area of 40 units in each case. However, a major saving comes with a hemisphere of the same volume which would be 9.1 units high and have a surface area of 525 square units, including the roof; a multistorey igloo would then have about one-third of the surface area of an 'H' or slab shape of equal volume.

The concept is simplified if we look at a large block of material and cut it in half, and by so doing create two new surfaces, and each time it is subdivided into smaller units more surfaces are created. The minimum

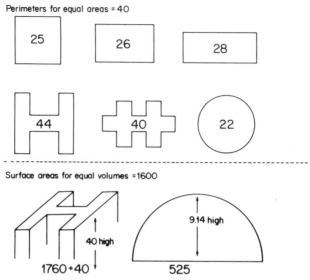

Figure 2 Relationships between perimeters and area, and between surface areas and volume

surface area, and hence minimum energy loss in a building of that shape, is achieved by creating the largest cube, or better still hemisphere, incorporating the total volume of material.

Clearly, it is not often practical to have a hemisphere, or even cubic configuration, for it may present very deep rooms if the cube is large. This objection is partly overcome when the cube is subdivided into several dwelling units sharing internal compartment walls and with a common roof, but inside an external cubic shell.

The discerning will note that when we have a terrace of cubic dwellings beside each other, or a row of cubic semidetached houses, these again begin to present a larger surface area–volume ratio than would be the case if all the units were housed in one gigantic cube, and if it embraced a whole terrace or even a whole town. There is, of course, an optimum size cube above which the difficulty becomes not the conservation of energy but the provision of light and ventilation and the dissipation of energy from the structure.

SURFACE CHARACTERISTICS

The surface of an igloo is smooth and without protrusions. If we wish to dissipate heat from a system, a well established method in engineering practice would be to provide it with fins, and providing a building with such fins if they are not formally isolated from the structure will result in

Figure 3   A view of an estate

Figure 4   A critical view of an estate

excessive unnecessary energy loss in winter and undesirable energy gain in summer. The illustrations above provide two different views of the same block of flats. The first (Figure 3) shows the letting agent's view and the second (Figure 4) gives the view of the 'energy auditor'; for comparison the heat-dissipating fins designed into an air-cooled engine are also shown (Figure 5). Design problems of this type are well known and, one assumes, will be avoided in the future, but less obviously a similar effect on a smaller scale is provided by some familiar concrete treatment shown in Figure 6, where the small fins provided on the concrete surface enhance the heat loss. To what extent a large exposed aggregate finish in concrete also contributes to heat loss has not, to my knowledge, been measured but may have a significant effect. The objective then must be to keep our surfaces as smooth as possible and without unnecessary projections.

COLOUR

It is not insignificant that snow is white, if it were any other colour it would more quickly melt in direct sunlight. This observation is not irrelevant in conventional construction.

In the next generation of structures, and in the improvements made to the existing stock of buildings, the level of thermal insulation will be considerably increased on current standards, and in consequence exterior

Figure 5 A design to dissipate energy

Figure 6   A design in concrete

weathering surfaces could be subjected to a much wider range of temperature. For example, the placing of roofing felts directly onto thermal
insulants of low thermal capacity results in the weathering surface reaching
a higher temperature in direct sunlight than a similar membrane laid
directly onto, say, a concrete substrate. Operating at a higher temperature
will result in more rapid deterioration of the surface materials where that
deterioration is chemical in nature, such as oxidation, and the general rule
of thumb is that a rise of about 10°C will double the rate of deterioration,
and thus halve the life of the component. In addition, rapid changes of
temperature can result in rapid movements known as thermal shock, with
consequent cracking of the construction materials. These effects can all be
reduced by the provision of a solar reflective treatment, but some of the
properties associated with colours are occasionally overlooked.

The three most relevant properties are reflectivity, absorptivity, and emissivity, and these properties are listed below for various colours. While there is a clear understanding of the absorptivity and reflectivity in materials, the property of emissivity is frequently not considered, and metallic finishes are used to provide a reflective surface for solar radiation. While a polished metal surface reflects 90% of the incident energy, just as a white surface would, the metallic surface is unable to reradiate readily even the small percentage of energy gained, due to its low emissivity (about 0.05). This should come as no surprise, for anyone who touches a polished metal surface exposed to direct sunlight would find that it gets very hot, despite its very highly reflective surface, where an adjacent white surface would be comparatively cool. Black surfaces, because of their high absorptivity and emissivity would be much more responsive than white to changes in radiant energy, and will be more subject to thermal shock with greater maximum and minimum temperatures than either white or metallic surfaces. However, depending on the conditions, the metallic surface may operate at a higher average temperature, although not reaching as high a maximum as the black surface. These effects of colour (Shown in figures 7 and 8) have been introduced here to indicate the connection between the energy balance of the structure and the durability of its components, and to note that the effects of colours on particular systems will best be determined by an experiment in the same climate and environment as the intended construction. For, quite apart from the complexity of the various factors involved, there is the additional consideration of whether the white surface will remain white, or the metallic surface remain polished in actual performance conditions, which may, therefore, cause the actual behaviour to differ considerably from the theoretical calculations.

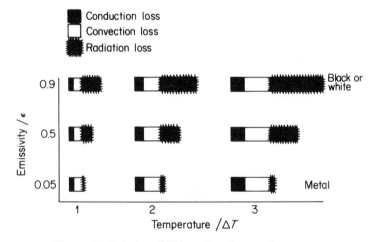

Figure 7   Relative abilities of surfaces to lose energy

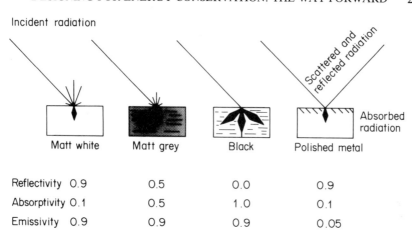

Figure 8   Input radiation balance

THERMAL CONDUCTIVITY AND THERMAL CAPACITY

Freshly fallen snow has a thermal conductivity of about 0.17 W/m K and compacted snow about 0.43 W/m K. Igloo snow is somewhere in between, say 0.3 W/m K and is used in blocks about 1 m thick. The material is of local origin and requires no transportation energy or manufacturing energy save that provided by the constructors' labour. This is an important point when we come to lightweight insulating materials, for the volumes involved are considerable and their transportation costs, which are volume- and not weght-related, contribute greatly to the overall cost of thermal insulation. It does provide a considerable cost advantage for those materials which are either of local origin, e.g. snow, wattle and daub, or thatch, or for those materials which can be transported in a dense form and expanded on site. Hence there is the advantage of those materials that can be foamed *in situ* or which can be expanded, like expandable polystyrene, on site with the appropriate equipment. There will doubtless be commercial interest in the future directed towards supplying dense materials which are processed in some way at the point of construction to provide thermal insulation.

Lightweight insulating materials generally owe their low density and insulating properties to the air entrained within their structure. This is achieved by containing the air in a lattice of fibre or cells and the insulating material inevitably has a large 'specific' surface area. By this is meant not the surface area which is determined by the shape of a block of material but the specific surface displayed by the sum of the surfaces of all the bubbles or fibres within the insulant. The large specific surface area has a profound effect on the fire behaviour of materials, increasing the rate of combustion and heat release and the rate of release of smoke and combustion products. This factor will no doubt be a severe constraint on the way in which many organic thermal insulants are used, and they will generally be confined to those applications where their surfaces are not exposed to a source of ignition.

COLD BRIDGES

While it may seem an odd way to make the point, there are no 'cold bridges' in igloo construction. By this is meant that there are no elements of construction of higher thermal conductivity than the building blocks, penetrating the skin of the structure. For example, there are no mortar joints or metal bars or solid concrete lintels such as we might have in a more conventional structure.

The microclimate close to cold bridges, because the adjacent air is at a low temperature, is one of higher relative humidity than elsewhere in the dwelling and mould is encouraged to grow here, and once established can spread by providing their own microclimates for survival.

Of course, it is much easier to deplore cold bridges than it is to eliminate them from a construction detail, but there should be increased emphasis in the future on the examination of prototypes in order to prevent cold bridges. Examinations of this nature can be carried out by the use of infrared scanning equipment such as the Aga Thermovision Camera, and a requirement for prototype structures to be examined in this way would be an eye-opener both to designers and prospective clients (Greater London Council, 1970).

SURFACE TEMPERATURE

Thermal conductivity is a most important parameter in considering energy loss through a structure, but it is not the most important when considering thermal comfort; here both thermal capacity and surface temperature play an important part.

The thermal resistance of the structure tells us something about the heat flow through the structural components, the thermal capacity tells us something about the energy necessary to heat up the components themselves, and the interior surface temperature tells us something about the comfort of the people in the dwelling. All these factors are influenced not only by the inherent thermal properties of the structural materials but also by their position in the structure, that is, whether the insulation is on the outside of the structure, within the thickness of the walls or on the inside. With the igloo, there is no such debate for the walls are isotropic in construction with the thermal conductivity uniform throughout the thickness of the wall; and of course it is not a structure which can be used to store energy within the walls for any excess energy in the form of heat would only cause the structure to melt.

Conventional buildings are unfortunately rather more complicated and we have the choice not only of what the thermal insulation values should be but of where the insulation should be placed in the structure, and these choices have given rise to a great deal of controversy of a kind that cannot easily be resolved because the parties on both sides of the argument are right to a degree but are considering different objectives and different data inputs.

The conclusions will differ according to the pattern of energy input to a structure, that is whether it is continuous or intermittent and whether the criterion adopted is one of energy efficiency for a given air temperature or one of comfort.

Structures of high thermal capacity are slow to respond to changes of energy input and this results in difficulties if they are intermittently heated. Thus structures incorporating thermal insulation as a sandwich between two heavy concrete layers will not perform well in conjunction with warm air heating systems used intermittently.

This combination will provide a suitable air temperature rapidly and will also be highly conservative of energy, but may well result in an uncomfortable environment caused by the cold surfaces of the enclosure and in mould growth and condensation.

This type of construction would, however, be satisfactory in a situation where the heating is continuous and has a large radiant element included, such that the surfaces themselves are heated before heat transfer to the air takes place and a desired air temperature is reached.

Of course, where the thermal capacity of the structure is very large, for example, a castle's wall, the system never reaches a steady equilibrium temperature.

To provide warmer surfaces in an intermittently heated structure of high thermal capacity, it is necessary to site the insulation on or near the inner surface of the structure, and this results in higher surface and air temperatures when the heating is on, and lower surface and air temperatures when it is off. This thermal behaviour contrasts with the smoother pattern where the insulation is within the wall structure, although the average temperatures may of course be about the same.

In order to describe the difference that the position of thermal insulation and the thermal conductivity and thermal capacity have on a structure, a parameter called the 'time constant' of the structure is used. The 'time constant' does not usually play a part in drawingboard calculations or in regulations, but is a most important concept in the design and in the matching of structure, heating, and climate, and can be regarded as the time taken for the temperature difference between the element of the structure being considered and its surrounding environment to change by approximately one-third.

COMFORT

From the foregoing discussion, it is clear that comfort is not directly related to energy consumption, and a brief explanation of what we need to be comfortable is useful.

Thermal comfort is generally concerned with radiant and convective heat exchange with our surroundings, and the contribution of conductive heat exchange can generally be discounted unless we are leaning against a hot or cold surface or are in wet clothes.

Apart from our clothing, which for this discussion we can take to be constant, the principal influences on radiant heat balance are the temperature, absorptivity, and reflectivity of the surroundings together with their configuration factor.

For a body in a room, the sum of the configuration factors of floor, ceiling, and walls would be unity. The configuration factor of each would be the solid angle which each subtended at the body divided by the total solid angle in a sphere. The need to concern oneself with the configuration factor when considering comfort in dwellings is small, but it does provide an explanation of why the closer one is to a cold surface such as a window the more significant its presence becomes.

A more important consideration is the temperature of the room surfaces and here the radiant energy exchange between the body and the surface is proportional to the fourth power of the difference of absolute temperature between them. This principle, known as the Stefan–Boltzman law, depending as it does on the fourth power of absolute temperature, is very sensitive to small temperature differences. Apart from the surface temperature, however, the reflectivity, absorptivity, and emissivity discussed earlier affect our comfort, and if walls are white or of polished metallic finish they will reflect back 90% or so of the incident radiation, but if black they will absorb it all. The igloo, with its white interior walls, scores again in this respect, although a skin covering is frequently used on the interior surface of the igloo.

This fur lining provides the low thermal capacity, high thermal resistance which thus warms up very quickly (i.e. has a short time constant) and substantially reduces direct radiant heat loss. However, if it were a minutely thin metallic surface it would be even better because of the low emissivity of the polished metal surface.

A minutely thin metallic coating is suggested, for otherwise the metal would conduct heat laterally throughout the area of the sheet and require a substantial heat input.

Thin metallic surfaces are used in this way in survival bags where a thin bag of plastics material with a metallic monolayer coating can keep a person warm by containing almost all the radiant heat by reflection and being unable to reradiate to its surroundings owing to its low emissivity. It should be noted that such a bag, even though able to keep a person warm in extremely low external temperatures actually provides little or no thermal insulation, that is resistance to conducted heat.

CONVECTION

Convected heat loss depends on the air temperature and flow rate of the air surrounding a body and also upon the humidity of the air. The humidity affects heat loss in two respects. First, the specific heat of air varies with its humidity, and hence the energy transferred to the air by heating to a particular temperature is increased with increasing humidity. Secondly, the amount of heat lost from the body through evaporative cooling will vary with the humidity of the surrounding air although at low humidities the body can compensate by reducing water loss.

Attention to air temperature is perhaps the consideration which designers are most aware of and heating systems are usually controlled to respond thermostatically to a preset air temperature and their output determined on the basis of air volume required to be heated and the U value of the structure.

VENTILATION AND EXTRACTION

While airflow rates affect thermal comfort, and on that basis alone still air (less than 0.1 m/s) would be preferred in cold conditions and allowance made to provide higher rates (1 m/s) in summer conditions, there are other considerations.

While ventilation plays a part in reducing summer temperatures, the principal needs of ventilation is to remove odours, carbon dioxide, and humid air, and this need may conflict with the desire to avoid draughts in winter.

Windows in exposed situations, such as high-rise buildings, are fitted with small ventilators but the airflow rates through them are so high that they produce an uncomfortable environment (or in other words a draught).

Figure 9 taken from wind velocity data shows that the air velocity of 0.1 m/s would be exceeded through a 2000 mm^2 ventilator in a high-rise building for about 95% of the time.

Sitting near this ventilator will be uncomfortable as often as not, but in an average room of say 40 m^3 the air change rate would be on average

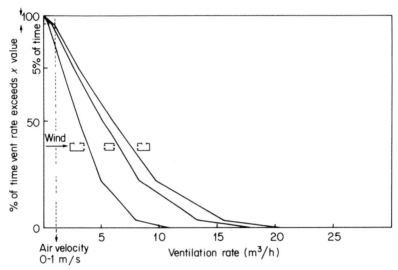

Figure 9   Ventilation through a 2000 mm² ventilator in a high-rise building

about one-eighth of an air change per hour, which is quite inadequate ventilation to remove odours, etc.

In elevated or otherwise exposed situations, ventilators must be carefully contrived if one is to achieve freedom from draughts combined with adequate ventilation rates.

The igloo heats air with a small blubber burner where the air enters through the entrance tunnel. The rate of ventilation is restricted by the total absence of any outlet save for a small hole in the centre of the dome to carry away combustion products from the cooking and to affect ventilation. The direction of airflow is thus regulated and generated water vapour from fuel and cooking extracted at the point of manufacture.

It is a considerable advantage to extract water vapour at the point of manufacture rather than to allow it to become distributed throughout the volume of a dwelling and then to remove it by ventilation or by air conditioning. Ventilation would obviously involve considerably greater volumes of replacement air to be heated to a comfortable temperature than that involved in replacement of the extracted water-vapour-laden air at the source of its production.

WATER

The moisture content of exposed masonry in the UK climate will of course depend on the density and type of masonry involved, but the pattern for water content shows a small daily cyclic change superimposed on a much larger seasonal cyclic change. In addition, there are large irregular changes caused by periods of heavy rainfall or dew.

The outer skin of a cavity brick wall may show daily changes of moisture content of 0.5–5% by weight and seasonal changes of 10–20% by weight with upper limits of about 25% by weight and lower limits of about 10% by weight.

It appears that the effect of the moisture in masonry that has been of principal concern regarding thermal properties is that of the change in thermal resistance with changes in moisture content. This is well documented and a relationship has been evolved from which the thermal resistance at any moisture content can be approximately deduced. Other effects have for some reason generally been neglected.

If the moisture content of the outer 100 mm skin of brickwork were to change by 0.5% by weight each day, what would the effect be? The weight of brickwork is about 1800 kg/m^3. The latent heat of vaporization of water is 2.5×10^6 J/kg, and to account for a 0.5% change requires $0.005 \times 2.5 \times 10^6 \times 180$ J/m^2 of brickwork. Since this occurs daily, the energy required is

$$\frac{0.005 \times 2.5 \times 10^6 \times 180}{60 \times 60 \times 24} = 26 \text{ W/m}^2$$

Thus, assuming only a 0.5% change by weight in moisture content, the energy involved is considerably greater than the design heat loss of 0.6 W/°C through the wall insulation, and even if this energy is largely provided by solar radiation and cooling the external air, there remains a probability that a great deal of energy is lost from a dwelling through evaporative cooling of the outer skin. The principle is not mysterious and has been used as a standard method of refrigeration in for example butter and wine coolers before modern refrigeration methods existed.

The chilling of the outer skin of a structure by evaporative cooling will increase the flow of energy through the structure whatever the thermal insulation installed, the heat flow being proportional to the difference in temperature between the inner and outer surfaces of the wall. For these reasons, the use of water-repellent treatments on walls or of impermeable outer skins would reduce the heat loss although the actual extent of the saving needs to be determined by experiment. Impermeable outer skins also require very careful detailing to prevent interstitial condensation within the thickness of the wall.

SYNTHESIS

The analysis of igloo-like structures is useful to provide an insight into a range of heat-conserving features and would no doubt lead designers to the beehive dwellings and public buildings found in 'primitive' societies.

It is of little use approaching the scientist with the wrong questions, such as 'what is the optimum window size for my building?' and receiving the answer 'zero'.

Energy saving must be recognized as a secondary requirement in buildings which have other primary requirements to fulfil. They have to function as schools or swimming pools or dwellings, and a synthesis of these primary requirements, with for example the lessons learned by the Greater London Council (1975, 1977a,b), from heat-conserving structures, is necessary.

We should consider shape and surface smoothness, water absorptivity, colour and emissivity, thermal capacity and heating systems, ventilation and extraction all as an integrated whole in building design. There is a danger that particular aspects such as thermal insulation will be singled out for isolated action and other areas neglected.

Prototypes of new construction should be monitored to search out cold bridges and areas of excessive heat loss and could even be certificated. A structure is usually a major capital investment and yet it comes without any prototype testing, any indication of its fuel consumption or of its maintenance requirements; would we accept a new motor car on that basis?

Insulation standards based on the heat loss per square metre of peripheral wall and window do little to regulate the shape of structures as indicated by the Greater London Council (1978), where an additional requirement to control the volumetric heat loss would go some way to influencing shape in an energy conservative manner.

A future standard of between 0.2 and 0.5 W/m² (minimum not average to avoid cold bridges) coupled with a volumetric heat loss of, say, 0.3–0.5 W/m³ would discipline the mind wonderfully towards the optimum shape if that is what is needed.

While the position of insulation will depend on the intermittency and type of heating and on the use to which a building is put, the presence of an insulating low-thermal-capacity inner lining, together with reflective surfaces of low emissivity, will enhance comfort levels.

Lightweight insulants prepared *in situ* from dense materials will reduce insulation costs and assist in the overall energy audit which takes into account the energy expended in producing the building including its insulation.

Many of these general requirements could be provided by double-skinned structures consisting of an inner sanctum of lightweight construction, perhaps foamed *in situ* between metal-foil-lined glass-reinforced cement or glass-reinforced gypsum shells. The outer weatherproof structure could be of glass and metal impermeable to water and providing a substantial air space between the inner and outer structures and an intermediate environment to reduce energy losses.

REFERENCES

Greater London Council, 1970, *Thermography*, GLC Development and Materials Bulletin no. 35.

Greater London Council, 1974, *Energy and Thermal Insulation*, GLC Development and Materials Bulletin no. 78.

Greater London Council, 1975, 1977a, 1977b, *Energy Conservation, GLC/ILEA Report*, GLC Development and Materials Bulletins no. 89 (1975), no. 109 (1977), no. 110 (1977) (three-part paper).

Greater London Council, 1978, *Energy Conservation in Dwellings*, GLC Development and Materials Bulletin no. 118.

Malthus, T. R., 1797, *Essay on Population*. Re-published, Dent, 1973.

Steffanson, V., 1921, *The Friendly Arctic*, Macmillan, London.

BIBLIOGRAPHY

Billington, W. S. (1967), *Buildings, Physics, Heat*, Pergamon, Oxford.

Bird, R. B., Stewart, W. E., and Lightfoot, E. N., 1960, *Transport Phenomena*, J. Wiley & Sons, Chichester.

Brinkworth, B. J., 1972, *Solar Energy for Man*, Compton Press, New York.

Central Policy Review Staff, 1974, *Energy Conservation*, HMSO, London.

Diamant, R. M. E., 1970, *Total Energy*, Pergamon, Oxford.

Fanger, P. O., 1972, *Thermal Comfort*, McGraw-Hill, New York.

Givoni, B., 1976, *Man, Climate and Architecture*, Applied Science Publishers, Barking.

O'Callaghan, P. W., 1978, *Building for Energy Conservation*, Pergamon, Oxford.

Porges, J., 1976, *Handbook of Heating, Ventilation and Air Conditioning*, Newnes-Butterworth, Sevenoaks.

Van Straaten, J. F., 1967, *Thermal Performance of Buildings*, Elsevier, Amsterdam.

Welty, J. R., 1974, *Engineering Heat Transfer*, J. Wiley & Sons, Chichester.

Energy Conservation and Thermal Insulation
Edited by R. Derricott and S. S. Chissick
© 1981 John Wiley & Sons Ltd.

CHAPTER 13

Energy and buildings at the Centre for Alternative Technology

R. W. Todd

National Centre for Alternative Technology, Machynlleth

INTRODUCTION

The Centre for Alternative Technology is a demonstration project which draws together ideas and methods forming sustainable alternatives to currently accepted, but shortsighted, approaches in various technological areas—mainly energy, food, and buildings. The Centre has a staff of 18 people; 16 people live on the site which is not connected to any mains services. The permanent exhibition at the Centre attracts more than 60 000 visitors each year. The bulk of this paper describes practical projects involving varying degrees of building insulation and the use of ambient energy sources, but first some comment on the philosophy underlying the work of the Centre is necessary to justify some of the design decisions.

The alternative technology movement is seeking a technology and lifestyle which is sustainable in the long term, taking into account finite energy, raw material, and food resources, the importance of fulfilling work for the individual, and our impact on the ecosystem. It is clear that the era of cheap energy is coming to an end due to depletion of fossil fuel, while a large fraction of the world still has an energy consumption per capita less than 10% of Britain's. It appears inevitable that world population will at least double before any stabilizing control can take effect and this increase will be mainly in the poor countries who are entitled to, and may ultimately demand, their fair share of the Earth's wealth of resources. If we plan to live within our fair share, the long-term future looks challenging but hopeful. If we continue to press for growth in our already excessive consumption, the future looks grim (Goldsmith, 1972; Schumacher, 1974; Thring, 1977; Todd, 1977).

We can make an important contribution to the developing world and to our own future stability by conserving fossil fuels as far as possible for use as chemical feedstock. The most environmentally acceptable way of doing this is by massive energy conservation measures and the widespread adoption of ambient energy systems. Short-term economic arguments are often used to reject such developments as solar heating; however, these arguments contain oversimplifications and must involve very uncertain guesses at the future to reach their conclusion. How, for instance, do you assess the 'value' of a solar heating system? Do you equate it to the cost of the energy it will save at current prices or try to guess the energy cost over the next twenty years? Or do you value it in terms of clean air, reduced nuclear risk, fossil fuels saved for more vital purposes in the future, or some security of energy supply in an uncertain future? It is clear that our motivation is rarely as simple as some economists might suggest.

Decisions made on an 'economic' basis inevitably reflect the decision maker's view of the future and his own interest. Thus it is quite easy for a nuclear energy engineer to show that—with his value judgements—wind power is of little significance, and likewise for the ambient–energy enthusiast to put forward a convincing argument that nuclear energy is a sure pathway to environmental and financial disaster. I suggest that what really matters is to have a vision of the kind of future we want and to work towards that; economics is merely a tool which can help us to achieve it most efficiently. In building design, this is particularly important as we are designing for a lifetime of the order of 100 years and economic guesses on this timescale are quite meaningless. Professor M. W. Thring states his challenging principle: 'What is essential for the long term survival of our civilisation is never economic and rarely politic.' However, short-term cost-effectiveness is important insofar as it is a measure of efficient use of materials (but not built-in obsolescence), reflects good system design, and encourages the early adoption of ambient energy devices (Thring, 1977; Chapman, 1975; BRE, 1975).

Designs at the Centre therefore take the use of insulation and ambient energy beyond the level generally considered economic and may also imply some reduction in comfort level below currently accepted design standards, but not necessarily below the comfort level in many homes at present. The Centre functions satisfactorily with a relatively small fossil-fuel consumption and no reliance on mains electricity. In the long term we aim to phase out fossil fuels completely and rely entirely on ambient energy sources; this will involve the use of biofuels, an area we are developing at present.

INSULATION OF BUILDINGS

The Centre has a wide range of building types. These include several old slate-walled cottages, a timber house, a highly insulated house built by Wates Built Homes Ltd., and two central buildings housing offices, bookshop, exhibition hall, eating facilities, library, and laboratory. Where cost and space allow, we have insulated walls to a U value of about 0.2 W/m^2 °C. Detailed monitoring of the thermal performance of some buildings is under way, and the others are being observed to gain experience of the different insulation techniques in a normal 'user' situation.

The cottages have all been rebuilt and various types of insulation have been incorporated. The original construction was solid slate block walls approximately 450 mm thick. All have approximately 150 mm of roof insulation. Two have been internally insulated, one with purlboard composite plasterboard/urethane foam sheets and the other with 75 mm glass fibre; the second is described in more detail later. In both cases, installation was carried out by people with an understanding of the principles involved (stopping air movement and the function of the vapour barrier), and we have had no problems with the buildings since completion. In both cases, the buildings respond quickly to heating and do not suffer from the wall condensation problem of similar uninsulated buildings.

A further small slate-walled cottage has been insulated externally using three different techniques on different walls. The east wall is treated with a commercial system comprising 100 mm Rockwool slabs covered by a paper-backed metal mesh attached to the wall by pegs passing through the insulation. The mesh is cement-rendered to complete the external surface. This system (by Cape Insulation Ltd) is straightforward around openings and appears to be very durable. The north wall was battened and clad in weatherboard; the 75 mm gap was filled with granulated polystyrene. Although the material cost of this method is higher, it is simple and well suited to do-it-yourself installation. The west wall was experimentally coated with 150 mm thick urethane foam sprayed directly onto the wall by Baxenden Chemicals Ltd. The bottom 450 mm was cement-rendered to guard against impact damage and the surface was then protected by a plastic paint-on surface and a final coat of Sandtex. The material cost of this method is quite high but it is very quick to apply: the total cost is similar to the other commercial system. The method yields a slightly bumpy surface not unlike a painted rough stone wall in appearance. It effectively seals the wall and prevents draughts. The surface can be damaged fairly easily but is quickly and easily repaired with a filler. Over its first year, no problems have been experienced. The high thermal capacity of the inside walls (approximately 10 kWh/°C) with a typical steady-state heat demand of 1 kW results in the cottage behaving very much like the uninsulated ones with intermittent heating. The benefits of the insulation are only apparent after about 24 h of heat input. This thermal characteristic is advantageous for making use of incidental and solar gains, but is inappropriate for intermittent fuelled heating.

The three-bedroomed timber house was constructed from a basic kit by Country Craft Homes Ltd. All external walls are insulated with 100 mm of glassfibre with 200 mm in the roof space; windows are double glazed throughout. A Pither anthracite stove provides direct heating of the lower floor, domestic hot water, and feeds two radiators on the upper floor. This building, together with the Wates house, has demonstrated the much-reduced need to distribute heating around the building when the exterior walls are well insulated. In each case, two heat emitters are capable of maintaining comfortable conditions throughout the building. The cost saving in not providing one or more heat emitters in each room at least partially offsets the cost of the additional insulation.

WATES CONSERVATION HOUSE

In the summer of 1975, Wates Built Homes were approached with the hope that they would take up the challenge of building a low-energy house at the Centre as a demonstration of what could be achieved with existing building technology. Wates responded enthusiastically and commissioned Peter Bond Associates to design the house. During preliminary discussion, the following decisions were made: to retain a building of fairly conventional appearance and size so as not to prejudice marketing possibilities, to limit the budget to £20 000, and to concentrate efforts on reducing energy demand rather than supplying energy from ambient sources. The last point ensured that the house design would be suitable for any location, initially in most cases connected to mains services, but in isolated situations (and perhaps more widely in the future) powered by ambient energy. The function of the house is partly to be a research tool and a means of gaining experience with large amounts of insulation and heat pumps but also to stimulate interest in low-energy housing among the public, building societies, etc.

The design employs a high degree of insulation in the walls, roof space, and beneath the ground floor, small quadruple glazed windows, mechanical ventilation, and inner walls of fairly high thermal capacity to take maximum advantage of incidental gains without local overheating and to smooth out the effect of diurnal temperature variations. The internal air design temperature was taken as 18°C. A heat pump is used for domestic hot water (DHW) supply and the design also attempts to minimize cooking energy and uses low-power lighting. Rain water from the roof is collected, filtered, and used for all purposes except drinking, with automatic reversion to mains water when necessary.

Cavity wall construction is used; the outer skin is brick and the inner Thermalite block. The 450 mm wide cavity is filled with 'Dritherm' glassfibre which was inserted during construction with little difficulty. The outer wall is buttressed internally and tied only at the buttresses to the inner wall. It is calculated that the dewpoint will occur on the outer surface of the insulation and since the insulation is layered in vertical planes, condensate will run downwards, together with any moisture penetration through the outer skin, to the base of the wall where drainage holes are provided to the outside. Inspection of the insulation, by removing bricks from the outer skin has shown that, even in the worst conditions, only the outer centimetre or so of insulation and the surface of the brick were damp. The rest of the insulation has remained dry. The top surface of the ground slab is laid to fall towards the outer skin, discharging any condensate via the same drainage holes. A polythene vapour barrier is also provided above the ground slab insulation; this also serves the function of stopping glass-fibre particles from being drawn into the building by the warm air heating system. The calculated U value for the walls is 0.075 $W/m^2 °C$.

A normal tiled pitched roof is used and, to accommodate 450 mm of roof insulation without excessive building height, part of the ceilings of the upstairs rooms are angled. A polythene vapour barrier is provided beneath the roof insulation and the roof space above the insulation is naturally ventilated. Some problems were experienced with condensation on the roofing felt. Quadruple glazing brings the window losses more into line with the

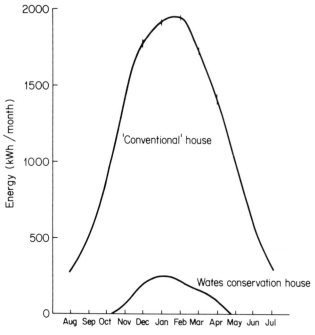

Figure 1 Space heating energy demand

rest of the structure. A fixed outer frame carries one sealed double-glazing unit and spaced at 200 mm, an inner openable frame carries a second similar unit.

Only one entrance is provided, thus avoiding through draughts, and this is fitted with double doors. Ventilation is provided by a 20 W extractor fan in the kitchen, fresh air being induced into the house by the slightly reduced internal pressure. The air change rate is one-quarter of the total house volume per hour and air is circulated round the house by the warm-air heating system fan. Additional automatically controlled fan ventilation is provided for the shower room and toilet.

The calculated specific heat loss rate for the house is 66 W/°C. The peak demand is 1.2 kW and the space heating requirement with incidental gains taken into account is only 8 kWh/day on average in December and January and a total of about 950 kWh over the heating season. This compares with 13 000 kWh over the heating season for an equivalent-sized conventional house. The heat demands of the conservation house and a comparable conventional house over a typical year, taking into account incidental gains, are shown in Figure 1. The house itself is shown in Figure 2.

Figure 2   Wates conservation home

Heating system

An air-to-air heat pump was chosen for space heating and the lower requirement enabled a small and cheap system to be employed. The compressor is the same type as used in a domestic deep freeze. The evaporator heat exchanger in the roof space is fed with outside air mixed with ventilation air from the kitchen extractor fan and the condenser is mounted with the compressor in a cupboard on the upper floor. Warm air is fed to all rooms via the underfloor spaces and returned via ceiling-level vents.

The refrigerant circuit can be reversed by a changeover valve for automatic defrosting of the evaporator and for summer cooling should this prove necessary. Experience of the hot summer of 1976 suggests that in this location cooling will not be needed as the long thermal time constant of the house provides a very stable temperature with little duirnal variation. A coefficient of performance (COP) of 2.5 was expected for this system.

In practice the heat pump was only capable of heating the building satisfactorily in fairly mild weather and, as soon as ambient temperature fell to a few degrees above zero, heat exchanger icing and a much reduced COP made the heat output inadequate. To check the thermal performance of the building, independently of the heat pump, a propane-fired gas boiler has been installed outside the house to feed two panel radiators inside. The heat input is measured by a water heatmeter. Tests are still under way, but preliminary results suggest that adequate whole-house heating can be achieved with around 1 kW heat input. We conclude that the house could in practice be heated for under £1 per week on average—one of the design objectives. A further observation is the reduced need to distribute heat so much in a well insulated building; two panel radiators maintain comfortable conditions over the whole house.

Hot water

A similar small heat pump is used to heat the 40 gallon hot water tank. A large volume of relatively low temperature (50°C) water is used to maximise the COP while maintaining an adequate hot water supply. The evaporator of this heat pump is immersed in a 60 gallon waste water tank below the ground floor. Waste water from the shower (no bath is fitted), all hand basins, and one kitchen sink is fed to this tank, and water from the bottom of the tank discharges to the drain as more water is added. By discharging water at a similar temperture to the cold water feed temperature, or colder if necessary, sufficient energy can be extracted to maintain the temperature of the hot water cyclinder. It has been estimated that heat lost from the water to the house is approximately balanced by the electrical energy supplied to the heat pump. A COP of 2 has been measured for this system. Electrical consumption is estimated at 4.5 kWh/day, 1640 kWh/yr, which is comparable with BRE estimates (Seymour-Walker, 1975). The performance of this system and the space heating system is monitored by a multichannel recorder. Typical data from these recordings are shown in Figure 3.

Cooking

A well insulated oven has been constructed based on a standard Belling oven surrounded by 150 mm of glass fibre. Tests have shown that energy use is reduced by a factor of about 2. An electric kettle and electric saucepans are provided as these are much more efficient than open ring cooking. So far tests are incomplete but we have successfully cooked a sponge cake with only 0.2 kWh! A modern version of the traditional haybox and an evaporative cooler are also fitted in the kitchen. A further experimental cooker has recently been fitted in the house, but is not at present operational. This is an electric storage AGA, supplied by Glynwed Ltd, which can store sufficient heat for several days' cooking. The energy can be fed intermittently at any rate up to about 8 kW. It can therefore be used on normal off-peak power or to store energy from wind or water turbines.

Lighting

Lighting in the house is provided by a 20 kWh storage battery fed by a Dunlite 2 kW aerogenerator mounted nearby. The system operates at 110 V dc. Despite poor siting of the aerogenerator, the system has provided a reliable and adequate supply for about three years so far. There have, however, been one or two easily rectified faults with the aerogenerator (Todd, 1979a).

Further work

A problem which has developed is the spalling of external brickwork and discoloration and flaking of the cement paint finishing surface. These faults can probably be blamed on the type of brick used and the incompatibility of the paint. However, there is the possibility that the high level of insulation has contributed because of the abnormal temperature cycle the bricks are exposed to and through the presence of condensate on the inner surface of the outer skin bricks.

Further tests on the walls, including brick temperature and moisture content, are under way together with more precise measurement of the ther-

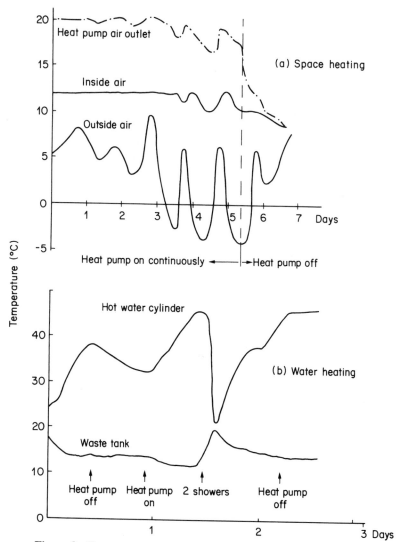

Figure 3 Temperature recordings from Wates conservation house

mal characteristics of the house. This work is being carried out in conjunction with the Department of Building Science, Sheffield University.

WIND-POWERED COTTAGE

In some situations wind power can offer advantages over solar power for the heating of buildings. Many existing properties are not well sited for solar heating systems and the space available for storage tanks is limited. A wind-power system using thermal (e.g. water tank) storage requires considerably less storage volume to achieve almost complete independence from other fuels because (a) the availability of wind energy matches the heating demand quite well and the maximum storage time required is much less than for solar power which, for a practicable collector size, requires several months storage to achieve a similar performance (Szokolay, 1975) and (b) the store can be operated up to much higher temperatures without loss of efficiency whereas high storage temperatures in solar systems imply increased collector losses and lowered efficiency.

Many isolated houses or groups of existing houses in areas with a fairly high mean windspeed (say above 5 m/s) could be heated by wind power. Such systems are, like solar systems, best combined with a high degree of insulation. The insulation is in itself cost-effective in the short term, has a long lifetime, and should be the priority investment. It also reduces the size of ambient energy system required. An existing cottage at the Centre has been modified and fitted with a wind-power system with the intention of providing a large percentage of the heating requirements and hot water for washing and a shower. The system is shown diagramatically in Figure 4.

Building structure

The cottage is at one end of a row of three old slate-walled cottages and was in need of considerable renovation. It was decided to insulate internally using 75 mm glass-fibre insulation covered by a polythene vapour barrier and plasterboard supported on 3 inch × 2 inch studwork, giving a

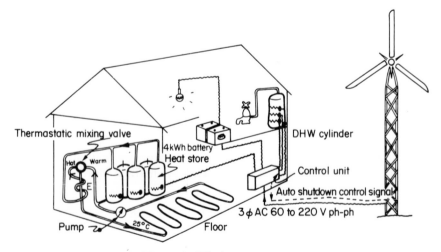

Figure 4   Wind-power system

U value of about 0.4 W/m^2 °C. This was the maximum insulation level consistent with acceptable loss of room size. Roof insulation of 80 mm was also installed. Windows are only single glazed but are small in area. A new ground floor was necessary as part of the renovation and this gave the opportunity to install a water-heated concrete floor. The 50 mm sand/cement screed is laid on 100 mm polystyrene block with a vapour barrier; 10 mm diameter nylon pipes are laid at 150 mm centres in the screed to provide a large-area low-temperature heating surface. A solid-fuel room and space heater is also provided.

Wind energy

The remainder of the design centred around the type of wind generator available. At the time the Swiss Elektro 5 kW machine was the largest well tried aerogenerator readily available and the three-phase AC permanent magnet alternator version of this machine was chosen. (A 10 kW version is now on the market.) It was erected on an exposed site 250 m from the cottage, on an 8 m guyed tower. Meteorological Office data and some earlier on-site measurements suggested an annual mean windspeed of around 11 mph (5 m/s) which, assuming the usual form of velocity duration curve for Britain (Golding, 1976) and an average overall C_p of 0.27* (the lower end of the range quoted by the manufacturers), indicated an

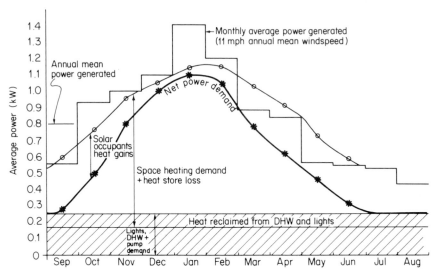

Figure 5   Power input and demand for wind-powered cottage

$$^*C_p = \frac{\text{electrical power output}}{\text{power in the wind}}$$

average power output of about 0.8 kW could be expected. The manufacturers' published power/windspeed curves indicate a considerably higher C_p but the more conservative value was taken as this seems more usual for machines of this size (Hengeveld, *et al.* 1978). Our subsequent measurements (Todd, 1979a) and a recent article (Carter, 1976) suggest that these high claimed C_p values are not achieved in practice. Figure 5 shows the calculated average power available for each month of the year, taking typical departures of monthly mean windspeed from annual mean windspeed and assuming that the same relationship between mean windspeed and average power output holds for monthly and annual means.

As the major part of the energy was required in the form of heat, thermal storage in water tanks was the obvious choice. The longest expected period of low windspeed (below the aerogenerator cutting-in speed) during the heating season is the main influence in choosing storage capacity, but for reliable energy supply it is not sufficient to size the storage equal to the longest expected low wind period. A plot of the probability of the power available exceeding any fraction of the long-term average power P_a for various values of storage time is necessary properly to assess the optimum storage capacity. An example of such a plot based on data given by Sørensen (1976) is shown in Figure 6. Using such a diagram, the relation between storage time (the time for which the store can supply the load with no wind) and the ratio of load of P_a for the required reliability of

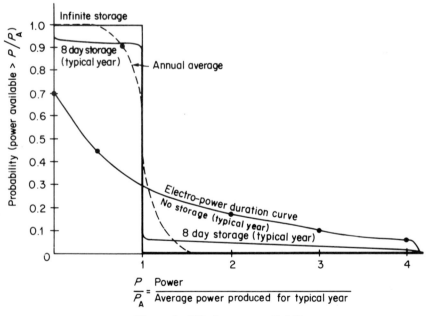

$$\frac{P}{P_A} = \frac{\text{Power}}{\text{Average power produced for typical year}}$$

Figure 6   Wind-power availability

supply can be established. In the example shown, eight days storage and load equal to P_a is assumed and the curve indicates a reliability of supply of approximately 90%.

However, this curve is for a particular machine and, with the lower cutting-in speed of the Elektro machine compared to that considered by Sørensen, the eight-day storage curve would be expected to rise to almost 1.0 as (power available)/P_a tends to zero. The problem is further complicated in a space heating application by the wide variation of load over the year, which suggests that the plot should be made for each month of the year. It should also include the effects of variations in annual mean power, also shown in Figure 6. A Markov model of the stochastic processes involved would be an appropriate way to tackle the analysis. It may enable use to be made of long-term wind and ambient temperature data without very lengthy computer model runs.

As inadequate data for this procedure were available for the site in question, a storage time (at midwinter load) of one week was chosen. The longest observed period with winds less than 7 mph—the Elektro cutting-in speed—during the last two winters was about four days and it was decided fairly arbitrarily to choose a seven-day storage time to take account of the possibility of low average windspeeds either side of such a four-day period. The storage time increases to about ten days at each end of the heating season. This storage capacity is similar to that suggested by several other authors (Golding, 1976; Steadman, 1975; Frost, 1974). The wind is being monitored on the site and further analysis will be performed when sufficient data are available.

Power demand

The power requirements of the cottage for lighting, and heating water are minimized by using low-power fluorescent tubes and localized lighting and by fitting a shower but no bath. With two occupants, the average power required for these functions is estimated as about 20 W for lighting and 220 W for hot water (17 gallons/day). A further 10 W average is required for a circulating pump. It is assumed that about one-third of the lighting and DHW energy can be reclaimed.

It was clear that the cottage could not be heated fully throughout with this size of aerogenerator so a compromise was reached in the design, heating the main room to about 18°C and the remainder of the house to about 12°C. The main room floor is heated directly and the heat store is positioned in the (ground floor) bathroom allowing heat to flow naturally from the ground floor to the upper floor. Because of the relatively low U valve of the outer walls and roof compared to the dividing floor, it is estimated that a temperature of at least 10°C should be maintained in the bedrooms. The heat loss coefficient for the main room, including ventilation and loss to other rooms, is calculated to be 56 W/°C and the average power demand for continuous heating is shown in Figure 5, together with the estimated average power produced per month and the lighting and water heating demand. Although this suggests that the design is satisfactory, it must be remembered that the power inputs and demands quoted are long-term averages for a particular month and there may be significant variations between the same month in different years. However, there are various energy economies which can be made to meet such variations, mainly on the space heating demand, such as reducing temperatures all round, reducing or using heat reclaim ventilation or more simply heating for only part of the day. Although the floor screed will have a cooling time constant of about eight hours, a significant saving could be achieved by only heating for part of the day.

Since the design of this system, measurement of windspeed has indicated an annual mean somewhat less than the assumed value of 11 mph. Many problems have also been encountered with the Elektro wind turbine (Todd, 1979a), the most serious being its poor output. Tests so far show only about 50% of the rated output at rated windspeed. The much-reduced power supply available has made proper operation of the system impossible, although much has been learnt and further tests are planned. The cottage heat store system will probably in the future provide part of the load of a much larger aerogenerator planned for the Centre. Work on a housing scheme based on this principle—a large aerogenerator feeding heat stores in a group of buildings—is under way in Hull (Liddel, 1978).

Heat store and controls

The heat store consists of five 60 gallon copper hot water cylinders, each fitted with an immersion heater. These are thoroughly insulated with about 150 mm of glass fibre. They operate under 4 m of water pressure head and the working temperature range is 25 to 105°C, giving a storage capacity of approximately 130 kWh. The tanks are coupled in parallel and feed a thermostatic mixing valve to ensure a water temperature of around 25°C is fed to the floor with any tank temperature above 25°C. A heat exchanger (E in Figure 4) is necessary to limit the maximum hot water input temperature to the mixing valve. A solid-fuel room and water heater is also provided.

The immersion elements, together with three more immersion elements in the domestic hot water cylinder, are switched in various series/parallel arrangements to match the load to the power available as the wind fluctuates and to give priority to domestic water heating when necessary. Its output can also be diverted to the Centre's main storage batteries to supplement electricity supply during the summer months when reduced rainfall limits the hydroelectric power available. The switching is achieved by electromagnetic relays in the Elektro control unit which has been substantially modified for this application. Room temperature is controlled by thermostatic switching of the circulating pump.

SOLAR-HEATED EXHIBITION AND OFFICE BUILDING

This building was designed in 1975 and aims at 100% solar space heating using long-term heat storage (Todd, 1978). Other influences on the design are the limited electrical power available on the site (particularly in summer when much of this must come from storage) and the availability of low-cost labour in the form of voluntary workers.

A building of this type presents a complex system design problem where many interdependent choices must be made. The individual components, such as the heat store, for example, cannot be optimized in isolation. Our approach was not one of seeking analytically an optimum solution but relied on general considerations, such as space availability, convenience of construction, other solar building designs and intuition in proposing a trial solution, and on analysis to test the solution and the effects of varying some major parameters. Whilst we feel the design is probably not too far from optimum given our constraints, the lack of adequate data and facilities prevented precise performance predictions.

Building design

The building is a conversion based on an old slate-walled workshop and now houses an exhibition hall and bookshop with an office above—200 m² floor area in total (Figure 7). To achieve a low specific heat loss rate, an internal perimeter wall has been added and the building has been insulated with between 150 and 200 mm of polystyrene all round. This was the first of many interrelated decisions. The resulting U valve of the walls is approximately 0.2 W/m² °C which ensures that the heat loss through them is only a small fraction, roughly one-quarter, of the total heat loss of the building. A further decrease in wall U value could not be justified unless the other heat losses, through windows, doors, and ventilation, could also be reduced. A compromise was necessary on window area, as maximum possible daylight was required into the downstairs exhibition area and the office to minimize the necessity for artificial lighting. The 50 m² of double-glazed window, part of which is fitted with insulating shutters of 40 mm translucent Okalux, account for just under half the total heat loss. The remainder is ventilation loss which we are minimizing by careful draught stripping, aiming at half an air change (230 m³) per hour in winter

Figure 7 Solar-heated exhibition and office building

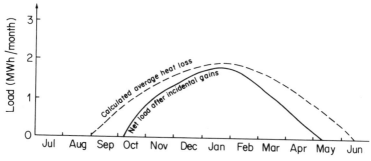

Figure 8 Space heating load

which comfortably exceeds the recommended level for 20 people (occupancy is usually much less than this). The calculated total heat loss coefficient is 250 W/°C. A low internal air temperature of 15.5°C was chosen for the winter months, and together with the fairly high mean radiant temperature, experience shows that this provides an acceptable comfort level except near doors and the glazed east wall where cold draughts are a problem. Despite the long cooling time constant of the building and heat emitter (about 20 h) and the warm-up time constant (about 4 h), a further significant saving, estimated to be 20%, can be made at very little cost by the use of night setback temperature control. The estimated annual demand curve for the building is shown in Figure 8. The solid line shows the demand when allowance is made for incidental heat gains (mainly solar gain through south windows). It would, of course, be possible to increase or decrease the demand with attendant savings or costs and to set these against the costs or savings of changing the energy output of the solar heating system. Rough calculations suggest that an increase (×2) in demand would bring an overall cost and space penalty, whereas reducing the demand (×0.5) would have little effect on the overall cost but may involve difficulties in achieving low ventilation rates, particularly in a building conversion. These comments apply only to our design and cannot be held to be generally true.

Collector and heat store

It was decided to use the existing south-facing roof surface of about 70 m² at 34° to the horizontal and an additional 30 m² of collector at 55° to the horizontal against the south wall. Calculations showed that the extra 30 m² would give a valuable increase in output of more than 60% over the worst of the heating season, for an area increase of only 43%. There was little scope for accommodating more collector area. The heat store was positioned outside the building for ease of access because of its experimental construction. This inevitably wastes some heat. Space was available for a maximum of 200 m^2, mostly below ground level.

We chose a double-glazed trickle-type collector based on Thomason's design, but incorporating some improvements. This was convenient as the remainder of the roofing was to be the same corrugated aluminium sheeting as used for the absorber, and the only significant additional costs for the roof collectors were for glass and glazing bars. The efficiency of this type of collector was known to be poor at high temperatures, but at the low temperatures anticipated in the heating season it was assumed that performance would be similar to a sealed absorber double-glazed collector. Under these conditions (around 30°C mean water temperature), the efficiency η was taken as

$$\eta = 0.64 - 3.2 \frac{\Delta T}{I}$$

where ΔT = mean water temperature − ambient temperature and I = insolation on collector plane in W/m^2, and at higher temperatures (around 50°C) the collector heat loss factor was assumed to be similar to a single-glazed collector:

$$\eta = 0.64 - 6 \frac{\Delta T}{I}$$

The collector flow rate is 0.52 l/s, a compromise between minimizing both temperature rise and pump power consumption.

The monthly output of the collector at several different temperatures was estimated in different ways using sunshine hours and total radiation data with estimated collector efficiencies (Duffie and Beckman, 1974) and using the results of a computer simulation (Everett, 1977) based on hourly Kew data. None of the methods were very satisfactory due to the nature of the local climate and the lack of local data in a suitable form. However, the curves presented in Figure 9 were derived for an average year and include the effect of shadowing by a hill to the south-east in midwinter. They have been used to estimate the performance of the system and to size the heat store.

To make best use of the collector, it is clear that its operating temperature must be kept to a minimum. Two separate heat stores are used to achieve this, a low-temperature store to smooth short-term fluctuations and a large inter seasonal store which can operate at high temperatures without forcing the whole system to do so during the heating season. The 30°C (mean water temperature) line in Figure 9 was therefore taken as an indication of the energy available via the short-term store for space heating (assuming the store is of adequate size).

A similar effect could be achieved with a single highly stratified store,

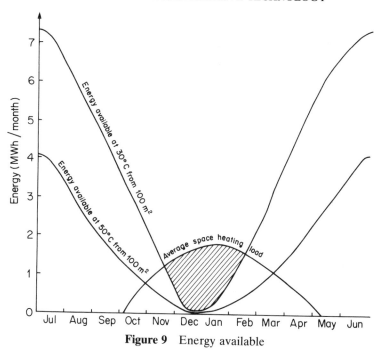

Figure 9   Energy available

but it was assumed in the design that long-term stratification would be difficult to achieve. To enable the heat to be used at low temperature, the building is heated via the ground floor sand/cement screed which has 10 mm nylon water pipes laid in it at 200 mm centres. This forms a large heat emitter (110 m²) with a low thermal resistance (from mean water temperature to internal air temperature) of approximately 0.0018°C/W. The maximum design heating load of 4 kW can therefore be provided by water at only 25°C. The low surface temperature (about 21°C maximum) reduces two of the usual underfloor heating problems—hot feet and the inability to adapt rapidly to incidental gains or changes in ambient temperature. In this case a rise of 1°C in internal air temperature due to increasing outside temperature reduces the output from the floor by about 800 W. As a result, the inside temperature will change by only about one-half of the outside change. This could be improved further by increasing the ratio of heating surface area to building heat loss coefficient. Figure 10 shows the whole heating system.

Sizing and design of long-term heat store

It can be seen from Figure 9 that, given an adequate short-term store, the building can be heated up to mid-November and after mid-February. The

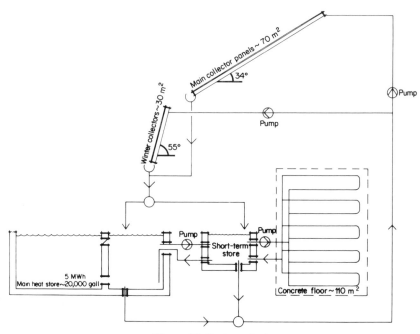

Figure 10 System design

energy deficit (shown hatched) of 2.9 MWh must be met by the main heat store. To find the necessary peak energy content of the store, it is necessary to make some allowance for losses from the store over the period before and during heat extraction. As a starting point in the design, the total losses (from September to February), were set roughly equal to the total energy required from the store in midwinter. This suggests a peak store content of at least 5 MWh depending on how much 'topping up' can be achieved after September.

It was decided to experiment with an unconventional low-cost water store design consisting of a hole in the ground lined with rigid polystyrene foam and a butyl rubber liner (see Figure 11). Groundwater buoyancy forces are not a problem on our particular site which is very well drained, but could be on other sites. However, raising the tank water level a little way above ground level could probably avoid any trouble. The temperature limits of the butyl rubber and the insulation (about 70°C), coupled with the poor collection efficiency at high temperatures, led us to aim for a maximum store temperature of about 65°C—lower if possible.

The minimum useful temperature of the store depends on the prevailing heating load as the store approaches exhaustion. As this should occur in February, the heating load could well be at its maximum 4 kW, so a minimum useful store temperature of 25°C was assumed. The system can,

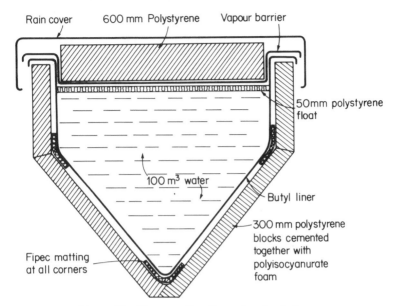

Rain cover   600 mm Polystyrene   Vapour barrier

50mm polystyrene float

100 m³ water

Butyl liner

300 mm polystyrene blocks cemented together with polyisocyanurate foam

Fipec matting at all corners

Figure 11   Cross section through a heat store

of course, still supply useful heat to the building down to water temperatures only slightly above internal air temperature, if any necessary
supplementary heat is supplied via a separate emitter—in our case a wood
stove.

This range of store temperature, 25 to 65°C, sets the volume of water
required at 107 m³ for 5 MWh. The measured volume of the store after
construction was approximately 100 m³. The store is 'V'-shaped to achieve
stable sides to the excavation; a shape which departs considerably from the
optimum area/volume ratio. It has roughly twice as much surface area as a
cube of equal volume and, retrospectively, this is probably the worst
feature of the design. A thickness of 30 cm of polystyrene all round, except
for the top which has 60 cm, yields an estimated heat loss coefficient
K of 20 W/°C. The top insulation is floated on a 2 inch layer of
polystyrene but is separated from it and the water by a vapour barrier.

The performance of the system was estimated by working backwards
from mid-February when the main store is expected to be exhausted,
calculating the store temperature a month earlier from heat input H_m, heat
extracted over the month E_m, and the losses, using:

$$T_{m-1} = \frac{(1 + 0.36K/S)T_m + E_m/S - H_m/S - 0.72(K/S)T_A}{1 - 0.36K/S} \tag{1}$$

where T_m = store temperature in month m, S = store capacity in KWh/°C, and T_A = average ambient temperature over the month. H_m is found from the mean collector temperature for the month and curves such as those in Figure 9:

$$H_m = f_m\left(\frac{T_{m-1} + T_m}{2}\right)$$

where $f_m(\mathord{|})$ is some function dependent on weather conditions during month m. This function can be incorporated into equation (1) if it is simple, otherwise T_{m-1} can be found iteratively, guessing an initial value for $(T_{m-1} + T_m)/2$.

The result of carrying out these calculations for our initial design (100 m^2, 100 m^3, and 20 W/ °C) is shown in Figure 12a, which indicates that the store can be reheated adequately starting mid-July at the latest and that the peak store content is 5.05 MWh corresponding to 68°C. This solution was finally adopted for the building.

Reducing collector area increases the energy deficit in midwinter and therefore calls for increased peak store content and/or a decrease in store losses. Figure 12b, c, and d show the effect of varying the store heat loss coefficient and volume to explore the possibility of reducing the collector area to 50 m^2. Although only a few permutations have been explored, the results are of some interest. After about mid-October, the curves are

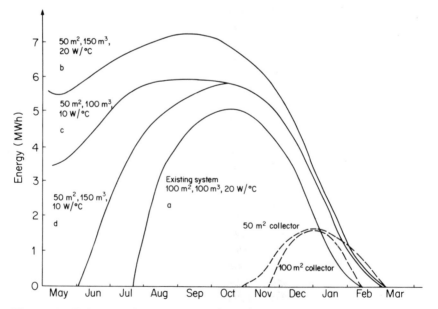

Figure 12   Estimates of store contents (———) and load met by store (– – – –) for various collector areas, heat store volumes, and store loss coefficients

essentially independent of store volume, depending mainly on energy extracted and on the heat loss coefficient. However, the ability to reheat the store depends quite critically on volume as shown by c and d.

Success hinges around having the store temperature sufficiently low by mid-August (working backwards) to be able to take advantage of the large amount of energy available in midsummer at a good collector efficiency. Curve c would also exceed the store temperature limit. The only viable option of these three, d, would call for an increase of about 1.3 times in tank liner area and an increase in volume of insulation of about 3.4 times if the store shape was kept the same. With our collector design, this would cost more than the saving made by reducing the collector area. However, with a store shape of lower surface area, some reduction in collector area below 100 m^2 would probably be advantageous. An insulated partition was fitted in the store, dividing it into 75 and 25 m^3. The two sections can be coupled by opening a valve or the smaller one can be operated on its own. This allows a reduction of heat loss and thermal capacity after the store is exhausted early in the year which could ease the problem of bringing the store back into use quickly should the temperature fall too low to be useful.

Control system

Twelve temperatures are monitored using semiconductor sensors: their outputs are amplified to $0 \rightarrow 5$ V ($0 \rightarrow 100°$C) and integrated circuit logic is used to provide control signals for the pump relays (Todd, 1979b). This approach allows switching thresholds and criteria to be changed easily and facilitates monitoring the system's operation. At present, the heat emitter temperature T_F is controlled by on/off switching of the pump based on a heat emitter temperature set point (T'_F) given by

$$T'_F = 21 - \frac{T_{ambient}}{3}.$$

Experience with the system

The 100 m^2, 100 m^3 system was constructed during 1976–77. The plumbing and control systems were not completed until mid-1977. The main store was heated to about 36°C and allowed to cool from September with no heat extraction or input to check the heat loss coefficient. This was found to be about 25 W/°C. Some rain leaked beneath the rain cover and this probably increased heat loss significantly. The collector used standard glazing bars from Bexley Glass Ltd with two 4 mm sheets of glass. Some problems were experienced with cracking at the bottom corners of the glass where it bears on small retaining clips. This occurred after the collector had reached high temperatures in the summer. The problem seems to have been cured by inserting soft packing material between the clips and glass. The other main problem with the collector has been leakage of water vapour into the building (the roof space beneath the collector is used as an office). This has caused considerable trouble with condensation on the colder north slope, particularly on cold sunny days. Careful sealing has reduced the problem but the experience would make us reject the trickle collector for such applications in future.

On first filling the store, the liner ballooned into a small crack in the insulation and burst. All joints in the insulation were sealed with sprayed polyisocyanurate foam and all corners where cracking was most likely were lined with Fipec matting. Since these modifications, no further trouble has been experienced. A further practical problem has been the shrinkage of the nylon floor pipes after being in use for some time. This amounted to about 20 mm on the width of the building and caused several water leaks at joints with the distribution pipes.

Although the system was designed assuming that stratification would not be maintained in the storage tanks, preliminary measurements indicate that stratification in the main store is good. During heating, the bottom temperature remained substantially unchanged until a volume of water equal to that of the tank had been circulated (several weeks). If stratification could be relied upon, a single store tank with inlets and outlets at different levels would probably suffice in place of the present system.

Since July 1978 the performance of the building has been monitored in some detail. Insolation, twelve temperatures, pump, and valve operation are all recorded on a multichannel chart recorder. During 1978, with only 70 m^2 of collector in use, the solar system provided all the space heat requirements between late February and late November. Owing to a combination of reasons (power failure, water loss, poor collector efficiency), the main tank failed to reach its design temperature by October and was therefore unable to supply sufficient heat for total heating midwinter. It has, however, maintained background heating and avoided very low internal night temperatures. A typical days operation is shown by Figure 13. More detail on the system's characteristics can be found in Whittaker (1979).

Electricity supply

The Centre is not connected to the National Grid, in order to make the point that renewable sources can provide an adequate electricity supply—particularly in an area like mid-Wales with its abundance of hills and streams. Providing a local supply creates jobs and keeps money in the area.

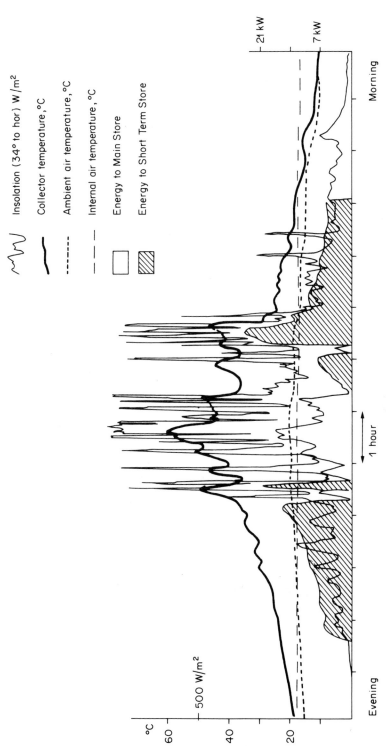

Figure 13 Operation of solar heating system on one autumn day

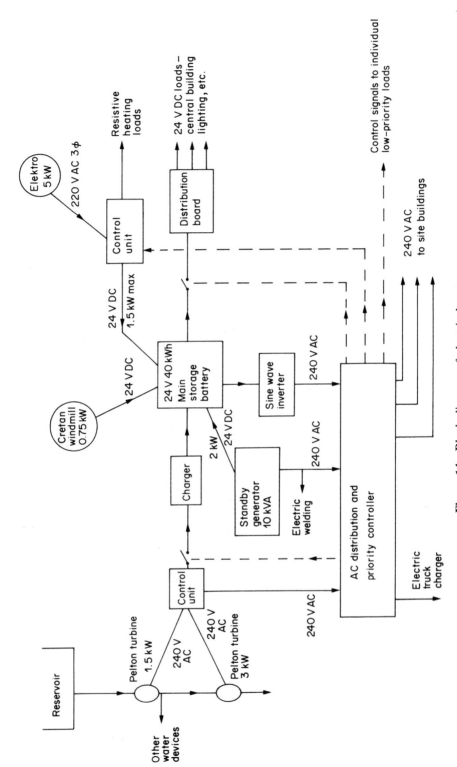

Figure 14 Block diagram of electrical system

At present our main electricity supply is used for electronic equipment, lighting, pumps, electric tools, and machines. We have installed two Pelton turbines which operate continuously during the winter and intermittently during the summer. They operate with heads of 30 and 45 m and produce a combined output of 4 kW. A small reservoir provides 500 kWh storage capacity. We also have 40 kWh of lead–acid battery storage. Electricity is distributed at 240 V AC, directly from the water turbine when available but otherwise from batteries via a static invertor. The batteries also accept power from two aerogenerators and feed low-voltage display lighting in the central buildings. In times of energy shortage, an automatic priority system sheds less important loads to conserve battery energy. A standby generator is occasionally used in long dry and windless periods. 'Off-peak' electricity is also used to charge the batteries of an electric vehicle (30 mph, 40 mile range) which provides local transport with a carrying capacity of half a ton. A block diagram of the system is shown in Figure 14.

DISCUSSION

The projects described serve a dual purpose. They stimulate public interest in energy conservation and the use of renewable energy sources by showing practical, if not yet fully developed, examples. They also are beginning to provide some useful data which are of great use in future design work at the Centre and elsewhere. Among the general public, we see much enthusiasm and a willingness to devote time and money to such projects—even those not offering a high economic return—provided sound advice based on practical experience is available. There is great scope for individual and local initiatives in these areas of energy supply and use where Government seems reticent to act.

REFERENCES

BRE, 1975, *Energy Conservation—a Working Party Report*, Building Research Establishment Current Paper. CP 56/75.
Carter, J., 1976, Testing an Elektro-Gemini, *Windpower Digest*, 1, no. 6.
Chapman, P., 1975, *Fuel's Paradise*, Penguin, Harmondsworth.
Duffie, J. A. and Beckman, W. A., 1974, *Solar Energy Thermal Processes*. Wiley-Interscience, New York.
Everett, R., 1977, private communication, Polytechnic of Central London.
Frost, L. N., 1974, Wind power in Ireland, *Technology Ireland*, December.
Golding, E. W., 1976, *The Generation of Electricity by Wind Power*, Spon, London.
Goldsmith, E., 1972, *Blueprint for Survival*, Penguin, Harmondsworth.
Hengeveld, H. J., et al., 1978, *Matching of Wind Rotors to Low Power Electrical Generators*, S. W. D., Amersfoort, The Netherlands.
Liddel, H., 1978, *A Low Energy Housing Development*, Hull College of Further Education.

Schumacher, E. F., 1974, *Small is Beautiful*, Abacus, Tunbridge Wells.

Seymour-Walker, K., 1975, *Low Energy Experimental Houses*, BRS Energy Research and Buildings, Building Research Station, Watford, HMSO.

Sørensen, B., 1976, Direct and indirect economics of wind energy systems relative to fuel based systems, *Proc. BHRA Conf. on Wind Energy Systems*.

Steadman, P., 1975, *Energy, Environment and Building*, Cambridge University Press, Cambridge.

Szokolay, S. V., 1975, *Solar Energy and Building*, Architectural Press, London.

Thring, M. W., 1977, A world energy policy, *Proc. Inst. Fuel Symp. on Potential for Power*, Southampton.

Todd, R. W., 1977, Low Energy Housing, *Proc. CICC Conf. on Ambient Energy and Building Design*, Nottingham.

Todd, R. W., 1978, A solar heating system with interseasonal storage, *Proc. UK—ISES Conf.* C15.

Todd, R. W., 1979a, *Experiences with Windpower*, Centre for Alternative Technology.

Todd, R. W., 1979b, *Control of Solar Space Heating System*, Centre for Alternative Technology.

Whittaker *et al.*, 1979, *European Solar Buildings*, Stephen George and Partners, London.

Energy Conservation and Thermal Insulation
Edited by R. Derricott and S. S. Chissick
© 1981 John Wiley & Sons Ltd.

CHAPTER 14

The insulation of houses: its impact on heating and ventilation design

S. J. Leach

Building Research Establishment, Garston

INTRODUCTION

The first part of this paper discusses the influence on the design and performance of building services of much higher levels of thermal insulation than are now current. This includes consideration of (a) the requirements to be met for fresh air supply and its relatively increased importance when conduction losses have been significantly reduced; (b) how the sizing of the heating system should be approached; (c) what are the requirements for the control of heat emission to maximize the benefits of and also avoid problems with incidental heat gains; (d) the possible future importance of low- temperature heat distribution; and (e) possible moisture problems associated with high levels of thermal insulation in roofs.

The second part of the paper is devoted to a description of the Building Research Establishment (BRE) experimental low-energy houses laboratories. These are being built with high levels of thermal insulation and will be used along with other studies to examine the consequences of this for services design and operation. The 'houses' include solar panels with seasonal heat storage, mechanical ventilation with heat recovery, and heat pumps, which are used for heat supply, heat recovery and heat management.

VENTILATION

The ventilation requirements of well insulated houses require separate consideration for 'summertime' and for 'wintertime' heating depending on whether or not the prime requirement is to cool the building to avoid the overheating which results from incidental heat gains mainly confined to the summer or whether or not the main need for ventilation is to supply human needs and to control moisture problems.

Summertime ventilation

The ventilation rate required in summer will depend on the thermal response of the building structure to sudden increases of heat input usually arising from sunshine through windows but also sometimes from other activities such as cooking. The thermal response depends on the thermal capacity of the inner surfaces of the house, and is a function of the density of the materials used together with the thermal resistance of the wall. Where thermal insulation is located on the inner surfaces of the outer walls, a rapid rise in temperature can occur. This can be countered to some extent by the use of heavyweight materials in the party wall (when present) and in the internal partition walls. Simplified design aids have been developed for use at the early stage of design which present the interaction between 'weight', glazing area, and the daytime ventilation rate. The basis of these 'summertime design aids' has been described by Milbank (1974) and Petherbridge (1976) and is intended mainly to be used to assist in producing office designs which will have acceptable thermal conditions without energy-consuming air conditioning. However, the principles behind these design aids apply also to house design.

Wintertime ventilation

A supply of fresh air is necessary in wintertime to ensure a comfortable, safe, and hygienic environment, but the heat loss to this air during the heating season will represent a substantial proportion of the total heat loss in the well insulated house. This indicates the need for (a) accurate knowledge of ventilation requirements, (b) the provision of adequate design methodology, and (c) control of gaps in construction to provide just the ventilation to meet these requirements by natural or mechanical means. It also suggests that mechanical ventilation with heat recovery may be worthy of consideration.

Ventilation is conveniently expressed in terms of 'fresh' or 'outside' air rate required to control a particular problem or meet a particular human need. The ventilation rates required depend upon activities within a house and its occupation level. Of the main reasons for supplying fresh air, the oxygen needed for breathing and for diluting the carbon dioxide from breathing leads to the least requirement and will not be further discussed.

The dominating requirements are for moisture and condensation control, together with the dilution of odours, for example from smoking. In certain circumstances, the supply of air to combustion appliances can dominate in particular rooms since an inadequate supply of oxygen can lead to the production of highly poisonous carbon monoxide and its ingress into an occupied space. These aspects are discussed and summarized by the Building Research Establishment (1977).

A simple statement of ventilation requirements like, say, three-quarters of an air change an hour (ACH) for the whole house cannot be made for the reasons given above, but it is clear that in most situations in the UK and for 'typical' use made of a house the requirement is in the range 0.5–1 ACH. For a ventilation rate in the middle of this range and for a conduction loss similar to that of the houses described later in this paper (U value for the opaque parts of the structure 0.35 W/m^2 °C), then the ventilation heat loss rate is about equal to the conduction loss.

It is interesting to examine how close are houses in practice to meeting this requirement. Studies by Warren (1975) of natural ventilation rates in modern housing have found a range of average wintertime whole-house ventilation rates (at the mean windspeed) with windows closed which spans a ventilation rate of 0.75 ACH, with the lowest value of 0.3 ACH and the highest value 1.35 ACH. This suggests that some houses today will be inadequately ventilated unless windows are opened in wintertime for long periods and that for most of the houses studied further reduction in natural ventilation rates by, say, draught stripping of windows and doors could lead to problems, particularly from condensation.

For the well insulated house in wintertime, the heat input from cooking can in most situations be the main source of incidental heat during the heating season. However, on most occasions this heat input will be much greater than the heat requirements of the kitchen and on many occasions will exceed the heat requirement of the occupied rooms of the house at the time of cooking. Mechanical ventilation with heat recovery is the only way of using usefully any but a small part of this incidental heat from cooking. Methods of recovery heat in fully mechanically ventilated houses are being studied in two of the BRE experimental low-energy houses laboratories to be described later and these studies will lead to a better assessment of the problems arising from incidental heat gains and also the possibilities of maximizing the contribution from incidental heat gains to a reduced energy consumption.

HEATING SYSTEM SIZING AND CONTROL

It is well known that the efficiency of fossil-fuel-fired boilers falls with decreasing applied load. Some typical examples for gas-fuelled boilers are shown in Figure 1, taken from a recent paper to the World Energy Conference by Brundrett et al. (1977). Similar results apply to oil- and coal-fired boilers. If a boiler is oversized for a particular application, the implication is that it operates on the lower part of the performance curve for a longer period than the correctly sized boiler. Bloomfield and Fisk (1977) have recently drawn attention to the sizing of the heat distribution system in comparison with the heat requirement of a house and concluded that an oversized heating system has little influence on the annual efficiency when the effects of intermittent use of housing is taken into account, but that it was important not to oversize the boiler with respect to the size of the heat distribution systems. The

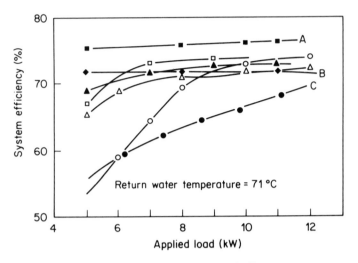

Figure 1 System efficiency versus boiler output

Symbol	Output (kW)	Bench efficiency (%)	Mode of operation
■	8.24	75.6	On/off
◆	14.7–4.7	76–76	Modulating
▲	14.7–5.0	77–75	Modulating
□	14.7	77	On/off
△	14.7	74.7	On/off
○	16.0–10.0	79–73	On/off
●	11.72–8.8	74–71.5	On/off

problems of reduced efficiency from boiler oversizing become important in a well insulated house for several reasons. First, the demand for domestic hot water is a much greater proportion of the total demand and, when the demand for heat by the storage cylinder is satisfied, leaves the whole heat output of the boiler available to satisfy a small demand for space heating. This suggests that sizing the boiler to be able to meet the maximum demand for hot water and space heating will lead to wasted energy, a result known for some time and reported by the Building Research Establishment (1960), but not always adopted in practice. Secondly, the size of the smallest boilers available today is significantly greater (around a factor of two) than the demand of the well insulated house. It is possible to manufacture boilers whose efficiency will be high at low loads and Figure 2 shows measurements made on an experimental boiler developed by British Gas which has ideal characteristics and shows an increasing efficiency with decreasing load. Until this boiler and similar appliances for other fuels are commercially available, there remains a

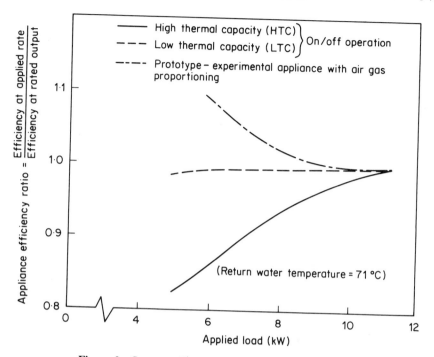

Figure 2 System efficiency ratio versus applied load

problem in achieving the full potential in energy saving from high levels of thermal insulation using fossil fuels in traditional ways.

Another feature of heating systems design which needs attention with high levels of insulation is the control of the heat distribution. It was pointed out in the previous section that incidental gains from the sun and elsewhere can lead to overheating or the need to open windows. To maximize the benefits from these incidental gains, the heating system must be able to switch off and cool rapidly in the parts of the house where these gains occur. This suggests rapid response, for example, from warm air on low-thermal-capacity radiators with either individual room thermostats or zoned heat distribution with thermostats as appropriate. This approach is not commonly adopted today and further R&D is needed before it would become a standard practice in view of the complex issues involved.

Before leaving the topic of heating, a further aspect of general importance is the question of the availability of fuels during the lifetime of a new house. Looking back over the last 60 years, and also trying to look into the future by reference to Cmnd. 7101 (Department of Energy, 1978), leads to the conclusion that over the life of a house there will be changes in the fuels used because of changes in the cost of the most important primary fuels, coal, gas, and oil, and the increased introduction of nuclear electricity. It is not possible to say when changes will take place, it is only possible to say that changes will take place. It is important therefore to retain flexibility so that the house will operate satisfactorily whatever happens. An important aspect for retaining flexibility is the use of low-temperature heat distribution systems, i.e. an oversized heat distribution system. As has been discussed above, this does not in itself lead to inefficiency, only to a small increase in capital cost. However, it would retain the option of introducing a later change without major upheaval or cost to, say, an electrically driven heat pump. Fisk (1976) suggests that it may therefore be worthwhile incurring the small extra cost during construction, to avoid a major cost later.

LOFT SPACE CONDENSATION

In a recent article Cornish (1977) has drawn attention to the need for increased loft space ventilation when the insulation level is increased. Studies in occupied houses were made where polythene vapour checks have been placed at ceiling level during construction. It was found that the vapour checks had little effect on the relative humidity within the roof space, and that water vapour was transferred into the roof space through holes around light drops, hatches, and wall/ceiling joints rather than by diffusion; this re-emphasizes the need to ensure that vapour checks are continuous and consequently airtight.

Investigations have also been made in pitched roofs of typical Scottish construction with timber or substitute timber sarking, of ventilation and airflow rates between dwelling space and loft space. It was found that for a wind speed of 4 m/s the ventilation rate was about 4 ACH and at this wind speed about 20% of the air entering the roof space was moist air coming from the room below. The effect of increasing the amount of insulation at ceiling level in typical two-storey terraced housing was studied and a comparison undertaken of the conditions in 20 pitched roofs with 25 mm of mineral wool at ceiling level and 20 similar roofs with 100 mm (see Figure 3). Average temperature in the well insulated roofs was 0.5°C lower than in the 'standard ' roof. This was sufficient to raise the relative humidity from 84 to 89%, with a consequent increase in the average moisture content of the roof timbers from 17.1 to 18.1%. This compares with an equilibrium moisture content of about 16% for timber under cover outside. In a second set of measurements with the outside temperature down to 1.6°C, the moisture content of the roof timbers in both sets of roofs increased on average by 1.1%. This difference of 1.1% moisture content, which was due to the lower external temperature, puts into perspective the difference of 1% ascribed to the incorporation of higher standards of ceiling insulation. All the information obtained from these field studies has been used in defining the parameters in the mathematical model,

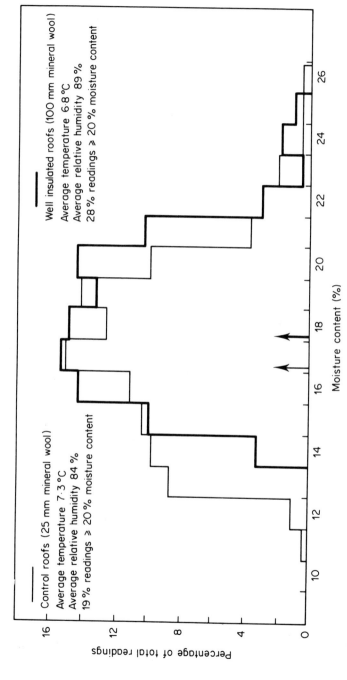

Figure 3 Effect on the moisture content of roof timbers of increasing the ceiling level insulation in terraced houses. Average external temperature is 6.2°C

and the results from the model are in good agreement with observed conditions in lightweight roofs. The model will be used to give design guidance on loft space ventilation for moisture control as a function of the insulation level.

EXPERIMENTAL LOW-ENERGY HOUSE LABORATORIES

In 1976 the insulation level of UK housing was raised substantially for new designs through an amendment to the Building Regulations to give a U value of 1.0 W/m^2 K for the opaque parts of the perimeter. BRE experimental low-energy house laboratories take this a stage further and incorporate the equivalent of 100 mm of insulating material in the walls and roof with 50 mm in the floor, giving a U value of about 0.35 W/m^2 K for the opaque surfaces. Three experimental houses are being built using different energy conserving services.

The construction materials used are traditional. The first two houses are timber-framed with brick cladding and 100 mm of insulation behind plasterboard. The third house is brick and blockwork with 60 mm of polystyrene insulation in a 110 mm cavity. Figure 4 shows the plan of two of the low-energy houses. The left-hand one is the 'heat recovery house' and the right-hand one the 'solar house'. The centre house is there for *ad hoc* energy studies.

The heat recovery 'house' uses mechanical ventilation based on the French 'Aldes' system with ducting for input and extract air and incorporating a static heat exchanger between them as shown in Figure 5. The heat exchanger will initially be the one supplied by 'Aldes' but later on other exchangers will be studied. Waste water from the bathroom and from the washing machine (but

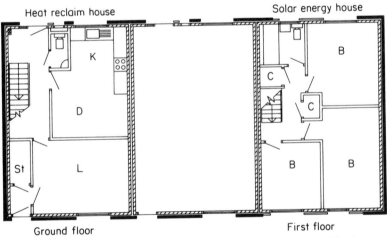

Ground floor   First floor

Figure 4   Plan of two of the BRE low-energy experimental houses

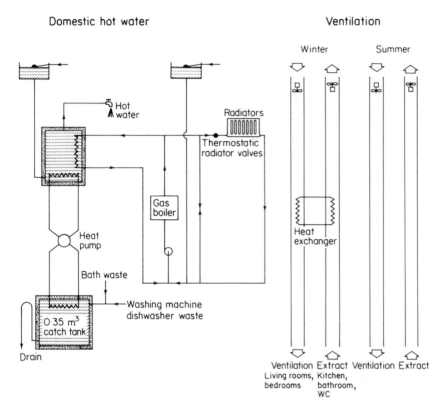

Figure 5 Heat reclaim house

not from the kitchen sink) will be put into an insulated dump tank and the heat returned to the hot water storage cylinder by a small heat pump immersed in a dump tank.

Figure 6 shows the energy system for the solar 'house' which has 22 m² of solar collector on the roof. The solar energy is stored in a 35 m³, well insulated tank of water. Heat pumps are used to assist the solar collector and to extract the stored heat and provide a useful temperature for distribution by fairly conventional systems. The only slightly unusual feature of the distribution system is that the radiators will be larger than usual and thermostatic control valves will be fitted to each radiator. In this house, heat pumps are used mainly to transport energy within the building from storage at a low temperature to storage at a higher temperature. Source temperatures from 5°C up to about 40°C will be encountered. Heat pump 2 is not used when the 35 m³ tank is above 40°C.

In the main storage tank, the water temperature may be raised to 60°C by solar heating during the summer using direct heat exchange. However, in

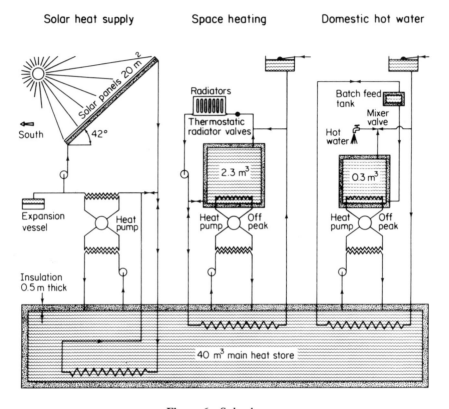

Figure 6   Solar house

order to be able to take full advantage of the capacity of the water, it will be necessary to use heat pump 1 when the solar radiation is inadequate.

The main tank provides the main supply of thermal energy for space heating and domestic water heating in the house. The heat distribution will be, as has been mentioned, by low-temperature radiators, and whilst the tank is hot enough heat will be taken directly through static heat exchangers for these circuits. As the temperature of the tank falls, heat pump 2 and heat pump 3 will be brought into operation.

Heat pump 2 runs on off-peak electricity and puts heat into the 1 m^3 tank. Similarly heat pump 3 is used to provide domestic hot water and has a heating rating of about 3 kW at the lowest evaporator temperature.

Figure 7 shows the heat pump 'house'. There is a mechanical ventilation system and an air solar preheater. Figure 8 shows the summer and winter operation modes. The main feature is the use of an air-to-air heat pump as the means of space heating. There is in addition mechanical ventilation with heat

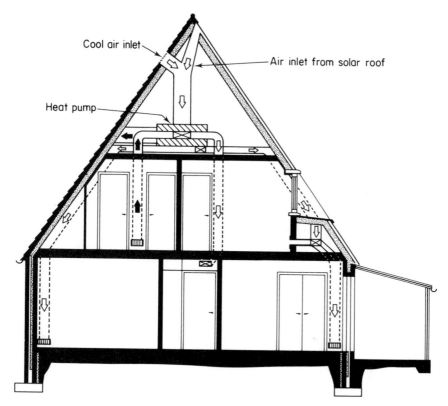

Figure 7   The experimental heat pump house

recovery by a second heat pump and the heating of the domestic hot water is provided by a third air source heat pump. The main supply of outside air to these machines is passed beneath the roof slope to provide prewarming when possible. In summer the roof air supply is used only to preheat the domestic hot water. The ventilation make-up is taken from an inlet in the north side of the roof in summer conditions. Further details are given by Leach (1977) and Seymour-Walker (1976, 1977).

Figure 9 shows the 'heat recovery' and the 'solar' houses at the stage in construction reached in March 1978. The heat recovery house is on the left and the solar house on the right. The first of these houses is operational and the period until the next heating season is being used to develop and install the instrumentation. The houses will not be occupied. The running of the houses will be automated and they will be operated according to both typical cycles of activity and also extreme cycles of activity, so that the limits of the services systems can be explored and the benefits from them and the thermal insulation evaluated.

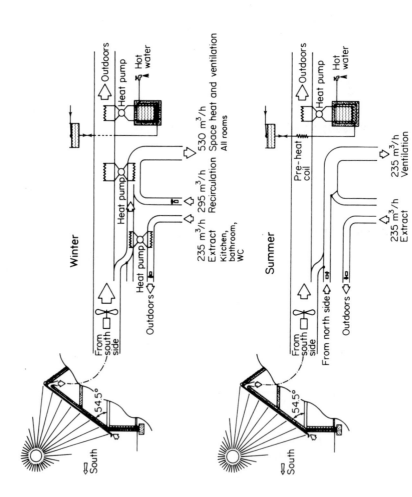

Figure 8 Heat pump house

Figure 9 'Low-energy houses' laboratories

ACKNOWLEDGEMENT

This paper is based on work of the Building Research Establishment and permission to use the work is acknowledged. The views expressed are those of the author alone.

REFERENCES

Bloomfield, D. P. and Fisk, D. J., 1977, Seasonal domestic boiler efficiencies and intermittent heating, *The Heating and Ventilating Engineer*, September, pp. 6–8.

Brundrett, G. W., Leach, S. J., Parkinson, M. J., Pickup, G. A., and Rees, N. T., 1977, Research into energy conservation in dwellings, *Proc. World Energy Conf.*, Istanbul, 19–23 September. Also *Coal and Energy Quarterly*, 1977, no. 14, pp. 19–30 and no. 15, pp. 14–27; and *Gas Engineering and Management*, 1977, December, pp. 430–441.

Building Research Establishment, 1960, Domestic heating and thermal insulation, *BRE Digest*, 133 (first series).

Building Research Establishment, 1977, Ventilation requirements, *BRE Digest*, 206, October, HMSO, London.

Cornish, J. P., 1977, Climates within roof spaces, *BRE News*, 41, Autumn.

Department of Energy, 1978, *Energy Policy, A Consultative Document*, Cmnd. 7101, HMSO.

Fisk, D. J., 1976, *Energy Conservation; Energy Costs and Option Value*, Building Research Establishment Current Paper CP 57/76.

Leach, S. J., 1977, Heat pump application in houses, *Electrowarme International*, Edition A 35, September.

Milbank, N. O., 1974, *A New Approach to Predicting the Thermal Environment in Buildings at the Early Design Stage*, Building Research Establishment Current Paper CP 2/74.

Petherbridge, P., 1976, Design aids and summertime energy for building services, *Building*, 26 March, pp. 103–104.

Seymour-Walker, K. J., 1976, BRE proposals for experimental low-energy houses, *Proc. 1976 Symp. of the International Council for Building Research Studies and Documentation (CIB)*, Garston, April.

Seymour-Walker, K. J., 1977, Experimental low-energy houses at the Building Research Establishment, *Proc. CICC Conf. on Ambient Energy in Building Design*, Nottingham, April, pp. 125–133.

Warren, P. R., 1975, Natural infiltration routes and their magnitude in houses—Part 1: Preliminary studies of domestic ventilation, *Proc. Conf. on Controlled Ventilation—Its Contribution to Lower Energy Use and Improved Comfort*, Aston Univ., September.

Energy Conservation and Thermal Insulation
Edited by R. Derricott and S. S. Chissick
© 1981 John Wiley & Sons Ltd.

CHAPTER 15

Insulation and condensation problems

V. I. Hanby

University of Nottingham

INTRODUCTION

The control of water is one of the primary requirements in the design of buildings. Water, in its three phases, is a principal agent in the deterioration of building structures via chemical, physical or biological processes. The unexpected presence of water may be revealed by a number of effects.

Many building materials expand as their moisture content increases, and contract as they dry out, often to less than their original size. This can result in the breakdown of the material itself or can lead to a splitting away from an adjacent material. Frost damage is not uncommon, particularly on roofs where the surface temperature may fall markedly below the ambient air temperature; blistering may also occur underneath membranes. High moisture contents may cause leaching and efflorescence which can carry away bonding materials, and electrolytic corrosion of metal structures can occur in damp areas. Biological deterioration is one of the most common symptoms observed, in particular wood rot and fungal growths on internal surfaces. In many cases, the origin of the water is difficult to identify, as there are several possible sources. The most common of these are rain penetration, capillary attraction of water from the ground, and internally generated moisture. This paper is concerned with problems resulting from the production of moisture inside buildings and how increased levels of insulation and design for reduced heat loss may be expected to affect the problem.

The moisture content of the air inside an occupied building is generally higher than that of the outside air which ventilates the building. This water vapour may have come from the occupants themselves, from their activities (e.g. cooking, washing) or from some process in an industrial building. Industrial condensation problems tend to be more avoidable than those in the domestic sector, possibly because of the more definite information available to the designer in the first instance. Most of the discussion in this paper is

351

concerned with condensation in dwellings. Despite the fact that the basic physics of the phenomenon is well understood, and a not-inconsiderable amount of material on this problem has been published (see, for example, Croome and Sherratt, 1972), it remains an irritatingly unpredictable problem in many instances, and one which is, regrettably, still frequently not considered by designers.

We can identify three main ways in which condensation of water vapour may occur in buildings.

(i) Condensation may occur on any surface which is below the dewpoint of the air that is in contact with it. This is frequently observed on windows where its effects may range from minor inconvenience to deterioration of window frames and the surrounding structure. The occurrence of high relative humidities on wall and roof surfaces can give rise to fungal growth and further deterioration.

(ii) Water vapour will diffuse from the region of high vapour pressure inside the building towards the lower vapour pressure outside, passing through the structure. The temperature gradient through the structure can cause internal condensation to occur (interstitial condensation).

(iii) Pressure differences across the building skin generated by wind or buoyancy forces cause air to leak through the structure. Where the flow is in an outwards direction, as the local temperature falls the possibility of condensation arises.

INSULATION AND CONDENSATION

To discuss the likely effects of high insulation levels on condensation we must first consider how the increased insulation level affects the internal environmental variables which influence the problem. It is necessary to consider insulation in a wider sense than that of reducing fabric losses, as ventilation also plays a pivotal role in determining both the heat losses from a building and the state of the air inside it.

The effect of structural insulation on heat loss pattern may be simply illustrated by taking as a model a 90 m^2 floor-area semidetached house, with single glazing and the walls insulated to the current Building Regulations standard of 1.0 W/m^2 °C and the roof to 0.6 W/m^2 °C. The ventilation is assumed to be 1.5 air changes per hour. This dwelling has a design heat loss of 276 W/°C. By insulating the walls to 0.3 W/m^2 °C and the roof to 0.2 W/m^2 °C (the 'insulated house'), the heat loss is reduced to 223 W/°C and the distribution of the heat losses is markedly changed (Figure 1). With ventilation losses possibly accounting for up to half the total, it is evident that some attention must now be directed towards this problem, but it must be done with care, as one of the functions of ventilation is to control internal moisture content.

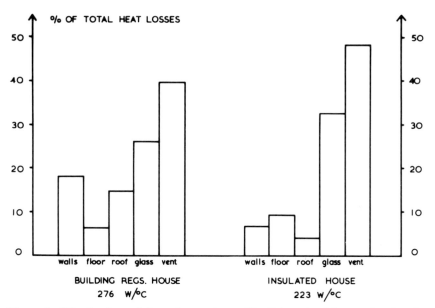

Figure 1 Distribution of heat losses in a 90 m² house with two levels of structural insulation

If we examine the internal air and mean surface temperatures in the two cases, when heated to a common environmental temperature (two-thirds mean radiant temperature plus one-third air temperature) by a largely convective source, we see that the insulated house has a higher mean surface temperature and a lower air temperature than the Building Regulations house. The inside environmental temperature is taken as 20°C and the outside air temperature 5°C.

	Mean surface temp. (°C)	Air temp. (°C)
Building regulations house	17.9	24.3
Insulated house	18.5	23.1

For the given ventilation rate and a typical rate of moisture input, the relative humidity would be about 4% higher in the insulated house, with the dewpoint increased by about 0.1 °C. The net effect of the insulation on surface condensation with these considerations would be to reduce the overall risk of its occurrence everywhere except on windows, where it may be marginally worse. It is shown later how the relative location of the insulation affects the problem.

VENTILATION AND CONDENSATION

It has already been mentioned that a building designed for low heat losses should have some kind of ventilation control. One and a half air changes per hour is not an unusually high figure yet several workers (Rosengren and Morawetz, 1976; Loudon, 1971) have suggested that an average rate of 0.5 air changes per hour is an adequate rate. The effect of a reduction in the ventilation rate on the state of the inside air is illustrated by Figure 2. Here the outside air is assumed to be at 5°C and 80% relative humidity (RH) and the inside air at 20°C. The moisture input is assumed to be 0.6 kg/h except during the hours 0700–0900 and 1600–1800, when it is taken as 1.2 kg/h. The curves illustrate the resulting fluctuations in internal relative humidity. The effect of ventilation rates on surface condensation in houses has been discussed by Loudon and Hendry (1972) who have shown the rates necessary to avoid such problems in a number of domestic situations.

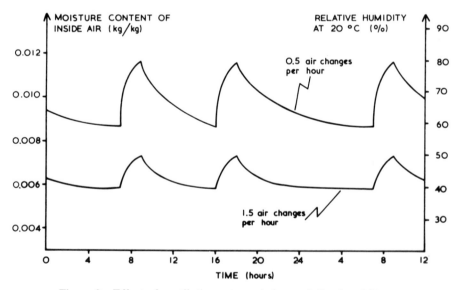

Figure 2   Effect of ventilation rate on indoor relative humidity

HEATING AND CONDENSATION

The previous section has raised the essentially nonsteady nature of the psychrometric and heat transfer processes which occur in occupied buildings. The

simple steady-state analysis presented earlier suggested that surface condensation problems might be reduced with better insulated wall surfaces for a
given level of thermal comfort. This only touches on part of the problem.
Many instances of condensation, particularly those arising in kitchens, have
been shown to arise from the variations in moisture content of the air and
surface temperatures resulting from intermittent occupancy.

Changing social patterns have led to the common situation where high
moisture inputs occur in the morning and evening and the heating is turned
off both during the night and during the unoccupied period in the day. This
trend has been facilitated by modern, fast-response heating systems and a
desire for reduced heating bills. If this variation in heat input is not matched
by the thermal response of the interior surfaces, the wall surface temperature
can lag behind the air temperature sufficiently to present a surface below the
dewpoint of the air for a significant period. It is the nature of the wall lining
which determines this speed of response rather than the overall U value of the
construction. Hence it is possible to have two walls of identical U value, one
of which responds rapidly to fluctuations in heat input by having the insulation near the interior of the wall (e.g. an insulating wallboard) and one which
displays a high thermal capacity by having the insulation nearer the outside
(e.g. a brick–brick cavity wall filled with urea-formaldehyde foam). Figure 3
illustrates a situation which could occur in the morning, when the heating is

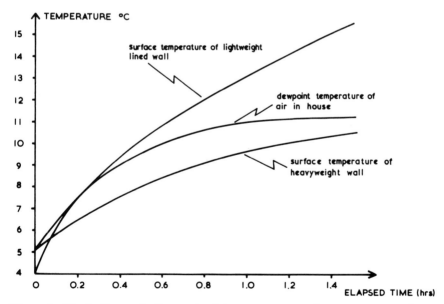

Figure 3 Illustration   of   the   potential   occurrence   of   temporary   surface
condensation in a dwelling

switched on and cooking initiated in both a fast- and a slow-response room. The curves were obtained using an electrical analogue.

To avoid this type of surface condensation, it is essential that the thermal characteristics of the structure are matched, as far as is possible, to the likely usage of the dwelling. Unfortunately the best location of the insulation, from an energy conservation point of view, is by no means a clearcut issue. If attention is confined to the heat input to the building, a lightweight lined, intermittently heated building will apparently use less energy because its mean internal temperature will be less than that of a similarly insulated heavier building (it is assumed that a given level of thermal comfort is to be maintained for a specified period in each case). This argument does not take into account heat gains from other sources, notably occupants, machinery, and solar gains. These adventitious heat gains become more significant, in terms of seasonal energy consumption, in better insulated buildings. A room can respond so quickly to solar gains, for instance, that it has to be cooled (by opening windows) and a useful source of heat may thus be wasted. These gains are most significant in the spring and autumn and act so as to shorten the heating season.

INTERSTITIAL CONDENSATION

The use of modern insulating materials has generated increased interest in the problem of interstitial condensation. Due to the porous nature of these substances, they tend not to have a high resistance to the diffusion of water vapour, but their high thermal resistance produces a steep temperature gradient across them, giving rise to a risk of interstitial condensation. This is particularly the case if the insulation is located near the inside of a structure. The principles of the calculation of the vapour pressure and temperature gradients through walls and roofs are adequately described elsewhere (British Standards Institution, 1975; Institution of Heating and Ventilating Engineers, 1970; Seiffert, 1970), but a few comments on the conclusions which might be drawn from such calculations would seem appropriate here. The steady-state calculations are based on values of resistance to diffusion of water vapour for common building materials, often expressed as a multiple of the resistance of air at the same temperature. These values, which are obtained in laboratory tests, may be in serious error when used in a practical calculation, as has been pointed out by Rodwell (1977) in relation to work on insulating wall linings.

The inside and outside conditions selected for the calculation are also important. Figure 4 shows the steady-state diffusion calculation performed for a brick–block cavity wall with the cavity filled with urea-formaldehyde foam. The calculations have been done for 'winter design conditions' of

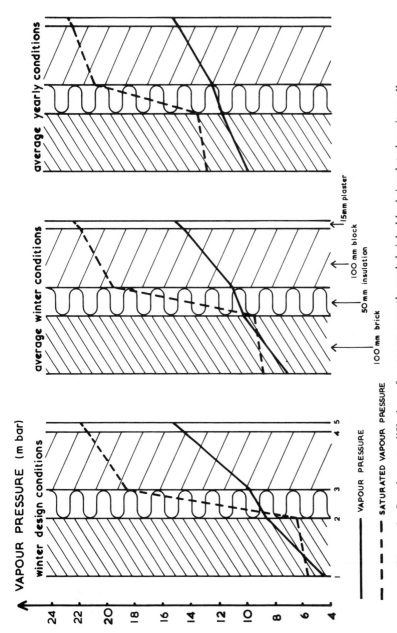

Figure 4 Steady-state diffusion of water vapour through brick–block insulated cavity wall

$-1°C$, 80% RH outside and 20°C, 65% RH inside; 'average winter conditions' of 5°C, 80% RH and 20°C, 65% RH; and for 'average yearly conditions' taken as 10.5°C, 78% RH and 20°C, 65% RH. The expected condensation zone is evident in the first two situations but the 'average year' shows no intersection of the curves. This illustrates that moisture accumulation during severe weather may dry out during the remainder of the year, particularly if the accumulation rate is low in relation to the moisture capacity of the relevant portion of the structure. If we assume that the inside and outside vapour pressures are those corresponding to the given conditions and that the vapour pressure at point 2 corresponds to the saturated vapour pressure, a balance can be constructed from the vapour pressure differences and the appropriate resistances, i.e. $5 - 2$ and $2 - 1$. This gives for 'winter design conditions'

Diffusion rate from room into wall	1.31×10^{-4} g/m^2 s
Diffusion rate from wall to outside	0.52×10^{-4} g/m^2 s
Net accumulation rate in wall	$\overline{0.79 \times 10^{-4}}$ g/m^2 s

Severe conditions such as these do not persist for long periods. For average winter conditions the corresponding net condensation rate is 0.225×10^{-4} g/m^2 s. If we assume that this moisture is deposited in the outer 25 mm of the insulation layer, then to change the moisture content of this layer by 1% would take 128 days at average winter conditions and 26 days at the winter design condition. Variations of moisture content in plastics insulation materials greater than 1% have been measured in protected membrane roofs (Hedlin, 1976) without ill-effects. The consequences of variation in moisture content of the materials involved is clearly an important consideration in any condensation risk assessment.

Of course, one would not normally expect interstitial condensation problems in a wall structure of this kind. It has been chosen simply as a familiar vehicle for illustrating certain principles. These techniques can yield valuable information on potentially troublesome structures, particularly those with a high diffusion resistance layer on the outside.

In view of the important roles of thermal and moisture capacity, the problem should ideally be treated as a nonsteady phenomenon. However, an accurate simulation will require a model which must take into account a number of other effects. For instance, the release of latent heat as condensation has been found to affect the simulation, and the increase in the moisture content of the material following condensation will modify the thermal resistance of the material and hence further affect the temperature distribution. An additional complication can occur in materials which have a significant number of small-diameter pores. A liquid surface in such pores will be curved, and hence vapour can exist in equilibrium with liquid at a pressure lower than

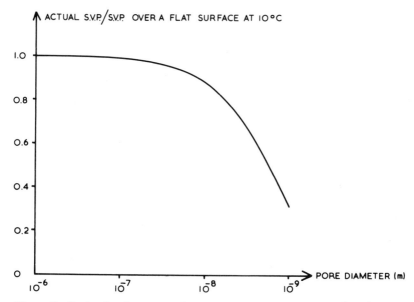

Figure 5   Reduction in saturated vapour pressure over a curved surface

that which would be required over a flat surface. Figure 5 shows the relationship between pore diameter and reduction in saturated vapour pressure. It can be seen that this effect is only significant in materials which have a considerable number of pores less than 10^{-8}m in diameter. Building materials that fall into this category include Fletton brick and concrete (Whiteley, *et al.*, 1977).

AIR LEAKAGE

Condensation may occur via air leakage through the structure caused by pressure differences set up by effects of wind and buoyancy—the agencies which promote natural ventilation. This is a likelier source of trouble than diffusion-controlled migration, although it tends to be considered less often. Much greater quantities of water can exfiltrate through walls by this means than can occur with diffusion, as has been pointed out by Baker (1973) and Ritchie (1976). The moisture condensing in the wall used as an example in the previous section could amount to 0.009 g/m^2 s, or two orders of magnitude higher than the diffusion-controlled rate. Although this problem is unlikely to be significantly affected by the degree of insulation, attention has recently been focused on a related problem which occurs in pitched roofs.

   The upwards leakage of air from the house into the loft space will increase the moisture content there depending on the leakage rate and the ventilation rate in the loft itself. Insulation of the ceiling and joists will tend to lower the temperature in the loft space and hence introduce a risk of condensation. The problem is complicated by the fact that nonsteady-state methods have to be used in its investigation, particularly so in respect of the consideration that the roof temperature is affected by radiation exchange and can be rather higher or lower than the ambient air temperature. A study of this problem has been briefly reported by the Building Reasearch Establishment (1978). The most effective way of avoiding the problem would appear to be to attempt to reduce the upwards leakage of air from the house into the loft by making any cracks or holes on the ceiling as small as possible. Baker (1973) has reported measured orifice areas of just under 18 000 mm^2 in a ceiling of 120 m^2, with actual pressure measurements indicating a total open area of around twice this figure.

CONCLUSION

The objective of this review has been to discuss some of the problems that reducing the heating requirement of buildings, particularly dwellings, may produce in the area of condensation control. It is apparent that the total thermal performance of the building is of great significance in this respect and that a clear understanding of the basic principles involved should assist those concerned with reducing energy consumption in buildings to avoid condensation problems. It is the opinion of this author that the question of interstitial condensation (diffusion-controlled) is a subject of overmuch concern, and that the solution sometimes advocated of locating the insulation towards the outside of the wall may result in surface condensation problems arising from the thermal inertia of the inner structure.

Although the basic physics of the condensation phenomenon are well understood, the application of these principles of thermodynamics to the building situation in detail is extremely complex. It is perhaps inappropriate at the current state of the art to spend too much time on calculations at the design stage when the combination of an understanding of the basic principles coupled with a good helping of commonsense is likely to prove more than adequate.

REFERENCES

Baker, M. C., 1973, *Moisture Problems in Built-up Roofs*, National Research Council of Canada Division of Building Research Technical Paper no. 390.
British Standards Institution, 1975, BS 5250:1975.
Building Research Establishment, 1978, Condensation in domestic pitched roofs, *BRE News*, **42**, 2–4.
Croome, D. J., and Sherratt, A. F. C., 1972, *Condensation in Buildings*, Applied Science Publishers, Barking.
Hedlin, C. P., 1976, *Moisture Content in Protected Membrane Roof Insulations—Effect of Design Features*, American Society for Testing and Materials Special Technical Publication 603.

Institution of Heating and Ventilating Engineers, 1970, *IHVE Guide*, A10.
Loudon, A. G., 1971, Building Research Establishment Current Paper CP 31/71.
Loudon, A. G., and Hendry, I. W. L., 1972, *Ventilation and Condensation Control, Condensation in Buildings*, Applied Science Publishers, Barking, p. 44.
Ritchie, T., 1976, Moisture degradation of masonry walls, *Proc. First Canadian Masonry Symp.*, University of Calgary, pp. 66–71.
Rodwell, D. F. G., 1977, Water vapour diffusion through plasterboard wall linings and damaged vapour barriers, *Building Services Engineer*, **45**, 69–74.
Rosengren, B., and Morawetz, E., 1976, The Termoroc House, *Proc. Conf. on European Solar Houses*, North East London Polytechnic.
Seiffert, K., 1970, *Damp Diffusion and Buildings*, Elsevier, Amsterdam.
Whiteley, P., Russman, H. D., and Bishop, J. D., 1977, *The Porosity of Some Building Materials*, Building Research Establishment Current Paper CP 21/77.

Energy Conservation and Thermal Insulation
Edited by R. Derricott and S. S. Chissick
© 1981 John Wiley & Sons Ltd.

CHAPTER 16

Thermal insulation and fire

E. G. Butcher and A. C. Parnell

Fire Check Consultants

INTRODUCTION

It has been obvious for many years that the use of combustible linings to ceilings and walls can add to the potential bonfire in the event of a fire, and most countries of the world have controlling legislation to limit the combustible nature of these surfaces.

With the increased awareness of the need to conserve energy, materials of a polymeric nature have become available which can be easily applied to walls and ceilings to improve the thermal insulation. Many of the statutory requirements controlling surface spread of flames refer to a test which, in origin, applied to the composite material as a whole, but chemists and physicists quickly developed surface coatings which would retard the spread of flames but which did not neccesarily inhibit the effect of such heat on the backing insulation. A classic example of this can be seen with fire-retardant paint finishes applied to polystyrene. The spread of flame along the surface is controlled by the fire-retardant nature of the finish, whereas the polymeric material is dripping to the floor. An approval certificate can, therefore, be issued for the surface quality without making any reference to the overall performance of the material and its ability to cause further combustion remote from the source. The Fire Research Station has published papers demonstrating that the use of this material on the ceilings need not be so damaging as long as it is totally adherent to the base strata without any air spaces (Malhotra, 1971).

Another material which has, until recently, been widely used to improve thermal insulation is expanded polyurethane, mostly in board form. This material has very good thermal insulating qualities and it is demonstrated later in this paper that this quality causes a rapid escalation of any heat source in the room at risk to a flashover condition when the whole room can be consumed by flames and the flame source, on reaching the polyurethane ceiling material, provides total combustion and very rapid disintegration of the material. However, the use of this material in traditional forms of construction normally provides additional fuel in the floor spaces, roof spaces, and roof linings, etc., which develops into a total conflagration. If the material were to be used to provide rapid ventilation for the fire and exhaustion of smoke to atmosphere, some merit could be considered for the use of exposed thermal insulating linings. However, the heat loss to atmosphere would probably negate any of the advantages of its use to insulate internally.

The use of thermal insulating materials, therefore, which are easily ignitable is generally deprecated by legislation throughout the world unless an insulating barrier of a fire-resisting material is provided on the exposed faces. The combined effect of this type of construction is discussed later in this paper with reference to thermal diffusivity.

NONCOMBUSTIBLE INSULATION

In the foregoing it has been shown that combustible thermal insulation material can provide a ready path for the rapid spread of fire. In this context it would be right to assume that if this thermal insulation were replaced with noncombustible insulation then the hazard associated would be eliminated.

Unfortunately the effect of thermal insulation in a building is a very complicated one and the hazard is not confined to the spread of fire through the body of the material. Other factors concerned with the thermal properties of the material are involved, and it can be shown that the fire hazard in a building can be enhanced even if the insulation is noncombustible.

FIRE SEVERITY

The factors which determine the severity of a fire in a building are several and their effects are interrelated in a complicated way. Figure 1 indicates this in very general terms. If the six boxes on the right-hand side of the figure are considered, it will be seen that the fire severity depends not only on the combustible material which is burning but also on features of the building in which it is contained.

Experimental confirmation of this statement is not lacking and Butcher *et al.* (1966), Thomas *et al.* (1967), and Thomas and Heselden (1972), for example, study this aspect of fire behaviour. All of this work has shown that changes in building geometry can have a major effect on fire severity. For instance, Butcher *et al.* (1966) show that in the particular room used for the fire tests, halving the area of the window opening had the same effect on the fire severity as doubling the amount of combustibles burning.

In this example, the larger window opening allowed greater heat loss from the fire and kept the maximum temperature of the fire lower. In the same way the heat lost through the walls and ceiling of the fire room can also contribute to the factors which control fire severity (see Figure 1).

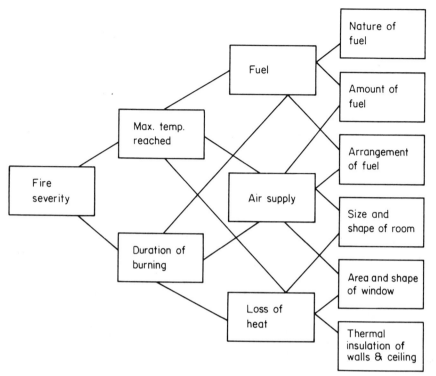

Figure 1   Factors that affect fire severity in buildings

IMPORTANCE OF THERMAL PROPERTIES

Experimental evidence of the importance, should a fire occur, of the thermal properties of the materials which form the walls and ceiling of a room has been accumulating over the years.

Tests were carried out at the Fire Research Station (1960, 1962) using different wall lining materials. In the reports of these tests it is stated 'that the thermal properties of the wall surfaces have a significant influence on the development of fire in rooms of small size. In this context the limits of size of a small room have not yet been determined.'

As far as is known, the additional work to correlate the adverse effect of the thermally insulating wall linings has not yet been done and it is suggested in this paper that the effect does not only apply to small rooms.

The results of these tests are given briefly in Table 1. In the tests the walls only were lined with the material stated, the ceiling in every case was non-combustible. The flashover time recorded in the third column is the time taken for the whole contents of the room to become involved in fire, the fire

Table 1   Effect of wall lining on 'flashover' time in a test room (Source: Building Research Establishment, 1976)

	Wall lining	Flashover time
Test 1	Dense noncombustible material (brick)	23 min 30 s
Test 2	Fibre insulating board with skim coat of plaster	12 min 0 s
Test 3	Hardboard with two coats of flat oil paint	8 min 15 s
Test 4	Sprayed asbestos[a]	8 min 0 s

[a]This material in sprayed form not used in the UK since 1970 (see Michaels and Chissick, 1979).

having been started in a single item of furniture. This event, the flashover, usually happens quite suddenly and the time taken for it to occur can be regarded as a measure of the rate at which the fire grows.

A repeat of Test 4 using the same coating which had been completely dried out during Test 4 gave a flashover time of 4 min 30 s.

The results of these tests show clearly that combustibility of the wall lining is not the only factor which can cause rapid growth of fire in a room.

Other experimental evidence is available. Also at the Fire Research Station, in the course of an extensive programme of large-scale fire tests, fires were staged in which the walls and ceiling of the test room were lined with mineral wood slabs to increase the thermal insulation, and the results were compared with those obtained with walls of vermiculite plaster and a ceiling of refractory cement slabs. These tests were reported by Butcher et al. (1966, 1968) and by the Fire Research Station (1968).

In the analysis of the results of this work, Heselden (1968) stated that temperatures in the compartment, gases, and the rates of heat transfer were only slightly higher in the lined experiments, and this is attributed to the fact that, in any case, only about 30% of the heat generated in the fire was transferred to the walls. However, a close inspection of the temperature rise curves for the compartment for the insulated and uninsulated conditions reveals that the rate of temperature rise for the insulated tests was about double that for the unlined tests ($160°C/min$ and $100°C/min$, compared with $88°C/min$ and $53°C/min$ respectively).

A similar conclusion, that increased insulation causes a faster rise of temperature in the fire room, can be extracted from the results, reported by Morris and Hopkinson (1976), of six fire tests staged in a mock-up of a two-storey dwelling in which four different kinds of ceiling material were used (plasterboard, fibre insulating board, polyurethane board, and polyisocyanurate board).

Interest in the problem is not confined to this country. In America Degenkolb (1976) published a report describing large-scale fire tests carried out in a specially erected building, which was 80 ft long, 24 ft wide, and 15 ft in height, with large door openings in one end wall and halfway along one long wall, by the Jim Walter Research Corporation of St Petersburg, Florida.

The building was constructed of a wooden framework with substantial-sized members, and it had an external cladding of 26 gauge metal sheet.

Fires were staged in the building as described above, without any additional lining, and then with insulation placed inside the wood framework and then in another test placed inside the building fixed to the wooden framework so that this latter was completely covered.

The lining materials used were 3 inch (75 mm) thick vinyl-faced fibreglass, 1 inch (25 mm) thick Styrofoam and 1 inch thick aluminium-foil-faced Thermax 600, an isocyanurate foam construction made by Celotex.

Degenkolb states that the results indicate that the placement of insulation, whether combustible or noncombustible, will have a profound effect on the burning characteristics of combustibles inside the building, be they contents or components of the building itself.

The experimental work cited above indicates that a fire problem exists in connection with the thermal insulation of a building. These do not by any means represent a comprehensive survey; many more examples could be found by a search of the literature. However, those described are sufficient to establish that there is sound experimental evidence to support the idea that the thermal properties of a wall or ceiling material are important in determining the severity of a possible fire in that enclosure.

The thermal properties of the wall and ceiling materials

For the purpose of the present discussion, the important thermal properties of the materials used for walls and ceilings are:

(i) the thermal conductivity (k), usually expressed in watts per metre (W/m);

(ii) the density (ρ), expressed in kilograms per cubic metre (kg/m^3); and

(iii) the specific heat (c) or the thermal capacity, expressed in joules per kilogram (J/kg).

Here it is worth noting that 1 watt (W) is equal to 1 joule per second (J/s). In heat flow calculations, the quantities are combined to form two other quantities, namely:

(iv) the thermal diffusivity, which is the thermal conductivity divided by the product of the thermal capacity and the density, and is expressed in square metres per second (m^2/s)

Thermal diffusivity $= k/\rho c$

(v) the thermal inertia, which is the thermal conductivity multiplied by the product of the thermal capacity and the density, and is in units of J^2/m^4 s or W^2 s/m^4

Thermal inertia $= k\rho c$

In the normal everyday heat flow conditions in a building, the insulation properties of a lining or wall material are often expressed by the U value, which is essentially the thermal conductivity k divided by the thickness of the material (see Building Research Establishment, 1972; Milbank and Harrington-Lynn, 1974).

However, the use of this steady-state quantity is not appropriate for the fire situation. In the latter the heating is rapid and the lining or wall material takes up heat itself but if its thermal capacity per unit volume (the product of the specific heat c and the density ρ) is low then it heats up quickly.

The rate of this temperature rise is proportional to the product of the three thermal properties listed above, or $k\rho c$, which is called the thermal inertia of the material.

In the rapid heating condition of a fire situation, the full thickness of the wall material is rarely heated; the depth of heating depends on the thermal conductivity k, on the thermal capacity per unit volume ρc, and on the duration of heating t.

Using standard heatflow analysis, as described by Carslaw and Jaeger (1959), the temperature rise of the insulant surface and the depth of heating in a fire situation can be calculated from the following equations:

$$\theta = \frac{2I}{\sqrt{\pi}} \sqrt{\frac{t}{k\rho c}}$$

and

$$L = \sqrt{\left(\pi \frac{k}{\rho c} t\right)}$$

where θ is the temperature rise of the surface, I is the heat radiation falling on surface from a fire, and L is the depth of heating.

Since the thermal properties are of such importance in the present discussion, it is of interest to consider their values for some materials likely to be used for walls and ceilings. These are given in Table 2.

Table 2 Physical properties

Material	Thermal conductivity, k (W/m °C)	Density, ρ (kg/m³)	Thermal capacity, c (J/kg °C)	Thermal diffusivity, $k/\rho c$ (m²/s)	Thermal inertia, $k\rho c$ (W² s/m⁴ °C²)
Plasterboard	0.16	950	837	2.0×10^{-7}	127 000
Fibre insulating board	0.057	300	1670	1.1×10^{-7}	28 600
Sprayed asbestos	0.075	240	816	3.8×10^{-7}	14 700
Foam polyurethane board (PU)	0.02	30	1260	5.3×10^{-7}	753
Expanded polystyrene (EPS)	0.038	16	1210	1.9×10^{-6}	735

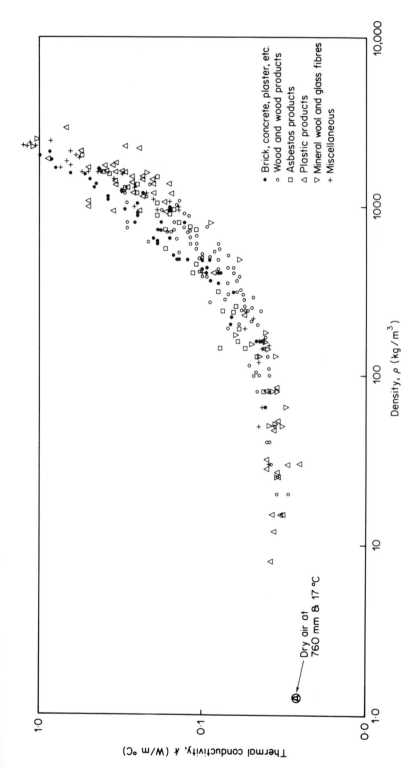

Figure 2 Relation between thermal conductivity k and density ρ

A glance at the formulae above shows that the rise in surface temperature depends on the thermal inertia, $k\rho c$. Table 2 shows that for the materials listed, this quantity varies over a range of nearly 200-fold compared with a variation in thermal conductivity of fourfold.

Thus the rise in surface temperature for a wall lining material is very sensitive to changes in their thermal properties, but the depth of heating, proportional to thermal diffusivity $k/\rho c$ is considerably less sensitive. In Table 2, $k/\rho c$ only varies over a tenfold range.

It is perhaps important to note that for very good insulants the value of k tends to a constant (approximately that for dry air) but the density ρ will continue to vary so that variation in the rate of rise of surface temperature for given heating conditions may also vary.

Figure 2 shows a plot of k against ρ for a range of values appropriate to those found in thermal insulation used in buildings.

Surface temperature of insulation material when used as a room lining and in a fire situation

In order to get an exact picture of the heating of the room surfaces from a fire, a complete heat balance of the room would need to be established which would take account not only of the heatflow, convective and radiative, into the walls and ceiling but also the heat loss back into the room from these surfaces. This is extremely difficult but for the purpose of developing the argument concerning the effect of thermal insulation on fire severity it is sufficient to use the simplified picture suggested below. All fires are different and so a truly realistic model is almost impossible to postulate, but that suggested below will indicate the difference which arises between the behaviour of the various materials, and it is comparative figures that are needed to illustrate the effect.

Suppose a small fire occurs more or less centrally in a room in which the distance from floor to ceiling is 3 m. Suppose the fire initially occupies a floor area 1 m diameter and that the flames, at a temperature of 900°C, are 1 m high. Then the radiated heat received at the ceiling immediately above the fire will be just over 6 kW/m². Suppose now that this fire grows slowly so that this level of radiation is the nett value falling on the ceiling during the first early stage of the fire, say the first 15 min.

Under this heating condition, the surface temperature of the ceiling above the fire will rise and a ceiling which is constructed of a good insulator will get hot more quickly than will a poorer insulator in the same position. In Table 3 the times to reach several temperatures are compared for several lining materials. It is suggested that the times given in Table 3 are quite startling. They show that over a relatively small fire on the floor of a room, the ceiling, if it were plasterboard, would reach a temperature of 600°C in 15 min, but that if

Table 3 Ceiling temperatures: Time for surface to reach given temperature when exposed to a small fire (net incident radiation = 6.4 kW/m²)

| Material | Time (s) to reach surface temperature of | | |
	200°C	400°C	600°C
Plasterboard (1.6 min)	98 (1.6 min)	390 (6.5 min)	(880 (14.7 min)
Fibre insulating board	22	88 (1.5 min)	198 (3.3 min)
Sprayed asbestos (dense)	11	45	102 (1.7 min)
Foam plastics board			
(PU or EPS)	0.6	2.3	5.2

the same ceiling had a thermal conductivity similar to that of foam plastics it would only take 5 s to reach that temperature.

If the ceiling were combustible, as fibre insulating board or foam plastics would be, then the ceiling would ignite in 1.5 min (if it were fibre insulating board) or 2 s if it were foam plastics.

If the fire were smaller than that suggested above, then the corresponding times to reach the various temperatures would be greater but the comparison between those for the very good insulator and the not-so-good insulator would be just as startling.

The heat transfer process from a fire to a building is complicated and the discussion here has taken it to be a simple process of radiated heat from the tips of the flames at the top of the fire. In fact, the ceiling would be heated by convection as well as by radiation so that a greater heatflow than that assumed above could result.

Mechanism of fire growth in a room

When a fire occurs in a room, the temperature of the walls and ceiling rises. Part of the heat received by the walls and ceiling travels outwards through the wall material and some of it is reradiated back into the fire room. If the walls and ceiling are constructed of good insulators, the heat flowing outwards through the material is small and consequently the surface temperature is hotter and the heat reradiated back into the room is greater.

The heat reradiated back into the room has an important effect on the fire growth in the room. Some of it is fed back into the fire, so making good some of the heat lost and maintains the rate of development of the fire. Some of it goes into the rest of the room, thus heating up items of furniture and other combustibles not yet involved in the fire. The higher the rate of this reradiation, the quicker these uninvolved items become ignited.

Using the model fire already suggested, it is possible to calculate how important is this reradiated heat component. For the present argument, a

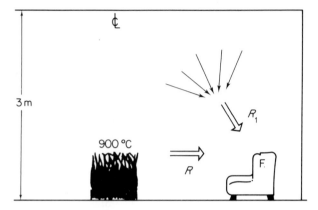

Figure 3 Heat re-radiated from ceiling to furniture at side of room after 2 min exposure. With ceiling of foam plastics R_1 = 7.5 × R; with ceiling of plasterboard R_1 = 0.25 × R.

piece of furniture 1 m high standing on the floor of the room 3 m from the centre of the fire is considered. The temperatures reached by the ceiling after the two-minute exposure have been taken and the result is shown in Figure 3.

The radiant heat received by the piece of furniture, both directly from the fire and by reradiated heat from the ceiling, after 2 min of exposure to the small fire that is being assumed, is illustrated in the figure.

Again, only the effect due to thermal insulation is considered and again the result is marked; indeed, it is frightening!

When the ceiling material has a thermal conductivity appropriate to that of a foam plastics board, the radiation received from the ceiling by a piece of furniture well away from the fire is 7.5 times that received direct from the fire, and this after only 2 min (see Figure 3). Corresponding values for the other insulating materials being considered are indicated in Table 4.

It is interesting to note how quickly this reradiation factor increases with time for an insulating ceiling surface and Table 4 shows the relation between

Table 4 Heat reradiated from ceiling: Heat reradiated = A × heat direct from fire. The figures given are values of A for various ceiling materials

	Value of A for exposure times		
Ceiling material	1 min	2 min	3 min
Foamed plastic board	4.5	7.5	10.5
Sprayed asbestos	0.5	1.1	1.4
Fibre insulating board	0.33	0.63	0.81
Plasterboard	0.23	0.24	0.35

Table 5   Time for piece of furniture at F to reach 300°C – ignition temperature

Ceiling material	Furniture material	
	Covered foam plastic	Plywood
Foam plastic (PU or EPS)	13 s	1.9 min
Sprayed asbestos	1.1 min	8.5 min
Fibre insulating board	2.1 min	12.8 min
Plasterboard	5 min	24 min

direct and reradiated heatflow for the four materials being considered and for exposure times of up to 3 min.

The values of A in Table 4 above indicate that with a very good insulator the reradiated heat very quickly becomes much greater than the heat received directly from the fire. For a good insulator the reradiated heat quickly becomes equal to the direct heat, but for a poor insulator, i.e. plasterboard, the reradiated heat is less than half the direct heat for at least 5 min.

It is now possible to carry this argument a little further and to estimate how quickly a piece of furniture (placed at F in Figure 3) will reach a temperature at which it might be expected to ignite.

The figures given in Table 5 give estimated times for this. It has been assumed that when the surface of the piece of furniture reaches 300°C, it will ignite and furniture items constructed (or covered) with two different materials have been used in the calculations. The first column gives the time for ignition of a piece of furniture of covered plastic foam for various ceiling materials and the second column gives times for the ignition of an article with plywood facing.

A glance at Table 5 will show that there is a marked difference in time of ignition depending on the material of which the furniture is constructed but, whatever furniture material is used, there is always a large difference according to the thermal properties of the ceiling surface.

The table shows that for a highly insulating ceiling surface the fire will spread rapidly from one piece of furniture to another, in times which can easily be less than 1 min so that the whole room will quickly become involved.

This possibility of rapid progressive ignition of the combustible items in the room indicates quite clearly that the importance of the thermal properties of the ceiling are just as great for large rooms as for small rooms, and it is suggested emphatically that there is no justification for the suggestion that the effect is only important in small rooms (see Fire Research Station, 1960, 1962).

Figure 4 illustrates the effect for those ceilings of different thermal conductivity.

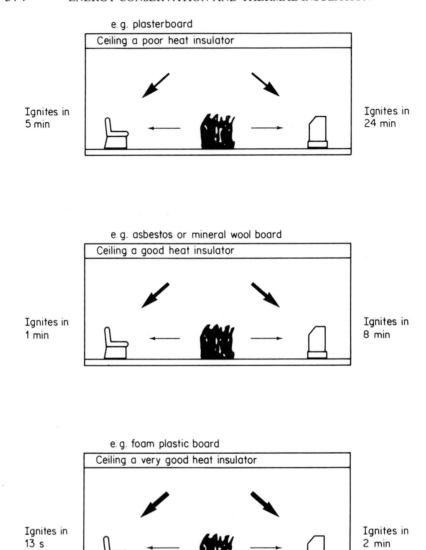

Figure 4   Ceiling material affects fire spread

Depth of heating

It has already been stated that, whereas for its normal purpose the heat insulation will be effective over its whole thickness, in the case of a fire the nonsteady-state condition applies and the lining material will only be heated to a limited depth.

In the fire situation, the depth to which the heat penetrates is determined by two factors, which are:

(a) The thermal diffusivity of the material, or $k/\rho c$; and
(b) the time (duration) of heating.

It is suggested throughout this paper that the thermal properties are important, indeed vitally important, in determing the rate of growth of a fire, that is to say the time taken for a small fire to flashover into a big fire. In considering safety of life, this initial growth period is of extreme importance. If the fire from its inception engulfs the whole room in less than a minute, or a similar short time, then escape of the occupants may be impossible. On the other hand, if this growth period is 5 or 10 min, then the escape or rescue of the occupants will be possible.

Once the fire has become large, the thermal properties of the wall and ceiling will still be important in that, because of the retention of heat, higher temperatures will result. But at this stage of the fire, heat will also be lost in other ways, by flow of hot gases out of the room, flow of cold gases into the room, etc. And so, by comparison with the earlier stages of the fire, at the full development time the increase of hazard by the thermal insulation effect is much less.

Consequently, for the purposes of discussing the critical thickness of the lining materials, heating times of 5–10 min will be considered, and on this basis the thickness values obtained in Table 6 have been achieved. A glance at this table will show that the better the thermal insulation (i.e. the lower the thermal conductivity) the greater the depth of heating. The reason for this is that the surface temperature rises much more quickly for the good insulator.

A second glance at the table gives the information that for all the materials listed, for the heating times used, the depth of material heated is 10 mm or more. It follows from this that:

(a) The use of a thin, noncombustible skin over the surface of a very good thermal insulator (whether combustible or not) will not prevent the surface of that thermal insulator from heating up very quickly (and if combustible reaching ignition temperature at a very early stage).
(b) If a plasterboard sheet (or material with similar thermal properties) is placed as the ceiling or wall lining and the good thermal insulator is then placed above or outside it, then the thermal properties of the plasterboard will determine the rate of growth of the fire and the good

Table 6   Depth of heating for various materials

Wall or ceiling lining material	Thermal diffusivity, $k/\rho c$	Depth of heating (mm) for heating time of	
		5 min	10 min
Plasterboard	0.201×10^{-6}	14	19
Asbestos cement board	0.298×10^{-6}	17	24
Asbestos insulating board	0.212×10^{-6}	14	20
Sprayed asbestos (dense)	0.383×10^{-6}	19	27
Hardboard	0.106×10^{-6}	10	14
Fibre insulating board	0.113×10^{-6}	10	15
Birch plywood	0.122×10^{-6}	11	15
Polyurethane (PU) board	0.531×10^{-6}	22	32
Polystyrene (EPS) board	$1.96 \ \times 10^{-6}$	43	61

insulant will have little or no affect on the fire but will be effective for its main purpose of good thermal insulation.

There is a recent and tragic example of the dangers associated with the use of a thin, noncombustible facing. A boiler room in an institutional-type building was lined with asbestos cement board as thermal insulation was placed behind it.

This arrangement satisfied all the Fire Precaution requirements commonly applied, but as a result of an accidental and very localized flame impingement on the asbestos cement board, which heated the surface for only 5 min or so, the fibre insulating board ignited, smouldered for six hours undetected, and then caused a disastrous fire.

The general conclusion to be reached from the discussion given in the last few pages is that a very good thermal insulator used as a wall or ceiling lining in a room can increase the fire hazard even if the insulant is noncombustible (Humberside County Council, 1977).

British Standard tests used for the control of lining materials

The dangers associated with the use of combustible linings as thermal insulation for walls and ceilings in buildings are to some extent recognized, and control is imposed by the use of British Standard tests (British Standards Institution, 1968, 1971).

These tests, the Fire Propagation Test and the Surface Spread of Flame Test, compare the combustion characteristics of various materials with one another. But in this respect they are not entirely satisfactory, since, although they may distinguish between 'very good' and 'very bad', they may not always put 'very good' and 'good' in the right order.

The Fire Propagation Test is a box test in which a small sample of the material (220 mm × 22 mm) is exposed to a specified heating condition and the temperature rise within the apparatus caused by the combustion of the test specimen is noted and averaged over the first 3 min of test, over the next 7 min, and finally over the last 10 min. Three indices of performance are thus obtained which can, if required, be compounded into an overall index. This test will to some extent measure how quickly and by how much heat is released by the combustion of the specimen.

The Surface Spread of Flame Test, on the other hand, is an open test in which a specimen of the material, 1 m long × 220 mm wide, is placed at the edge of and at right-angles to a radiant panel 900 mm square. A gas flame provides pilot ignition for the first minute of the test and the distance of flaming along the specimen after the test has lasted 10 min is noted. According to the distance travelled by the flaming, a standard of performance is allocated.

The Fire Propagation Test is used to measure Class O and the Surface Spread of Flame Test designates materials as Class 1, 2, 3 or 4. The limits set to obtain the classification in both tests are necessarily arbitrary, and in addition there is the slightly anomalous situation that because different tests are involved a material can be correctly graded as both Class 1 and Class 0.

In both these tests the material should be tested in its complete form, that is to say the surface material must be assembled over its correct substrate, and the future use of the material over a substrate of different thermal properties would invalidate the test result. This fact may not be generally appreciated and the substitution of a better thermal insulant as a substrate behind a thin surface material could create a completely different degree of hazard.

The main shortcoming of the tests is that they concentrate on the combustion properties without taking sufficient account of the important thermal characteristics. They do not, therefore, give a realistic assessment of the way wall and ceiling lining materials affect the growth of fire in the compartment, nor do the tests in any way represent a real fire situation.

They are, nevertheless, the only criteria used by Building Control and Fire Prevention Officers to decide on the suitability of a wall or ceiling lining material for a particular building, and it is clear that they are not by any means satisfactory for this purpose.

There are examples in existence where the use of combustible boards has been made possible under the Building Regulations 1976 by the application of a noncombustible facing material to the board. Aluminium foil and asbestos paper have both been used for this purpose. With an applied surface of this sort, the combustible board will obtain a sufficiently high rating in the Surface Spread of Flame Test to be acceptable for use in many buildings.

Even if this device reduces the danger due to spread of fire by combustibility, an assumption which is extremely doubtful, the hazard of accelerated fire growth due to the thermal conductivity will still be present, and this example is a clear illustration of the shortcomings of the BS Test.

CONCLUSION

The foregoing discussion makes it possible to give suggestions as to how the fire hazard associated with the use of thermal insulation in buildings can be minimized.

Combustible insulation

In this case measures must be taken to ensure that fire cannot spread uninterrupted over large areas of combustible insulation. If it is used in a cavity, void or duct, then fire spread within these spaces must be reduced by means of adequate fire stopping; if it is used as a lining material, then it should if possible be disposed in limited areas with space between to provide a fire break, and preferably it should not be used to line ceilings.

Noncombustible insulation

For this material the danger lies, as explained already, in the thermal properties of the material. The insulation properties for normal everyday use can be achieved and the fire hazard minimized by arranging the insulation in two layers.

The surface of the room should be lined with a material of relatively high thermal conductivity, say plasterboard, and the low thermally conductive material can then be placed behind it. This arrangement allows a required U value to be obtained but, provided the thickness of the facing material is of the order shown in Table 6, then the increase in fire hazard will be small.

REFERENCES

British Standards Institution, 1968, *Fire Propagation Test for Materials*, BS 476 (part 6): 1968.
British Standards Institution, 1971, *Surface Spread of Flame Test for Materials*, BS 476 (part 7): 1971.
Building Research Establishment, 1972, Standard *U*-values, *BRE Digest*, 108, HMSO, London.
Building Research Establishment, 1976, Fire hazards of lightweight insulation, *BRE News*, 38, p. 10, HMSO, London.
Butcher, E. G., Bedford, G. K. and Fardell, P. J., 1968, Further experiments on temperatures reached by steel in building fires, *Proc. Symp. no. 2* held at Fire Research Station, Borehamwood, 1967, Paper no. 1, HMSO, London.
Butcher, E. G., Chitty, T. B. and Ashton, L. A., 1966, *The Temperature Attained by Structural Steel in Fires*, Fire Research Technical Paper no. 15, HMSO, London.

Carslaw, H. S. and Jaeger, J. C., 1959, *Conduction of Heat in Solids*, 2nd edn, Clarendon Press, Oxford.

Degenkolb, J. G., 1976, Will energy conservation have an effect on fire protection of buildings?, *Building Standards*, Sept–Oct.

Fire Research Station, 1960, Development of fire in rooms partially lined with combustible materials, *Fire Research*, 1959, p. 21, HMSO, London.

Fire Research Station, 1962, Development of fire in rooms lined with combustible material, *Fire Research*, 1961, p. 28, HMSO, London.

Fire Research Station, 1968, Behaviour of structural steel in fire, *Proc. Symp. no. 2* held at Fire Research Station, Borehamwood, 1967, HMSO, London.

Heselden, A. J. M., 1968, Parameters determining the severity of fire, *Proc. Symp. no. 2* held at Fire Research Station, Borehamwood, 1967, Paper, no. 2, HMSO, London.

Humberside County Council, 1977, Report of the Committee of Inquiry into the fire at Wensley Lodge, Hessle, 5th January 1977, HCC.

Malhotra, H. L., 1971, Expanded polystyrene linings for domestic buildings, Fire Note No. 12, HMSO, London.

Michaels, L., and Chissick, S. S., eds, 1979, *Asbestos*, vol. 1, John Wiley & Sons, Chichester.

Milbank, N. O. and Harrington-Lynn, J., 1974, *Thermal Response and Admittance Procedure*, BRE Current Paper CP 61/74, Building Research Establishment, Garston.

Morris, A. W. and Hopkinson, J. S., 1976, *Fire Behaviour of Foamed Plastic Ceilings used in Dwellings*, BRE Current Paper CP 73/76, Building Research Establishment, Garston.

Thomas, P. H. and Heselden, A. J. M., 1972, *Fully Developed Fires in Single Compartments* (CIB Report no. 20), Fire Research Note 923, Fire Research Station, Borehamwood.

Thomas, P. H., Heselden, A. J. M. and Law, M., 1967, *Fully Developed Compartment Fires—Two Kinds of Behaviour*, Fire Research Tech. Paper no. 18, HMSO, London.

Energy Conservation and Thermal Insulation
Edited by R. Derricott and S. S. Chissick
© 1981 John Wiley & Sons Ltd.

CHAPTER 17

Thermal insulating blockwork

J. H. Hampshire

*TAC Construction Materials Ltd, Widnes**

It is necessary to dwell a little on the history of the concrete block industry, the product range, and manufacturing techniques in order both to set the scene for the reasons why one particular product could be expected to perform better than another and to highlight the considerable advance made within the industry in developing products to meet more stringent thermal insulation standards.

This chapter is therefore divided into three main sections:

1. Brief history of the concrete block industry.
2. Products available and manufacturing techniques.
3. The changing standards and products to meet the required thermal insulation values.

1. BRIEF HISTORY OF THE CONCRETE BLOCK INDUSTRY

The term 'Breeze Block' is synonymous with the use of concrete blocks in domestic dwellings and, despite the fact that both product quality and technical performance have advanced dramatically, the term is still in use.

The breeze block was manufactured on simple hand-operated block machines from coke dust, or breeze, as a cheap low-quality block. It was not until the war that dense natural aggregates were used in some quantity and this led to the establishment of relatively large block plants in various parts of the UK, mainly under the control of quarry and pit owners.

Demand for dense blocks slackened in the domestic market while keeping its share of the agricultural and industrial markets, so block manufacturers turned to the use of clinkers, a by-product from chain gate power stations.

The introduction of British Standards for aggregates, the general widening of the scope of the Building Regulations, and the installation of more sophisticated machinery all tended to enhance the image of the industry, but the

*Since writing this paper, J. H. Hampshire has transferred to Concrete Masonry Group, Wrexham.

block was still regarded as being little more than a cheap substitute for the common brick.

The late 1950s saw the large-scale production of aerated concrete. Here was an industry that was capital-intensive with a relatively expensive product which had to be sold on performance and quality. It was also around this time that the artificial lightweight aggregate industry began to be formed with the sintered PFA (pulverized fuel ash), pelletized sintered clay, and expanded shales, etc. Thus by the 1960s the modern block industry was taking shape, backed by the large companies, improved technology, and sophisticated plant and equipment.

2. PRODUCT RANGE AND MANUFACTURING METHODS

Concrete blocks can be divided into two main product groups (Figure 1):

(a) aerated concrete, and (b) aggregate concrete.

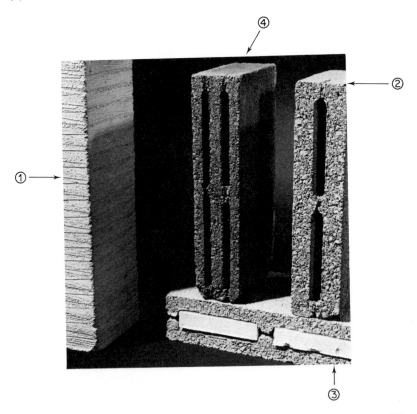

Figure 1 Concrete blocks (broad classification): 1, aerated (totally solid); 2, lightweight aggregate (slotted or solid); 3, organic filled; 4, clinker/furnace bottom ash

Aerated, or 'gas', concrete is in a class of its own, with all the competitors manufacturing blocks by very similar techniques and marketing products with similar physical properties.

The aggregate, or 'dense', concrete blocks, however, can be broadly subdivided depending upon the type of aggregate used, as:

(a) manmade lightweight aggregates, examples being sintered PFA, clay, foamed blastfurnace slag, etc.;
(b) reclaimed materials, examples being clinker and furnace bottom ash; and
(c) dense natural aggregates, examples being limestone, granite, etc.

There are other materials used, for example perlite, graded wood particles, etc.

The aerated concrete block manufacturers, because of the very nature of the process and basic raw materials, concentrate on manufacturing very lightweight units of rectangular solid section. The aggregate blocks on the other hand are available in a much wider variety of forms, lightweight and dense, facings and commons, solid, cellular, and hollow, and special sections.

2.1 Manufacturing process

Aerated concrete

Aerated concrete is made by introducing a foaming agent into a wet mixture of fine silica and a cementitious binder, usually cement and lime. The foaming agent used for aeration is usually powdered aluminium which in the presence of water reacts with the lime constituent of the binder to generate innumerable minute bubbles of hydrogen which remain stable whilst the initial set takes place. The foamed mix is cast into a large 'cake' which whilst still plastic is cut by wires through the horizontal and vertical planes into the required block size. The blocks are then placed in an autoclave and subjected to high-pressure steam. On removal from the ovens the blocks are ready for despatch and use.

Aggregate production

As stated previously the main aggregate groups are the manmade lightweight aggregates, reclaimed materials, and the dense natural aggregates.

The dense reclaimed aggregates have been used for many years and are essentially quarried, crushed, and screened or simply screened, the quality and properties of the aggregate being dependent upon the particular source.

Manmade lightweight aggregates on the other hand are a much more controlled product having a consistent high quality and are manufactured on sophisticated large-scale plants. An example of a manmade lightweight aggregate is a sintered pulverised fuel ash.

The basic raw material (pulverised fuel ash) is blended with a fuel and pelletized. The aggregate is produced by a basic sintering process which provides a rapid increase in temperature of the raw material to $1100–1300°C$, which is just below the complete vitrification temperature. Expansion occurs due to the generation of gases within the pyroplastic material and, on cooling, the glasslike cellular structure is retained in the sinter cake. The cake is then crushed, screened, and graded.

Aggregate block production

Aggregate blocks are produced on a block machine or 'press'. The blended aggregate and cement are mixed with water and fed into the press. The mould box on the press is filled with the concrete mix and vibration promotes compaction of the concrete. The green blocks are immediately stripped from the mould and cured.

The blocks can be produced on fully automatic plants incorporating low-pressure steam curing or by a technique called egg-laying, where blocks are dropped onto concrete and air-cured before being picked up and stacked to await despatch.

3. THE CHANGING STANDARDS, AND PRODUCTS TO MEET THE REQUIREMENTS

Whilst specifically dealing with thermal insulation, it must be remembered that the concrete block has also to fulfil many other performance requirements, i.e. structural performance, fire protection, sound insulation, etc. These functional performance requirements are often conflicting; for example, good insulation against sound calls for high densities, whereas thermal insulation calls for low densities.

3.1 Meeting a thermal transmittance coefficient of 1.0 $W/m^2\,°C$

1974 Amendment to the Building Regulations

Amendments to the Building Regulations were announced in January 1974, and these included 17 main items where the regulations were to be amended and, of these, part 'F' Thermal Insulation, is of particular relevance.

The main change to part 'F' was a reduction in the thermal transmittance coefficient (U value) of the external wall of dwellings from 1.7 $W/m^2\,°C$ to 1.0 $W/m^2\,°C$ and the introduction of an average overall value of 1.8 $W/m^2\,°C$

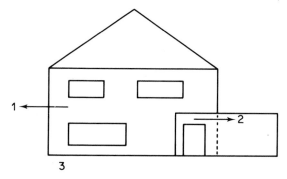

Figure 2 The thermal transmittance coefficients of the external wall of dwellings incorporated in the Building Regulations (1976): 1, U value = 1.0 W/m² °C; 2, U value = 1.7 W/m² °C; 3, average U value = 1.8 W/m² °C

for the perimeter including windows and partially ventilated spaces (Figure 2).

The amendments came into force on 31 January 1975 and incorporated into the Building Regulations (1976).

Compliance with the Building Regulations

The Department of the Environment (1974) issued advice which related not only to thermal insulation but also to the requirement for sound insulation performance for flanking walls and identified a deemed-to-satisfy provision of a mass of 120 kg/m² of wall (Figure 3).

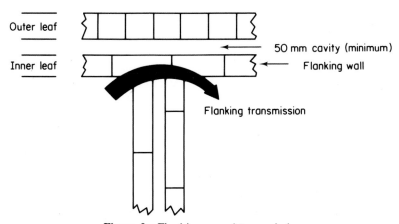

Figure 3 Flanking sound transmission

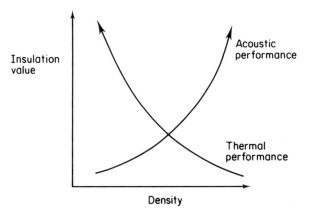

Figure 4   The conflicting requirements of thermal and
acoustic performances

This caused problems in that generally better thermal performance meant lower weights, while better sound insulation meant greater mass/weight (Figure 4).

A guidance note was also issued by the Department of the Environment (1976) to part 'F' of the Building Regulations on the interpretation of the rules and actually illustrating the standard methods of calculation of U values for various structures, including those using slotted blocks in the inner leaf. Reference to the thermal insulation note, Part 'F', shows that the statutory regulations for the thermal transmittance value of 1.0 W/m^2 °C could be satisfied by compliance with:

(a) deemed to satisfy (schedule II), and
(b) determining the U value by calculation.

The crux of the thermal performance of a block is the thermal conductivity of the material from which the block is manufactured. The dependence of thermal conductivity on density for the general class of masonry materials is given by an empirical curve, as expostulated by Arnold (1969), the Building Research Establishment (1972) and the IHVE (1975). This so-called 'standard curve' represents the behaviour of many traditional masonry materials, so that it is reasonable to use it to obtain data necessary for the calculation of the U-value of a structure. Nevertheless, it must be remembered that it is possible to produce concrete and other masonry materials with lower k values for a particular density than that predicted by the standard curve.

The aerated concrete blocks could meet the 1.0 W/m^2 °C in a traditional brick:cavity:block construction using the deemed-to-satisfy schedules. However, the aggregate block manufacturers, if using K values from the standard BRE curve, would have to resort to complicated slotted blocks or organic foam fills into standard blocks.

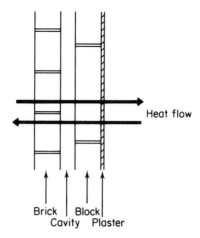

Heat flow

Brick | Block
Cavity Plaster

Figure 5 Diagrammatic representation of the thermal transmittance
of a traditional cavity wall

Manufacturers of manmade lightweight aggregates amassed considerable
evidence to show that their materials performed better than suggested by the
standard BRE curve; this work however has not been published, but has been
communicated privately by the Association of Lightweight Aggregate
Manufacturers.

The Department of the Environment, concerned that all aggregate block
manufacturers would claim substantially better performance, brought in more
stringent requirements for proving better values:

(1) For materials only claiming up to 15% better thermal conductivity than
the standard curve, one certificate required.
(2) For materials claiming better than 15%, then four certificates required,
two from each of two different laboratories.

Due to this, the manmade lightweight aggregate concrete blocks were able to
satisfy the requirements by producing solid or single slotted blocks with certification showing lower K values than the standard curve.

Calculation of U values in structures containing blocks

The thermal transmittance (U value) of a structure is best illustrated by
reference to the most common construction (Figure 5). It is found by adding
the resistance of all the elements in the structure:

inside surface,
plaster,

block,
cavity,
brick, and
outside surface,

and obtaining the reciprocal, i.e. $U = 1/R$. All the resistances are standard and are listed by the Building Research Establishment (1972), except for the block which needs to be computed separately (IHVE, 1975).

It must be noted that in computing the resistance values certain broad assumptions are made, i.e. the heat travels in a straight path with no heat bridging occurring, the effects of mortar joints are ignored (except in the calculation of sound insulation performance), and that the materials condition out to steady moisture contents dependent upon the exposure conditions.

As stated previously, better thermal performance is generally achieved by lowering the material density. The aerated concrete blocks of low density (in masonry materials terms), and coupled with the very nature of the material, i.e. closed pore structure, give good thermal insulation performance. Figure 6 illustrates such a block.

The sintered lightweight aggregates, foamed slag, and furnace bottom

Figure 6   Tacbloc GC aerated concrete block (with enlarged section showing aerated texture)

ashes, by their very nature and manufacturing process, also contain minute air cells which are a contributory factor in their good thermal performance. Figure 7 illustrates a lightweight aggregate block.

Figure 7 Tacbloc GP lightweight aggregate block (with enlarged section showing honeycombed texture

Products developed and being used to meet the U values of 1.0 W/m² °C

These are given below. The reader is asked to consult Figure 1, as well as the following list:

(1) aerated concrete (totally solid blocks);
(2) lightweight aggregate concrete (solid or single slotted);
(3) clinker/FBA (furnace bottom ash) (solid, single or double slotted);
(4) clinker/FBA (organic foam-filled cavities); and
(5) dense (polystyrene slabs, inserted into slots within the block).

Typical constructions which will meet a U value of 1.0 W/m² °C

Two such construction types are shown.

Double leaf and cavity

Brick: cavity: block: plaster

Render: block: cavity: block: plaster

Brick: insulated cavity: block: plaster

Brick: partly insulated cavity: block: plaster

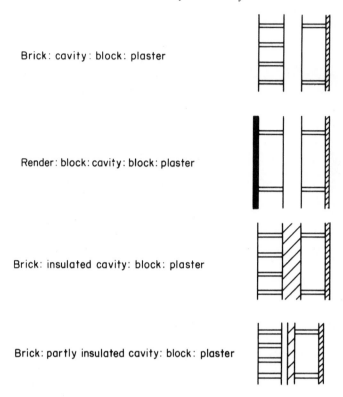

In the examples shown on cavity constructions, the blockwork is normally 100 mm thick and the plaster shown is lightweight. Improved thermal performance can be achieved by the use of dry-lining systems.

Single leaf and cladding

Tile hanging on battens: block: plaster

Weatherboard on battens: block: plaster

3.2 Meeting a thermal transmittance coefficient of 0.6 W/m² °C

In the Building Regulations (First Amendment) (1978), the U value of 0.6 for nondomestic buildings became law in the UK (1 June 1979) and it is suggested that the value 1.0 for housing is only an intermediate step and a lower value will later be introduced for housing. Part 'FF' refers. (Department of the Environment (1979).)

It could be argued that the concrete block industry as a whole should forget about designing and modifying products to meet more stringent thermal insulation standards and go about their main purpose of producing structural elements, offering fire protection, sound insulation, etc., and leave insulation improvement to the insulants like mineral wool, glass fibre, formed plastics, etc., and specially designed dry-lining systems. However, it is possible to attain U values of 0.6 and still keep to traditional constructions.

Typical constructions which will meet a U value of 0.6 W/m² °C

One construction type is considered.

Double leaf and cavity

Brick: cavity: block[1]: dry lining

Brick: cavity: block[2]: dry lining

Brick: cavity: block[3]: plaster

Render: block: cavity: block[4]: plaster

Brick: cavity[5]: block: plaster

The following points should be noted:
(1) A block thickness in excess of 100 mm is required to achieve the required resistance value.
(2) Slotted blocks of manmade lightweight aggregate with low thermal conductivity can be used.
(3) Foam-filled blocks can be used.
(4) The 100 mm block thickness can be kept by using slotted or foam-filled blocks made from lightweight aggregate.
(5) The cavity can be totally or partially filled with insulant.

In all cases the thermal performance can be improved by the use of dry-lining systems.

ACKNOWLEDGMENT

The author would like to thank the Directors of TAC Construction Materials Ltd for permission to publish this paper, and particularly Block Division for the photographs of Tacblocs.

REFERENCES

Arnold, P. J., 1969, Thermal conductivity of masonry materials, *J. Inst. Heating and Ventilation Engineers*, **37**, August.
Building Regulations, 1976, HMSO, London, 1977.
Building Regulations, 1978, First Amendment, HMSO, London, May 1978.
Building Research Establishment, 1972, Standardised *U* values, *BRE Digest*, 108, London.
Department of the Environment, 1974, *Part 'F' and 'G': Thermal and Sound Insulation* (Technical Information Note), Welsh Office.
Department of the Environment, 1976, *The Testing of Building Materials in Relation to Part 'F' of the Building Regulations*, Note BRA/668/68, September.
Department of the Environment, 1979, Part FF: Conservation of fuel and power in buildings other than dwellings (Guidance Note) HMSO, London.
IHVE, 1975, Thermal and other properties of building structures, *IHVE Guide A3*, The Institution of Heating and Ventilation Engineers, London.

Energy Conservation and Thermal Insulation
Edited by R. Derricott and S. S. Chissick
© 1981 John Wiley & Sons Ltd.

CHAPTER 18

Cellulose fibre insulation in the USA and its application in the UK

W. D. McGeorge and P. P. Yaneske

Environmental Design Services, Glasgow

1. GROWTH AND DEVELOPMENT OF CELLULOSE FIBRE INDUSTRY IN THE USA

The following description is based on a report prepared by the Brookhaven National Laboratory (1978) for the US Department of Energy.

In the 1800s, thermal insulation was composed of naturally occurring materials, largely vegetable in origin and based principally on wood and its derivatives. Patents for cellulosic fibre insulation were taken out round about this time. The product, however, did not become firmly established in the marketplace until the 1950s.

The first large-scale manufactured insulation products were based on rock and slag wools and began operating during the late 1920s and early 1930s. The 'Battle of the Fuels' in the 1950s of electric heating versus fossil-fuel heating was a major stimulus to the insulation industry. The obvious advantages of using recycled newsprint together with fire-retardant chemicals to form cellulose fibre insulation caused the rapid development of the fibre industry during the 1950s. The capital project costs of a cellulose fibre manufacturing plant are a good deal less than those of a glass fibre plant of similar manufacturing capacity; also the energy used during the manufacturing process of cellulose fibre is less than that used in the production of glass fibre.

The current size of the cellulose fibre industry in the USA is difficult to assess since there are many small manufacturers whose numbers tend to fluctuate. An OBRA (Office of Business Research and Analysis) report (Brookhaven National Laboratory, 1978) estimates approximately 250 manufacturers as of June 1977 with a production capacity of 0.77 million tonnes. This represents a threefold increase in the number of manufacturers over the previous year. An ICF report (Brookhaven National Laboratory, 1978) gives a more conservative production capacity of 0.27 million tonnes and indicates a twofold increase in the number of manufacturers over the previous year.

Even the more conservative estimates of the growth and production capacity of the cellulose fibre industry still demonstrate the rapid inroads which this product has made into the insulation manufacturing market since the 1950s. The OBRA report indicates that the US glassfibre manufacturing capacity is 0.71 million tonnes, a figure of similar magnitude to that given for cellulose fibres.

1.1 Future development of cellulose fibre in the USA

Since cellulose fibre insulation comprises essentially two types of ingredient, namely newsprint and chemical retardants, its growth is of necessity linked to the future availability of these materials.

There have been recent reports of a boric acid shortage (Brookhaven National Laboratory, 1978), boric acid being the most common chemical in current use as a flame retardant for cellulose fibre insulation. Presently, boric acid is used in a ratio of approximately 1.4 kg of boric acid to 18 kg of newsprint, so that the current annual demand is about 48 000 tonnes. With the expected increase in cellulose output, the boric acid demand could increase fourfold or more by 1980. The current boric acid capacity is approximately 183 000 tonnnes/year with US Borax and Chemical accounting for 65% of this amount. US Borax plans a 30% expansion, expected to be completed by 1980. Other increases in availability could result from imports, particularly from Turkey, Russia, and South America.

Even with the increase in boric acid supply, there could still be future problems with availability due to the much higher rate of increase in demand predicted for cellulose insulation. Developments of new flame retardant formulations requiring less boric acid should alleviate potential supply problems. Several chemical companies, including US Borax and Chemical, are now in the process of investigating different formulations requiring less boric acid.

The ICF report indicates that, except for short-term fluctuations, waste paper for cellulose insulation will be available.

In summary, the US cellulose fibre market has developed from almost zero in the 1950s to become a substantial component of the current output of insulating materials, with a particularly marked growth rate during the 1970s.

Currently there are four trade organizations representing the US cellulose fibre industry, with a possible fifth organization in the process of being formed. These associations are as follows:

(a) National Cellulose Insulation Manufacturers Association, which is the longest established of the trade associations;
(b) Society of International Cellulose Manufacturers Association;
(c) Manufacturers Association of Cellulose International; and
(d) Society of American Cellulose Insulation Manufacturers.

Negotiations are currently taking place to attempt to form an amalgam of these associations so that the industry is represented by a single body.

2. GROWTH AND DEVELOPMENT OF CELLULOSE FIBRE INDUSTRY IN THE UK

It was noted by an international business group that even after the energy crisis of 1973 inspired by OPEC (Organisation of Petroleum-Exporting Countries) there was no cellulose fibre industry in the UK despite the remarkable growth rate of cellulose fibre in the USA based on the material's cost-effectiveness to both consumer and producer.

This group set about establishing a cellulose manufacturing industry in the UK with the first factory being established in Livingston in the Scottish central industrial belt. Prior to the establishment of the factory, marketing and technical studies were carried out which confirmed the feasibility and appropriateness of cellulose fibre insulation for use in the UK market.

A technical advice agreement was negotiated with one of the most successful cellulose insulation manufacturers in the United States of America. Under this agreement, which gives the UK company, Diversified Insulation Ltd, the exclusive European rights, a full consulting service was supplied to assist in the manufacturing, sales, and installation of cellulose fibre in the UK.

In August 1977 the decision was taken to proceed with the project. Key items of manufacturing equipment were imported from the USA, the remainder of the equipment being manufactured in Scotland. Financial backing for the project was partly received from the Scottish Development Agency and from the Industrial and Commercial Finance Corporation.

Diversified Insulation Ltd, the first UK company to manufacture cellulose fibre insulation, has been fully operational since June 1978 and is now approaching maximum manufacturing capability of 30 000 tonnes per annum.

Encouraged by the success of this venture, other manufacturing plants have been established by CIBCO Insulation, J. & J. Maybank, and Thermo-ken which, together with Diversified Insulation Company's plans for further expansion, would indicate that the accelerated growth rate of cellulose fibre observed in the USA will shortly be paralleled in the UK.

Since cellulose fibre manufacture has only recently been established in the UK, there are no indications of newsprint shortages; however, chemical shortages could parallel the USA experience.

The cellulose fibre manufacturing industry in the UK is represented by the Thermal Insulation Manufacturers Association.

3. MANUFACTURING PROCESSES

Cellulose fibre insulation is manufactured from recycled newsprint and cardboard, or virgin wood, which is pulverized to fibre form and treated during processing by various chemicals to provide flame retardancy and to inhibit fungal attack. Typical of the chemicals used are borax, boric acid, and aluminium sulphate added to the material in proportions which amount up to 25% by weight.

Owing to a difference in the supply and demand situations for newsprint and paper feedstock between the UK and the USA, manufacture in the UK is likely to be based on selected newsprint only. Another difference is that cellulose fibre insulation is available in loose-fill and spray-on forms in the USA, whereas in the UK it is currently available only as a loose-fill.

3.1 Dry process

The original manufacturing technique for cellulose fibre insulation was a dry process and today the dry process is still the most widely used. It consists of feeding the paper feedstock, in most cases newsprint, into one or more mills, where it is shredded and pulverized into a fibrous bulk material and blended with a dry chemical additive. The final product is then conveyed to a bagging machine for packaging.

The process stems directly from agricultural technology which has been refined into sophisticated on-line production systems. Such systems include the following basic components:

(1) newsprint feeder and conveyor belt;
(2) hammermills;
(3) chemical hopper and feeder; and
(4) bagger.

In a single mill operation, paper stock is hand fed into the mill to allow for sorting of undesirable foreign materials. A typical single mill operates with up to 150 horsepower motors and may function with either fixed or swinging hammers or work as a cutter and shredder, forcing the final product through a screen with approximately 6 mm openings. The dry chemical is usually fed by an auger directly into the outlet of the mill where it is blended mechanically and forcefully onto the fibres. The final product is then conveyed pneumatically to the bagging machine and packaged.

This type of operation has been used by between 75 and 100 plants in the USA and can achieve production rates as high as 4 tonnes per hour, though about half this rate is usual for long-term operation. Though the final product is satisfactory in its thermal performance, due to the lack of balanced feed rates for the paper and the chemicals it is unlikely that such a process can offer the quality control factors necessary to meet the specification most countries are likely to adopt with regard to, for example, fire performance and corrosiveness.

Multiple mill processes have been designed, with many variations, to cope with the problems of quality control, increased output, layout, and material handling. The process employed by the Diversified Insulation Company Ltd at their Livingston factory in Scotland is a sophisticated multiple mill process and its layout is shown diagrammatically in Figure 1.

Sorted and selected newsprint is fed by conveyor into a shredder where it is broken down into strips, some 70 mm square, by the use of cutters and thence to the first hammer mill, where the paper is further reduced, emerging through the final screen some 10 mm square. This 'confetti' is then conveyed pneumatically to a roof-mounted cyclone where the fine dust is driven off. From the cyclone, the material then enters a hopper or holding bin which is used to achieve a uniform feed rate to the final hammer mill in the process.

The dry chemicals, in this case the component mix of borates with a small amount of aluminium sulphate, are proportioned and blended in a chemical mixer and then ground in a chemical mill to the consistency of fine talcum power. This finely ground material, which is readily dispersed

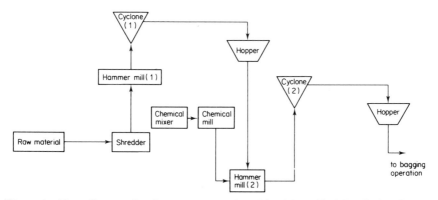

Figure 1   Flow diagram for dry process employed by Diversified Insulation Co. Ltd, Livingston, Scotland

and easily blended into the cellulose fibres, is introduced simultaneously with the ground paper from the hopper at controlled rates into the final hammer mill by mechanical means. Some processes may introduce the chemicals by pneumatic means or gravity feed. This is the most critical stage of the process, and human supervision must be combined with mechanical or electrical controls to maintain the consistent quality required in the final product.

In the Diversified Insulation Company's process, the treated cellulose fibre emerging from the final hammer mill is conveyed pneumatically to a second cyclone where any fines are removed. The material then falls into a bagging hopper from which it is fed by augers into bags at a controlled pressure. The density of the bagged material and therefore its potential area of coverage on installation is controlled at this stage. Any spillage from the bag-filling operation is conveyed pneumatically back to the first hammer mill. The bags are then weighed, conveyed through a horizontal stitcher, and passed through rollers which shape them to the required dimensions, ready for storing or loading.

3.2 Wet processes

Two methods available for the manufacture of cellulose fibre insulation can be classified as wet processes. The first process parallels the manufacture of the dry material except that liquid dispensing equipment replaces such items as bins, augers, and mills used in handling the dry chemicals. Instead, an aqueous solution of the chemicals is sprayed, sprinkled or misted into the cellulose fibre material, normally after it leaves the first mill and before it enters the final one. The removal of the excess moisture from the material depends on the evaporation induced by the air movement and heat produced in the final milling process. Clearly the amount of chemical added and its evaporation characteristics are critical to the properties of the final material, and either too little or too much chemical may result in an unusable end product.

The second process uses conventional paper-making techniques and equipment. The paper stock is first reduced to a slurry by a pulping process, then some 50 to 60% of the water is squeezed from the pulp by compression, and finally the material is dried and fluffed prior to bagging. There is a wider range of opportunities to introduce chemicals, dyes or other modifications here than in other processes.

Wet processes appear to offer the potential, as yet unproven, advantages of better chemical dispersion into the material with improved fire-retardant characteristics, control of corrosiveness, and resistance to leaching. Nevertheless, because of the relative, and perhaps deceptive, simplicity and low cost of the dry process, very few plants currently use a wet process, and therefore little technical information is available on which to base comparative evaluation.

3.3 Cellulose fibre insulation for spray-in applications

The technique of using bonding agents to bind sprayed cellulose fibre to substrates has been known for many years. Material intended for spray-on applications is manufactured in a similar manner to that described for the dry process. However, since the material is usually left as an exposed finish, its flame-retardant characteristics are vital to its performance. To this end, more chemical, typically boric acid, is added.

3.4 Chemical treatments

Since the macerated newsprint or other feedstock provides the thermal insulating properties, the primary purpose of the chemical treatment is to provide the cellulose fibre insulation with fire-retardant properties. It is also desirable that the treatment should inhibit fungal attack and provide vermin resistance. Clearly, it is important to know how the addition of chemicals affects the toxicity, corrosiveness or moisture absorbency of the final product. For the most part, the chemicals used in the treatment of cellulose fibre insulation are nondurable, that is they are water-soluble.

Many chemicals have been used, either singly or in combination, in attempts to provide an acceptable fire-retardant performance at minimum cost. The behaviour of treated cellulose fibre is complex, and it has been found that, while one combination of chemicals may produce a high resistance to flame spread but permit severe afterglow, changing the balance of the chemicals may not only alter the results but may also reverse them (Brookhaven National Laboratory, 1978).

Common chemicals used in the treatment of cellulose fibre are: boric acid, borax, aluminium sulphate, ammonium sulphate, calcium sulphate, mono- and diammonium phosphate, sodium carbonate, calcium carbonate, and hydrated alumina. With respect to fire retardance, resistance to fungal growth, corrosiveness, and moisture absorbency, the borates are believed to be the best all-round chemical treatment for cellulose fibre insulation. The optimum fire-retardant properties are obtained with a 1:1 ratio of boric acid to borax, but expense and availability have forced lower boric acid rates to be used. The sulphates are generally inferior in their fire-retardant properties but are often used to extend the boric acid. In particular, aluminium sulphate is an antivermin proofing agent. Ideally, the use of sulphates should be limited to around 2% by weight of the insulation.

Phosphates have been used for many years as effective flame retardants for textiles. Although the ammonium phosphates may be used in a similar manner to the borates, they are associated with problems of corrosiveness, moisture absorption, and fungal growth, which they do not inhibit. The carbonates are reported to be used to alleviate the corrosiveness of sulphates by neutralizing their acidity. Hydrated alumina has been suggested for use in a similar manner to borax, but its effectiveness is as yet unproven.

In the USA, a number of proprietary chemical mixtures are available, both in dry or liquid forms and with borate or phosphate bases. More detailed information on the chemicals currently used in the treatment of cellulose fibre insulation is available in a report prepared by Volkman *et al.* (1978). In the UK, no proprietary mixes are as yet available and, for example, the Diversified Insulation Company prepares its own chemical formulation based on the borates.

3.5 Health and Safety

The manufacture of cellulose fibre insulation involves the storage of flammable paper feedstock, the use of chemicals and machinery with moving parts, and the generation of dust and noise, especially in the dry process. It is therefore important to observe basic health and safety measures.

Newsprint should be stored so as to minimize the risk of spontaneous combustion, and the large amount of combustible dust likely to be generated by the manufacturing process must be continually removed. While the chemicals used in the treatment of the cellulose fibre are generally of low to moderate toxicity, protective clothing in the form of overalls, gloves, and possibly eye goggles should be worn when handling the chemicals or the finished product. Recommended threshold limits for volume concentration of the chemicals should be observed and engineered by such methods as enclosure, local exhaust, and ventilation, and if appropriate, dust masks should be used. Spilled chemicals should be immediately disposed of by thorough dilution with water.

The hand-loading of paper onto conveyor belts, the changing of hammermill screens, the bagging operation, and the general maintenance of equipment with moving parts can produce hazardous conditions if machinery is not properly guarded or provided with safety isolation mechanisms. Machines need to be prevented from 'walking' under vibration or tipping over, and must be laid out so that the unwary are not tripped or snagged.

As might be expected from the nature of the process, high levels of noise are commonly generated. The most practical method of dealing with this problem is to ensure that protective ear muffs are worn by all personnel involved directly in the manufacturing process.

4. SPECIFICATIONS AND QUALITY CONTROL FACTORS

At the time of writing, there are no British Standard Specifications directly applicable to cellulose fibre insulation. However, current and future products for the UK market will benefit considerably from the experience accumulated in the USA. It is therefore most unlikely that the kind of problems that have arisen in the USA from poorly made cellulose fibre insulation, which has led to fires, corrosion of metallic elements in contact with the material, and inflated claims for thermal resistance, can arise in the UK. Indeed, concern in the USA over quality control procedures has led to increasingly stringent specifications being introduced over recent years, so that similar problems are unlikely to arise in the future.

The American Society for Testing and Materials (1973) has issued a standard which deals with materials, packaging and marking, and a number of physical requirements important as quality control factors such as density, thermal resistance, burning characteristics, and corrosiveness. The various trade associations in the USA (see section 2) require members to test their products according to procedures contained in ASTM Standards and Federal Specifications. Federal Specification HH-I-515D was introduced in June 1978 and incorporates both modifications to the original test procedures described in ASTM C739–73 and a new requirement with respect to resistance to fungal attack which have increased the stringency of the test procedures to the point that many of the cellulose fibre insulation products manufactured up to that time failed to meet the new requirements. Federal Specification HH-I-515D lays down requirements for the following:

(1) settled density;
(2) starch content;
(3) thermal resistance;
(4) moisture absorption;
(5) odour emission;
(6) corrosiveness;
(7) fungal resistance;
(8) critical radiant flux;
(9) smouldering combustion; and
(10) marking of bags.

In addition to any tests the manufacturer may carry out or have carried out on his behalf, qualification tests must be conducted on the first nine items listed above at a laboratory chosen by the Federal Supply Service of the General Services Administration in order to qualify as an approved material within the meaning of the specification.

Although there are differences in approach and testing procedures between the UK and the USA, any future UK standard or certification will need to deal in some way with the items listed above. The element of independent testing is also important. Attention should also be paid to the quality control test procedures employed by the manufacturer, since regular testing of the finished product is desirable to prevent possible substandard material reaching the installers. Small-scale tests for settled density, starch content, burning characteristics, pH value, and odour emission can be carried out on the manufacturer's premises at frequent intervals. Larger-scale tests and tests requiring more complicated apparatus and usually a long timescale, such as the determination of thermal resistance, corrosiveness, and resistance to fungal attack, can be carried out by commercial laboratories or by local university or polytechnic departments where the necessary equipment and expertise exists to deal with the complex properties of cellulose fibre insulation.

5. PROPERTIES AND PERFORMANCE

5.1 General

Treated cellulose fibre is usually of a light-grey colour in appearance and consists of a fibrous mass in which the individual fibres are typically 'C'-shaped and give the material a tendency to clump by their interlocking action. Small pieces of paper may be visible, but this is quite normal. The material is pleasant to handle.

5.2 Density

The recommended installed density for cellulose fibre loft insulation is 40 kg/m³. Typical applied densities for blown-in or poured-in insulation lie in the range 35 to 50 kg/m³, with the higher values referring to the poured-in insulation. Even higher values are found in wall applications.

The density may increase through settlement after installation. For loft insulation installed at a density in the above range, settlement is typically between 2.5 to 5% of the original thickness.

5.3 Thermal transmission

Although thermal insulating materials such as cellulose fibre do not have thermal conductivity as a defined physical property, the concept of an apparent thermal conductivity has had currency as a practical way of defining their thermal performance. Measurements of the apparent thermal conductivity of cellulose fibre have been carried out in the USA, usually on small samples around 1 inch (25.4 mm) thick and 12 inches (30.5 mm) square at a mean temperature of 75°F (23.9°C), using guarded hotplate or heatflow meter methods. A comparison of such measurements has been carried out and reveals a wide divergence in the values of apparent thermal conductivity reported (Brookhaven National Laboratory, 1978). However, for loose-fill cellulose fibre loft insulation at a typical applied density of 2.5 lb/ft³ (40 kg/m³), North American sources commonly quote design values of about 0.27 Btu in/ft² h °F (0.039 W/m² °C) for the apparent thermal conductivity or, correspondingly, of about 3.7 ft² h °F/Btu in thickness (0.65 m² °C/W) for the thermal resistance (ASHRAE, 1972; US Department of Housing and Urban Development, 1975; Egan, 1975: Canadian Office of Energy Conservation, 1976).

As with any fibrous thermal insulating material of relatively low density, the mechanisms of heat transfer within cellulose fibre insulation form a very complicated system in which radiation effects play a significant role. As a consequence, the thermal properties of the material are to some extent thickness–dependent, and should strictly be measured both at the density and the thickness at which it is intended to install the insulation. The thermal resistance of cellulose fibre increases with increasing thickness but decreases with increasing density.

Recognition that thickness affects the measurement of the thermal transmission properties of many insulating materials has led to the recent development of large-scale guarded hotplate and heatflow meter equipment for measurements on materials at greater thickness and to developments in the guarded hotbox technique where large sections of building components containing insulation can be tested.

In the UK, the Agrément Board has developed apparatus for determining the thermal transmittance and conductance of building components containing insulation which is designed to satisfy the criteria of ASTM C236-6 (American Society for Testing and Materials, 1966) and BS 874:1973 (British Standards Institution, 1973). In this apparatus, sections of loft panel including the installed insulation, applied as in practice, have been tested and the actual thermal conductance of the sections plus insulation measured. By using standard values for surface, air space, and roof resistances, standard U values for a typical roof construction of various slopes can be computed. This has been done for both glass fibre quilt insulation with a measured thickness of 101 ± 8 mm and measured density of 9.5 ± 0.8 kg/m³ and for blown cellulose fibre insulation with a measured thickness of 93 ± 8 mm and measured density of 37.9 ± 3.2 kg/m³, and the results are shown in Table 1 (Agrément Board, 1979a,b). It can be seen that, for a nominal thickness of 100 mm, blown cellulose fibre provides about 30% more thermal resistance than the glass fibre quilt under the conditions of the test.

Table 1 Standard U values for a typical roof construction obtained using the Agrément Board Thermal Transmittance Apparatus (Issued by permission of the Diversified Insulation Company Limited)

Pitch of roof (deg)	Standard U value (W/m² °C)	
	Glass fibre quilt	Treated cellulose fibre
20	0.77 ± 0.02	0.56 ± 0.02
25	0.77 ± 0.02	0.57 ± 0.02
30	0.78 ± 0.02	0.57 ± 0.02
35	0.79 ± 0.02	0.57 ± 0.02
40	0.79 ± 0.02	0.58 ± 0.02

5.4 Fire resistance

Cellulose fibre insulation is manufactured from an organic material, normally used newsprint, and is by definition combustible and must, therefore, be treated chemically to provide resistance to fire. The basic principle involved is that the installed insulation should not add to the fire hazard that already exists in a traditional structure. In the case of loft insulation, the treated cellulose fibre should certainly be no less fire safe than the wooden members supporting the roof structure.

Although the occurrence of fires that have spread from the loft to the rest of the building is very limited, it is desirable in the interests of increased safety that cellulose fibre should exhibit neither sustained flaming nor smouldering combustion after exposure to potential igniting sources. Examples of possible ignition sources in the roof are a discarded lighted match or cigarette, a carelessly used blowtorch, and overloaded, and consequently overheated, electrical wiring.

However, at the time of writing, there exist no accepted tests in the UK that can establish the fire performance of a loose-fill, organically based insulation for horizontal application in a satisfactory manner. The test methods given in ASTM C739 and Federal Specification HH-I-515D cannot simply be transferred to the UK where the conditions in roof spaces are different, both in terms of services in the roof space and in terms of environment, especially with respect to humidity and moisture levels. In any case, it is widely recognized that these test methods are not completely adequate.

A number of tests have been carried out both by the Agrément Board and by the Fire Research Station on samples of cellulose fibre treated with a borate-based fire retardant in order to assess its behaviour in fire conditions. The conclusion reached by the Agrément Board (1977) as a result of these tests was that installation of this particular cellulose fibre insulation would not result in a significant increase in the fire hazard which exists in the normal loft situation, with the caveat that the material was not recommended for use in completely sealed spaces where, if subject to burning, the normally insignificant evolution of toxic products in the form of smoke, carbon monoxide, and carbon dioxide would increase through lack of available oxygen.

Nevertheless, the tests used to arrive at this assessment cannot be regarded as definitive, and adequate test procedures await development.

5.5 Corrosiveness

The potential corrosiveness of cellulose fibre insulation derives from its chemical treatment. If the chemical treatment is properly formulated, adequately dispersed into the fibre by the manufacturing process, and the final product is subjected to satisfactory quality control procedures, then there appears to be no significant risk of corrosion occurring in materials with which the insulation might come into contact.

There are no general or comprehensive methods by which to test for the potential corrosiveness of building insulation products, but ASTM C739 does contain a particular test procedure for cellulose fibre insulation. Corrosion tests in accordance with ASTM C739 have been carried out on mild steel, shim steel, and copper specimens (Heriot-Watt University, 1978) using the product Shelter Shield manufactured in the UK by the Diversified Insulation Company. In addition, samples of mineral wood and glass fibre mat were also tested for comparison. All three insulating materials proved to be noncorrosive to the test metals as defined by ASTM C739. It is interesting to note that while all three insulating materials were inert to copper, only Shelter Shield remained completely inert to mild steel. All three materials produced slight surface etching of the shim steel.

Further tests were carried out at Heriot-Watt University (1979) in which the pure chemicals used in the production of Shelter Shield and mixed in the correct proportions were tested for corrosiveness against galvanized iron, mild steel, copper, and aluminium in the absence of any fibre. After 14 days in a controlled environment of 45°C and 95% relative humidity, none of the metal specimens showed perforations or pinholes, and only in the case of mild steel was there a weight loss which did not exceed 1.69% in any of the specimens tested.

5.6 Resistance to fungal growth

Cellulose fibre insulation contains some 70% by weight of readily hydrolysable carbohydrates that could be attacked by 'dry rot' and 'wet rot' fungi. The material depends on its chemical treatment to inhibit fungal growth. On advice obtained from the Princes Risborough Laboratory, samples of Shelter Shield, the only cellulose fibre insulation in production in the UK in June 1978, were tested for resistance to fungal attack by a consultant mycologist at Glasgow University (1978).

For 'dry rot', the material was subject to attack by *Serpula lacrimans* strain 5, a strain of intermediate decay ability and *Serpula lacrimans* strain 42A, a very active strain. For 'wet rot', *Coniophora puteana* strain 11, which is a standard test strain in many European laboratories, was used. The results were that, after several weeks incubation in conditions likely to be more favourable for the growth of these fungi than found in practice, no fungal growth was found on any sample of Shelter Shield, that the fungi failed to grow into the material from an outside nutrient source, and that the Shelter Shield fibres completely inhibited fungal growth when added to cultures in nutrient solutions.

These results are specific to the product Shelter Shield and should not be assumed for other products unless backed by test results. Some uncertainty is reported by the Brookhaven National Laboratory (1978) on resistance to fungal growth of cellulose fibre manufactured in the USA.

5.7 Moisture absorption

Both ASTM C739 and Federal Specification HH-I-515D contain tests which place a limit of 15% on weight gain from moisture absorption. Any properly treated cellulose fibre should pass these tests. Though no equivalent test procedures exist in the UK as yet, the Agrément Board (1977) has tested samples of cellulose fibre with a borate-based chemical treatment and concluded that it does not absorb excessive amounts of moisture and that, as a consequence, its thermal performance will not be significantly affected by moisture in normal roof conditions.

5.8 Water vapour permeability

Cellulose fibre insulation has a high permeability and presents no significant resistance to the passage of water vapour when installed at densities of 35 to 60 kg/m^3.

5.9 Odour

Cellulose fibre insulation has no odour.

5.10 Effects of Ageing

Some settlement may occur over long periods of time depending on environmental conditions and the original installed density. The effects of age on thermal performance and fire resistance, if any, are not established.

6 INSTALLATION AND APPLICATION OF CELLULOSE FIBRE INSULATION

Although, as previously discussed, cellulose fibre can be supplied for spraying application, the main use of cellulose fibre has been as a blown or poured loose-fill insulation.

Since cellulose fibre is one of the few types of loose-fill insulation that can be blown in excess of its own specific density and reportedly will not settle significantly, this makes it an ideal infill material for use either vertically in external timber-framed walling or horizontally as a loft insulant.

In the USA, wood-framed external wall construction constitutes the largest form of residential construction, and this in turn represents the major area of the USA cellulose fibre market (see Figure 2).

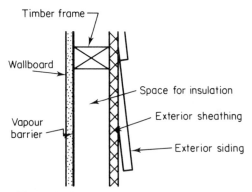

Figure 2   Typical external timber-frame wall as used in the USA (insulation omitted)

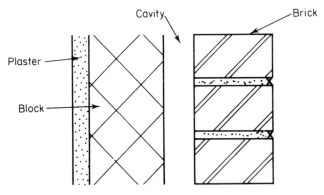

Figure 3   Typical external wall as used in the UK

In the UK, although timber-framed construction is gaining ground, external walling in the residential sector is constructed traditionally by brick cavity walling, or solid stone or solid brick in the case of older buildings (see Figure 3).

Currently, cellulose fibre is not installed in external walls in Britain, although the inner leaf timber-frame construction is similar to the timber-framed construction of the USA (see Figure 4).

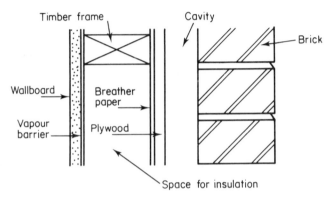

Figure 4   Typical external timber-framed wall as used in the UK (insulation omitted)

6.1 Methods of installing cellulose fibre in external walls

Installation must be made by pneumatic means. The normal installation equipment consists of a pneumatic machine incorporating a hopper into which the contents of bags of compressed cellulose fibre are deposited (as shown in Figure 5). The fibres are agitated by metal tines and are then blown at a controlled density through flexible plastic hosepipes (50 to 65 mm diameter) to the point of installation (as shown in Figure 6).

A minimum of two operatives are required, one to load the hopper and the other to carry out the installation, the flow of cellulose fibre through the delivery hose being controlled by means of a remote control switch operated from the hose delivery point.

There is little difference between new construction and retrofit in the methods of filling the external wall void. In new construction, careful design can ensure that means of entry are as accessible as possible and that difficult-to-fill cavities are avoided. However, the general principles are the same for new construction as for retrofit.

The following procedure is recommended by the National Cellulose Insulation Manufacturers Association, Inc. (NCIMA). Holes are drilled 15 mm to 50 mm in diameter, depending on the timber-framed construction, in each

Figure 5   Operative loading blowing machine

wall cavity. The vertical distance between the holes and top or bottom plate should not exceed 600 mm and the vertical distance between the holes should not exceed 1500 mm. Walls with shingle or lapped cladding should have the holes drilled as near the shadow line as possible. Walls with brick cladding should have holes 15 to 20 mm in diameter drilled in the mortar joints. All holes should be filled with suitable plugs or material.

Filling with a fill tube is occasionally adopted. When using this method, only one entry hole per cavity is necessary. The fill tube should be inserted far enough to reach within 450 mm of the plate farthest from the point of entry. Fill tube size will depend upon the size of hole which can be drilled

Figure 6   Operative blowing cellulose fibre in place

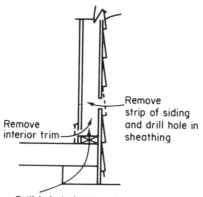

Remove
interior trim

Remove
strip of siding
and drill hole in
sheathing

Drill hole in bottom plate

Figure 7 Alternative points of entry
for cellulose fibre fill tube (ref.
NCIMA—National Cellulose Insulation
Manufacturers Assoc. Inc.). External
timber-framed wall—floor level

(see Figures 7 and 8). After the entry holes are drilled, all cavities should
be checked by optical probes for fire stops or other obstructions in order
to ascertain that the free flow of the insulation material is uninhibited. In
existing buildings, weak areas in either internal or external faces of wall

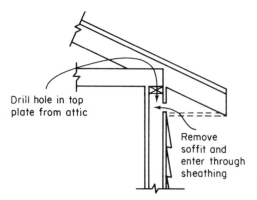

Drill hole in top
plate from attic

Remove
soffit and
enter through
sheathing

Figure 8 Alternative points of entry for cellulose fibre fill tube. External timber-framed wall—roof level

should be reinforced, and any areas of moisture penetration or dampness should be attended to prior to blowing, which is normally carried out from the bottom hole working upwards and entirely filling the void between standards with cellulose fibre. Care should be taken to ensure that wall cavities used as air ducts for heating or air conditioning are not filled with insulation. Finally, in new construction, the small cavities around door and window frames should be filled with insulation, prior to the installation of the interior covering.

From the foregoing description, it can be seen that methods of installing cellulose fibres in timber-framed walls in the USA could also be used in timber-framed dwellings in the UK, although the British method of separating the inner and outer leaves with a cavity would make exterior blowing in a retrofit situation slightly more difficult.

In general, cellulose fibre is not recommended for use in timber-framed construction below ground level, e.g. in a basement location, nor should cellulose fibre be used in the cavity wall situations shown in Figures 3 and 4.

6.2 Methods of installing cellulose fibre in lofts

The next most important use of cellulose fibre insulation in the USA and the current sole use of cellulose fibre in the UK is as loft insulation.

Where ceilings are accessible, the preferred method of installation is by pneumatic means previously described for external wall insulation. In the USA, a pouring quality of cellulose fibre insulation is available for use on a 'do-it-yourself' basis. This variant of cellulose fibre is not currently available in the UK, although its production requires only a modification to the degree of compression employed after manufacture and during packaging.

Where ceiling voids are enclosed, e.g. in the case of flat roofs, installation must be by pneumatic means using a fill tube method. In installations of this type, the void must be completely filled by inserting a fill tube into the void between joists or roof members and withdrawing it as each void is filled, the air setting on the blowing machine being similar to that for vertical wall filling.

In the large majority of cases where ceilings are accessible, blown cellulose fibre is a rapid method of covering the loft area entirely, either between joists or between and over the tops of joists depending on the thickness of insulation required.

When using a pneumatic process, the air setting on the application machine should be such that the least amount of air possible is used to convey the insulation to the hose delivery point, while achieving the manufacturers recommended density.

In new construction, where individual vents are used in the soffit, the rafter space, immediately in front of and on either side of the vent, should be provided with an airway (see Figure 9). All other spaces should be completely blocked with an insulation batt (see Figure 10). Where continuous strip soffit ventilators are used, airways should be provided every third rafter space and all other spaces completely blocked with an insulation batt.

In an existing ceiling the opening into the soffit area may be blocked by batt insulation cut and fitted between joists (see Figure 11). Care should be taken if soffit ventilation is provided to allow a gap of at least 25 mm between the underside of the roof boarding and the cellulose fibre insulation installed at ceiling level. Alternatively, the provision of a permanent airway as illustrated in Figure 9 could be considered.

In general, care should be taken to contain the blown cellulose fibres only over the area requiring insulation. Rigid batt insulation should be placed round access hatches to contain the blown material. Blocking should

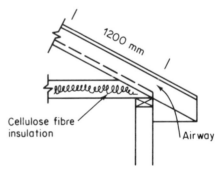

Figure 9   Construction of airway at eaves level

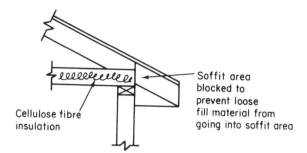

Figure 10 Blocking cellulose fibre at eaves level

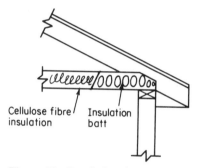

Figure 11 Insulation batt at eaves level

also be placed round light fittings recessed into the ceiling, otherwise heat build-up could occur which could cause failure of the lighting element. As a general rule a minimum clearance of 75 mm should be maintained between cellulose fibre insulation and heat-emitting elements such as chimney flues, central heating flues, etc., although other requirements may be imposed by building regulations and codes of practice.

Cold air and combustion air intakes and exhausts must not be blocked nor should blown fibres be used in a situation where there is a likelihood that fibres could be drawn into a heating or air-conditioning system.

In common with all loft insulants, attention should be given to ensure that adequate ventilation of the roof space is maintained. In the USA where thicknesses in excess of 200 mm are not uncommon, mechanical ventilation of the roof space has been found to be the most expedient method of ensuring an adequate airflow.

References

Agrément Board, 1977, Assessment Report No. 212, August.

Agrément Board, 1979a, Test Report No. 271, January.

Agrément Board, 1979b, Test Report No. 272, January.

American Society for Testing and Materials, 1966, *Thermal Conductance and Transmittance of Built-up Sections by Means of the Guarded Hot Box*, ASTM Standard C236–66.

American Society for Testing and Materials, 1973, *Standard Specification for Cellulosic Fiber (Wood—base) Loose—fill Thermal Insulation*, ASTM Standard C739–73.

ASHRAE, 1972, *ASHRAE Handbook of Fundamentals*, 1972 ed. chap. 20, p. 361.

British Standards Institution, 1973, *Methods for Determining Thermal Properties, with Definitions of Thermal Insulating Terms*, BS 874: 1973.

Brookhaven National Laboratory, 1978, *An Assessment of Thermal Insulation Materials and Systems for Building Applications*, Brookhaven National Laboratory Report BNL–50862 for the US Department of Energy, June.

Canadian Office of Energy Conservation, Energy, Mines and Resources, 1976, *Keeping the Heat In*.

Egan, M. D., 1975, *Concepts in Thermal Comfort*, Prentice-Hall, New Jersey, USA.

Glasgow University, 1978, *Report on the Resistance to Decay of Samples of 'Shelter Shield' Insulation Material by 'Wet Rot' and 'Dry Rot' Fungi*, prepared by the Department of Botany, Glasgow University for Environmental Design Services, October.

Heriot-Watt University, 1978, Department of Building Technical Report No. JMS/13/78, September.

Heriot-Watt University, 1979, Department of Building Technical Report No. JMS/1/79, January.

US Department of Housing and Urban Development, 1975, *In the Bank Or Up The Chimney*, US Government Printing Office, Washington DC, April.

US Government, 1978, Federal Specification HH-I-515D.

Volkman, J., Billau, R. L., and Manley, F., 1978, *Cellulose Insulation: A Look Beyond The Newsprint*, Report by the National Center for Appropriate Technology, Butte, Montana, USA.

Energy Conservation and Thermal Insulation
Edited by R. Derricott and S. S. Chissick
© 1981 John Wiley & Sons Ltd.

CHAPTER 19

Foamed phenolic thermal insulation materials

A. Barnatt

Lankro Chemicals Ltd, Manchester

INTRODUCTION

Phenolic foams were reported as early as 1909, when Baekeland noticed the evolution of gases during the preparation of mouldings from phenol-formaldehyde resins (Baekeland, 1909). Little else was reported until the 1940s, when it is thought that phenolic foams were used in Germany to fill ship bulkheads and as substitutes for balsawood. These early materials were of limited value as insulants, due mainly to their poor mechanical properties, but their outstanding performance in fire situations was undoubtedly recognized.

A surge of interest was noticed in the 1960s (Benning, 1969), probably following an increased interest in the behaviour of plastics insulants in fire, and doubtless assisted by a great amount of polyurethane, polyisocyanurate, and polystyrene technology available at that time.

By the 1970s phenolic foams became available which overcame the early criticisms of inadequate physical properties and difficult processing, and also showed excellent performance in fire situations. New resins were developed as a result of an increased knowledge of foam technology, the influence of acid hardeners and surfactants became more fully understood, and machinery manufacturers were increasingly aware that the production of phenolic foam required machinery other than that used for polyurethane and polyisocyanurate.

On the basis of these developments, the world market for phenolic foams increased from less than 1000 tonnes in 1968 (Lowe *et al.*, 1978) to about 35 000 tonnes in 1978 (Norman, 1978). However, about 25 000 tonnes per annum (tpa) are currently used in Russia, no doubt influenced by their specification of these materials for insulation of public buildings, and hence the Western Europe market is about 10 000 tpa.

415

This comparatively slow growth of phenolic foam in Western Europe has undoubtedly been influenced by the very rapid growth of polyurethane and polystyrene (with its emphasis on cost/performance, from an insulation point of view), and a lesser emphasis on behaviour in potential fire situations than is the case today.

Modern society increasingly demands that insulating materials show improved safety performance in fire situations, and it is in these circumstances than phenolic foams have a contribution to make. They would not claim to be the universal insulating material, and currently available foams are unlikely, at least in the short term, to compete with polyurethane, polyisocyanurate, and polystyrene on a purely cost/insulating performance basis. However, their performance in a fire situation should do much to ensure a more rapid growth than their history would suggest.

Phenolic foams find an increasing use in Canada for built-up roofing, either over a steel deck, concrete or wood. In all cases an adhesive and vapour barrier are used between the substrate and foam insulation, which is then normally covered with felt and asphalt. In the UK, profiled steel and aluminium cladding, preinsulated with phenolic foam lining boards, has been successfully used for cladding several public sector buildings, including those of regional Electricity Generating Boards and British Rail.

The concept of a phenolic foam building panel has proved attractive to designers on account of its fire performance ratings, and this type of structure is currently used in, for example, the American Express Building (Brighton), London Bridge Office Block (London), and the Wellington Centre (Aldershot). It has also been used for the information kiosks at the recent Montreal Olympics and for on-site accommodation cabins in France (Norman, 1978). As a floor insulant, it is used in several large London office blocks.

Profiled phenolic foam is used to a significant extent for the insulation of fermentation vessels and pipework in breweries, where it has the added advantage that, unlike previously used polyisocyanurate foam, it does not cause stress cracking at welded joints.

At subzero temperatures, the insulation value of phenolic foam approaches that of polyurethane foam, and so the two materials become similarly cost-effective from an insulation point of view. The fire performance advantage of the phenolic foam clearly makes it a material to be considered very seriously for the insulation of, for example, liquid-gas vessels.

It can therefore be appreciated that the applications for phenolic foam are increasing at an encouraging pace. As successful case histories build up, it is increasingly specified for fire risk situations. Criticisms of early material are shown to be not applicable to modern foam, and the foundation for large-volume use is nearing completion.

INSULATING PROPERTIES

Since phenolic foam is primarily intended as an insulating material, it is pertinent to consider it in the overall spectrum of such materials (Building Products of Canada, 1977) (Table 1). Cellular materials are generally better insulators than fibrous materials, because they impede the flow of air more efficiently. Plastic cellular materials in particular are low-cost, excellent insulators, and phenolic foams belong in this category. Their main competitors may in one sense be regarded as polystyrene, polyurethane, and polyisocyanurate foams.

Phenolic foam, at its current state of development, is a better insulator than glass wool or mineral wool, about the same as polystyrene, and worse than polyurethane or polyisocyanurate foams (Table 2). It is usually marketed at densities in the range 30–40 kg/m^3, which is about the same as polyurethane and polyisocyanurate, but about twice that of polystyrene foam. Consequently, about half the weight of polystyrene foam produces an equivalent insulation value, while polyurethane and polyisocyanurate have an approximate 40% weight advantage on account of their superior insulating properties.

At current UK prices, it is difficult for phenolic foam to compete with other cellular plastic insulants on a purely cost/performance basis, although it has been suggested that this situation could well change if production levels increased to those of polyurethane and if future raw material prices reflect the 1977–78 trend (Norman, 1978) (Table 3). Nevertheless, for present economic conditions and state of technology, phenolic foams must be regarded mainly as an insulating material to be used in potential fire applications, and it is in these circumstances that most future growth is envisaged. They are therefore likely to complement, rather than compete with, polyurethane and polyisocyanurate foams.

Fully cured phenolic foam, under ambient conditions, normally contains about 10% (w/w) of water, and can absorb significantly greater quantities under more humid conditions. Although this has little effect on physical properties of the foam, it adversely influences the K factor to such an extent that, in common with many other cellular plastic insulating materials, phenolic foam must be protected from moisture in service if it is to remain economically viable (Lowe et al., 1978) (Figure 1). The moisture gain by phenolic foam in protected membrane roof structures has been investigated by the National Research Council of Canada (Hedlin, 1977).

A further factor which influences the K factor of phenolic foam is the temperature at which it is used. At temperatures of the order of $-90°C$, and below, the K value of phenolic foam is approximately the same as that of polyurethane and polyisocyanurate foam, and thus cost/performance figures are very similar. Possible low-temperature insulation, in potential fire situations, is a future growth area for phenolic foam (Figure 2).

Table 1 The spectrum of common insulating materials. (Reproduced by permission of Building Products of Canada)

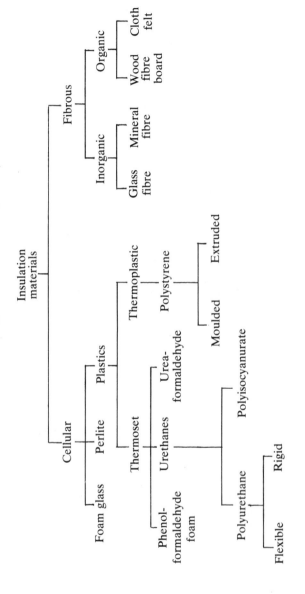

Table 2 K factors of common insulating materials

Material	K factor (W/m $°$C)
Polyurethane foam (aged)	0.023
Polyisocyanurate foam (aged)	0.023
Polystyrene foam	0.032
Phenolformaldehyde foam	0.032
Glass wool	0.040
Mineral wool	0.040

Table 3 Cost comparisons projected to 1980–81 (UK) of common insulating materials. (Reproduced by permission of Cape Insulation Ltd.)

Insulants compared	Nominal density (kg/m³)	K factor (W/m $°$ C)	Cost index, at equivalent production levels
Phenolic foam	35	0.031	100
Polyurethane (fire-retarded)	35	0.023	110
Polyisocyanurate (rigid)	35	0.023	130
Expanded polystyrene	32	0.035	90
Glass/mineral wool	58	0.040	110
Foam glass	125	0.045	150

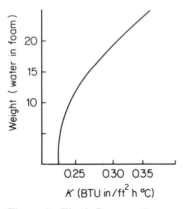

Figure 1 The influence of water content on K factor. (Reproduced courtesy of *The European Journal of Cellular Plastics*, Technomic Publishing Company, 265 Post Road West, Westport, Connecticut 06880, USA)

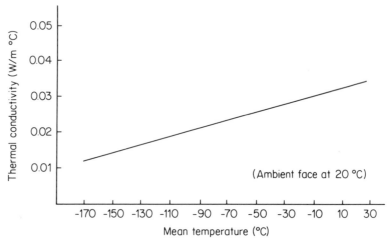

Figure 2   The variation of K factor with temperature for open-celled phenolic foam

CELL STRUCTURE

It is interesting that little difference in the K value/temperature relationship is apparent between so-called 'closed-cell' foams blown with pentane and so-called 'open-cell' foams blown with fluorocarbons such as monofluorotrichloromethane, and this must lead the technologist to question open/closed-cell terminology in the context of phenolic foams.

It has become common practice to use the terms open- and closed-cell in the context of polyurethane technology, where a high closed-cell content (as measured by the ASTM D2856–70 air pycnometer method) is usually associated with a low K value, low water absorption, and low water vapour transmission. This is not necessarily the case for phenolic foam, and much evidence exists to the contrary.

For example, it has been shown that the water absorption of phenolic foam (as measured by immersion of a 5 cm cube of foam under a 6 mm head of water for 7 days) is only slightly influenced by closed-cell content (Lowe et al., 1978) (Figure 3). However, the water absorption of phenolic foam can be influenced to a very great extent by the nature of the surfactant used in its manufacture, strongly suggesting that water absorption is a function of surface chemistry rather than closed-cell content.

Similarly it has been shown that the water vapour permeability of phenolic foam is not influenced by closed-cell content (Dyke, 1977) (Table 4). It is interesting to note that the water vapour permeability of the phenolic foam outlined in Table 4 is of the same order as that of polyisocyanurate foam.

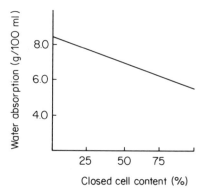

Figure 3 The influence of closed-cell content on water absorption. (Reproduced courtesy of *The European Journal of Cellular Plastics*, Technomic Publishing Company, 265 Post Road West, Westport, Connecticut 06880, USA)

These findings must lead one to question the closed-cell measurements obtained using ASTM D2856–70 for this type of material. It is commonly known that a polyurethane or polyisocyanurate foam with a high closed-cell content (according to ASTM D2856–70) retains the blowing agent used during its manufacture, and this results in a low K factor (approximately 0.023 W/m °C. This is not the case for currently available phenolic foam, for which the K factor (approximately 0.032 W/m °C) seems little influenced by closed-cell content, and is significantly higher than that

Table 4 The relationship of closed-cell content on water vapour permeability of 25 mm samples and water uptake (1 week immersion of a 50 mm cube) for phenolic foam

Closed-cell content (%)	Water uptake (vol. %)	Water vapour permeability at 25°C and 75% RH (perm. ins.)
6	5	5
42	7.3	5.4
70	9.4	6.0
80	7.1	4.8
85	7.6	5.2
90	7.6	4.8
Isocyanurate (90 + closed cell)	–	5.9

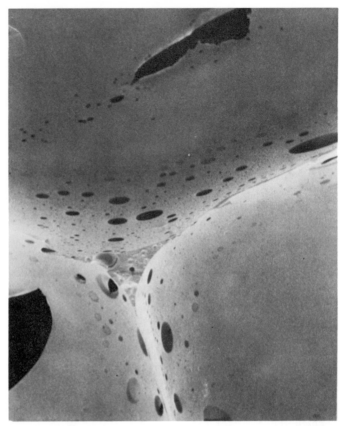

Figure 4 Phenolic foam at × 5500 magnification. (Reproduced courtesy of *The European Journal of Cellular Plastics*, Technomic Publishing Company, 265 Post Road West, Westport, Connecticut 06880, USA)

expected from a foam which has retained its blowing agent. It is not unusual to find phenolic foams with *initial K* values which approximate to that of polyurethane and polyisocyanurate, and such foams are relatively easy to prepare. However, foams in which this low *initial K* value does not gradually increase to about 0.032 W/m °C do not seem to be currently available.

The apparent discrepancy between closed-cell content (according to ASTM D2856–70) and *K* value has caused workers to investigate the structure of phenolic foam using electron microscopy at magnifications up to 5500 (Lowe *et al.*, 1978) (Figure 4). The structure of the phenolic foam under consideration (batch-produced from a 'slow' chemical system) was likened by Lowe *et al.* to individual balloons lightly squashed together with

Table 5 Closed-cell measurements before and after drift

Before drift (%)	After drift (%)
94	73
81	61
30	24

individual surfaces and interstices between the various elements. On a micro-scale nearly all the cells had one or more holes of about 1 micron (1 μm, 0.001 mm) diameter, and this was the case for foams which were both open- and closed-cell by ASTM D2856–70.

It was postulated by Lowe et al. that the interstices between individual cells may be responsible for the phenomenon of 'drift' occasionally noticed during closed-cell determinations. The original determination takes about 15 min, and this can be followed by a gradual drift to an equilibrium situation over a further 15 min. Examples of closed-cell measurements before and after drift are given in Table 5. Possibly the drift is due to a slow diffusion round the cells to fill the interstices gradually.

A possible cause of the microholes in the cell walls is expansion of blowing agent which was dissolved in the cell wall when it was still in a semi-fluid state. Another possibility is that they are caused by elimination of formaldehyde vapour, formed if polymerization occurs via formation of a benzylic ether (Saunders, 1973) (Figure 5).

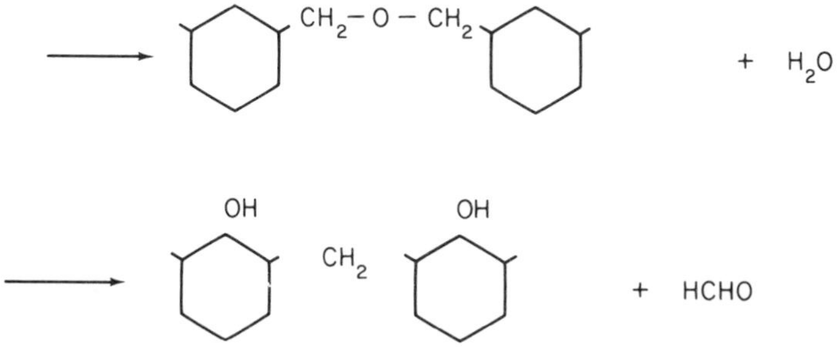

Figure 5 Polymerization mechanism relating to phenolic foam

Much work is currently in hand throughout Europe and the USA to develop phenolic foams with structures such that the blowing agent is retained within the cells. If fluorocarbon 11 or 113 blowing agents are used, the resulting foams should have K values of the same order as polyurethane and polyisocyanurate foams. The economics of phenolic foams are thus vastly improved, and the potential market similarly influenced. Such approaches tend to use significantly higher viscosity resole resins than is the case for currently available foams, and the chances of developing low K factor material in the short/medium term are judged to be high. However, until these foams are available, there seems little point in differentiating between currently available open- and closed-cell structures (as measured by ASTM D2856–70). Additionally, a phenolic foam with a truly closed-cell structure should ideally be defined as a material which gives the same K value characteristics as polyurethane or polyisocyanurate.

FIRE PROPERTIES

Phenolic foams are likely to take an increasing share of the plastic insulant market on account of their improved behaviour in a fire situation. Like all organic materials they burn under sufficiently severe conditions, but with more difficulty and considerably less smoke emission than, for example, polystyrene, polyurethane or polyisocyanurate foams.

Many tests are available for the evaluation of fire properties on a laboratory scale, and certain phenolic foams are classed as nonflammable on PV65 2259 A, B, and C of France's CSTB official test organization, as Class 0 nonflammable according to French merchant marine standards, as Class I according to Dutch standard NEN 1076, nonflammable according to ASTM D1692–59T, and 'acceptable for use as an insulant in cargo spaces' according to a certificate that is granted by the Marine Surveyors of the UK Department of Trade (*Modern Plastics International*, 1974a).

Other phenolic foams are reported to achieve Class I on BS 476 (part 7), a calorific heat flux of less than $0.4/\text{cal}/\text{cm}^2 \text{s}$ on NEN 1076 (Holland), and a burnthrough time of 30 min for a $1000°\text{C}$ propane pencil torch on a 25 mm thick sample (APA Foam Products, 1976).

Another laboratory technique rapidly gaining popularity for testing phenolic foam is critical oxygen index (COI) (Fennimore and Martin, 1966), which is defined as the percentage of oxygen in a mixture of oxygen and nitrogen that will just support combustion. In particular, critical oxygen index provides a useful guide to the punking tendency of a phenolic foam; this being the afterglow which persists after a piece of foam has been heated by a flame to $800–1000°\text{C}$, and the flame removed. A bad foam will smoulder through completely after a period of time, whereas no afterglow will be noticed in a correctly formulated modern foam. Phenolic foams of COI less than about 35 very frequently punk, whereas foams with COI greater than 35 have been found satisfactory in service.

A recent technique which could well be used for testing future phenolic foams is temperature index, which is a measure of COI at elevated temperatures.

Further small-scale testing has been carried out by the Construction Industry Scientific and Technical Centre (CSTB) in France (Hognon, 1976), who compared rigid cellular polyvinylchloride (PVC), polyurethane foam, polyisocyanurate foam, polystyrene foam, and phenolic foam in a series of experiments judged to give an indication of the behaviour of these materials in a real fire situation. Part of the CSTB programme involved measurement of:

(a) the higher calorific value (*HCV*), which is defined as the quantity of heat released by complete combustion of unit mass of the test material; and

(b) the volatile calorific value (*Q*), which is defined as the heat released by combustion of the gases liberated by thermal degradation of unit mass of the test material.

HCV is measured by burning the test material in oxygen under pressure, whereas *Q* is measured by burning the test material at atmospheric pressure and igniting gases of thermal decomposition by means of a pilot light. It follows that *Q* is numerically less than *HCV*, and is also more related to conditions in a real fire situation. Table 6 summarizes *HCV* and *Q* values obtained by CSTB, where the higher calorific value of expanded polystyrene and lower calorific value of phenolic foam can readily be seen.

Further CSTB work measured the rate of heat generation of these materials, suggesting that this may give an indication of potential escape time in a real fire situation, and results were of the same order as those

Table 6   *Q* and *HCV* values of common insulating materials. (Reproduced by permission from Hognon, 1976)

Material	*Q* (kcal/kg)	*HCV* (kcal/kg)	*Q/HCV*
Phenolformaldehyde foam	1389	5160	0.27
Rigid cellular PVC	1517	5457	0.28
Polyisocyanurate foam	1628	6290	0.26
Fire-retarded polyurethane foam	2360	5960	0.40
Non fire-retarded polyurethane foam	2658	6540	0.41
Fire-retarded expanded polystyrene	3947	10050	0.39

obtained for Q and HCV measurements, i.e.

Phenolic foam	
Rigid cellular PVC	Increasing
Polyisocyanurate foam	rate of
Non fire-retarded polyurethane foam	heat
Fire-retarded polystyrene foam	generation

Although much work has been carried out on phenolic foam using laboratory-scale fire tests, it is now widely recognized that such tests do not necessarily relate to the performance of the test material in a large-scale fire situation. It is for this reason that there has been a move in recent years towards fire testing on a larger scale, with the object of approaching real fire conditions, and this type of test has shown the advantages of phenol formaldehyde to be:

(a) improved flame resistance over, for example, polystyrene or polyurethane; and

(b) very low smoke emission in a fire situation.

Large-scale testing of phenolic foam has been carried out at Queen Mary College (London) (1975) in their test chamber (Figure 6). The chamber consists of a building (4 m × 3 m) which opens onto a corridor (1 m × 12 m). Steel doors with adjustable vents are fitted at one end of the corridor and at the outside entrance to the test chamber, and the doors are used to adjust the ventilation to the sample in the test chamber. Measurements can be taken on thermocouples, gas analysers, and smoke detectors.

Phenolformaldehyde and polyisocyanurate foams, both faced with thin paper/foil, were compared on this test rig, and it was reported that: 'The PF sample showed a greater resistance to heat penetration, although there was a tendency for the foam to crack and flake off in the flame zone. The isocyanurate was damaged to a greater extent and collapsed during the experiment. The level of smoke generated from the PF foam was significantly lower than that from the isocyanurate.'

It is in this lack of smoke in a fire situation that phenolic foams are outstanding, as can be seen from the results of the QMC work (Figure 7) and from XP2 Smoke Chamber experiments (Figure 8).

Further work on smoke evolution has been carried out by the Fire Safety Centre at the University of San Francisco (Hilado et al., 1978). In particular, phenolic foam was compared with other common plastic materials according to ASTM E162, and shown to be significantly better than polystyrene and PVC, but worse than red oak or fir plywood (Table 7).

In the American Underwriters Laboratory Tunnel Test, sprayed phenolic foam at 0.8 lb/ft^3 density showed a rating of 25, smoke density value of 2,

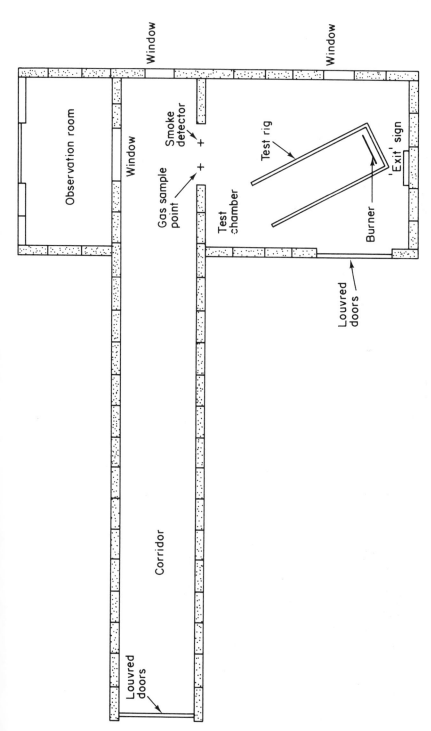

Figure 6 Queen Mary College Fire Test Facility. (Reproduced by permission of QMC Industrial Research Ltd)

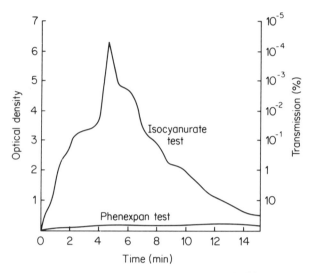

Figure 7 Smoke Obscuration data of Phenexpan phenolic foam and isocyanurate foam

and fuel contribution of 15 (*Modern Plastics International* 1974b). More recently, phenolic foam has achieved a flame spread index of 15, fuel contribution of 15, and developed smoke of 20, according to ASTM E84 (Building Products of Canada, 1977).

On one occasion, a demonstration classroom was constructed of triangular panels, measuring 2.3 m on each side, consisting of glass-reinforced polyester skins with 50 mm of phenolic foam between. Large-scale testing carried out before construction of the classroom involved using similar

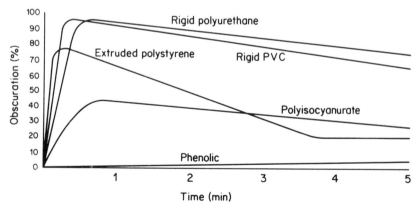

Figure 8 XP2 Chamber smoke generation characteristics

Table 7   Smoke evolution data of common plastic materials, red oak, and plywood

Material	Thickness (10^{-3} inch)	Smoke deposit (mg)
Red oak	750	0.3
Fir plywood (exterior)	250	0.3
PVC	147	28.9
PVC (FR)[a]	147	10.5
Polystyrene	66	23.0
Polymethylmethacrylate (FR)[a]	125	40.6
Phenolic (laminate)	63	1.1
Unsaturated polyesters (resin only)	125	5.7–9.7
21% glass	62	15.9
27% glass	85	18.4
27% glass (FR)[a]	95	22.3

[a]FR = fire-retarded.

panels as roofing on an asbestos-walled structure, and lighting wood fires inside it such that an inside temperature of 800°C was achieved in 20–30 min. After 35 min the structure showed no signs of external damage (*Modern Plastics International*, 1976).

CHEMISTRY AND TECHNOLOGY

Phenolic foams are usually manufactured by the acid-catalysed polymerization of phenolformaldehyde resins in the presence of a blowing agent and a cell control agent. The phenolformaldehyde resins are normally resole resins, prepared by reacting 1 mole of phenol with 1.5 to 3.0 moles of formaldehyde under alkaline conditions.

There are many variants within the scope of the reaction outlined in Figure 9. For example, the degree of *ortho/para* substitution may be varied with reaction conditions, part of the phenol may be replaced by alkylphenols, chlorinated phenols or substituted phenols such as *bis*phenylolpropane. A part of the formaldehyde may be replaced by other aldehydes such as acetaldehyde or glutaraldehyde, and nitrogen-containing compounds such as urea or melamine may be incorporated in the structure in place of part of the phenol.

It is widely accepted that the polymerization reaction is complex, involving many competing reactions. However, it is generally thought that the methylol groups of the resole initially condense, with elimination of water, to give benzylic ethers, which in turn cleave under acidic conditions to form methylene bridges between the aromatic nuclei (see Figure 5).

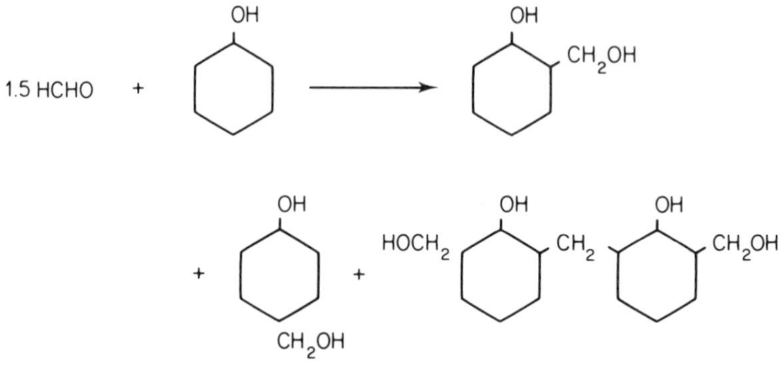

Figure 9   Formation of resole resins

Most resole resins in commercial use contain 10–20% water in order to give products of workable viscosity. Any further reduction usually leads to a rapid increase in viscosity, although it could well be that the technical advantages, from the point of view of improved K value and physical properties, likely to be obtained from foams manufactured using high-viscosity resoles may in future lead to development of new processing techniques.

A significant problem with water-soluble resoles is that of limited storage stability, which generally worsens as viscosity increases. A storage stability of 4–12 weeks at 20°C is usual; with significant improvements at lower temperatures. Current resole developments suggest a move towards less stable systems, which must in turn place additional demands on manufacturing schedules, and restrict the transport of resin over other than short distances prior to use by the foam manufacturer. This in turn may lead to local manufacture of resole and foam throughout the world.

A basic formulation for the manufacture of phenolic foam is:

Constituent	Parts by weight
Resole resin	100
Blowing agent	12
Surfactant 11	1
Acid hardener	7–30 (depending on hardener type)

Resoles used in phenolic foam production are typically characterized by viscosity, reactivity, water content, free phenol content, and free formaldehyde content. Of these, it is probably true to say that the most important are viscosity and reactivity, since these are influenced to a large extent by the other parameters. To make phenolic foam in a satisfactory manner, it is necessary to use a resole with the correct balance between viscosity and reactivity, and it is the task of the resole manufacturer to ensure that this is obtained in the correct manner. For example, both reactivity and viscosity can be influenced by molecular weight and/or water content and/or free phenol content, and a correct balance must be achieved between all of these factors.

Optimum resin characteristics are very much influenced by the type of foam manufacturing process to be used. In general, low-viscosity, high-reactivity resoles (typically 1500 cP at 25°C) are used for continuous processes such as lamination and spraying, whereas higher-viscosity, lower-reactivity resoles (typically 3000 cP at 25°C) are used for discrete processes such as production of discrete blocks and *in situ* void filling.

Many blowing agents for phenolic foam have been investigated, and these may be generally described as materials which decompose under heat or acid conditions to give gaseous products, or vaporizable liquids. Examples of acid-sensitive materials include sodium carbonate, ammonium carbonate, sodium nitrate, sodium sulphite, and sodium bicarbonate. Diazonium salts are an example of heat-sensitive material. Some systems, mostly of Russian origin, use hydrogen liberated by the reaction of aluminium powder with the acid catalyst.

Nearly all modern phenolic foams are blown using vaporizable liquids, usually pentane, saturated fluorocarbons or, sometimes, methylenechloride. It was originally thought that pentane produced foams with smaller and more regular cells than the saturated fluorocarbons, but work carried out in recent years has shown that this is not the case. Much pentane is still used throughout Europe, where its flammability problem can be outweighed by its low cost. Nevertheless, the continued use of pentane seems unlikely of foams of low K value are to be developed, since these require a fluorocarbon blowing agent retained in a truly closed-cell foam to ensure a K value of the same order as polyurethane or polyisocyanurate foam. In particular, mixtures of fluorocarbon 11 and fluorocarbon 113 are in increasing use, with a tendency to use fluorocarbon 113 alone.

The properties of phenolic foam, particularly regarding cell structure, water vapour transmission, and water absorption, are greatly influenced by the type of surfactant used in its manufacture, and it is in this area that major advances have been made during recent years. Recommended surfactants have included silicone oils, nonylphenylethoxylates, polyglycol ethers of sorbitan mono-oleate, and castor-oil ethoxylates. Silicone oils are used less frequently in modern foams, since they almost invariably produce open-celled material, which is against all current development trends.

The acid catalyst may be selected from mineral acids such as hydrochloric acid and sulphuric acid, organic acids such as toluenesulphonic acid and phenolsulphonic acid, and polymeric hardening materials such as sulphonated Novolaks. All of these acids, or mixtures of them, are, or have been, used commercially, although hydrochloric acid is becoming less popular on account of the possible formation of the highly toxic dichlorodimethylformal, together with a possible adverse influence on the corrosive properties of foam made from it. Its frequent use in early phenolic foams could well explain their initial poor reputation regarding acidity and corrosion.

Early phenolic foams were often friable materials, and it became common practice to add modifiers to improve foam flexibility. Such additives may cure along with the resole under the influence of the acid catalyst, or they may be unreactive materials. Examples of the former are ureaaldehyde condensation products, epoxides and Novolaks, whereas examples of the latter are polyesters and polyacrylonitriles.

Finely divided fillers such as talc, mica, wood flow, and carbon black have been used to improve foam structure, but these have the disadvantage of increasing foam density.

Resole resin viscosity may be reduced by additives such as polyvinylacetate, polyvinyl alcohol and polyvinylbutyral, with consequent simplification of mixing during the foaming process.

Many phenolic foams require external heat to achieve a satisfactory state of cure, and this may restrict on-site applications. Also, it can be difficult to obtain an even temperature distribution in large blocks by external heating alone, since the foam itself is a good insulator. One solution to these drawbacks is to include in the reaction mixture a substance which reacts exothermically with the curing agent. Examples of this type of material include phosphorus pentoxide, boron oxide, and calcium carbonate; the latter also acts as a blowing agent when it reacts with the acid catalyst to form carbon dioxide. Another solution, claimed to be more practical, is to use liquid unsaturated nuclear oxygen containing heterocycles such as pyran and furan derivatives (Ciba Geigy, 1970).

As produced, most phenolic foam contains up to 30% water, due to water initially contained in the resole resin and that formed during the condensation polymerization reaction. Much of this is lost during foam storage, over a period which is influenced by such factors as storage conditions and foam dimensions. In addition, the blowing agent used in foam production is usually lost. The combined effect of water and blowing agent is that the weight of fully cured and stabilized phenolic foam is about 75% of the weight of the chemicals used in its manufacture. Such a significant factor must be taken into account during economic considerations of the foaming process.

Typical physical properties of a fully cured phenolic foam are given in Table 8.

Table 8   Typical physical properties of phenolic foam

Density (kg/m^3	35
Thermal conductivity by BS 4370 (part 2) (W/m $^\circ$ C)	0.032
Compressive strength by BS 4370 (part 1) (kPa)	150–200
Cross break strength by BS 4370 (part 1) (kPa)	200
Maximum service temperature for continuous operation ($^\circ$C)	150
Water uptake by ASTM D2842-69 (% v/v)	7
Moisture vapour transmission by BS 4370 (μg/Nh)	20
Shrinkage at 130°C for 7 days (%)	1.5
Closed-cell content by ASTM D2856-70T (%)	85

MANUFACTURING TECHNIQUES

Phenolic foams are usually manufactured, on machine scale, from three-component (acid catalyst/blowing agent/all other components) or two-component (acid catalyst/all other components) systems. Machinery operates on the same basic principles as that for polyurethane, in that the components are pumped from storage tanks, in a predetermined ratio, to a mixing head, which discharges mixed chemicals onto a conveyor system or into a mould. However, there are important differences between polyurethane and phenolic foam machines, as outlined below.

(i) Phenolic resins are of higher viscosity than resins for polyurethane foams, and so require larger-diameter pipework and larger pumps.

(ii) Many hardening agents are corrosive, and hence require special metals for the construction of storage tanks, pumps, lines, and mixer.

(iii) Provision sometimes has to be made for inflammable blowing agents such as pentane.

(iv) Higher chemical temperatures are frequently needed for resole resins.

Laminate is produced on a continuous basis, using a variety of facing materials. It is preferable that at least one of the facings allows water to escape from the foam either during manufacture or subsequent storage, since the insulation properties of the foam will otherwise be impaired. This type of production is in use in Canada, where laminate is manufactured at an average rate of 250 m^2/h at thicknesses of 30–100 mm and densities of 35–100 kg/m^3.

Considerable production of laminate is thought to be carried out in the Eastern Sector, for which the equipment illustrated in Figure 10 is used. This laminator produces foam up to a rate of 4.5 m/min at a maximum trimmed width of 1.25 m. Thickness can vary between 2 and 10 cm. Foam chemicals are dispensed through a metering/mixing unit capable of handling liquids of 30 000 cP at up to 16 kg/min total chemicals output. The

1000/214   PHENOLIC LAMINATOR CONVEYOR PRESS

1000/215   PHENOLIC LAMINATOR 3 COMPONENT METERING UNIT

Figure 10   Laminator for phenolic foam production. (Courtesy Viking Engineering Ltd)

complete machine is designed in order that inflammable blowing agents such as pentane may be used. The conveyor system, 16 m long, is similar to that used for polyurethanes, and will withstand a maximum working pressure of 0.5 kg/cm².

Foam-filled panels are also manufactured on a discontinuous process. Panel facings are generally loaded into multidaylight presses with temperature-controlled platens, and each panel then filled with foam chemicals through one or more injection points. Long sections are sometimes filled using a probe, which is withdrawn from the mould at a rate proportional to that at which the chemicals are injected.

It has been reported that phenolic foam has been spray applied to the exterior of a pavilion in France (*Modern Plastics International*, 1978), but the lack of further reports are possibly an indication that more development work is needed in this area.

Most phenolic foam slabstock, certainly in Europe, is manufactured as discrete blocks, usually about 8 ft × 4 ft × 2 ft. Chemicals are machine dispensed into a suitable mould, which may have a device for ensuring that the block has a flat top, and the filled mould is placed into an oven for 1 to 2 h. The foam is then removed from the mould and preferably maintained at 15–20°C for about 3 days before cutting. Such treatment minimizes the chances that the foam block will crack due to thermal shock combined with an incomplete cure.

Development of a process for continuous production of slabstock is well advanced in Europe, and should become commercially viable within the next 12 months. This type of manufacture reduces the high waste factor often associated with discrete block production, mainly due to skin formation and consequent trimming losses.

CONCLUSION

Over recent years, phenolic foams have developed to such an extent that they are now technically and commercially viable insulants for those applications which require increased fire resistance over, for example, polyurethane and polyisocyanurate. Processing equipment has developed at a similar rate, and this has coincided with a hardening of the attitudes of government and local authorities towards the performance of insulants in fire. For these reasons, modern phenolic foam can be regarded as a new type of insulating material, with a high probability of becoming a much-higher-volume product than at present.

REFERENCES

APA Foam Products, 1976, *Phenexpan Phenolic Foam, the Realistic Solution to Insulation Fire Hazards*, Trade literature.

Baekeland, L. H., 1909, *Ind. Eng. Chem.*, **1909** (1), 149.

Barnatt, A., Dyke, R., and Hillier, K., 1978, Phenol formaldehyde foams, Paper presented by A. Barnatt to *Studiedagen Kunstofschuimen*, Utrecht, Holland, February.

Benning, C. J., 1969, *Chemistry and Physics of Foam Formation*, vol. 1, John Wiley & Sons, New York, pp. 423–445.

Building Products of Canada, 1977, Trade literature.

Ciba Geigy, 1970, *Foamed Materials and Their Preparation*, Patent BP 1335973.

Dyke, R., 1977, private communication on behalf of Diamond Shamrock (Europe).

Fennimore, C. P., and Martin, F. J., 1966, *Modern Plastics*, **43**, 141.

Hedlin, C. P., 1977, Moisture gains by foam plastic roof insulation under controlled temperature gradients, *J. Cellular Plastics*, Sept/Oct, 313.

Hilado, C. J., Cumming, H. J., and Machado, A. M., 1978, Screening materials for smoke evolution, *Modern Plastics International*, July, 53.

Hognon, B., 1976, *Contribution Possible des Principaux Isolants Synthétiques à l'Aggravation des Dangers en Cas d'Incendie*, Report of Cahiers du Centres Scientifique et Technique du Batiment.

Lowe, A. J., Barnatt, A., Chandley, E. F., and Dyke, R., 1978, Phenol formaldehyde foams, *European J. Cellular Plastics*, **1**, January, 42.

Modern Plastics International, 1974a, July, pp. 14–17, Upgraded phenolic foams aim for bigger slab of insulation markets.

Modern Plastics International, 1974b, August, p. 52, It's the comeback trail for urea and phenolic foams.

Modern Plastics International, 1976, June, p. 8, Phenolic foam is bubbling again.

Modern Plastics International, 1978, May, p. 18, The fire challenge: phenolics and urethanes meet head on.

Norman, E. G., 1978, Rigid phenolic foam insulation, Paper presented to *Building Construction Group Conf.* at the Building Research Station, UK.

Queen Mary College, 1975, *Comparative Fire Testing of Phenexpan Phenolic Foam and an Isocyanurate Foam*, QMC Report 00/440 (1975).

Saunders, K. J., 1973, *Organic Polymer Chemistry*, Chapman and Hall, London, p. 291.

Energy Conservation and Thermal Insulation
Edited by R. Derricott and S. S. Chissick
© 1981 John Wiley & Sons Ltd.

CHAPTER 20

Foamed polyurethane insulation

A. R. D. Lambert

Bayer UK Ltd, Richmond

INTRODUCTION

For too many years insulation has been ignored in the United Kingdom due mainly to the relatively cheap energy which has been afforded by our indigenous coal supplies. This fuel source was also fundamental as a foundation for gas and electricity industries which have served us well into the third quarter of the twentieth century.

Exploration for oil and its introduction as an industrial and domestic fuel has been economic and competitive until the most recent developments in the Middle East, where political activity has brought about a drastic change in the basic cost and transportation of this all-important commodity. Future revolutionary moves in politically sensitive areas can only cast further gloom and concern over the industrialized nations who have designed their environments around oil and its byproducts.

But maybe we should not be too quick to grumble at the rising fuel costs. Perhaps we have been too slow in considering ways to use the energy not only sparingly but economically, and too slow in looking at the many cases where modifications to our original design or concept of an energy use would have meant that less fuel could well have provided more power.

There are some people who are of the opinion that we, in the British Isles, have been slow in realizing that our climatic environment is either changing or was never as temperate as we would have ourselves believe. Whilst the theoretical requirements to keep our island moderate exist, i.e. the Gulf Stream and other warmth afforded by the vast sea masses that surround us, it is very cold comfort indeed when life grinds to a halt just because we have had a 'freak' snowfall of 12 inches or that the temperature has dropped to $-6°C$ due to unusual easterly winds blowing from Russia.

It is true that insulation now serves a major role in the design and construction of most buildings, but it may be interesting to discover how and when this

437

came about. Industrial premises, business houses in particular, became very aware of the need to insulate perhaps as a result of direct action by employees who refused to work below a comfortable temperature. It is perhaps ironic that these same employees may well have returned home to spend their leisure time in conditions which might have been worse.

However, it is now estimated that over 50% of all private houses have some form of central heating. But the amount of effective insulation in these dwellings leaves a lot to be desired. It is only in the last decade that it has been made a regulation that a minimum U value for the roofs of domestic dwellings be specifically achieved.

Technical insulations have been required for many years, as many installations and applications could never have functioned without them. Natural materials were obvious media, and mineral fibres, cork, and glass were most commonly used.

But the development of expanded plastics and rubber products presented the engineer and architect with reliable and consistent products with improved insulation values. Among these developments was rigid polyurethane foam.

EARLY HISTORY AND DEVELOPMENT

It was in response to the significant development of Nylon® by Dupont in the USA that polyurethane was initially synthesized in Germany by Bayer who were looking for alternative synthetic fibres. Early research was continually spoiled by bubbles of gas within the polymer samples. It was established that this was caused by the presence of water which generated carbon dioxide to be formed as a side reaction.

Very quickly, this foamed plastic was being investigated and developed in its own right by Otto Bayer and his coworkers for uses in the engineering and construction industry. The high strength-to-weight ratio of this foam was a contributory factor to its applications in the German war effort where it was used for reinforcement of aircraft wings, etc.

At the end of the Second World War, polyurethane foam technology was acquired by the Allies and considered well worth developing further due to the shortages of rubber from the Far East. Research had reached a stage by this time where flexible polyurethane foams were possible, and this lent itself ideally for the production of mattresses and other upholstered items much needed by the furniture, bedding, and transport industries now in need of a rubber substitute. In this way, rigid foam development was halted in favour of flexible and semiflexible foams.

However, in the late 1950s serious interest was shown from the refrigeration industry, and by the mid-1960s rigid urethanes were to be found in the construction and technical insulation world. This was undoubtedly encouraged by successful research into the utilization of less hazardous starting products to enable an *in situ* or on-site application to be made. Hartmoltopren® was now established as an insulation material in Germany and Western Europe.

PRODUCTION AND MANUFACTURE

As with most chemical processes, the most effort was required in offering industry facility which was akin to their present facilities and 'know-how', and therefore the rubber and allied industries were first to become involved.

The polyurethane process is based upon the reaction of a polyglycol with a polyisocyanate to result in a thermosetting plastic. The nature of the type of starting materials has a direct effect on the flexibility or stiffness of the final plastic material, and this can be quite finely controlled. If water is also present, then carbon dioxide gas is also produced, which will cause the plastic to foam. When controlled by catalysts, a stable plastic foam can be manufactured. With flexible foam, the gas is allowed to permeate from the foam prior to 'setting-up'. But rigid foam encapsulates the gas and uses these bubbles to achieve its good insulation characteristics. For several years the relatively good insulation value, or conversely the poor conductivity value, of the carbon dioxide gas was sufficient, but, with the introduction of fluorocarbons as blowing agents in the reaction, the insulation characteristics were clearly improved.

These fluorocarbons replaced the water in the reaction and did not use any of the polyisocyanate in the reaction. This immediately was of interest. But these low-boiling-point liquids soon vaporized with the exothermic heat generated in the chemical reaction to produce a plastic foam containing thousands of bubbles of fluorocarbon gas.

The first commercially produced rigid foams utilized the technology and equipment in the existing flexible foam factories. But due to technical difficulties, early manufacture was not of a continuous nature. Large 'loaves' of rigid polyurethane foam were thus produced and subsequently cut into slabs or sheets for further installation as insulation materials. Before long, very intricate parts were being cut for pipe sections or profiled slabs for internal roofing.

This conversion process proved to be very expensive and wasteful and soon alternative methods of manufacture were under consideration. Pipe sections were being moulded and profiled sheets were soon to be replaced by the on-site technique of applying the polyurethane foam by spray application (Figure 1). *In situ* pouring of the material was also developed to satisfy many requirements in the building and shipbuilding industries.

This on-site work was made possible by the utilization of portable processing machines which adapted the 'know-how' from the factory-based equipment.

Polyurethane foam processing machines are basically a means to pump, formulate mix, and dispense two or three liquid components. The raw materi-

Figure 1   Spray insulation of roof

als must not come together until the very last moment. Within seconds of the materials being mixed, a foaming reaction takes place which produces a rigid but lightweight foamed plastic which can be handled within several minutes. The processing machine must be capable of accurately pumping and formulating, and the mixing must be intimate. The mixing head must be easy to use in the case of portable equipment, and the entire equipment needs to be handled by fully trained operators. In the interest of health and safety, the operators require protective clothing, and in many cases, a means of avoiding inhalation of polyurethane particles or fumes. This is best afforded by fresh-air-fed facemasks.

Because of the problems experienced with on-site and *in situ* work during the 1960s, there has been a return to the use of ready-made products produced in the relatively safer environment of the factory. This has meant the development of further processes to facilitate the production of composite elements. Such production utilizes a double band or laminator which can apply rigid or flexible facings to the foam.

PHYSICAL PROPERTIES

Rigid urethane foam can be produced over a wide range of densities, and this is usually the base for comparing the mechanical properties of one foam with another.

(a) Density

A rigid polyurethane foam is normally produced at densities of 30–200 kg/m^3 (2–12 lb/ft^3).

Most slab foam production is produced to give a free rise density but in many cases foams are moulded and the density is subsequently higher, and the surface can show a slight skinning tendency.

(b) Mechanical properties

According to the density polyurethane foams have the following properties.

(i) Comprehensive strength

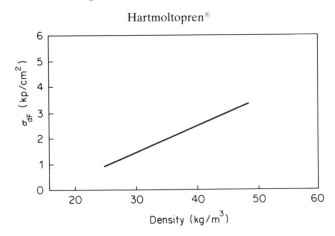

Hartmoltopren®

(ii) Tensile strength

Tensile strength and elongation at break (measured in accordance with DIN 53455)

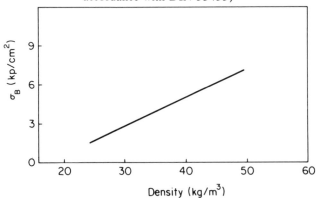

(iii) Shear modulus

Shear Modulus (measured in accordance with DIN 53427)

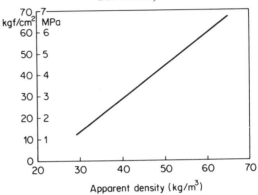

(iv) Elongation at break

Elongation at break, 22°C, 55% RH. Test specimen as per DIN 53455 Form I, 25 mm web but 10 mm thick

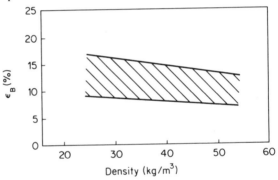

(v) Flexural strength

Flexural strength as per DIN 53423, 22°C, 55% RH (minimum values)

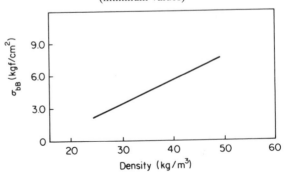

(vi) Coefficient of thermal conductivity

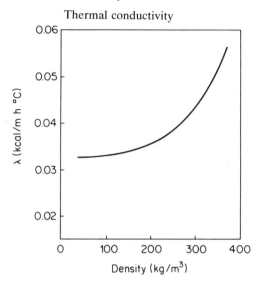

Thermal conductivity

(vii) Dimensional stability

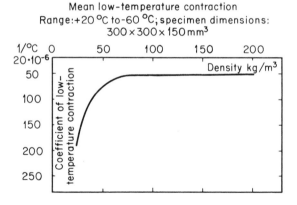

Dimensional Stability

Mean low-temperature contraction
Range:+20 °C to-60 °C; specimen dimensions:
300 × 300 × 150 mm³

(viii) Water absorption

The water absorption, measured on samples immersed for 24 h, is about 1 vol.% with fluorocarbon-blown foam, and 3% where carbon dioxide is present.

(ix) Chemical resistance

Rigid polyurethane foam resists fresh water and sea water, dilute acids and base solutions, aliphatic hydrocarbons (e.g. propane), normal petrol, diesel fuels, and mineral oils.

(x) Weather resistance

The material should normally incorporate a facing for external applications. Exposed polyurethane foam has shown good resistance, but a darkening in colour is observed together with some likelihood to become brittle.

(xi) Adhesion

Rigid polyurethane foam bonds excellently to most surfaces as long as they are free of dust, grease or moisture. Typical facing materials are: paper, board, roofing-felt, wood, hardboard and chipboard, stone, concrete and asbestos cement, steel, and aluminium, as well as plastic facings.

(xii) Temperature performance

Operating temperatures of $-200°C$ to $+100°C$ are common, with some special products suitable up to $+150°C$. The material can even stand hot bitumen for short periods.

APPLICATIONS

(a) Refrigeration

As mentioned previously, the refrigeration industry was among the first to realize the insulation potential of rigid polyurethane foam. In both industrial and domestic appliances, units had been using cork or mineral wool and then glass fibres. Plastic foams, initially polystyrene and then polyurethane, were applied as slabs, whereupon the improved U value (resistance to thermal conductivity) meant that, for the same size outer cabinet of a refrigerator or deep freezer, the effective storage space inside could be increased due to the thinner section of insulating material required. If we consider refrigerated appliances such as refrigerators, deep freezers, and display cabinets, the use of polyurethane brought about a new and improved method of construction.

Such units are basically metal boxes or wrappers with a plastic or metal box inside. Slab materials previously contributed little or nothing to the strength or building of the appliance. But the possibility of *in situ* foaming using rigid polyurethane afforded the best insulation together with good bonding to the various substrates used, therefore producing a sturdy and robust cabinet with the minimum of mechanical fixings (Figure 2).

In use, cabinets insulated with polyurethane have been proved reliable and

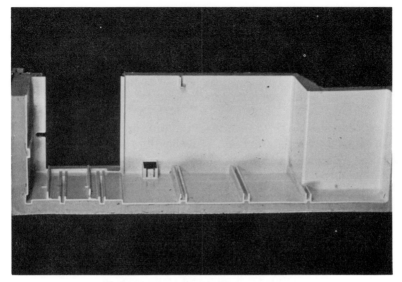

Figure 2   Cross section of a refrigerator

durable. The superior physical properties are highlighted by the good resistance to moisture ingress which normally occurs with cork or fibrous materials. The insulation remains consistent and does not pack down like fibrous insulation.

(b) Cold rooms and stores

As with refrigerated appliances, rigid polyurethane now plays a major part in the design and construction of cold rooms and cold stores.

Most cold rooms are produced as modules which are assembled on-site to stand freely within an existing store or shop. Once again the reduced wall thickness offered by the improved U value increases storage space, and the lightweight foam assists easy and speedy construction.

In the fabrication stage, the elements which are produced to make up the modules are easily fixed together and doors can be cut to suit the final design after the panels are foamed. Roofs and floors can be reinforced by utilizing a polyurethane foam of higher density without significant change in U value.

In some cases, existing rooms can be made into cold rooms by the *in situ* pouring of polyurethane around the original walls should they be strong enough and accessible. This was done in particular at butchers and fish shops. Such cold stores utilizing approximately 120 mm thick polyurethane have a service temperature range of $-18°C$ to $-20°C$.

Cold stores are usually purpose-built constructions where the entire building is to be maintained at low temperature. In many cases, a building shell is erected and insulated panels are fixed to the inner walls and roof. This is a common method in the UK, but in Europe many stores are produced by applying insulated panels as cladding to a frame structure.

This method is often cheaper but the insulated panels need to be thicker and often of a higher density to withstand movement which is possible due to the panel experiencing a greater temperature difference from the outer to inner face.

In practice, panels up to 10 m long by 1 m wide are produced in the factory and can be readily transported and assembled on site with the minimum of mechanical handling. In combination with metal facings, the rigidity of polyurethane-insulated panels have been proved to span larger areas unsupported.

Once again, the bonding characteristics of the foam are an advantage when manufacturing cold store panels, as the locking mechanism and sealing gaskets are often foamed in place.

(c) Transport and containers

As a follow-up to its use in the previous applications, it was predictable that polyurethane would be used to insulate transport specializing in movement of refrigerated goods. Insulated boxes were fixed to the chassis of heavy goods vehicles to deliver frozen foods or dairy products (Figure 3), and road tankers were insulated with polyurethane to maintain raw materials at the tempera-

Figure 3 Insulated delivery truck

tures industry requires for processing or to facilitate the easy loading and discharging of temperature-sensitive products. The cylindrical shape of most tankers encouraged the earlier use of pour- or spray-in-place polyurethane. However, today, curved sections are often made from moulds in factories. Again the rigid foam was sought after not only for its insulation value but also for its contribution in stiffening the entire composite structure.

This is very much the case with the insulated International Standards Organization containers (see Figure 11) which we have come to accept as the favoured methods of moving cargoes other than bulk. These insulated boxes are constructed in different ways but are virtually all insulated with rigid polyurethane. This was decided at an early stage when the various methods of transporting containers were considered in respect of the various climates for which they must be suitable and the need to conserve as much cargo space as possible, coupled with the need to stack the units up to five high; a light-weight sturdy insulated box was specified, and polyurethane as insulant and construction material met this requirement.

The actual construction used slab polyurethane, pour-in-place, and in one instance an internal spray system. The *in situ* method presented problems due to the sheer size of the boxes. Other *in situ* foaming pressures which were generated (Figure 4) or were of robust enough construction to require no further support. Because the steel frames of the containers are designed to take the majority of the load, one of the facing materials is usually of light-weight construction.

A large jig to facilitate *in situ* foam insulation of an ISO container is obviously extremely expensive and the speed of demoulding determines how many jigs are necessary. This method has been utilized in the United Kingdom whereupon the entire container was filled at one time. However, this method is only economical where large numbers of identical containers are being produced by a single manufacturer. Experience has shown that fabrication from slab polyurethane, produced and profiled to suit the container constructor's own requirements, has given the best guarantee of a satisfactorily built and insulated box.

Railways have sought to utilize rigid polyurethane for the obvious insulation of rolling stock to carry refrigerated cargo. Not so obvious, however, is the early use of the material to insulate and later assist in construction of personnel wagons. *In situ* urethane was used in roofs and bulkheads of carriages to improve insulation and help cure the drumming effect transmitted from the wheels through the two-skin construction. In Germany, a complete train was built using glass reinforced plastics reinforced with polyurethane, which ran daily transporting personnel from the local town to a large industrial complex.

Currently, we can find urethane in most aircraft, ships, and automobiles, often fulfilling a multipurpose function.

Figure 4 Stationary mould for refrigerator production

The German automotive giant, Volkswagen, uses rigid polyurethane within the bulkhead of the rear seat and engine in its long-running 'Beetle'. This affords sound and heat insulation. Others use it to stiffen frames or fill cavities which all too often become traps for water and therefore promote rust formation.

A contribution has also been made to the 'space age'. Urethane has played a major role in insulating the vehicles against the wide extremes of temperature that can be experienced outside of the Earth's protective atmosphere. Not only is the craft insulated with polyurethane, but all the important liquified-gas tanks which provide the life-support and motive power are protected with the urethane plastic. Coincidentally, polyurethane in another form is used in foaming the solid propellants which are the main fuel source for some rocket designs.

(d) Construction industry

In this application area, there is probably the most potential growth which has continuously fallen short of the predictions made by optimistic engineers and technologists in the plastics industry. World usage has increased from approximately 5000 tonnes in 1965 to approximately 150 000 tonnes ten years later. Whilst there is a pronounced increase in usage, it is much less than estimated usage predicted year after year. Why is this?

Perhaps it is the conservative attitude of an industry which has been working with traditional materials for hundreds of years. If we consider how long we have had to develop ways in which to apply stone, timber, and metals, maybe we should be satisfied with the use of plastic materials in general in the building industry considering the 50 years of their development and the somewhat shorter involvement of polyurethanes.

The first inroads were made into building with prefabricated units where comparative low weight, coupled with good mechanical properties and speedy construction on site, proved to be an advantage over some other materials. The excellent insulation value was known and promoted to the industry but it was some time before architects and engineers began specifying the material.

Factory-made building modules incorporating rigid polyurethane were soon to find a place in the market. Panels with steel, aluminium or GRP facings were speedily produced and fitted to construction frames as cladding panels. Sophisticated jointing systems were developed to prevent thermal bridges being formed. The comparatively light weight of such panels led to easy handling and erection on site. But these composite elements were still considered non-load-bearing and investigations were made to improve the structural properties.

Inorganic filling materials were considered and the most suitable were

Figure 5 Bungalow from Dumont & Besson

expanded clay and glass spheres. Such aggregates were already known in the manufacture of lightweight concrete and the K factors were reasonable. The difficulty was in determining how much filler to add and then to ensure a good flow of polyurethane throughout the panel to obtain good bonding of the aggregate. The overall density of these filled foams are approximately 200 kg/m^3 with the compressive strength increased to 0.8 MPa; whereas the K factor is 0.064 W/m K.

Complete bungalows (Figure 5) have been constructed from modules made in this manner and, in some instances, two storeys have been permitted. The same construction has been adopted to produce sanitary units (Figure 6), where the plumbing and bathroom furniture are constructed together with the polyurethane-filled panel in the factory and then transported to the site. Such panels have played a major role in the general modernization of older properties. In some cases, complete bathroom cells are then assembled into the partially finished buildings, usually apartment blocks, etc.

Probably the largest amount of rigid polyurethane for the building industry is used in the insulation of roofs (Figure 7). Continental buildings are predominantly flat-roofed and lend themselves to being insulated from the outside. Slabs of rigid polyurethane with various facing materials are laid onto hot

Figure 6   Sanitary modular unit

Figure 7   Roofing slabs for roof insulation

bitumen and then sandwiched between more bitumen and bitumenized felts beneath a layer of fine gravel to form a screed. The facing materials are usually impervious to water and the slabs are often applied in two layers to avoid jointing problems. The slabs themselves are produced continuously on a laminating machine within the factory.

In the United Kingdom the majority of polyurethane laminate has been used for ceiling board, where the underside of the roof is traditionally insulated. Such laminates are usually thin and only where rigid facings are used, possibly for wall insulation, is the polyurethane of any substantial thickness.

Rigid polyurethane for cavity wall insulation is only practised at the time of building. The general flow characteristics and pressures involved during foaming make it necessary for one to be assured of complete filling by special techniques. In some cases granulated polyurethane has been blown into cavities in the same way as mineral fibres. Whilst this improves the insulation values significantly, it is not as good as 'foamed *in situ*'.

More recently, polyurethane-filled window sills (Figure 8) and polyurethane-filled aluminium window frames have improved the U values of such elements and the problem of internal condensation has been eliminated. Such elements are now possible from high-density polyurethane, which

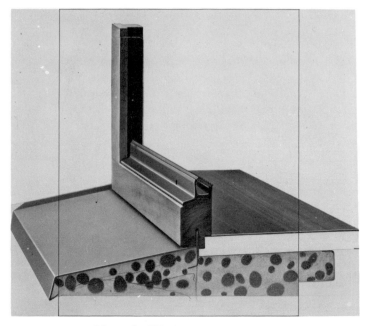

Figure 8   Window sill from Risse

exhibits the same properties and dispenses in many cases with expensive and heavy metal frames.

(e) Shipbuilding

This industry was a very early user of urethane. Being extremely conscious of wasting any cargo space, the possibility of thinner insulation was investigated earnestly by naval architects who required to insulate ships' provision rooms and holds. In the late 1950s early 1960s a number of fully refrigerated vessels were built using urethane as the major insulant. Such ships carried meat, vegetables or fruit in bulk. In most cases the urethane was applied *in situ* by pouring behind shutters or spraying.

Following the relatively successful operation of such ships, the relatively new bulk liquid/gas vessels were subsequently insulated. Here, there was a marked advantage in applying 4–6 inches of polyurethane as opposed to 12 inches of mineral fibre or balsawood. In many instances, access to the areas of insulation was so limited that fixing of conventional materials would have proved very difficult, if not impossible. It should also be noted that rigid urethane can also act as a secondary barrier in the case of tank rupture. Many other materials would not meet the requirement without some form of reinforced coating.

Currently, the holds of many large container ships have used polyurethane for insulation. Some have been sprayed but the trend is to apply factory-made laminates incorporating aluminium foil and glass cloth as facings to the ship's side. It is often to be found as insulant for brine pipes or ventilation trunking where it can be applied as ready-made sections or foamed *in situ*. Other properties of polyurethane make it even more attractive to the marine industry. Its closed-cell structure gives it great impermeability to water and it is therefore used for buoyancy or anticorrosion in ships' rudders, tank tops, and lifeboats.

(f) Technical insulations

Oil refineries and chemical plants use rigid polyurethane for tank insulation and pipe insulation (Figure 9); liquid-gas tanks are insulated by sections or slab or, where of an intricate design, by *in situ* application. In some instances, large areas of tank frames have been sprayed automatically to keep crude oils warm enough to facilitate easy transfer.

Pipe sections are often cut from slab in the factory or formed in moulds ready for fitting. *In situ* foaming of pipes (Figure 10) has proved uneconomic unless the diameter of the pipe is big enough to eliminate overspray or a 'pipe-in-pipe' method is used. The latter method has proved the most successful and when this is done within a factory-controlled environment, sufficient confidence can be placed in these units for use in district heating schemes.

Figure 9 Various pipe insulations

Figure 10 *In situ* filling of pipes

This method of providing heating to large numbers of dwellings or factories is used to great advantage in Scandinavia and on the Continent. The preinsulated pipes are taken to the site where they are assembled in trenches, and the joints are either *in situ* foamed or premoulded units fitted around them before a protective cladding is applied.

RIGID POLYURETHANE FIRE RETARDANCY

Polyurethane foam is an expanded organic material which, given the right ignition temperature and initial flame source, will burn. The degree to which a rigid urethane is easily ignitable or continues to burn is governed by the density and chemical make-up. The former is self-explanatory, but the latter criterion is the one most researched into by our material manufacturers and processors alike.

Rigid polyurethane is normally produced from starting products of the methylene di-isocyanate (MDI) type, and, whilst a foam can be made to burn, the surface spread of flame would be far greater than the combustion rate of the entire piece. The rate of burning can be significantly slowed down by the addition of known 'fire retardants'. These are usually organic phosphates. If these are added to the base materials, then combustion is severely restricted and in many cases the foam will 'self-extinguish' in a few seconds. This is provided that no other fire source is available to keep the foam burning.

Laboratory tests are only an indication of fire retardancy and larger-scale tests are very necessary to attempt to assimilate actual conditions.

In very few cases is polyurethane foam left without a final coating or protective layer. Most applications, as previously described, use the material together with other conventional materials as facings to produce composite elements or structures. When the material is used in this way, with expert design and correct use, the fire problems are no greater than with other materials. The important matter is one of education. We should be aware that organic materials can burn, and therefore take the necessary precautions to eliminate fire source and direct contact with the foam.

International standards are set to which building materials must be evaluated, and the plastics industry in general has made major steps to improve the rating of previously easily combustible materials.

New types of urethane foam have shown a great improvement on surface spread of flame, achieving a Class 1 to BS 476 (part 7). But more important is the burnthrough time of composite panels, for these give a representative result under actual fire conditions.

The products of combustion are also an important discussion point. One hears of so many gases being produced that the layman is confused and perhaps notices products that are known to be poisonous irrespective of the percentage amount that is produced. Like all organic materials, the major product of combustion is carbon monoxide (CO) and this, together with the reduced oxygen content due to the fire requiring this to support combustion, is responsible for the majority of deaths in fires. Carbon monoxide can be absorbed into the bloodstream to form carboxyhaemoglobin which poisons the body, enough to kill. Other gases produced are the oxides of nitrogen and hydrogen cyanide (HCN). This final gas is indeed lethal, but the amounts produced are considerably less than the CO which has already been explained. The HCN is therefore possibly an overkill medium. Ironically, the very materials that are used to inhibit fire spread are those which break down to give the poisonous fumes.

But this situation must be considered in the correct context and in direct comparison with other known materials. For it has been shown that greater quantities of such gases are generated by cork, wood, and wool when evaluated on a weight-for-weight basis.

The fire problem is very real, but it can be tolerated, and correct use of the material is possible to minimize the fire risks.

SUCCESSFUL PROJECTS

(a) Refrigerators and deep freezers

The firm of Robert Bosch, Giengen/Benz, West Germany, are well known for their range of refrigerated appliances. They have utilized rigid polyurethane to reduce insulation thickness and expediate construction and production times.

A typical appliance has maximum capacity but low exterior dimensions (500 mm × 500 mm × 850 mm). The inner liner is thermoformed acrylonitrile butadiene styrene, whilst the base, rear wall, and top are of aluminium/paper foil. Rigid polyurethane at a density of 30 kg/m³ and wall thickness of 20 mm is injected to bond these materials plus the coated steel sheet wrapper. The top is finished as a work surface using a melamine resin board which is stuck to the aluminium foil.

Changing from glass fibre to rigid polyurethane permitted Bosch to increase the internal capacity of the refrigerator whilst the outer dimensions were standard. The versatility of polyurethane produces a robust unit, reinforces the ABS, and permits savings to be made on the outer wrapper. Only the side walls need to be from coated steel sheet, as the urethane will bond and stiffen the aluminium/paper foil at the base, rear, and top. Production times are such that in one instance up to 3000 units per day may be produced from a plant in Italy.

Tyler Refrigeration International GmbH of Schwelm, West Germany, produce freezers and cooling cabinets for supermarkets and caterers. Urethane was chosen for its low weight, long service life, maximum insulating efficiency, and low insulator thickness. Service temperatures range from 0°C to +8°C for cooler to −18°C to −23°C for deep freezers.

When polyurethane is cast between two steel plates, a self-supporting structure is produced. In this instance, foam density is 40 kg/m³ and thickness varies from 53 mm for coolers to 65–100 mm for freezers.

(b) Cold stores

Factory-produced sandwich panels utilizing plastic-coated galvanized sheet and stainless-steel sheet are typical of constructions recommended by Romakowski KG, of Unterthureim, West Germany. A wall thickness of 100 mm at foam density 60 kg/m³ is sufficient to maintain a processing area at +10°C, a cooling area at 0°C, and a deep-freeze area at −28°C. The K value for such a system is 0.21 W/m K.

(c) Transport

Prefabricated bodies for regrigerated lorries are made by Clark Schumann of

Kichheim-Bolander, West Germany (Figure 11). The body comprises two side walls, one front end, one rear end, roof, and floors. Each element incorporates a Snap-Lock system for simple construction which enables one vehicle to be produced in one week by two semiskilled workers. Prior to using rigid polyurethane foam, two skilled workers needed three weeks to achieve the same production rate.

The exterior facings are aluminium sheet over an aluminium frame with a rear frame of steel profiles. Interior facings of roof, end wall, side, and rear door is GRP or bonded plywood. Rigid polyurethane is poured *in situ* to thicknesses of 60–100 mm. Service temperatures are −18°C for frozen foods and 0°C for chilled goods.

Refrigerated ISO containers 20 ft long having a service temperature range of −30°C to +70°C are manufactured by Industriewerke Transportsystem GmbH, Lubeck, West Germany. Polyurethane of density 40 kg/m^3 is used to produce sandwich panels of 55–175 mm depending on its use for side wall, end wall or roof. The unladen weight of such a box is 2400 kg and the maximum heat emission from the entire surface is 21 W/K.

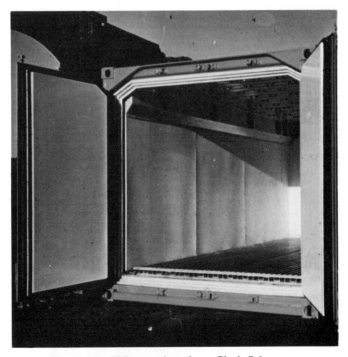

Figure 11   ISO container from Clark Schumann

(d) Building

Flat-roof insulation is best undertaken by laying slabs made continuously on a factory-based laminator. Chemische Werke Worms-Weinsheim GmbH, West Germany, used polyurethane as the basis of their development for 'Tropenia' panels. The construction was with bitumenized board on one side of the polyurethane and glass roofing mat or aluminium foil on the other. Tension-released channels are formed in one side of the board. The material is light to handle and can be cut easily whilst not crumbling or forming much dust. At a density of 30–35 kg/m^3 and compressive strength of 2.0 kgf/m^2 it is strong enough to be walked upon. An early project was to prevent 15 000 m^2 of roof forming condensation within the cellular concrete. By the use of this product, the solution was economically achieved. Installation time is said to be 50% of that required for conventional materials.

Cladding panels such as 'Isowand' from Hoesch in West Germany (Figure 12) are constructed with two galvanized steel sheets, either stove-enamelled or plastic-coated on one side, and rigid polyurethane to 35 or 120 mm thick is applied. Considerable savings are made on heating costs. Special attention is made to the jointing system to prevent thermal bridging.

Figure 12   Hoesch building panels

Figure 13   Cross section of Dumont & Besson panels

Dumont & Besson in France have been utilizing aggregated filled polyurethane since 1969 for building elements (Figure 13).

Rigid polyurethane lightweight concrete sections are produced in the following manner: PVC feed pipes are inserted into a wooden supporting frame. The entire cavity is then filled with blown glass or similar expanded granules with a grain size of about 10–30 mm. After that, the prepared frame is placed in a jig. Finally, the individual sections are foam-filled using a foaming machine. Wall coverings consisting of wooden facings or plasterboard can be foam-bonded directly, thus eliminating the need for interior plastering after assembly.

Sections for a detached house with a floor area of about 120 m^2 are transported to the building site by lorry complete—inclusive of heating and plumbing installations, rafters, etc.

About 60–65 kg of rigid polyurethane are needed per cubic metre of wall volume, i.e. with a wall thickness of 7 cm, 4.1–4.5 kg of the reaction mix are required for each square metre.

Perhaps one of the most interesting projects in the building industry would be rigid polyurethane for emergency housing. Bayer AG in cooperation with the West German Red Cross developed a method of foaming polyurethane 'igloos' for victims of natural disasters, such as earthquakes and floods (Figure 14).

The technique is to inflate a rubber hemisphere on a turntable and to coat with release agent prior to rotating and applying sprayed rigid polyurethane

Figure 14   Constructing a foam igloo

foam. After several minutes an igloo is produced and the rubber ballon deflated. Windows and doors are easily cut into the finished dome. During recent earthquakes in Turkey, Chile, and Nicuragua approximately 500 dwellings were produced at each time. The equipment was simple to transport and the chemicals required were not bulky when one considers that they are expanded 30 times in order to produce a foam.

(e) Shipbuilding

Two major areas are of interest today in the marine industry. They are the container ships and bulk liquid/gas tankers.

Overseas Containers Ltd and Associated Container Lines were the pioneers in the use of rigid polyurethane for both their vessels and containers. In the late 1960s several large vessels were laid down in both the UK and West Germany. Some vessels were initially insulated with spray foam throughout the hold and under the hatch covers, but a change was later made to factory-made slabs attached to the ships' sides. Only 50 mm of polyurethane was applied, which was considerably less than the thickness of conventional insulation. The ships were so designed that every available space was accounted for and the insulation could not have been thicker.

Bulk liquid/gas carriers really needed the good insulation properties of polyurethane to reduce the necessary foam thickness. Specialized vessels were built to carry liquid ethylene for ICI from the UK to Holland. These ships have one tank which is insulated with 6 inches of polyurethane with an epoxy resin/glass secondary skin as a protective barrier. These vessels have been in service for over 10 years, transporting their cargoes daily across the North Sea.

(f) Technical insulations

District heating schemes employing the insulated pipe-in-pipe method have been relatively successful in Scandinavia and on the Continent. Meier-Schenk of Switzerland are among several companies active in this field offering their 'Pan-Isovit' system.

The service pipe may be of black steel for heating or galvanized or copper for sanitation. They may be welded sections or seamless, and are fitted inside a seamless extruded rigid polyethylene sleeve. Rigid polyurethane at a density of 80–90 kg/m^3 is injected into lengths of 6–12 m. Diameters vary between 26.9 and 609.6 mm.

The pipe is laid into the ground and anchored at strategic points, joints are made good, and are insulated with preformed sections or foamed *in situ*. The permissible temperature range is from $-200°C$ to $+130°C$. Large blocks of apartments have been serviced with hot water heating by these district heating schemes. A steam-generating station is either purpose-built or excess energy from electricity power stations is diverted through these lines. Some schemes have been in existence for several years. The success of the scheme depends upon the quality of construction and the ability to detect any leaks which may occur in the line. The latter is a subject of earnest investigation and development at this time.

FUTURE TRENDS

Conservation of our natural resources and a more responsible attitude with regard to the use of energy is of the utmost importance over the next 25 years. Promises and prophecies are made concerning energy from atomic power and solar source, but these will also be only truly economic if we improve our efficient usage.

Insulation must be a prime consideration and polyurethane foam can play a major role. In buildings there are strong indications that urethanes will be used in many of the versatile applications discussed earlier. Roofing board is likely to increase markedly as more construction firms gain first-hand experience from the manufacture of such plastic foams as well as the civil engineering aspects of such projects. Recent acquisitions of rigid polyurethane processors by internationally recognized building concerns must indicate a growing confidence in the material as well as confirming the advantages to be gained by the use of plastics.

In West Germany, plans have long been under consideration for the building of 'polyurethane city'. In such a development the material is widely used to construct a town which will be relatively self-sufficient regarding energy supplies. District heating schemes would provide heating from the electrical generating station which in turn would be complimented by hot water from solar cells. The general construction would incorporate polyurethane roofing and wall panels together with between-floor insulation of polyurethane onto concrete and timber to offer the 'burolandschaft' effect. All ventilation trunking would be constructed either from aluminium/foam laminates or be sprayed polyurethane onto conventional metal fabrications. Windows are of course double-glazed with polyurethane insulation inside the aluminium frames or, in the case of high-rise blocks, solid polyurethane window frames are preferred because of the low maintenance costs.

At the same time, rooflight frames are of high-density polyurethane and ventilation pipes and roof gulleys are produced from similar materials. They have the advantage of not absorbing water and therefore do not crack during cold spells.

Bathrooms are either of the modular construction incorporating plumbing and attached furniture or are complete cells as made by Kunbau in West Germany. To improve the overall appearance of an extremely functional concrete, steel, and polyurethane foam construction, decorative facades can be made from polyurethane coated GRP/polyurethane foam laminates.

The use of many of the plastics items considered may be of a higher initial cost. But the prime reason for their use is to save energy and cut the costs of maintenance which proves to be soaring each year.

The world of polyurethane is clearly extending vastly and it is still only 35 years old. Such a material must play an important role in building a secure future for the next generation, and we should not be parochial in our thoughts. Surely the exploration of other worlds indicates that there will be a requirement to use tools and materials which are today described as synthetic or plastic. The development of mankind has moved in what often seems at the time to be strange. But viewed over the centuries, there is a clear pattern of man's ability and interest to widen his horizons.

The infancy of space research has already shown clearly that the use of plastics is an important contribution to its success. The first settlements in space or on foreign worlds are unlikely to be made of wood, stone or steel, etc. We must realize that if there are good reasons to use perhaps more expensive materials in order to make things possible or viable outside of our environment, then perhaps a use closer to home should be encouraged strongly.

Energy Conservation and Thermal Insulation
Edited by R. Derricott and S. S. Chissick
© 1981 John Wiley & Sons Ltd.

CHAPTER 21

Foamed glass thermal insulation materials

H. Robert

Pittsburgh Corning Europe N.V., Brussels

I. WATER VAPOUR DIFFUSION AND INTERSTITIAL CONDENSATION

A. Introduction

Water vapour as a gas is invisible, although a good number of people are misled by saying that 'the water in the kettle is boiling because there is steam coming out of it'. In fact, what we see in this case is not steam or vapour but condensed water particles. For this reason, it is so difficult to make people understand the importance of water vapour diffusion in the building and industrial fields.

In the building field many people say: 'We never bothered about thermal insulation and never thought of water vapour diffusion in old buildings but they are still standing.' We should not forget that our way of life and the methods of construction have changed completely— even over the last 50 years. In the earlier days we did not have central heating systems, but we had a fireplace and a chimney in every room. When we sat around the fireplace it was warm in front of us, while we froze at the back. Our doors and windows did not fit as they do today. Together with the chimney, we had a natural ventilation system whereby dry air was taken in from the outside in winter and a great deal of humid air was evacuated via the chimney. The ceilings in our rooms were very much higher and we had a far greater volume for the same floor surface (or number of people in the room). There was no central hot water supply in the building. We did not have fully automatic washing machines or dishwashers and last, but not least, we did not have humidifiers. With our modern comfort, we introduced the problem of water vapour diffusion.

463

B. Air composition

The air in which we live daily is a mixture of gases, and it is composed of:

(a) 21 vol.% of oxygen;
(b) 78 vol.% of nitrogen; and
(c) 1 vol.% of other gases, such as CO_2, water vapour, argon and the other rare gases such as H_2, He, Ne, Kr and Xe.

C. Partial pressure of water vapour in the air

We have already forgotten the age of the steam engine, of railway locomotives moving trains only by vapour pressure in the pistons of the engine.

Each gas present in a close space exercises a pressure against the walls of this space and acts as if it was completely alone in occupying this space, even if other gases are present as well. As the other gases act in exactly the same way, they also exercise their full pressure against the same walls. For this reason we call this gas pressure a partial pressure (relevant to the gas), a fact that has been known for centuries. This partial pressure varies rapidly with the temperature of the gas. It increases in an exponential way with temperature. This is even better understood when considering the general gas law (Boyle–Mariotte and Gay Lussac):

$$PV = xRT$$

where P = gas pressure in mbar (or psi), V = gas volume in m^3 (or cu. ft), x = gas quantity in kg (or lb), R = gas constant in m kgf/kg K (relative to each gas), and T = absolute temperature in K (kelvins). In this law and elsewhere, we note that kg represents kg mass, kgf represents kg force, and that

$$T(K) = 273 + t \ (^{\circ}C)$$

($^{\circ}$C) degrees = Celsius or centigrade) and

$$R = \text{gas constant} = \frac{R_a}{M} = \frac{\text{general gas constant}}{\text{molecular weight of gas considered}}$$

with R_a = 848 m kgf/kMol K. From this we see that if the volume V remains constant and if the gas is trapped, the quantity x of gas remains constant as well. R itself is a constant, so the pressure P has to increase with increasing temperature T.

The partial pressure of water vapour in the air is given in saturated water vapour pressure tables and is expressed in the following units: kgf/m^2; gf/cm^2; psi; mm Hg = torr (from Torricelli); in Hg; N/m^2 = newton/m^2 = Pa = pascal; bar; and mbar.

Various units of force and pressure

1 N is that force that gives to 1 kg mass an acceleration of 1 m/s^2, i.e.

$$1 \text{ N} = 1 \text{ kg} \times 1 \text{ m/s}^2 \tag{1}$$

1 kg force (1 kgf) is that force that gives to 1kg mass an acceleration of 9.81 m/s^2, i.e.

$$1 \text{ kgf} = 1 \text{ kg} \times 9.81 \text{ m/s}^2 \tag{2}$$

If we now multiply equation (1) by 9.81, we obtain

$$9.81 \text{ N} = 1 \text{ kg} \times 9.81 \text{ m/s}^2 \tag{3}$$

and we notice that the left-handed side of equation (3) is identical to that of equation (2). Therefore we can write:

$$1 \text{ kgf} = 9.81 \text{ N}$$

By definition, a pressure of 1 N/m^2 is called a pascal (Pa), so

$$1 \text{ N/m}^2 = 1 \text{ Pa}$$

By definition

$$10^5 \text{ Pa} = 1 \text{ bar}$$

or

$$100\ 000 \text{ Pa} = 1 \text{ bar}$$
$$100 \text{ Pa} = 1 \text{ mbar}$$

It unfortunately happens that we have to translate from one pressure unit to another, and for this we refer to Table 1. We also have, for example, that

$$1 \text{ g/cm}^2 = 10 \text{ kg/m}^2 = \ldots$$

and

$$1 \text{ mm Hg} = 13.595 \text{ kg/m}^2 = \ldots$$

D. Relative humidity in the air at a temperature t

We can give two definitions of relative humidity, and it is always useful to remember both.

First definition of relative humidity ϕ

The relative humidity (RH) indicates the degree of saturation in water vapour of air at a temperature t. That is to say, it is the ratio of the actual

Table 1   Conversion tables

(1) Length

Units	mm	m	in	ft
1 mm	= 1	= 0.001	= 0.0394	= 0.00328
1 m	= 1000	= 1	.= 39.37	= 3.281
1 in	= 25.4	= 0.0254	= 1	= 0.0833
1 ft	= 304.8	= 0.3048	= 12	= 1

Example: 5 m = 5 × 3.281 ft = 16.41 ft.

(2) Surface area

Units	m^2	sq. in	sq. ft
1 m^2	= 1	= 1.550	= 10.76
1 sq. in	= 6.45×10^{-4}	= 1	= 6.94×10^{-3}
1 sq. ft	= 0.0929	= 144	= 1

Example: 3 sq. in = $3 \times 6.45 \times 10^{-4}\, m^2$ = $19.35 \times 10^{-4}\, m^2$

(3) Volume

Units	litre or dm^3	m^3	cu. in
1 l or 1 dm^3	= 1	= 0.001	= 61.02
1 m^3	= 10^3	= 1	= 6.102×10^4
1 cu. in	= 0.0164	= 16.39×10^{-6}	= 1
1 cu. ft	= 28.32	= 28.32×10^{-3}	= 1.728
1 imperial gallon	= 4.546	= 4.546×10^{-3}	= 277.3

Units	cu. ft	imperial gallon
1 l or 1 dm^3	= 0.0355	= 0.220
1 m^3	= 35.32	= 220
1 cu. in	= 57.87×10^{-5}	= 36.08×10^{-4}
1 cu. ft	= 1	= 6.234
1 imperial gallon	= 0.1604	= 1

Example 2 m^3 = 2 × 220 imperial gallon = 440 imperial gallon.

(4) Speed

Units	km/h	mph
1 km/h	= 1	= 0.6214
1 mph	= 1.6094	= 1

Example: 60 km/h = 60 × 0.6214 mph = 37.28 mph.

(5) Mass–weight

Units	kg (mass)	g (mass)	lb (mass)
1 kg	= 1	= 1000	= 2.205
1 g	= 0.001	= 1	= 2.205×10^{-3}
1 lb	= 0.4536	= 453.6	= 1

Example: 2.2 lb = 2.2 × 0.4536 = 1.00 kg.

(6) Density

Units	kg/m^3	lb/cu. ft
1 kg/m^3	= 1	= 0.0624
1 lb/cu. ft	= 16.019	= 1

Example 8 lb/cu. ft = 8 × 16.019 = 128 kg/m^3.

(7) Pressure–Stress

Units	Pa (N/m^2)	kgf/m^2	atm
1 Pa	= 1	= 0.1020	= 1.020×10^{-5}
1 kgf/m^2	= 9.81	= 1	= 10^{-4}
1 atm or 1 kgf/cm^2	= 0.981×10^5	= 10^4	= 1
1 lbf/sq. ft	= 47.883	= 4.882	= 4.882×10^{-4}
1 lbf/sq. in	= 6.895	= 703.1	= 7.03×10^{-2}

Units	lbf/sq. ft	lbf/sq. in
1 Pa	= 2.09×10^{-2}	= 1.45×10^{-4}
1 kgf/m^2	= 0.2048	= 14.22×10^{-4}
1 atm or 1 kgf/cm^2	= 2048	= 14.22
1 lbf/sq. ft	= 1	= 0.695×10^{-2}
1 lbf/sq. in	= 144	= 1

Example: 2 atm = 2 × 9.81 × 10^4 = 10^4 Pa.

Table 1   Conversion tables

(8) energy

Units	kcal	J	kWh	Btu
1 kcal	= 1	= 4.1868	= 11.63×10^{-4}	= 3.97
1 J	= 2.39×10^{-4}	= 1	= 2.778×10^{-7}	= 9.48×10^{-4}
1 kWh	= 860	= 3.6×10^6	= 1	= 3.413
1 Btu	= 0.252	= 1.055	= 2.93×10^{-4}	= 1

Example: 200 J = $200 \times 2.39 \times 10^{-4}$ = 478×10^{-4} = 0.0478 kcal.

(9) Thermal conductivity

Units	kcal/m h °C	W/m °C	Btu in/ft² h °F
1 kcal/m h °C	= 1	= 1.163	= 8.064
1 W/m °C	= 0.86	=	= 6.93
1 Btu in/ft² h °F	= 0.124	= 0.144	= 1

Example: 0.036 kcal/m h °C = 0.036×1.163 = 0.042 W/m °C.

(10) Coefficient of heat transfer

Units	kcal/m² h °C	W/m² °C	Btu/ft² h °F
1 kcal/m² h °C	= 1	= 1.163	= 0.205
1 W/m² °C	= 0.86	= 1	= 0.176
1 Btu/ft² h °F	= 4.882	= 5.678	= 1

Example: 0.60 W/m² °C = 0.60×0.86 = 0.52 kcal/m² h °C.

amount of water vapour present in the air at a given temperature to the maximum amount of water vapour that air at that temperature can hold.

The symbol used is ϕ; it is expressed in per cent; it has no unit as it is a ratio; and it is normally written as a decimal.

Example

If we are given that the RH is 70%, this implies that

$\phi = 0.7$ (0.70 = 70/100 = 70%

The symbol ϕ alone has no meaning; it has to be referred to the temperature of the air. Why?

We have seen in section I.B that the air is a mixture of gases: let us suppose that we represent the gas particles by circles, as indicated below.

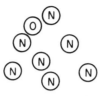

It is known (from the molecular weights) that N_2 (nitrogen) molecules and O_2 (oxygen) molecules are much greater than H_2O (water vapour) molecules. So we can fill the gaps between the N_2 and O_2 molecules with the smaller H_2O molecules. As all gases expand when their temperature is increased, the particles occupy a greater volume, i.e. they have more freedom to move and the spaces between them increase as well. As the increase in intermolecular space is relatively greater than the increase in volume of the water vapour molecules, we can lodge more H_2O in the free space.

Therefore it is clear that air at a given temperature can only contain a limited quantity of water vapour as a gas, but that it can take a greater quantity (also limited) if its temperature is higher. So we always have to indicate the relative humidity for a given temperature.

Second definition of relative humidity ϕ

The relative humidity is the ratio of the real vapour pressure P_r present in the air to the saturation vapour pressure P_s present in the air at the same temperature, i.e.

$$\phi = P_r/P_s \tag{4}$$

This last definition is very important and allows us to determine the dewpoint temperature corresponding to a given air temperature and relative humidity, using saturation vapour pressure tables.

Tables 2 and 3 give water vapour pressure in saturated air, Table 2 above water and Table 3 above ice, expressed in Pa (pascals). Table 4 allows direct reading of the dewpoint temperature corresponding to a given air temperature and a known relative humidity.

E. Absolute humidity

Absolute humidity is the real quantity of water vapour present in the air. It is normally expressed in the following units: g/kg of dry air; lb of moisture/lb of dry air; and sometimes in g/m^3 of air or lb/cu. ft of air.

Table 2 Saturation water vapour pressure p_s in $P_a = (N/m^2)$ according to GOFF formulas (ASHRAE Handbook, 1977, Ch. 5, p. 5.2)

+	0.0	0.1	0.2	0.3	0.4	0.5	0.6	0.7	0.8	0.9
0	611.1	615.6	620.1	624.6	629.1	633.7	638.3	642.9	647.6	652.3
1	657.0	661.7	666.5	671.3	676.1	681.0	685.9	690.9	695.8	700.8
2	705.9	710.9	716.0	721.1	726.3	731.5	736.7	742.0	747.2	752.6
3	757.9	763.3	768.7	774.2	779.7	785.2	790.8	796.4	802.0	807.7
4	813.4	819.1	824.9	830.7	836.5	842.4	848.3	854.3	860.3	866.3
5	872.4	878.5	884.6	890.8	897.0	903.3	909.6	915.9	922.3	928.7
6	935.1	941.6	948.2	954.7	961.3	968.0	974.7	981.4	988.2	995.0
7	1001.8	1008.7	1015.7	1022.7	1029.7	1036.7	1043.9	1051.0	1058.2	1065.4
8	1072.7	1080.0	1087.4	1094.8	1102.3	1109.8	1117.3	1124.9	1132.6	1140.3
9	1148.0	1155.8	1163.6	1171.4	1179.4	1187.3	1195.3	1203.4	1211.5	1219.7
10	1227.9	1236.1	1244.4	1252.8	1261.2	1269.6	1278.1	1286.6	1295.2	1303.9
11	1312.6	1321.3	1330.1	1339.0	1347.9	1356.9	1365.9	1374.9	1384.0	1393.2
12	1402.4	1411.7	1421.0	1430.4	1439.9	1449.3	1458.9	1468.5	1478.2	1487.9
13	1497.6	1507.5	1517.3	1527.3	1537.3	1547.3	1557.4	1567.6	1577.8	1588.1
14	1598.5	1608.9	1619.3	1629.9	1640.5	1651.1	1661.8	1672.6	1683.4	1694.3
15	1705.2	1716.2	1727.3	1738.5	1749.7	1760.9	1772.3	1783.6	1795.1	1806.6
16	1818.2	1829.9	1841.6	1853.3	1865.2	1877.1	1889.1	1901.1	1913.3	1925.4
17	1937.7	1950.0	1962.4	1974.8	1987.4	2000.0	2012.6	2025.4	2038.2	2051.1
18	2064.0	2077.0	2090.1	2103.3	2116.5	2129.8	2143.2	2156.7	2170.2	2183.8
19	2197.5	2211.2	2225.1	2239.0	2252.9	2267.0	2281.1	2295.3	2309.6	2324.0

	0	1	2	3	4	5	6	7	8	9
20	2338.4	2353.0	2367.6	2382.2	2397.0	2411.8	2426.8	2441.8	2456.9	2472.0
21	2487.3	2502.6	2518.0	2533.5	2549.1	2564.7	2580.5	2596.3	2612.2	2628.2
22	2644.3	2660.5	2676.7	2693.1	2709.5	2726.0	2742.6	2759.3	2776.1	2793.0
23	2809.9	2827.0	2844.1	2861.4	2878.7	2896.1	2913.6	2931.2	2948.9	2966.7
24	2984.6	3002.5	3020.6	3038.8	3057.0	3075.4	3093.8	3112.4	3131.0	3149.8
25	3168.6	3187.5	3206.6	3225.7	3244.9	3264.3	3283.7	3303.2	3322.9	3342.6
26	3362.5	3382.4	3402.4	3422.6	3442.8	3463.2	3483.7	3504.2	3524.9	3545.7
27	3566.6	3587.6	3608.7	3629.9	3651.2	3672.6	3694.1	3715.8	3737.5	3759.4
28	3781.4	3803.5	3825.7	3848.0	3870.4	3893.0	3915.6	3938.4	3961.3	3984.3
29	4007.4	4030.6	4054.0	4077.4	4101.0	4124.7	4148.5	4172.5	4196.5	4220.7
30	4245.0	4269.5	4294.0	4318.7	4343.5	4368.4	4393.4	4418.6	4443.9	4469.3
31	4494.8	4520.5	4546.3	4572.2	4598.3	4624.5	4650.8	4677.2	4703.8	4730.5
32	4757.3	4784.3	4811.4	4838.6	4866.0	4893.5	4921.1	4948.9	4976.8	5004.8
33	5033.0	5061.3	5089.8	5118.4	5147.1	5176.0	5205.0	5234.2	5263.5	5292.9
34	5322.5	5352.2	5382.1	5412.1	5442.2	5472.5	5503.0	5533.6	5564.3	5595.2
35	5626.2	5657.4	5688.8	5720.3	5751.9	5783.7	5815.6	5847.7	5880.0	5912.4
36	5944.9	5977.6	6010.5	6043.5	6076.7	6110.0	6143.5	6177.2	6211.0	6245.0
37	6279.1	6313.4	6347.9	6382.5	6417.3	6452.2	6487.3	6522.6	6558.1	6593.7
38	6629.4	6665.4	6701.5	6737.8	6774.2	6810.9	6847.3	6884.6	6921.8	6959.1
39	6996.5	7034.2	7072.0	7110.0	7148.2	7186.6	7225.1	7263.8	7302.7	7341.8

Table 3 Saturation water vapour pressure P_s in $P_a = (N/m^2)$ above ice, according to GOFF formulas (ASHRAE Handbook, 1977, Ch. 5, p. 5.2)

	0.0	0.1	0.2	0.3	0.4	0.5	0.6	0.7	0.8	0.9
- 0	611.11	606.10	601.12	596.18	591.28	586.41	581.58	576.79	572.04	567.31
- 1	562.63	557.98	553.36	548.78	544.24	539.73	535.25	530.81	526.40	522.02
- 2	517.68	513.37	509.09	504.84	500.63	496.45	492.30	488.18	484.10	480.04
- 3	476.02	472.03	468.06	464.13	460.23	456.36	452.52	448.70	444.92	441.17
- 4	437.44	433.74	430.07	426.43	422.82	419.24	415.68	412.15	408.65	405.18
- 5	401.73	398.31	394.92	391.55	388.21	384.89	381.60	378.34	375.10	371.89
- 6	368.70	365.54	362.40	359.29	356.20	353.13	350.09	347.07	344.08	341.11
- 7	338.16	335.24	332.34	329.47	326.61	323.78	320.97	318.18	315.42	312.68
- 8	309.95	307.26	304.58	301.92	299.29	296.67	294.08	291.50	288.95	286.42
- 9	283.91	281.42	278.95	276.49	274.06	271.65	269.26	266.88	264.53	262.19
-10	259.88	257.58	255.30	253.04	250.80	248.57	246.37	244.18	242.01	239.85
-11	237.72	235.60	233.50	231.42	229.35	227.30	225.27	223.25	221.25	219.27
-12	217.30	215.35	213.41	211.49	209.59	207.70	205.83	203.97	202.13	200.31
-13	198.50	196.70	194.92	193.15	191.40	189.66	187.94	186.23	184.54	182.86
-14	181.19	179.54	177.90	176.28	174.67	173.07	171.49	169.91	168.36	166.81
-15	165.28	163.76	162.26	160.76	159.28	157.81	156.36	154.91	153.48	152.06
-16	150.66	149.26	147.88	146.51	145.15	143.80	142.46	141.14	139.82	138.52
-17	137.23	135.95	134.68	133.42	132.17	130.93	129.71	128.49	127.28	126.09
-18	124.90	123.73	122.56	121.41	120.27	119.13	118.01	116.89	115.78	114.69
-19	113.60	112.53	111.46	110.40	109.35	108.31	107.28	106.26	105.24	104.24
-20	103.25	102.26	101.28	100.31	99.35	98.40	97.45	96.52	95.59	94.67
-21	93.76	92.86	91.96	91.08	90.20	89.33	88.46	87.61	86.76	85.92
-22	85.08	84.26	83.44	82.63	81.82	81.03	80.24	79.45	78.68	77.91
-23	77.15	76.39	75.64	74.90	74.17	73.44	72.72	72.00	71.30	70.59
-24	69.90	69.21	68.52	67.85	67.18	66.51	65.85	65.20	64.55	63.91
-25	63.28	62.65	62.03	61.41	60.80	60.19	59.59	58.99	58.40	57.82

Table 4

Ambient air temperature (°C)	% Relative humidity												
	40	45	50	55	60	65	70	75	80	85	90	95	100
	Dew point temperatures in °C												
50	32.7	34.8	36.7	38.5	40.1	41.6	43.0	44.4	45.6	46.8	47.9	49.0	50
49	31.8	33.9	35.8	37.6	39.2	40.7	42.1	43.4	44.7	45.8	47.0	48.0	49
48	30.9	33.0	34.9	36.6	38.3	39.7	41.1	42.4	43.7	44.9	46.0	47.0	48
47	30.1	32.1	34.0	35.7	37.3	38.8	40.2	41.5	42.7	43.9	45.0	46.0	47
46	29.2	31.2	33.1	34.8	36.4	37.9	39.2	40.5	41.7	42.9	44.0	45.0	46
45	28.3	30.3	32.2	33.9	35.4	36.9	38.3	39.6	40.8	41.9	43.0	44.1	45
44	27.4	29.4	31.3	33.0	34.5	36.0	37.3	38.6	39.8	40.9	42.0	43.1	44
43	26.5	28.5	30.4	32.0	33.6	35.0	36.4	37.6	38.8	40.0	41.0	42.1	43
42	25.6	27.6	29.4	31.1	32.6	34.1	35.4	36.7	37.9	39.0	40.1	41.1	42
41	24.7	26.7	28.5	30.2	31.7	33.0	34.5	35.7	36.9	38.0	39.1	40.1	41
40	23.9	25.8	27.6	29.3	30.8	32.2	33.5	34.8	35.9	37.0	38.1	39.1	40
39	23.0	24.9	26.7	28.3	29.9	31.2	32.6	33.8	35.0	36.1	37.1	38.1	39
38	22.1	24.0	25.8	27.4	28.9	30.3	31.6	32.8	34.0	35.1	36.1	37.1	38
37	21.2	23.1	24.9	26.5	28.0	29.4	30.7	31.9	33.0	34.1	35.1	36.1	37
36	20.3	22.2	24.0	25.6	27.1	28.4	29.7	30.9	32.0	33.1	34.2	35.1	36
35	19.4	21.3	23.1	24.6	26.1	27.5	28.7	29.9	31.1	32.1	33.1	34.1	35
34	18.5	20.4	22.1	23.7	25.2	26.5	27.8	29.0	30.1	31.1	32.1	33.1	34
33	17.6	19.5	21.2	22.8	24.2	25.6	26.8	28.0	29.1	30.2	31.2	32.1	33
32	16.8	18.6	20.3	21.9	23.3	24.6	25.9	27.1	28.2	29.2	30.2	31.1	32
31	15.7	17.7	19.4	20.9	22.4	23.7	24.9	26.1	27.2	28.2	29.2	30.2	31
30	15.0	16.8	18.5	20.0	21.4	22.7	24.0	25.1	26.2	27.2	28.2	29.2	30
29	14.1	15.9	17.6	19.1	20.5	21.8	23.1	24.2	25.2	26.3	27.2	28.2	29
28	13.2	15.0	16.7	18.2	19.6	20.9	22.1	23.2	24.3	25.3	26.2	27.2	28
27	12.3	14.1	15.7	17.2	18.6	19.9	21.1	22.2	23.3	24.3	25.3	26.2	27

Table 4—continued

Ambient air temperature (°C)	% Relative humidity												
	40	45	50	55	60	65	70	75	80	85	90	95	100
	Dew point temperatures in °C												
26	11.4	13.2	14.8	16.3	17.7	19.0	20.1	21.3	22.3	23.3	24.3	25.2	26
25	10.5	12.3	13.9	15.4	16.7	18.0	19.2	20.3	21.3	22.3	23.3	24.2	25
24	9.6	11.4	13.0	14.4	15.8	17.1	18.2	19.3	20.4	21.4	22.3	23.2	24
23	8.7	10.5	12.1	13.5	14.9	16.1	17.3	18.4	19.4	20.4	21.3	22.2	23
22	7.8	9.6	11.2	12.6	13.9	15.2	16.3	17.4	18.4	19.4	20.3	21.2	22
21	6.9	8.7	10.2	11.7	13.0	14.2	15.4	16.4	17.5	18.4	19.3	20.2	21
20	6.1	7.8	9.3	10.7	12.1	13.3	14.4	15.5	16.5	17.4	18.4	19.2	20
19	5.2	6.9	8.4	9.8	11.1	12.3	13.4	14.5	15.5	16.5	17.4	18.2	19
18	4.3	6.0	7.5	8.9	10.2	11.4	12.5	13.6	14.5	15.5	16.4	17.2	18
17	3.3	5.0	6.6	7.9	9.2	10.4	11.5	12.6	13.6	14.5	15.4	16.2	17
16	2.4	4.1	5.7	7.0	8.3	9.5	10.6	11.6	12.6	13.5	14.4	15.2	16
15	1.6	3.2	4.7	6.1	7.4	8.5	9.6	10.7	11.6	12.5	13.4	14.3	15
14	0.6	2.3	3.8	5.2	6.4	7.6	8.7	9.7	10.6	11.6	12.4	13.3	14
13	− 0.2	1.4	2.9	4.2	5.5	6.6	7.7	8.7	9.7	10.6	11.4	12.3	13
12	− 1.0	0.5	2.0	3.3	4.5	5.7	6.7	7.8	8.7	9.6	10.5	11.3	12
11	− 1.8	− 0.4	1.0	2.4	3.6	4.7	5.8	6.8	7.7	8.6	9.5	10.3	11
10	− 2.6	− 1.1	0.0	1.4	2.7	3.8	4.8	5.9	6.8	7.6	8.5	9.3	10
9	− 3.4	− 2.0	− 0.7	0.5	1.7	2.8	3.8	4.8	5.8	6.7	7.5	8.3	9
8	− 4.2	− 2.8	− 1.5	− 0.4	0.7	1.9	2.9	3.9	4.8	5.7	6.5	7.3	8
7	− 5.0	− 3.6	− 2.4	− 1.2	0.1	0.9	2.0	2.9	3.8	4.7	5.5	6.3	7
6	− 5.8	− 4.4	− 3.2	− 2.0	− 1.0	− 0.0	1.0	2.0	2.9	3.7	4.5	5.3	6
5	− 6.6	− 5.2	− 4.0	− 2.9	− 1.8	− 0.9	0.0	1.0	1.9	2.7	3.5	4.3	5
4	− 7.4	− 6.0	− 4.8	− 3.7	− 2.7	− 1.7	− 0.8	0.0	0.9	1.8	2.6	3.3	4
3	− 8.2	− 6.8	− 5.7	− 4.5	− 3.5	− 2.5	− 1.6	− 0.8	0.0	0.7	1.6	2.4	3
2	− 9.0	− 7.7	− 6.5	− 5.4	− 4.3	− 3.4	− 2.5	− 1.7	− 0.9	− 0.2	0.6	1.3	2

	1	2	3	4	5	6	7	8	9	10	11	12	
1	− 0.3	− 0.4	− 1.0	− 1.8	− 2.5	− 3.4	− 4.2	− 5.2	− 6.2	− 7.3	− 8.5	− 9.8	1
±0	− 0.6	− 1.2	− 1.9	− 2.6	− 3.4	− 4.4	− 5.1	− 6.0	− 7.0	− 8.1	− 9.3	−10.7	0
− 1	− 1.6	− 2.2	− 2.9	− 3.6	− 4.4	− 5.2	− 6.0	− 7.0	− 8.0	− 9.0	−10.3	−11.6	− 1
− 2	− 2.6	− 3.2	− 3.9	− 4.6	− 5.4	− 6.2	− 7.0	− 7.9	− 8.9	−10.0	−11.2	−12.5	− 2
− 3	− 3.6	− 4.2	− 4.9	− 5.6	− 6.3	− 7.1	− 8.0	− 8.9	− 9.9	−10.9	−12.1	−13.5	− 3
− 4	− 4.6	− 5.1	− 5.9	− 6.6	− 7.3	− 8.1	− 8.9	− 9.9	−10.8	−11.9	−12.9	−14.4	− 4
− 5	− 5.7	− 6.2	− 6.8	− 7.5	− 8.3	− 9.0	− 9.9	−10.8	−11.8	−12.8	−13.9	−15.3	− 5
− 6	− 6.6	− 7.2	− 7.8	− 8.6	− 9.2	−10.1	−10.9	−11.8	−12.7	−13.8	−14.9	−16.2	− 6
− 7	− 7.6	− 8.2	− 8.8	− 9.5	−10.3	−11.0	−11.8	−12.7	−13.7	−14.8	−15.9	−17.2	− 7
− 8	− 8.5	− 9.1	− 9.8	−10.5	−11.2	−11.9	−12.8	−13.7	−14.6	−15.7	−16.7	−18.1	− 8
− 9	− 9.5	−10.1	−10.8	−11.5	−12.1	−12.9	−13.8	−14.6	−15.6	−16.6	−17.7	−19.0	− 9
−10	−10.5	−11.1	−11.7	−12.5	−13.2	−13.9	−14.7	−15.6	−16.5	−17.5	−18.6	−19.9	−10
−11	−11.6	−12.1	−12.8	−13.5	−14.1	−14.9	−15.7	−16.5	−17.5	−18.5	−19.6	−20.8	−11
−12	−12.6	−13.2	−13.8	−14.4	−15.1	−15.9	−16.7	−17.5	−18.5	−19.5	−20.5	−21.8	−12
−13	−13.6	−14.1	−14.8	−15.4	−16.1	−16.8	−17.6	−18.5	−19.4	−20.4	−21.5	−22.7	−13
−14	−14.6	−15.1	−15.7	−16.4	−17.2	−17.8	−18.6	−19.5	−20.3	−21.4	−22.4	−23.5	−14
−15	−15.6	−16.1	−16.7	−17.4	−18.1	−18.8	−19.5	−20.4	−21.4	−22.3	−23.2	−24.6	−15
−16	−16.5	−17.2	−17.7	−18.4	−19.1	−19.7	−20.5	−21.4	−22.2	−23.1	−24.3	−25.5	−16
−17	−17.5	−18.1	−18.7	−19.4	−20.0	−20.7	−21.5	−22.3	−23.1	−24.1	−25.1	−26.3	−17
−18	−18.5	−19.1	−19.7	−20.3	−20.9	−21.7	−22.4	−23.1	−24.2	−25.0	−26.1	−27.1	−18
−19	−19.5	−20.0	−20.6	−21.3	−22.0	−22.6	−23.3	−24.2	−25.0	−25.9	−26.9	−28.2	−19
−20	−20.5	−21.1	−21.6	−22.3	−22.9	−23.6	−24.3	−25.2	−26.1	−26.9	−27.8	−29.0	−20
−21	−21.5	−22.1	−22.7	−23.2	−23.9	−24.6	−25.4	−26.2	−26.9	−27.8	−28.8	−30.0	−21
−22	−22.4	−22.9	−23.6	−24.2	−24.8	−25.5	−26.2	−27.2	−27.8	−28.7	−29.6	−30.8	−22
−23	−23.4	−24.0	−24.6	−25.2	−25.8	−26.7	−27.2	−28.1	−28.9	−29.7	−30.7	−32.1	−23
−24	−24.5	−25.0	−25.5	−26.2	−26.7	−27.4	−28.1	−28.9	−29.9	−30.7	−31.8	−33.1	−24
−25	−25.5	−26.1	−26.6	−27.0	−27.7	−28.5	−29.0	−29.9	−30.8	−31.8	−32.9	−34.0	−25
−26	−26.5	−26.9	−27.4	−28.1	−28.7	−29.4	−30.2	−30.9	−31.8	−32.8	−33.8	−34.9	−26
−27	−27.4	−28.1	−28.5	−29.1	−29.9	−30.4	−31.2	−32.0	−32.9	−33.9	−34.9	−35.9	−27
−28	−28.5	−28.9	−29.6	−30.2	−30.7	−31.5	−32.2	−33.1	−33.9	−34.8	−35.7	−36.7	−28
−29	−29.6	−30.2	−30.7	−31.3	−31.9	−32.7	−33.4	−34.2	−35.0	−35.8	−36.6	−37.6	−29
−30	−30.6	−31.1	−31.7	−32.3	−33.0	−33.6	−34.4	−35.1	−35.8	−36.6	−37.5	−38.4	−30

Note

Now we shall consider the following. In winter, we often hear people say: 'It is cold but it is dry.' By this, I want to draw your attention to a misleading fact. When the temperature is low outside, the air can contain very little water vapour; its *absolute* humidity is *very low*, but its *relative* humidity is *very high*, at least $\phi = 0.80$ at $-10°C$. Table 5 expresses the absolute humidity content of saturated air at various temperatures.

F. Dewpoint or dewpoint temperature

Let us consider air at a given temperature t and a relative humidity ϕ. When we now lower its temperature, it contracts in volume, the voids between N_2 and O_2 particles get smaller, and the relative humidity ϕ increases. If we continue to reduce the temperature, we reach a point at which the air is saturated ($\phi = 1$). This temperature is called the saturation temperature t_s. For this condition, $P_r = P_s$, so $\phi = 1$ at t_s, the saturation temperature or dewpoint temperature.

If we know the conditions of the air, as determined by temperature t and relative humidity ϕ, we can determine t_s. The dewpoint or saturation temperature t_s is the temperature at which *condensation* of water vapour *starts*. Dewpoint temperatures (t_s) for various temperatures and relative humidities are given in Table 4.

G. Psychrometric chart or Mollier diagram

On this chart (see Figure 1) we find a lot of interesting information in condensed form. Let us first get acquainted with scales and units as well as with the curves drawn on this chart.

Each point in the area under the saturation curve determines a climatic condition given by the air temperature and its relative humidity.

(a) Isotherms

These are the vertical lines that link up all points where the air temperature is the same. They end up on a temperature scale, represented by the lower bottom scale.

(b) Isobars

These are the horizontal lines that link up all points where the partial pressure (real of saturated) of water vapour in the air is the same.

Table 5 Absolute humidity present in saturated air (expressed in g/m^3) as a function of the air temperature (in $°C$)

$t(°C)$	$\psi(g/m^3)$	$t(°C)$	$\psi(g/m^3)$	$t(°C)$	$\psi(g/m^3)$	$t(°C)$	$\psi(g/m^3)$	$t(°C)$	$\psi(g/m^3)$
−20	0.88	1	5.19	21	18.33	41	53.77	61	135.9
−19	0.97	2	5.56	22	19.42	42	56.50	62	141.9
−18	1.06	3	5.95	23	20.56	43	59.35	63	148.1
−17	1.16	4	6.36	24	21.77	44	62.33	64	154.5
−16	1.27	5	6.79	25	23.04	45	65.40	65	161.1
−15	1.39	6	7.25	26	24.37	46	68.67	66	168.0
−14	1.51	7	7.74	27	25.76	47	72.03	67	175.2
−13	1.65	8	8.26	28	27.23	48	75.33	68	182.6
−12	1.80	9	8.81	29	28.76	49	79.18	69	190.2
−11	1.96	10	9.39	30	30.36	50	82.98	70	198.1
−10	2.14	11	10.00	31	32.04	51	86.93	71	206.2
−9	2.33	12	10.65	32	33.80	52	91.03	72	214.6
−8	2.53	13	11.34	33	35.65	53	95.30	73	223.4
−7	2.75	14	12.07	34	37.58	54	99.74	74	232.4
−6	2.99	15	12.84	35	39.60	55	104.30	75	241.8
−5	3.25	16	13.65	36	41.71	56	109.1	76	251.4
−4	3.53	17	14.50	37	43.92	57	114.1	77	261.4
−3	3.83	18	15.39	38	46.22	58	119.3	78	271.7
−2	4.15	19	16.32	39	48.63	59	124.7	79	282.3
−1	4.49	20	17.29	40	51.14	60	130.2	80	293.3
−0	4.84								

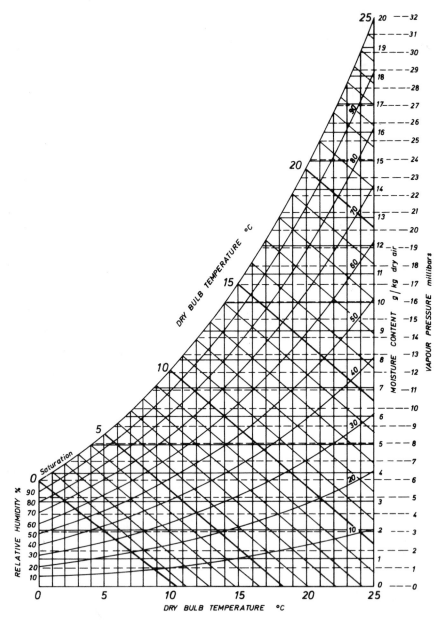

Figure 1   Psychrometric chart or Mollier diagram

(c) Relative humidity curves

These link up all the points where the relative humidity is the same. The top
curve represents a 100% relative humidity and is called the 'saturation curve'.

(d) Absolute humidity

This is indicated in the first column (when reading from left to right) and is expressed in grams of water vapour per kilogram of dry air (g/kg dry air).

(e) Partial pressure (either real or saturated)

The partial pressure of water vapour in the air can be read in the second column and is expressed in grams per square centimetre (g/cm^2).

(f) Specific volume

This is indicated by slightly sloping lines that end up at the upper bottom scale. The specific volume of 1 kg of air, including its humidity, for given conditions of temperature and relative humidity, is expressed in cubic metres per kilogram of air mixture (m^3/kg air mixture).

(g) Dry-bulb temperature

This is the temperature that we can read from a normal thermometer. The lower bottom scale is a dry-bulb thermometer temperature scale.

(h) Wet-bulb temperature

This is the temperature read on a thermometer whose bulb is surrounded by wet cotton. The wet-bulb temperature is always lower than the dry-bulb, except when the air is saturated in water vapour, when the wet- and dry-bulb temperatures are the same.

When water passes from the liquid to the gaseous state, we say that it evaporates. This change of state of aggregation requires energy, in the form of heat, called the latent heat or evaporation heat.

The water retained in our wet cotton evaporates too and withdraws the necessary heat from the thermometer. As the rate of evaporation depends upon the state of saturation of the surrounding air, the quantity of water that evaporates is limited to the quantity of vapour that can be taken up by the surrounding air. This explains why the difference in temperature between the dry- and wet-bulb readings is also proportional to the relative humidity present in the air.

The wet-bulb temperature isotherms on the graph are lines inclined at approximately 45° that end up on a sloping temperature scale in front of the saturation curve.

H. Use of the psychrometric chart

The reader is referred to Figure 2.

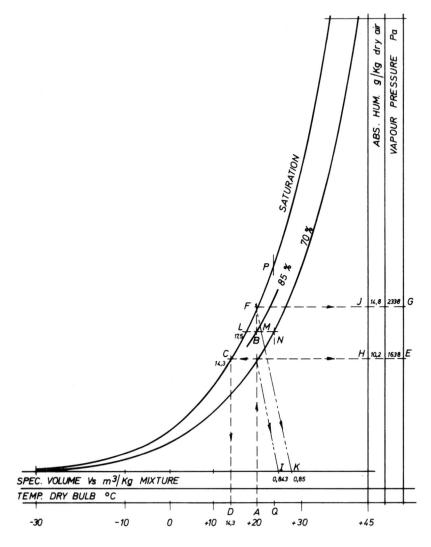

Figure 2   Use of psychrometric chart

Example 1

Given

Dry-bulb air temperature, 20°C
Relative humidity, $\phi = 0.70$

To find

(a) t_s.
(b) P_r.
(c) P_s.
(d) Check ϕ.
(e) Absolute humidity ψ in g/kg dry air and in g/m^3 air mixture.
(f) Absolute humidity of 20°C saturated air in g/m^3.
(g) Knowing (c) and (d), determine the quantity of humidity present in 1 m^3 of air at 20°C and $\phi = 0.7$.

Solution

(a) On Figure 2, start at the bottom scale at 20°C at A, and follow the isotherm until it reaches relative humidity $\phi = 0.7$ at B. B is the point that represents the given air conditions on the chart. Then follow the isobar to the left until it meets the saturation curve at C. There we can read the temperature $t_s = 14.3$°C.

We can also read the same temperature on the bottom scale at D when we drop along the isotherm passing through C.

(b) From our point B, we now travel to the right and find a direct reading on the vapour pressure scale of 16.7 g/cm^2 at E which is the P_r value.

(c) To find P_s, we start from the intersection point F of our 20°C isotherm with the saturation curve and follow the isobar on the right, and again we can read directly on the pressure scale the value of 23.8 g/cm^2 at G, which is our P_s value.

(d) $\phi = P_r/P_s$ and so

$$\phi = \frac{16.7}{23.8} = 0.7$$

(e) The isobar passing through B again allows a direct reading on the absolute humidity scale of 10.2 g/kg dry air at H.

We now note that, although we have found ψ in g/kg, it is practically impossible to imagine a quantity of 1 kg of dry air, but it is far easier to imagine a volume of 1 m^3. Hence we wish to determine ψ in g/m^3.
From our point B we now draw a line parallel to the specific volume lines and find at I a direct reading on the specific volume scale of 0.843 m^3/kg. (Note that this is per kilogram of air mixture.)
To find the absolute humidity in g/m^3 air mixture, we first have to transform our previous reading in g/kg dry air into g/kg air mixture. This is very

easy: we have 10.2 g of water vapour per kilogram of dry air. When we mix both, we can say that we have 10.2 g of water vapour per (1 kg + 10.2 g) of air mixture or 10.2/1.0102 g/kg air mixture.

Now our problem is solved in a few lines. From the specific volume we learn that

$$0.843 \text{ m}^3 \text{ weighs } 1 \quad \text{kg}$$

$$1 \quad \text{m}^3 \text{ weighs } \frac{1}{0.843} \text{ kg}$$

and from the previous operation we learn that

$$1.0102 \text{ kg contains } 10.2 \quad \text{g}$$

$$1 \quad \text{kg contains } \frac{10.2}{1.0102} \text{ g}$$

$$\frac{1}{0.843} \text{ kg contains } \frac{10.2}{1.0102} \times \frac{1}{0.843} = 12 \text{ g/m}^3$$

(f) We can now do exactly the same for 20°C saturated air. Starting from F to the right along the isobar, this gives us a reading of 14.8 g/kg dry air at J.

We find the corresponding specific volume by drawing a line parallel to the specific volume curves from point F and read at K a value of 0.85 m³/kg. The absolute humidity in this case is:

$$\frac{14.8}{1.0148} \times \frac{1}{0.85} = 17.2 \text{ g/m}^3$$

(g) It is most interesting to note that air at 20°C with $\phi = 0.7$ can only take up another (17.2 − 12) = 5.2 g/m³ until saturation is reached. In other words, to evacuate 5 g of moisture by a stream of air at 20°C and $\phi = 0.7$, at least 1 m³ of air is required. Even 5 g is on the high side, as it is almost impossible to saturate the passing air completely. This is important when dealing with roof, wall, and floor constructions where humidity occurs.

Example 2

Given

Assume we have an inside air temperature of 20°C and an insulation thickness has been determined to satisfy a required U value. In winter conditions, between Foamglas* and concrete, a temperature of 17.5°C is obtained.

---

*Foamglas®—foamed glass insulation manufactured by Pittsburgh Corning.

To find

(a) What maximum relative humidity can we allow on the inside in order to avoid interstitial condensation?
(b) Suppose the room is fully air conditioned so that the relative humidity remains constant. Up to what temperature can we raise the inside air to avoid interstitial condensation?

Solution

(a) In this case, using Figure 2 we read the dewpoint temperature along the saturation curve at point L, draw a horizontal line to the right (isobar), and find the intersection with the 20°C isotherm at point M, where we read the maximum allowable relative humidity of 85% ($\phi = 0.85$).

(b) In this case, we continue the horizontal line LM until it reaches the $\phi = 0.7$ curve at N, and read off the temperature to which this corresponds: 23.2°C at points P or Q.

As vapour diffusion is such a serious problem in buildings, cold storage, and industrial applications, we should be aware of its importance to architects, engineers, and owners of buildings.

Example 3

Given

Inside conditions, 20°C and $\phi = 0.70$
Outside conditions, -15°C and $\phi = 0.80$

To find

Determine the difference in water vapour pressure between inside and outside.

Solution

We find by direct reading (on Figure 1 that

P_r inside $= 16.7$ g/cm^2
P_r outside $= 1.2$ g/cm^2

and so we obtain that

pressure difference $= 15.5$ g/cm^2 or 155 kg/m^2
$= 31.75$ lb/ft^2
$= 1520.5$ Pa

Thus we see that moisture is pressed into the construction under these conditions at more than 30 lb/ft^2 = 1500 Pa.

I. The use of tables giving the partial pressure of water vapour in saturated air

Although not as complete in information as a psychrometric chart, a table of partial pressure values (see Table 2) is often useful, so we shall now examine how to use such tables.

In Table 2, the first column gives the temperature in degrees Fahrenheit (°F). In the second column, the corresponding saturation water vapour pressure is expressed in pounds per square inch (lb/in^2).

Example

(a) Inside air conditions are 20°C (68°F), and ϕ = 0.70, and we wish to determine the dewpoint temperature t_s.

We know from equation (4) that $\phi = P_r/P_s$ that ϕ is given (= 0.70) and that we can read P_s from Table 2, i.e. P_s = 0.3390 lb/in^2 for 20°C (68°F). So, from our equation (4), we can calculate P_r:

$$P_r = \phi \times P_s = 0.7 \times 0.3390 = 0.23730 \text{ lb/in}^2$$

Now in the tables we have to find the temperature at which the water vapour pressure in saturated air is 0.2373 lb/in^2. Thus

at 58°C we read 0.23849
at 57°C we read 0.23006
i.e. a difference of $\overline{0.00843}$

So we have to find the exact temperature by interpolation. We have 0.23730 − 0.23006 = 0.00724. Thus the dewpoint temperature (t_s) corresponds to:

$$57 + \frac{0.00724}{0.00843} = 57 + \frac{724}{843} = 57.9°F$$

or

$$(57.9 - 32) \times \tfrac{5}{9} = 14.4°C$$

(b) The temperature at the edge on the warm side of the Foamglas layer is 17.5°C (63.5°F taken from Table 1 or calculated: 17.5 × $\tfrac{9}{5}$) + 32 = 63.5). For this temperature, in Table 2 we find (by interpolation) a p_s value

$$P_s = \frac{0.28488 + 0.29505}{2} = 0.28997 \text{ lb/in}^2$$

or, say, $P_s = 0.290$. This value corresponds to the real vapour pressure that can be allowed with respect to our inside temperature.

Working out equation (4), we find

$$\phi = \frac{P_r}{P_s} = \frac{0.290}{0.339} = 0.8$$

(P_s for 68°F, 20°C, corresponds to 0.339 lb/in^2) which is the maximum allowable relative humidity at 20°C (68°F) to ensure that the dewpoint falls in the Foamglas.

(c) If $\phi = 0.70$, what would then be the maximum allowable inside temperature in order to avoid condensation below the Foamglas insulation?

If ϕ remains constant at 0.70, we know $P_r = 0.290$ lb/in^2 (from our temperature of 17.5°C (63.5°F). Then equation (4) gives us:

$$P_s = \frac{P_r}{\phi} = \frac{0.290}{0.7} = 0.414 \text{ lb/in}^2$$

which is our new P_s value that corresponds to a temperature of 23.3°C (74°F) or by interpolation 73.9°F.

Table 3 is more complete for temperatures above 30°C, but the partial vapour pressure is given in millibars, where

$$1 \text{ mbar} = 100 \text{ N/m}^2 = 100 \text{ Pa} = \frac{100 \times 212}{9.81 \times 1550}$$
$$= 90 \text{ 145 lb/in}^2$$

Table 5 gives the values of absolute humidity in the air expressed in g/m^3 air mixture at saturation.

If we have to determine the absolute humidity at 20°C and $\phi = 0.7$, we have to multiply the absolute humidity at saturation by 0.7 (the rate of saturation of the air at that temperature).

Example

If we are given that absolute humidity of saturated air at 20°C is 17.29 g/m^3, then absolute humidity ψ at 20°C and $\phi = 0.70$

$$\psi = 17.29 \times 0.7$$
$$= 12.103 \text{ g/m}^3$$

J. Simplified method of determining the insulation thickness required to avoid condensation of water vapour in the supporting structure

Let us consider the roof construction shown in Figure 3. We shall look at the problem from the inside going to the outside. (In general, it is better to adopt a uniform method for calculations, graphs, etc. In choosing this method, we should try to remain as logical as possible. As heat always travels from the warm to the cold side, we should also start at the warm side and go towards the cold side.)

In order to keep the problem as simple as possible, we will not consider screeds to falls in the present example. Thus we have, from the inside:

0.015 m gypsum plaster coat
0.10 m reinforced concrete
x m Foamglas cellular glass
0.01 m built-up roofing
0.04 m gravel

The air conditions are as follows:
Inside conditions, $+20°C$ and $\phi = 0.70$
Outside conditions, $-10°C$ and $\phi = 0.80$

To satisfy the above conditions, the dewpoint temperature should be in the Foamglas cellular glass at the warm side, next to the reinforced concrete.

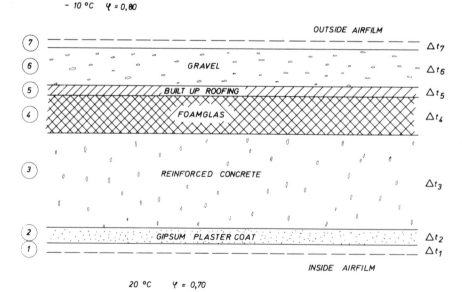

Figure 3   Roof construction

(a) Determine the dewpoint temperature for 20°C and $\phi = 0.7$ using Table 4.

We found a value of 14.4°C but for practical applications this is rounded off to 14.5°C. (Often the theoretical t_s values are rounded up to the next half degree centigrade)

(b) The allowable temperature drop between inside air and the temperature reached between reinforced concrete and Foamglas cellular glass is $20 - 14.5 = 5.5$°C.

The quantity of heat that passes through a construction is proportional to the temperature drop (inside air temp. − outside air temp.) and is inversely proportional to the thermal resistance of the construction, or

$$Q = \frac{\Delta t}{R_t}$$

The same quantity of heat travels through each layer that belongs to this construction. This allows us to write:

$$Q = \frac{\Delta t}{R_t} = \frac{\Delta t_1}{R_1}$$

where Δt_1 is the temperature drop in the first layer, and R_1 its thermal resistance. If the quantities Δt_1, R_t, and R_1 are known, we can determine

$$\Delta t_1 = \frac{\Delta t}{R_t} \times R_1$$

The total temperature drop is therefore

$$\Delta t_1 + \Delta t_2 + \Delta t_3 + \cdots + \Delta t_7$$

and that between inside air and Foamglas is

where

$$\Delta t_1 + \Delta t_2 + \Delta t_3 \tag{6}$$

Δt_1 = temperature drop in inside air film
Δt_2 = temperature drop in plaster coat
Δt_3 = temperature drop in concrete
Δt_4 = temperature drop in Foamglas cellular glass
Δt_5 = temperature drop in built-up roofing
Δt_6 = temperature drop in gravel
Δt_7 = temperature drop in outside air film

This temperature drop is given for one layer (as mentioned above) by:

$$\Delta t_i = \frac{\Delta t}{R_t} \times R_i$$

where Δt_i is the temperature drop in considered layer i; Δt is the temperature drop between inside and outside air; R_t is the total thermal resistance of construction/air/air; and R_i is the thermal resistance of layer i under consideration.

The thermal resistance R is given in general for a solid layer by

$$R = \frac{L}{k} = \frac{\text{thickness of layer in metres}}{k \text{ value of material}}$$

and has units m^2 °C/W, and for the air film by

$$R = \frac{1}{\alpha}$$

in the same units.

For our present construction we have:

No. of layer, i	Material	k value (W/m °C)	Thickness, L (m)	Thermal resistance, $R = L/k$ (m^2 °C/W)
1	Inside surface resistance	—	—	0.120
2	Plaster coat	0.70	0.015	0.021
3	Reinforced concrete	2.00	0.10	0.050
4	Foamglas	0.048	x	$x/0.048$
5	Built-up roofing	0.19	0.01	0.053
6	Gravel	0.80	0.04	0.050
7	Outside surface resistance	—	—	0.043
				$R_t = 0.337 + x/0.048$

We also have the following temperature drops:

$$\Delta t_1 = \frac{30}{R_t} \times 0.12$$

$$\Delta t_2 = \frac{30}{R_t} \times 0.021$$

$$\Delta t_3 = \frac{30}{R_t} \times 0.05$$

so that the allowable drop between inside air and Foamglas (i.e. 5.5°C) is from equation (6):

$$5.5 = \frac{30}{R_t} \times 0.191$$

which gives

$$R_t = \frac{30}{5.5} \times 0.191 = 1.042$$

Thus from the total R_t in the table, we obtain

$$R_t = 1.042 = 0.337 + \frac{x}{0.048}$$

and

$$x = 0.048(1.042 - 0.337) = 0.048 \times 0.705 = 0.034 \text{ m}$$

or a thickness of 4 cm of Foamglas cellular glass.

We shall plot a temperature graph and examine the maximum relative humidity that we can allow on the inside. With 4 cm of Foamglas cellular glass we get for this layer ($i = 4$):

$$R_4 = \frac{0.04}{0.048} = 0.833 \text{ m}^2 \, °\text{C/W}$$

and for the total

$$\begin{aligned} R_t &= 0.12 + 0.021 + 0.05 + 0.833 + 0.053 + 0.05 + 0.043 \\ &= 1.170 \end{aligned}$$

The temperature drops in each layer are:

$$\Delta t_1 = \frac{30}{1.170} \times 0.12 = 3.1 \left\{ \quad 20 \right.$$

$$\Delta t_2 = \frac{30}{1.170} \times 0.021 = 0.5 \left\{ \quad 16.9 \right.$$

$$\Delta t_3 = \frac{30}{1.170} \times 0.05 = 1.3 \left\{ \quad 16.4 \right.$$

$$\Delta t_4 = \frac{30}{1.170} \times 0.833 = 21.3 \left\{ \quad 15.1 \right.$$

$$\Delta t_5 = \frac{30}{1.170} \times 0.053 = 1.4 \left\{ \quad -6.2 \right.$$

$$\Delta t_6 = \frac{30}{1.170} \times 0.05 = 1.3 \left\{ \quad -7.6 \right.$$

$$\Delta t_7 = \frac{30}{1.170} \times 0.043 = 1.1 \left\{ \begin{array}{l} -8.9 \\ -10.0 \end{array} \right.$$

$$\overline{\quad 1.170 \quad} \quad \overline{\quad 30.0 \quad}$$

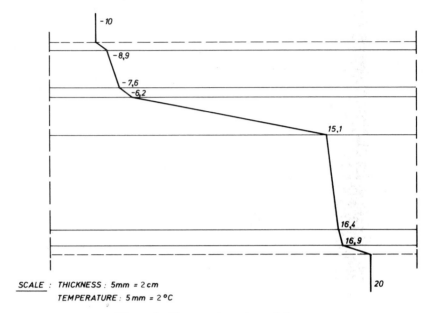

SCALE : THICKNESS : 5mm = 2 cm
TEMPERATURE : 5 mm = 2 °C

Figure 4   Temperatures in each layer

The last column of values is obtained by subtracting the temperature drop in the inside air film to find the inside surface temperature, then subtracting the temperature drop in the plaster coat to find the temperature between plaster coat and concrete, etc. These results we now plot on a graph (see Figure 4).

The saturation vapour pressure at 15.1°C is 1716.0 Pa, so

$$\phi = \frac{1716}{2338.7} = 0.73$$

which represents the maximum allowable relative humidity in the room for 20°C inside air temperature.

The corresponding U value for 4 cm Foamglas cellular glass is:

$$U = \frac{1}{1.170} = 0.85 \ \text{W/m}^2 \,°\text{C}$$

Semigraphical method

We can also apply a more graphical method, the so-called triangular method (see Figure 5).

Again, we calculate the thermal resistance of each layer (with the excep-

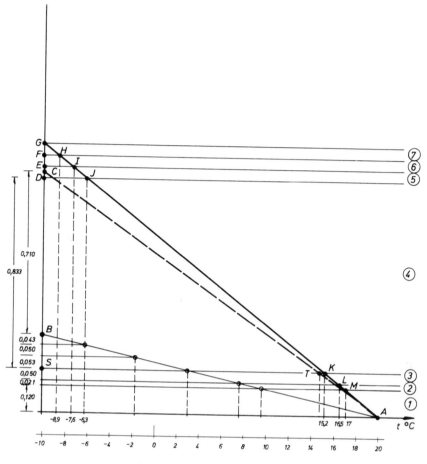

Figure 5   Triangular method

tion of the insulation layer which is not known), and plot those values on the ordinate of an R–t (thermal resistance–temperature) graph starting with the inside air film resistance at O. On the abscissa we plot a temperature scale with the outside air temperature at O and the inside air temperature on the right.

We now draw horizontal lines through the points we have plotted on the ordinate. These lines represent the thermal resistances of the various layers in the construction.

For clarity, we did not draw the lines above the concrete layer but gave each layer a number.

When we link up points A and B, the intersection with the horizontal lines gives us the temperatures at these planes; we only have to read them on the bottom scale.

Now we plot the dewpoint temperature (14.5°C) on the graph and draw a vertical line t_s. We wanted the dewpoint to be between concrete and Foamglas, i.e. at point T, which is the intersection with the t_s isotherm.

We now have to draw a line from A through T which meets our ordinate at C. The distance between B and C is the graphical representation of the missing thermal resistance (71 mm on the scale), i.e. 0.71 m² °C/W. This value equals $L/0.048$ for Foamglas, so

$$L/0.048 = 0.71$$

gives

$$L = 0.048 \times 0.71 = 0.034 \text{ m}$$

This corresponds to our previous result. This is a theoretical result and we now determine the practical thickness that should be applied. The nearest standard thickness being 4 cm, we determine the corresponding thermal resistance:

$$\frac{L}{k} = \frac{0.04}{0.048} = 0.833 \text{ m}^2 \text{ °C/W}$$

This value is now plotted on the graph, starting at point S (on top of the concrete) to point D. On top of D, we again plot the thermal resistances of the built-up roofing (DE), gravel EF), and outside air film (FG).

If we connect points G and A, the intersection of the hypotenuse AG with the horizontal lines that represent the layers gives us the corresponding temperatures on the temperature scale (points HI, J, K, L, and M).

In this case we read from inside to outside: 20, 17, 16.5, 15.2, −6.3, −7.6, −8.9 and −10°C, which correspond fairly well with our previous results.

This triangular method is often used by roofing contractors.

K. The Glaser method

1. Aim of the Glaser method

By this method we can examine the problem of interstitial condensation in more detail. It allows us to determine:

(a) whether interstitial condensation occurs or not;
(b) the quantity of water vapour diffusing into the construction;
(c) the quantity of water vapour passing through the construction;
(d) the quantity of water vapour passing through the construction;
(e) the quantity of water vapour that condenses in the construction;
(f) the zone where this condensation occurs;

(g) the zone where we have water in liquid form;

(h) the zone where we have ice;

(i) the required minimum resistance of the vapour barrier to be applied on the warm side to reduce condensation to an acceptable value;

(j) the maximum inside temperature or relative humidity that can be allowed to avoid interstitial condensation;

(k) the minimum conditions on the inside or outside for temperature and relative humidity for which interstitial condensation danger exists;

(l) humidity balance over the year for
   (i) winter period (2 months),
   (ii) summer period (3 months), and
   (iii) eventually, intermediate seasons—spring and autumn (7 months).

2. Definitions with respect to water vapour diffusion

Unfortunately, many different units are used in the various countries when dealing with water vapour diffusion. We can divide them into two main parts:

(a) those that work with the permeance of the materials; and

(b) those that use the resistance of materials against the diffusion of water vapour.

Let us consider the first group and then the second group.

(a) Permeance

This idea was first developed in the USA. Permeance is a physical property of materials to let water vapour pass; its unit is a perm, i.e.

$$1 \text{ perm} = \frac{1 \text{ grain}}{\text{ft}^2 \text{ h (in Hg)}} \qquad \left(1 \text{ grain} = \frac{1}{7000} \text{ lb}\right)$$

Thus 1 perm is the passage of 1 grain of water vapour per square foot of a material, having a *given thickness*, per hour if there is a real partial pressure difference of one inch of mercury column between both sides of the material.

This property is also known as a perm rating and is mainly used for sheet form materials. Therefore the thickness of the material is not considered in the unit (or dimension).

This idea was then brought to Europe where the unit was adapted to a metric perm, i.e.

$$1 \text{ metric perm} = \frac{1 \text{ gram}}{\text{m}^2 \text{ 24 h (mm Hg)}}$$

Apart from the units, its definition was exactly the same. According to SI units, a perm rating is expressed in gram/m^2 h mbar.

(b) Permeability

For other materials, the idea of the thickness was introduced and we obtained the perm inch rating. This was then called the *permeability* of a material. In the metric system we got exactly the same, but transformed it into a perm centimetre and perm centimetre rating or permeability, i.e.

permeability = permeance × unit thickness

It is easy to remember the difference between permeance and permeability. Permeability has more letters than permeance, so it is permeance multiplied by something, and that something is either the inch or the centimetre.

(c) Water vapour diffusion resistance factor, μ-value

The μ value indicates the resistance against water vapour diffusion of a layer of material in comparison with the resistance against water vapour diffusion of an air layer of the same thickness, under the same conditions of temperature and atmospheric pressure. As this μ value is a ratio, it has no dimensions.

(d) Specific resistance against water vapour diffusion

This resistance is defined as ρ value, where

$$\rho = \frac{\mu_s R_D T_m}{\delta}$$

and s is the thickness of the material (in m), R_D is the gas constant of water vapour (in kJ/kg K), T_m is the mean temperature (in K), and δ is the diffusivity of water in the air (in m^2/h). The units of ρ are

$$\frac{m \times kJ \times K \times h}{kg \times K \times m^2} = \frac{kJ \, h}{kg \, m}$$

Notes

(i) If we consider the general gas law, we find for water vapour under atmospheric pressure at 0°C (known as normal conditions, n.c.) that

$$PV = XR_D T_m$$

where $P = 101\ 325\ \text{N/m}^2$ (atmospheric pressure), $V = 22.4141 = 0.0224$ m^3 (volume of 1 mol gas at n.c.), $x = 18$ g (1 gram-molecule of water vapour, H_2O) $= 0.018$ kg, and $T_m = 273$ K ($0°C$ expressed in K).
We can rewrite this as

$$R_D = \frac{PV}{xT_m}$$

$$= \frac{101\ 325 \times 0.022\ 414}{0.018 \times 273.15}$$

$$= 462\ \text{N m/kg K}$$
$$= 0.462\ \text{kJ/kg K}$$

(ii) Diffusivity of water vapour in the air: in order to understand better what diffusivity means, we can transform its units of m^2/h into (m/h) $\times$ m. We know that m/h stands for a speed and m for a distance. So we can simply define diffusivity as the speed at which water vapour penetrates (diffuses) into the air (over a depth of 1 m).

(iii) In order to simplify calculations, the value of

$$R_D T_m/\delta = N$$

is given in tables (see Table 6). The units of N are then kJ h/kg m^2. In this case ρ becomes

$$\rho = \mu s N$$

(e) Quantity of water vapour diffusing through a layer (or layers) of material

The quantity that passes is always proportional to the difference in pressure and is inversely proportional to the resistance. This is a law that we also meet in electricity and in heat transfer.
Let us call g this quantity that gets through, then we have:

$$g = \frac{\Delta p}{\rho} = \frac{\text{real pressure difference between inside and outside}}{\text{total resistance of layer(s)}}$$

Units are as follows: Δp is expressed in Pa (N/m^2), and ρ is expressed in kJ h/kg m, so g is expressed in N kg m/m^2 kJ h. As 1 J = 1 N m, we obtain for g:

$$\frac{\text{N kg m}}{\text{m}^2\ \text{kN m h}} = \frac{\text{kg}}{\text{m}^2\ \text{h}}$$

Table 6 $N = R_D\, T_m/\delta$ (k J h/kg m²)

$t(°C)$	N	$t(°C)$	N	$t(°C)$	N	$t(°C)$	N
+0	$1.519.10^6$	+10	$1.475.10^6$	+20	$1.434.10^6$	+30	$1.397.10^6$
+0.5	$1.517.10^6$	+10.5	$1.473.10^6$	+20.5	$1.432.10^6$		
+1	$1.515.10^6$	+11	$1.471.10^6$	+21	$1.430.10^6$		
+1.5	$1.512.10^6$	+11.5	$1.469.10^6$	+21.5	$1.428.10^6$		
+2	$1.510.10^6$	+12	$1.467.10^6$	+22	$1.426.10^6$		
+2.5	$1.508.10^6$	+12.5	$1.465.10^6$	+22.5	$1.425.10^6$		
+3	$1.506.10^6$	+13	$1.463.10^6$	+23	$1.423.10^6$		
+3.5	$1.504.10^6$	+13.5	$1.461.10^6$	+23.5	$1.421.10^6$		
+4	$1.501.10^6$	+14	$1.458.10^6$	+24	$1.419.10^6$		
+4.5	$1.499.10^6$	+14.5	$1.456.10^6$	+24.5	$1.417.10^6$		
+5	$1.497.10^6$	+15	$1.454.10^6$	+25	$1.416.10^6$		
+5.5	$1.495.10^6$	+15.5	$1.452.10^6$	+25.5	$1.414.10^6$		
+6	$1.493.10^6$	+16	$1.450.10^6$	+26	$1.412.10^6$		
+6.5	$1.490.10^6$	+16.5	$1.448.10^6$	+26.5	$1.410.10^6$		
+7	$1.488.10^6$	+17	$1.446.10^6$	+27	$1.408.10^6$		
+7.5	$1.486.10^6$	+17.5	$1.444.10^6$	+27.5	$1.407.10^6$		
+8	$1.484.10^6$	+18	$1.442.10^6$	+28	$1.405.10^6$		
+8.5	$1.482.10^6$	+18.5	$1.440.10^6$	+28.5	$1.403.10^6$		
+9	$1.479.10^6$	+19	$1.438.10^6$	+29	$1.401.10^6$		
+9.5	$1.477.10^6$	+19.5	$1.436.10^6$	+29.5	$1.399.10^6$		
−0	$1.519.10^6$	−10	$1.566.10^6$	−20	$1.616.10^6$	−30	$1.669.10^6$
−0.5	$1.521.10^6$	−10.5	$1.569.10^6$	−20.5	$1.619.10^6$	−30.5	$1.672.10^6$
−1	$1.524.10^6$	−11	$1.571.10^6$	−21	$1.621.10^6$	−31	$1.675.10^6$
−1.5	$1.526.10^6$	−11.5	$1.574.10^6$	−21.5	$1.624.10^6$	−31.5	$1.678.10^6$
−2	$1.528.10^6$	−12	$1.576.10^6$	−22	$1.627.10^6$	−32	$1.681.10^6$
−2.5	$1.531.10^6$	−12.5	$1.579.10^6$	−22.5	$1.629.10^6$	−32.5	$1.684.10^6$
−3	$1.533.10^6$	−13	$1.581.10^6$	−23	$1.632.10^6$	−33	$1.686.10^6$
−3.5	$1.536.10^6$	−13.5	$1.584.10^6$	−23.5	$1.635.10^6$	−33.5	$1.689.10^6$
−4	$1.538.10^6$	−14	$1.586.10^6$	−24	$1.637.10^6$	−34	$1.692.10^6$
−4.5	$1.540.10^6$	−14.5	$1.589.10^6$	−24.5	$1.640.10^6$	−34.5	$1.695.10^6$
−5	$1.543.10^6$	−15	$1.591.10^6$	−25	$1.643.10^6$	−35	$1.698.10^6$
−5.5	$1.545.10^6$	−15.5	$1.594.10^6$	−25.5	$1.645.10^6$	−35.5	$1.701.10^6$
−6	$1.547.10^6$	−16	$1.596.10^6$	−26	$1.648.10^6$	−36	$1.705.10^6$
−6.5	$1.550.10^6$	−16.5	$1.599.10^6$	−26.5	$1.651.10^6$	−36.5	$1.707.10^6$
−7	$1.552.10^6$	−17	$1.601.10^6$	−27	$1.653.10^6$	−37	$1.710.10^6$
−7.5	$1.554.10^6$	−17.5	$1.604.10^6$	−27.5	$1.656 10^6$	−37.5	$1.712 10^6$
−8	$1.557.10^6$	−18	$1.606.10^6$	−28	$1.659.10^6$	−38	$1.715.10^6$
−8.5	$1.559.10^6$	−18.5	$1.609.10^6$	−28.5	$1.661.10^6$	−38.5	$1.718.10^6$
−9	$1.561.10^6$	−19	$1.611.10^6$	−29	$1.664.10^6$	−39	$1.721.10^6$
−9.5	$1.564.10^6$	−19.5	$1.614.10^6$	−29.5	$1.666.10^6$	−39.5	$1.724.10^6$
						−40	$1.727.10^6$

(f) Pressure drop in a layer

The pressure drop in a layer is always directly proportional to the resistance of this layer. This can be concluded from subsection (e). When we now consider various layers, each having a specific resistance ρ, and when we call ρ_1 the resistance of the first layer, ρ_2 the resistance of the second layer, etc., it is clear that as $g = \Delta p/\rho$, this is the same ratio for each layer, as the same quantity passes through each of them.

In other words (see Figure 6a), the quantity that is passing through the total construction (here composed of three layers) passes through each layer, so we can write:

$$g = \left[\frac{\Delta p}{\rho} = \frac{\Delta p_1}{\rho_1}\right] = \frac{\Delta p_2}{\rho_2} = \frac{\Delta p_3}{\rho_3} = \cdots \tag{8}$$

(a)

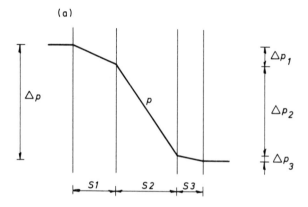

(b)

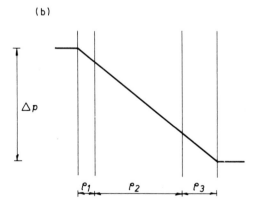

Figure 6   Variation of layer thickness and resistance
with pressure

When we now consider that part of equality (8) in the large brackets, we see that the only unknown factor is Δp_1, as Δp is the total pressure drop between inside and outside, ρ is the total resistance (was determined for each layer), ρ_1 is the resistance of the first layer. So,

$$\Delta p_1 = \Delta p/\rho \times \rho_1$$

and this we can repeat for each layer.

When we plot these results on a pressure–layer thickness $(p-s)$ graph, we obtain a broken line or p curve, indicating the real pressure variation inside the construction as a function of the thickness(es) (see Figure 6a).

If we now plot the pressure as a function of the resistance, we obtain a straight line, as the ratio of $\Delta p/\rho$ is a constant and equals g (see Figure 6b).

(g) The Glaser diagram

In the Glaser diagram we plot on the ordinate the temperature in °C, the partial pressure of water vapour for saturated air, and the actual partial water vapour pressures corresponding to inside and outside conditions. On the abscissa, we plot the ρ values of the various layers (see Figure 7).

Given

For a built-up construction, we have:
  Inside temperature, t_i
  Inside relative humidity, ϕ_i
  Outside temperature, t_o
  Outside relative humidity, ϕ_o
  Air film resistance inside, $1/\alpha_i$
  Air film resistance outside, $1/\alpha_o$
  Thickness, K and μ values for each layer

Question

Examine whether or not interstitial condensation takes place.

Solution

The reader is referred to Figure 7.

(1) We determine the temperature gradient through the given construction. This is already our temperature curve for our Glaser diagram.

(2) We determine the mean temperature t_m in each layer.

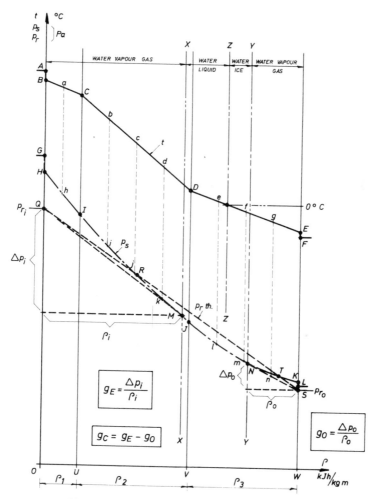

Figure 7 Glaser diagram, winter conditions

(3) We calculate the ρ value for each layer. For this we read off the N values in Table 6 corresponding to the mean temperatures that we obtained in (2).

(4) We determine the scale for the ρ values and plot these on the abscissa of the Glaser diagram and draw vertical lines through the endpoints of the ρ values (points O, U, V, W).

(5) We choose a scale for the temperature, and plot the temperature gradient on the upper part of the Glaser diagram (line ABCDEF).

(We note that it is easier to draw a temperature scale on the right side of the graph in order to keep the left side clear for the pressure values. This temperature scale should not start at the origin, but should be drawn on the upper part of the diagram independently of the origin of abscissa and ordinate (point O).

(6) A saturation partial pressure of water vapour in the air corresponds to each point of the temperature curve. This saturation pressure is found from Tables 2 or 3 (P_s values).

We choose a scale for the pressure and start by plotting the P_s values that correspond to the temperatures reached between the different layers, as well as for the inside and outside temperatures (points A, B, C, D, E, and F) on the t curve and points G, H, I, J, K, and L on the P_s curve). These points may not allow us to draw the P_s curve with sufficient precision, in which case we have to determine the intermediate points.

For the determination of intermediate points, we can apply two methods:

(i) We take full degree temperatures on the temperature graph, find the corresponding P_s values in Tables 2 or 3, and plot these values on the ordinate passing through the intersection of our temperature curve with the horizontal line that we have drawn through the chosen full temperature (and representing the corresponding isotherm). As we are marking on millimetre paper, it is not always easy to apply this method as there are too many vertical lines.

(ii) A simpler method (applied here) consists of indicating equidistant points on the temperature curve (points a, b, c, d, e, f, and g) that correspond to the vertical 5 mm lines of our paper. We then read the corresponding temperature in tenths of degrees (°C) on the temperature scale, look up the corresponding P_s value in Tables 2 or 3 and plot it on the corresponding 5 mm line. This gives us the points h, i, j, k, l, m, and n of the P_s curve. Once we have a sufficient number of points of the P_s curve, we can draw this curve on the graph.

(7) We now determine the actual vapour pressures corresponding to the inside and the outside air conditions, going out from inside t_i and ϕ_i to outside t_o and ϕ_o.

This gives us points Q and S.

(8) We link up points Q and S by a straight line, which represents the theoretical variation of the actual partial pressure of water vapour through the construction, as a function of the resistance against water vapour diffusion of each layer.

We notice that this line crosses the P_s curve at points R and T. The easiest interpretation of this would be that the saturated pressure is reached between these points (condensation area). But this interpretation is not correct as the actual vapour pressure can never be higher than the saturation vapour pressure for a defined temperature.

According to Glaser, the variation of the actual vapour pressure as a function of the resistance of the construction against water vapour diffusion follows the tangent from point Q to the P_s curve, then follows the P_s curve, and leaves it again along the tangent drawn from point S to the P_s curve. This means that condensation only occurs between points M and N.

(9) When we now draw vertical lines through the points M (x–x line) and N (y–y line), we indicate the area in the construction where condensation occurs.

Now we should go a little step further. We draw a horizontal line through 0°C on the temperature scale, which meets the temperature curve at T. When we now draw a vertical line (z–z) through T, we divide the condensation area in two parts. This allows us to analyse what happens in the construction:

(i) From the warm surface up to the x–x axis we have water vapour in gaseous form, which is harmless.

(ii) From the x–x axis up to the z–z axis we have water vapour in liquid form, so we get condensation. The results are:
— Increase of the k value, and thus increase of the U value, causes an increase in fuel consumption, and makes the x–x axis (dewpoint area) to move towards the warm side. The situation gets worse. The effects become greater with time. Increased danger of corrosion of metal parts in this area (nails, electric cables, etc.).

(iii) From the z–z axis to the y–y axis: area where snow and/or ice is formed. The results are:
— Same as (ii), but four times more important.
— If sufficient water vapour gets in to build a thin ice skin in the y–y plane, it acts like a vapour barrier on the cold (wrong) side, which increases the amount of condensation that occurs as less vapour can escape to the cold side.

(iv) From y–y axis to cold side: here again we have water vapour in a gaseous state, which is harmless.

Thus we make the following conclusions: It is important to note that interstitial condensation takes place inside the construction, and it remains hidden from an observer on the warm side as well as on the cold side. If the effects become visible (humidity), a state of saturation of the construction is reached, and then it is too late. A big problem then emerges: how do we get the construction dry again and how can we prevent condensation in the future?

(10) The quantity of water vapour that diffuses into the construction is given by:

$$g_E = \Delta p_i / \rho_i$$

expressed in kg/m² h. The quantity that diffuses out of the construction is given by:

$$g_o = \Delta p_o / \rho_o$$

expressed in kg/m² h, as this quantity is proportional to the pressure difference and inversely proportional to the resistance.

We call Δp_i the pressure drop on the warm side of the condensation area (towards the inside), ρ_i the total resistance for the same area, g_E the vapour flow in, Δp_o the pressure drop on the cold side of condensation area (towards the outside), ρ_o the total resistance for same area, and g_o the vapour flow out.

The values Δp_i and ρ_i are measured on the graph; as the scales used for plotting the graph are known, we can easily determine the actual values of Δp_i and ρ_i and so obtain g_E. It is the same for the values Δp_o and ρ_o.

The quantity of water vapour that remains in the construction is given in kg/m² h by:

$$(g_E - g_o)$$

We now make the following remarks.

(i) At 0°C and an atmospheric pressure of 760 mm Hg, water vapour has a density of 0.77 kg/m³. At the same temperature, water has a density of 999 kg/m³, and ice has, at 0°C, a density of 917 kg/m³.

In order to obtain a better idea of the importance of these values, we shall simply determine the volumes of water vapour, water, and ice for a mass of 1 kg at 0°C and 760 mm Hg (atmospheric pressure):

1 kg of water vapour has

a volume of $\dfrac{1}{0.77}$ = 1.3 m³

1 kg of water = $\dfrac{1}{999}$ = 0.001 m³

1 kg of ice = $\dfrac{1}{917}$ = 0.00109 m³

ICE ⟵——— WATER ———⟶ VAPOUR
      × 1.1   × ~ 1300
(ca. 10% in-   (ca. 1300 times increase in   crease in volume)
volume)

(ii) When water vapour is transformed into water, i.e. when it condenses, a latent heat of 2.256 MJ/kg is set free. When water is transformed into ice, i.e. when it freezes, 0.33 MJ/kg are set free.

In relation to the mass of the structure, these quantities of heat are insignificant and do not have a practical influence on the temperature in the construction.

(11) We now consider the required minimum resistance of the water vapour barrier to reduce condensation (see Figure 8).

If we continue the tangent from point S (the actual partial vapour pressure on the cold side) until it crosses the horizontal line through point Q (the actual partial vapour pressure on the warm side), we find a ρ_m value (horizontal length between points V and Q).

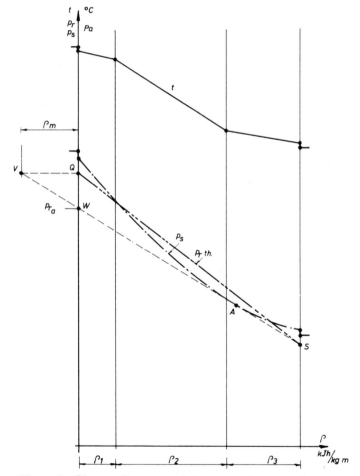

Figure 8 Glaser diagram—minimum resistance of water vapour barrier

This is seen on the scale as the value of the missing resistance against water vapour diffusion. As $\rho = \mu s N$, we can determine the value μs. The N value corresponds in this case to the surface temperature on the warm side. Now we only have to find a suitable material that gives us this μs resistance with an acceptable thickness of the foil.

(12) At the same time we can consider the point W, where the tangent SA meets the inside surface. The P_r value at point W represents the maximum allowable actual water vapour pressure in order to avoid interstitial condensation. For this we can examine two values:

  (i) If t_i (air temperature on the warm side) remains constant, we can determine allowable ϕ_i:

$$\phi_i = P_r/P_s$$

    where $P_s = P_{ra}$ (value found at point W) and P_s is the saturation pressure corresponding to t_i.

 (ii) We can also keep ϕ_i constant and determine t_i:

$$\phi = P_r/P_s$$

    so,

$$P_s = P_r/\phi$$

    In Tables 2 or 3 we then find the corresponding t_i value (allowable for this P_s value).

(13) We can then apply exactly the same Glaser method for conditions in any season. The following important points should be considered.

  (i) In winter and intermediate seasons, we have marked differences between inside and outside conditions. This means that we have great actual partial pressure drops. As the quantity of water vapour diffusing in is proportional to this pressure drop, we will find a high value for this vapour flow.

    On the other hand, in summer there are only slight temperature differences between inside and outside conditions and quite often we are faced with a change in the direction of water vapour flow. In this case, we have relatively small quantities of water vapour that diffuse out of the construction. This explains why it is easier to get condensation into the construction than to get it dry.

 (ii) In the precise study, we considered k values for dry materials, and, in fact, started with a dry state of construction. Nevertheless, we should not forget that when dealing with concrete and masonry, it normally takes three to four years before these materials have reached a state of equilibrium whereby humidity no longer gets out.

In practice, all materials have their natural degree of humidity, which is expressed as a percentage by volume or by weight. This humidity content also has an influence on the k values of the materials.

(14) Summer conditions: this is the period when humidity leaves the construction. Having determined the area where interstitial condensation occurs in winter, we consider the middle of this area as the plane in the construction where saturation is reached (or present). This simplifies calculations (see Figure 9).

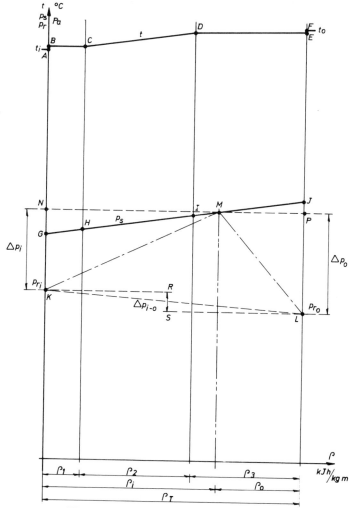

Figure 9 Glaser diagram, summer conditions

We again apply the same procedure:

(i) determine the temperature variation through each layer;
(ii) calculate the corresponding ρ values;
(iii) plot a ρ–t graph;
(iv) determine the P_s curve; and
(v) indicate P_r values on the inside and outside.

We now have humidity transmitted from the middle of the construction towards the inside and towards the outside, and simultaneously also a diffusion of water vapour (gas) from the side where we have the higher P_r value towards the lower P_r value.

As mentioned above, the method is somewhat simplified and the real mechanism of switching over from condensation to drying-out phases is somewhat different and more complicated in practice (see Figure 9).

When condensation stops, the drying-out phase starts, but very slowly, as the actual vapour pressure drop between the limits of the area where condensation occurred and the outside vapour pressure is very small.

Suppose we have prepared the graph as mentioned above (Figure 9). In this diagram, we have taken P_{ro} lower than P_r, although t_o is higher than t_i, only to demonstrate that these conditions may occur in practice.

The quantity of water vapour leaving the construction towards the inside is given by:

$$g_1 = \frac{\Delta p_i}{\rho_i} = \frac{(N - K)}{(M - N)} \tag{9}$$

The quantity of water vapour leaving the construction towards the outside is given by:

$$g_2 = \frac{\Delta p_o}{\rho_o} = \frac{(P - L)}{(M - P)} \tag{10}$$

But we still have the normal diffusion from inside to outside, given by:

$$g_3 = \frac{\Delta p_{i-o}}{\rho_{i-o}} = \frac{\Delta p_{i-o}}{\rho_T} = \frac{(R - S)}{(N - P)} \tag{11}$$

This means that the quantity of humidity leaving the construction equals:
$$g_1 + g_2 - g_3 \tag{12}$$

(15) We can apply the same study for the intermediate seasons, and make a balance over the year. Our conclusion then is that more humidity must get out of the construction in summer and eventually during the intermediate seasons than the quantity that condenses during the winter period. If this condition is not met, a finite quantity remains in the construction each year and saturation will be reached after a number of years, due to accumulation.

(16) The general conditions applied in Germany according to Dr Rer. Nat. Cammerer of the Cammerer Institute, Munich, in examining the risks for interstitial condensation are:

Winter

$$t_o = -10°C \qquad \phi_o = 0.80$$
$$t_i = +20°C \qquad \phi_i = 0.50$$

Summer

$$t_i = t_o = +12°C \qquad \phi_i = \phi_o = 0.70$$

These conditions are only valid for dwellings. For office buildings, fully air-conditioned buildings, industrial buildings, schools, etc., conditions are taken according to the real situation and local climatic conditions, which are generally obtained from meteorological stations in the neighbourhood.

L. The dewpoint curve method

The reader is referred to Figure 10. This method, although less accurate than the Glaser method, is easier to understand.

For this graph, we do not plot ρ values against saturation vapour pressure or actual vapour pressure. Instead, on the ordinate we only plot the temperature and the dewpoint temperature, while on the abscissa we put the real thickness of each layer to scale (or vice versa, thickness on ordinate and temperature on abscissa).

Once we have determined the ρ values for each layer, say we have three layers with values ρ_1, ρ_2, and ρ_3, we also know the total specific resistance ρ of the construction. Knowing the actual inside and outside vapour pressures (given by inside and outside air conditions), we can determine the pressure drop in each layer.

The pressure drop in each layer is always directly proportional to the total pressure drop and is inversely proportional to the total specific resistance (cf. temperature drop in a layer). This allows us to write:

$$\Delta P_1 = \frac{\Delta p}{\rho} \times \rho_1 \left\{ \begin{array}{l} P_{r\,i} \rightarrow t_{s1} \\ \\ \end{array} \right.$$

$$\Delta P_2 = \frac{\Delta p}{\rho} \times \rho_2 \left\{ \begin{array}{l} P_{r\,1/2} \rightarrow t_{s2} \\ \\ P_{r\,2/3} \rightarrow t_{s3} \end{array} \right.$$

$$\Delta P_3 = \frac{\Delta p}{\rho} \times \rho_3 \left\{ \begin{array}{l} \\ P_{r\,o} \rightarrow t_{s4} \end{array} \right.$$

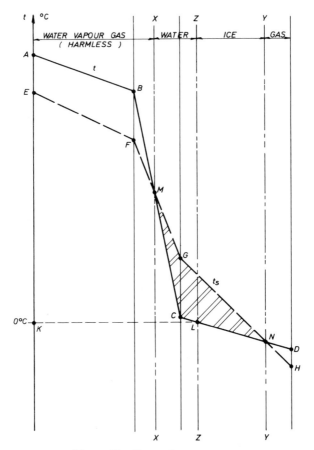

Figure 10   Dewpoint temperature

From these values, we obtain the actual vapour pressure values inside, and at each interface of the layers, i.e. P_{ri}, $P_{r1/2}$, $P_{r2/3}$, and P_{ro}. To each of these values corresponds a dewpoint temperature t_{s1}, t_{s2}, t_{s3}, and t_{s4} (taken from Tables 2 or 3) which allows us to plot the dewpoint curve (t_s), i.e. points E, F, G, and H.

Comparing the dewpoint temperature curve t_s with the actual temperature curve t, i.e. points A, B, C, and D, through the construction, we know that condensation occurs when the actual temperature is lower than the dewpoint temperature at the same location in the construction.

The points M and N (on Figure 10), where the dewpoint curve crosses or meets the real temperature curve, correspond to the points R and T on the Glaser graph (see Figure 7).

So this method only gives an idea of the area in which condensation occurs, and does not enable us to determine the quantity of water vapour that condenses in the construction. We can indicate the 0°C isotherm and the planes x–x, y–y, and z–z as in the Glaser method. Nevertheless, this method does indicate what really happens in the construction itself.

II. ANALYSIS OF A CONVENTIONAL BUILT-UP ROOF

A. Survey

A roof construction is a composite of various layers, each satisfying a different function (see Figure 11). Not enough people dealing with, or specialized in the building field, are aware of this.

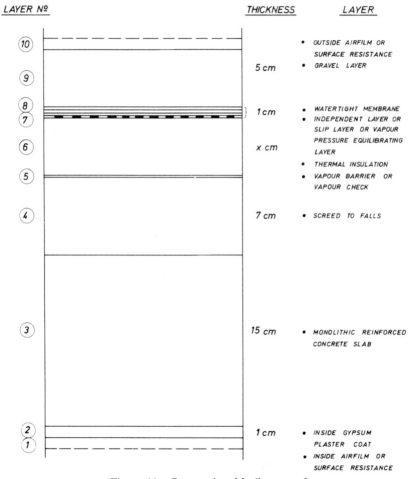

Figure 11 Conventional built-up roof

So let us first analyse the problem, leading to the various requirements that we should put forward for each layer and the construction in general. This analysis is very general in nature and applies for any roof. From inside to outside we have the following layers and their aim(s) and/or function(s).

1. Inside air film

From the thermal point of view, it symbolizes the heat transmitted by radiation and convection from inside ambient conditions to the roof surface.

2. Gypsum plaster coat

The functions of this are (a) to obtain a smooth ceiling for aesthetic reasons to hide the imperfections of the underside of the concrete slab, and (b) to improve light reflection by being often kept white.

3. Supporting structure

An example of this is a monolithic concrete slab. This structure forms the support for the roof of the building: mechanical resistance against roof load, snow and wind load, pedestrian traffic or vehicle traffic.

4. Screed to falls

Its function is to speed up the evacuation of rain water or melting ice or snow to the drains, and to avoid water standing on a watertight membrane (ponding). The watertight membrane consists of a relatively thick layer, often put down as three-layer thick roofing felts fully adhered with hot bitumen, including overlapping of each lane of each layer and staggering. However, there is still doubt about perfect watertightness and for this reason it is preferable to get rid of the water before it can find a pinhole to penetrate into the layers below.

5. Vapour barrier

As in winter vapour pressure is higher on the inside of the building than on the outside, vapour transmission takes place from the inside towards the outside.

The only function of a vapour barrier is to create resistance against the passage of water vapour, as it is known that condensation takes place when the dewpoint is reached. As the most important temperature drop automatically takes place in the thermal insulation (greatest thermal resistance), the dewpoint will normally fall in the insulation layer. So condensation usually occurs in this layer.

As the k value of a thermal insulation is about 0.050 W/m °C, and the k value of water reaches 0.58 W/m °C, whereas the k value of ice is 2.3 W/m °C (four times worse than that of water), it is understandable that everything is done to avoid condensation in the thermal insulation. (We only enjoy wearing a pullover when the pullover is dry.)

In order to reduce heat flow (heat loss), we use a layer of thermal insulation. Neverthless, we cannot reduce this heat flow to nought. Heat flow and the flow of water vapour are very similar. The driving force in the first case is a difference in temperature, and in the second case a difference in pressure.

The function of a vapour barrier is only to reduce the flow of water vapour, but it does not reduce this flow to nought either.

(I should like to remark here that also an independent layer of roofing felt is applied under the vapour barrier to protect the same against local damage. This independent layer is often interpreted as a vapour pressure equilibrating layer, which is false.)

6. Thermal insulation

Why do we apply thermal insulation in buildings? The answers are as follows:

(a) To *save energy*, not only in winter but also in summer when we are dealing with buildings that are air conditioned.

(b) To *protect the supporting structure* of our buildings against the temperature variations that occur on the outside, from day to night and summer to winter.

(c) To *increase our comfort*, by
   (i) creating a pleasant ambient air temperature (18 to 21°C);
   (ii) controlling relative humidity of the air (40 to 60% for temperatures of 21 to 18°C);
   (iii) limiting temperature drop between ambient air temperature and surface temperature on the inside of our buildings (to 3 to 5°C);
   (iv) providing for good ventilation (20 to 30 m³/person h) whereby:
      — airspeed in room <0.2 m/s, and
      — airspeed at defusor $<_2$ m/s.

(d) To *control condensation* in constructions. Less condensation should get into the construction during the winter and the intermediate seasons than the quantity that leaves the construction during summer and intermediate periods. By intermediate periods, we mean spring and autumn.

It depends upon the inside and outside climatic conditions whether condensation or drying occurs during the intermediate periods.

(e) It is advisable to keep the mass of the building on the inside, as this takes advantage of the heat accumulated in the mass and makes heating and air conditioning run more smoothly, as the control instruments are not activated by each temperature variation that occurs on the outside.

Mass affects thermal attenuation and dephasing, and is also important from the acoustic point of view.

7. Independent layer, slip layer or water vapour pressure equilibrating layer

The aim of this layer is to take up the differential movement between the watertight membrane and the thermal insulation to reduce the risks of eventual rupture due to fatigue of the watertight membrane.

This independent layer (or spotmopped layer) was also meant to avoid blistering due to high vapour pressure coming from the inside or underside, or caused by a temperature rise due to solar radiation.

In this independent layer, pressure would be divided and, if all goes well, the pressure could escape on the sides of the roof (roof edge) or via small chimneys installed on the roof. If an insulating material is applied that is not gastight, there is no need for a vapour pressure equilibrating layer, as in this case the volume of the insulation represents a sufficient volume to take up the air (gas) expansion without the risk of forming blisters.

We can conclude that only the first reason for the existence of such a layer is valid.

8. Watertight membrane

This avoids any rainwater penetration into the thermal insulation, or any penetration of melted ice or snow.

9. Gravel layer

This is compulsory in France. It has several functions:

(a) protection of the watertight membrane against ultraviolet radiation from the sun;

(b) mechanical protection of the watertight membrane against accidental foot traffic;

(c) protection of the watertight membrane against rapid temperature changes (thermal inertia of mass);

(d) ballasting the watertight membrane against wind uplift action; and

(e) reducing the free surface of humidity with outside air—thus, evaporation of humidity present on top of the watertight membrane is slowed down.

As we know that the specific heat of water is the highest for all materials and that the heat of evaporation is also relatively high, it helps to moderate temperature changes on top of the watertight membrane.

10. Outside air film

The presence of this has the same effect as the 'inside air film' (as explained above).

B. Behaviour and efficiency of some important layers

Let us now discuss in more detail the behaviour and the efficiency of some important layers.

1. Vapour barrier

(a) The main requirement for this layer is that it should be more vapourtight than the watertight membrane. Unfortunately, this condition is seldom satisfied unless high-quality vapour barriers are used. Only those having a metal foil should be considered as high-quality vapour barriers.

(b) The vapour barrier is only a rather thin membrane; the thickness of the metal foil normally used is 0.1 to 0.08 mm. It is manufactured in sheets or rolls and applied with about 5 to 10 cm overlap. Its weakest point is at the overlap joints because the metal foil is not continuous and the effectiveness depends entirely on the tightness of the joints, that is on the type of adhesive used and on the workmanship of the people applying it.

(c) It is important that the surface on which the vapour barrier is applied is even, as this foil is very easily damaged.

(d) As small movements in the supporting structure may occur, the vapour barrier should be fully embedded in hot bitumen or applied on top of a spotmopped independent layer.

(e) Depending upon the thermal insulation chosen, the thermal expansion and contraction of this material should be considered if it is adhered to the vapour barrier.

(f) Special care should be taken at the roof edges where the vapour barrier should be continued and flashed against the wall.

From the above, we can understand that the requirements for the effectiveness and successful operation of a vapour barrier are extremely severe.

If we want this effectiveness we have to pay for quality and application care, yet we still do not have any long-term guarantee of perfect performance.

2. Slip layer under vapour barrier

Often, for high-quality application, a slip layer, or spotmopped independent layer, is applied under the vapour barrier. It consists of a spotmopped roofing felt. Often this slip layer is called a vapour pressure equilibrating layer, but this is completely wrong. The reason why this layer is wrongly considered as a water vapour equilibrating layer is that for some time (and this still happens today) short pipes were installed on the roof to evacuate what was considered to be excess vapour (see Figure 12A)

Instead of improving the situation, it made things worse. A direct thermal short circuit was created, the supporting structure was locally cooled, and condensation occurred in the concrete as well as along the pipe.

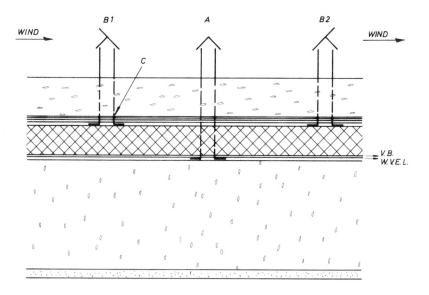

V.B. = VAPOUR BARRIER
W.V.E.L. = WATER VAPOUR EQUILIBRATING LAYER

Figure 12   Short pipes to evacuate vapour

3. Water vapour pressure equilibrating layer

The origin of this layer is found in problems of blistering that occurred on roofs. This was because the vapour barrier did not work properly or it was less tight than the watertight membrane on top of the insulation. This caused an accumulation of water vapour and condensation on the insulating material. When the outside temperature rises, the dewpoint plane travels to the outside as well, and condensed water starts to evaporate more easily. This is accompanied by an important increase in volume (gas) and pressure of this gas against the watertight membrane, which causes blisters. In order to evacuate or balance out this gas pressure, a loosely laid roofing felt which has small gravel added on the underside and holes of about 2 cm diameter (see Figure 13) is put on top of the insulation material. When the second layer of felt is applied with hot bitumen, the hot bitumen gets through the holes of the loosely laid felt and adheres to the insulation layer, resulting in a spotmopped adhered felt layer. The only aim

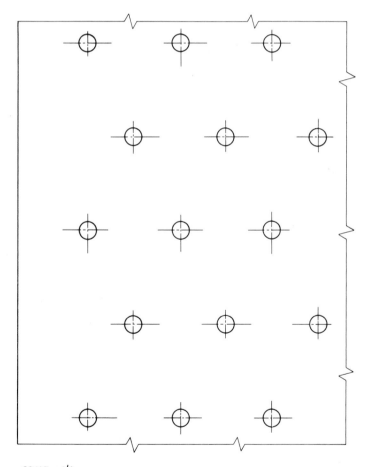

SCALE : 1/4

Figure 13 Spotmopped roofing felt or independent layer

of this layer is to avoid blisters between insulation and built-up roofing, but it does not avoid blistering if humidity or air are trapped between the first and the second or the second and the third roofing felts.

In order to improve this system, small pipes are put on the roof, passing through the built-up roofing and ending in the vapour pressure equilibrating layer (see Figure 12B). This again creates two new important problems:

(a) The joint between pipe and built-up roofing must be made watertight. This always remains a weak point due to the difference in thermal movement between metal and roofing, mainly in winter when the bitumen is cold.

(b) When humidity conditions are such that the actual vapour pressure in the atmosphere is higher than the actual vapour pressure present in the insulation, moisture migrates into the roof.

Notes

When using an insulating material that is not water vapour tight, its volume is normally sufficient to allow vapour pressure to be spread and to avoid blistering.

Some people believe that this system of small pipes allows thermal insulation that became wet to dry out again, which is very doubtful. To make something dry, a stream of nonsaturated air is required.

(i) If we suppose the air passing through to be at 20°C and have 70% relative humidity, it has an absolute humidity of 12 g/m³, and when saturated only 17 g/m³ (see Table 5).

This means that 1 m³ of air is required to evacuate 5 g of humidity, and 1 m³ is, in this case, an enormous volume compared to the thickness of the gravel layer of say 3 mm.

For 100 m² of roof, the total volume is only 100 × 0.003 = 0.3 m³ or 300 litres as an absolute maximum, and when we take off the volume of the gravel (say 50%) there only remain 150 litres.

If the air temperature is lower, conditions are even far worse.

(ii) How can the air in this thin space get to move? We only get an air flow if we have a pressure difference between inlet B1 (Figure 12) and outlet B2. The temperature at B1 and B2 is exactly the same, as both openings are exposed in the same plane under the same conditions. We might only get a temperature difference if a number of these small chimneys is exposed to sunshine and others are situated in the shade; this creates a small pressure difference, but even this is not sufficient to make the air move in the thin layer, as the resistance due to friction is far greater. If we now suppose that the small chimneys are equipped with directable openings, whereby one is put in the direction of the prevailing winds and the other in the opposite direction, the air is blown in with a pressure difference equal to the wind pressure plus 40% of the wind pressure to be added for wind suction on the other side. Still, it is practically impossible to make the air move in the thin layer, due to internal friction.

Our conclusion therefore must be that no real drying-out process can take place.

4. Movement in the insulation layer

When we plot the temperature graph for winter conditions and again for summer conditions, we can calculate the mean temperature in the various layers, and from there the thermal movement for each layer.

When we also consider the size of the insulation blocks or slabs, we can determine the relative movement in the joint, when we also take account of the movement in the supporting layer.

In dealing with plastic foams, this movement should be taken into consideration. When the watertight membrane is fully adhered to the insulation material, we shall see later that the elongation of the roofing felt is far greater than 400%, which can lead to serious damage of the watertight membrane above the joints between insulation slabs owing to conditions of fatigue caused by these continuous movements and accompanying stresses.

Since plastic foams do not withstand high temperatures, the hot bitumen is applied on the roofing felt prior to its application to the insulation layer. It is only adhered when the hot bitumen has cooled sufficiently. This means that the joints between the insulation slabs are not filled with bitumen but remain open; so in this joint area the roofing felt has the minimum resistance.

In winter, when the cooling process starts from the outside, first the bituminous layers are cooled, whereby the bitumen becomes harder and more brittle. As this layer is under stress in these conditions, very small hairline cracks occur in the weakest area above the joints of the insulation slabs. In these cracks the first condensation takes place, as this is the coolest part, together with capillary action.

When the outside temperature goes down further, the freezing point in this area is reached, and the condensed water becomes ice and takes up a 10% greater volume, which means that the hairline cracks get larger. So we do not only have an erosion of our watertight membrane from the outside or topside but also from the underside, which will seriously reduce the effectiveness of the watertight membrane in time. In practice, condensation may already have occurred underneath the watertight membrane when the outside air temperature reaches +10 to +12°C. This outside temperature is met during a longer period over the year in general.

Notes

This danger of fatigue caused by movement will occur less when a spot-mopped independent layer is applied (see Figure 13), as the relative movement is spread over the distance between the spots of bitumen. In this case we are also dealing with an air layer between insulation and watertight membrane. This can never be dry air. Air always has a relative humidity, so the danger of condensation is a real one.

Furthermore, any independent layer is an air layer where water can circulate freely beyond control. In the case of accidental roof puncture, water gets in on one side of the roof and runs freely to the lowest point of the roof structure following the path of minimum resistance. Gradually the total roof becomes saturated and when finally the damage is noticed, it is already too late; repair has become almost impossible without taking down the total roof insulation, watertight membrane, and gravel layer.

The work is obviously considerable, and could be avoided entirely by the use of a totally inert cellular glass insulation which is water vapour tight and watertight.

III. PRACTICAL EXAMPLE 1: compact built-up roof

A. Built-up roof

The built-up roof comprises, from inside to outside:
(1) inside air film,
(2) plaster coat,
(3) reinforced concrete slab,
(4) screed to falls,
(5) Foamglas cellular glass thermal insulation,
(6) multilayer bitumous roofing felt,
(7) gravel, and
(8) outside air film.

The dimensions and L and K values are given in Table 7.

B. Requirements

These are as follows:
(a) required U value = 0.60 W/m^2 °C;
(b) for t_i = 20°C and ϕ_i = 0.50, and for t_o = −5°C and ϕ_o = 0.85, the dewpoint should be in the insulation layer.

C. Procedure

(a) Determine for each known layer its thermal resistance L/K.
(b) Calculate the required $1/U$ value.

(c) Determine the required $1/U$ value to have t_s between screed to falls and Foamglas, for a given outside temperature in winter.

(d) Take the largest required $1/U$ value and determine the corresponding thickness of the thermal insulation; the practical thickness will be the next thicker standard thickness.

(e) Determine the temperature drop across each layer, as well as t_m for each layer.

(f) Plot the t graph across the construction.

(g) Determine for each layer the μ or μs value, and calculate the corresponding ρ value.

(h) Determine inside and outside actual partial vapour pressure, as well as Δp (total pressure drop).

(i) Calculate the actual vapour pressure drop across each layer and the corresponding actual vapour pressure change across the construction.

(j) Determine the dewpoint temperature values corresponding to these vapour pressures.

(k) Plot the t_s curve across the construction.

(l) Check for interstitial condensation.

(m) If condensation occurs, determine the quantity of water that enters during the two-months winter period and the quantity of water vapour that leaves the construction during the three summer months. Make up the balance; normally for safety reasons, twice the quantity that gets in during winter should get out during summer.

D. Solution

(a) We shall fill out Table 7A.

(b) For $U = 0.6$ W/m^2 °C we require a $1/U$ value $= 1/0.6 = 1.667$ m^2 °C/W.

(c) As t_s should be in A (see Figure 14), $t_s = 9.3$°C (direct reading from Table 4), so

$$t_i - t_s = 20 - 9.3 = 10.7°C$$

This should be equal to

$$\Delta t_1 + \Delta t_2 + \Delta t_3 = \Delta t_4$$

where

$$\Delta t_1 = \frac{20 - (-5)}{R_t} \times 0.106$$

$$\Delta t_2 = (25/R_t) \times 0.022$$

$$\Delta t_3 = (25/R_t) \times 0.107$$

$$\underline{\Delta t_4 = (25/R_t) \times 0.055}$$

$$\overline{10.7} = (25/R_t) \times \overline{0.290}$$

Table 7

1 No. of layer	2 Material	3 L (m)	4 k (W/m°C)	5 L/k ((m²°C)/W)	6 Δt (°C)	7 t (°C)	8 t_m (°C)	9 μ	10 s (m)	11 μs (m)	12 N (kJ h/kg m)	13 $\rho = \mu s N$ (N h/kg)	14 Δp (Pa)	15 P_r (Pa)	16 t_s (°C)	17 Remarks
1	Inside surface resistance	—	—	0.106												
2	Plaster coat	0.01	0.46	0.022												
3	Reinforced concr. slab	0.15	1.40	0.107												
4	Screed to falls	0.07	1.28	0.055												
5	Foamglas® T2		0.044	—												
6	BU roofing	0.01	0.19	0.053												
7	Gravel	0.05	0.80	0.063												
8	Outside surface resistance	—	—	0.045												
9																
10																
	Total			0.451												

$P_r = \phi \times P_s$

$P_r = 0.50 \times 2338.4 = 1169.2$ Pa
$P_{ro} = 0.80 \times 401.73 = 321.4$ Pa

$\dfrac{\Delta p}{\rho} = \qquad$ W/m²°C

Inside air temperature $t_i = 20°C$ Relative humidity $\phi_i = 0.50$
Outside air temperature $t_o = -5°C$ Relative humidity $\phi_o = 0.80$
Temperature difference inside–outside

Thermal transmittance $U = \dfrac{\quad}{\quad} = \qquad$ W/m²°C
Actual heat flow $Q = U\Delta t = \qquad$ W/m²
Allowable $\phi_i = \qquad$ (to keep t_s in Foamglas)

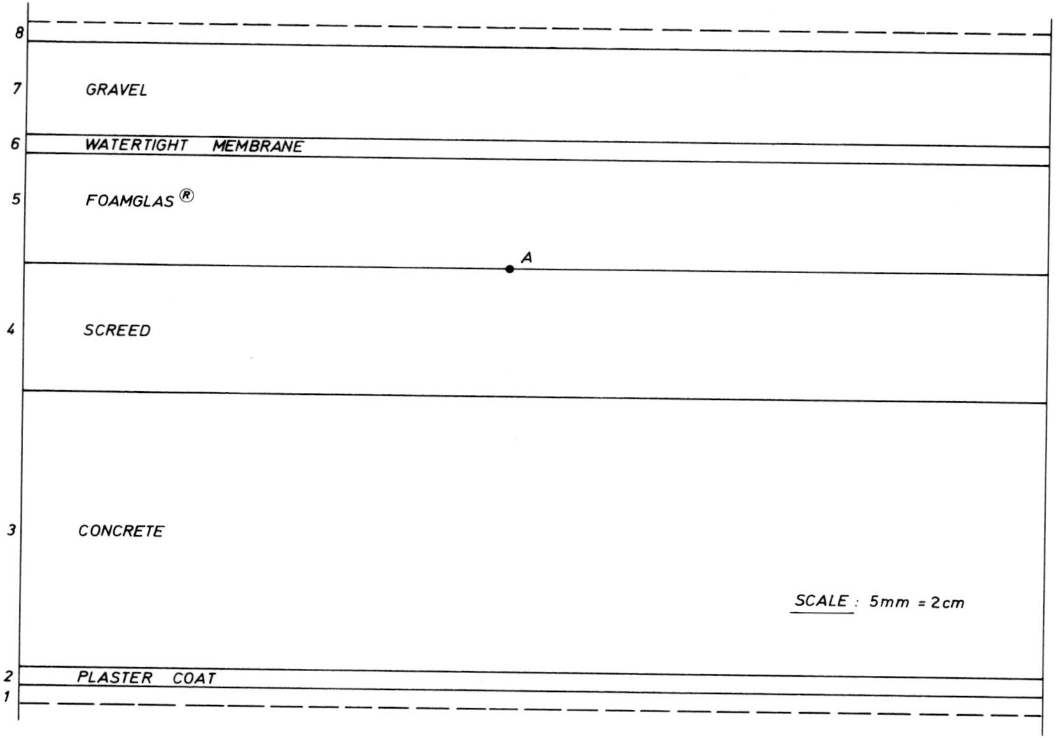

8

7   GRAVEL

6   WATERTIGHT MEMBRANE

5   FOAMGLAS ®

      A

4   SCREED

3   CONCRETE

2   PLASTER COAT

1

SCALE : 5mm = 2cm

Figure 14 Compact roof construction

or

$$R_t = \frac{25}{10.7} \times 0.290 = 0.678 \text{ m}^2 \, °\text{C/W}$$

As this is smaller than 1.667 m² °C/W, the required insulation thickness is dictated by the required U value.

Without insulation, we have a total resistance (Table 7a) of 0.451 m² °C/W. Thus we need to add $1.667 - 0.451 = 1.216$ m² °C/W. Or since $L/k = 1.216$ and as $k = 0.044$ W/m °C, this gives us a thickness value

$L = 0.044 \times 1.216 = 0.054$ m

of Foamglas, which means 55 mm.

(d) We now fill out Table 8. We calculate L/k for Foamglas

$$L/k = 0.055/0.044 = 1.250 \text{ m}^2 \, °\text{C/W}$$

This leads to

$$R = 1/U = 1.701 \text{ m}^2 \, °\text{C/W}$$

and

$$U = 1/1.701 = 0.588$$

From equation (5) we can thus write for $Q = \Delta t/R = U \Delta t$ that

$$Q = 0.588 \times 25 = 14.697$$

This at once our constant factor to calculate the temperature drop in each layer.

(e) We calculate the temperature drop in each layer as

$$\Delta t_n = \Delta t/R \times R_n$$

with Δt_n the temperature drop in layer n, Δt the temperature drop between inside and outside, $R = 1/U$ the total thermal resistance, and $R_n = L_n/k_n$ the thermal resistance of layer n.

Note

As the sum of the temperature drops in the layers here equals 25.1°C due to rounding errors, we correct the largest resistance or temperature drop and write 18.3°C for Δt_5 instead of 18.4°C which was the result of $(25/1.701) \times 1.250$ (this sum always has to be checked).

We determine the mean temperatures in the various layers, except in the air layers.

(f) See t graph of Figure 15

Table 8

No. of layer	Material	L (m)	k (W/ m °C)	L/k (m² °C/ W)	Δt (°C)	t (°C)	t_m (°C)	μ	s (m)	μs (m)	N (kJ h/ kg m)	$\rho = \mu s N$ (N h/kg)	Δp (Pa)	P_r (Pa)	t_s (°C)	Remarks
1	Inside surface resistance	—	—	0.106	1.6	20	—	—	—	—	—	—	—			
2	Plaster coat	0.01	0.46	0.022	0.3	18.4	18.25	6	0.01	0.06	1.441×10^6	0.086×10^6	0.01	1169.20	9.3	
3	Reinforced concr. slab	0.15	1.40	0.107	1.6	18.1	17.30	34	0.15	5.10	1.441×10^6	7.349×10^6	1.02	1169.19	9.3	
4	Screed to falls	0.07	1.28	0.055	0.8	16.5	16.10	20	0.07	1.40	1.446×10^6	2.024×10^6	0.28	1168.17	9.3	P_s for 15.7°C
5	Foamglas® T2	0.055	0.044	1.250	18.3	15.7	6.55	70000	0.055	3850	1.490×10^6	5736.5×10^6	795.46	1167.89	9.3	= 1783.6 Pa
6	BU roofing	0.01	0.19	0.053	0.8	-2.6	-3.00	—	—	240	1.533×10^6	367.92×10^6	51.02	372.43	-5.8	P_s for 20°C
7	Gravel	0.05	0.80	0.063	0.9	-3.4	-3.85	1.2	0.05	0.06	1.537×10^6	0.092×10^6	0.01	321.41	-7.5	= 2338.4 Pa / 1783.6
8	Outside surface resistance	—	—	0.045	0.7	-4.3	—	—	—	—				321.40	-7.5	$\phi_i = \dfrac{2338.4}{1783.6} = 0.76$
9						-5.0										
10	Total			1.701	25.0							6113.971×10^6	847.8			

$P_r = \phi \times P_s$

$\dfrac{\Delta t}{R} = \dfrac{25}{1.701}$

Inside air temperature $\quad t_i = 20°C$
Outside air temperature $\quad t_o = -5°C$
Temperature difference inside–outside $\quad \Delta t = 25°C$

$P_r = 0.50 \times 2338.4 = 1169.2$ Pa
$P_{r_o} = 0.80 \times 401.73 = 321.4$ Pa
Relative humidity $\phi_i = 0.50$
Relative humidity $\phi_o = 0.80$

$\dfrac{\Delta p}{\rho} = \dfrac{847.8}{6113.971 \times 10^6}$

Thermal transmittance $U = \dfrac{1}{1.701} = 0.588$ W/m² °C

Actual heat flow $Q = U\Delta t = 14.7$ W/m²
Allowable $\phi_i = 0.76$ (to keep t_s in Foamglas)

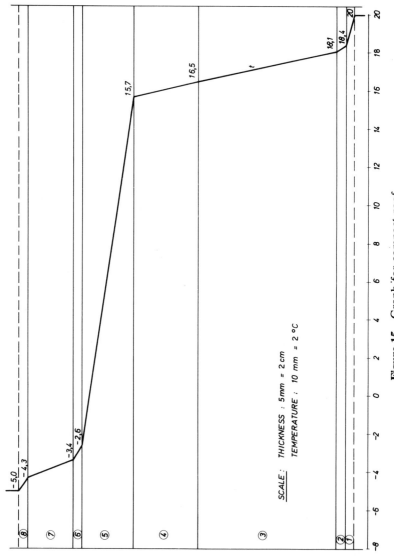

Figure 15 Graph for compact roof

(g) Now we fill out the μ values in Table 8 using the values in Table 9, and the μs values taken from Table 10 with the N values found in Table 6 and corresponding to the mean temperature in each layer and we complete the ρ values.

(h) We now determine P_{ri} and P_{ro}, as well as Δp. We know that $\phi = P_r/P_s$. We find P_s in Table 3 corresponding to t_i and t_o. Then we know that

$$P_{ri} = 0.50 \times 2338.4 \quad = 1169.2 \text{ Pa}$$

and

$$P_{ro} = 0.80 \times 401.73 = \underline{321.4 \text{ Pa}}$$
so
$$\Delta p = \overline{847.8 \text{ Pa}}$$

Table 9 Water vapour diffusion resistance factor 'μ', according to J. S. Cammerer, Germany

Material	Density kg/m^3	μ-value
Brickwall	1600 to 1800	9
Clinker masonry	2000	120
Ceramic tiles		300
Concrete	2300	34
Sandstone	2250	22
Sandlime brick	1900	16
Lime mortar plaster		10
Cement mortar plaster		20
Gypsum mortar plaster		6
Pumice concrete	780	3
Cellular gas concrete		6
Pumice concrete board	900	5
Broken brick concrete		5
Asbestos cement board	1800 to 1850	65
Hard fibreboard		62
Lightweight woodwool building slab		6
Pine wood (4% to 8% humidity by weight)	μ decreases with	170
Beech wood (10% to 50% humidity by weight)	increasing humidity	36
Epoxy resinboard, glassfibre reinforced		200 000
Expanded natural cork	100 to 150	15
Expanded natural cork, bitumen bound	150 to 230	5
Polystyrene foam	15 to 50	40
Polyurethane foam	40 to 150	40
PVC foam	25 to 35	200
Urea formaldehyde foam	12	1.7
Soft woodfibre board		4
Mineral fibre mats and slabs (with organic binders)	300	6
Monoblock (mineral fibre board with mainly organic binders)	220	3
Cellular glass layers (bitumen bound)		70 000

Table 10   Water vapour diffusion resistance: practical μs values (according to Verbia/Wisda, Olten, Switzerland)

(1) Bituminous roofing felts and vapour barriers

Material	Applied with	Thickness, s (m)	μs (m)
F 3 (felt reinforced)	1 layer hot bitumen	0.0030	57
V 60 (glass fibre reinforced)	1 layer hot bitumen	0.0035	80
A 60 (asbestos reinforced)	1 layer hot bitumen	0.0035	80
J 2 (jute reinforced)	1 layer hot bitumen	0.0040	90
F 60 (felt reinforced)	1 layer hot bitumen	0.0040	90
J 3 (jute reinforced)	1 layer hot bitumen	0.0045	100
G 3 (glass fabric reinforced)	1 layer hot bitumen	0.0041	100
J 4 (jute reinforced)	1 layer hot bitumen	0.0051	120
J 5 (jute reinforced)	1 layer hot bitumen	0.0065	150
ALU 10 B (with 0.1 mm Al foil)	1 layer hot bitumen	0.0037	150
ALU 10 B (with 0.1 mm Al foil)	2 layers hot bitumen	0.0052	180

(2) Watertight membranes (BU roofing)

Material	Applied with	Thickness, s (m)	μs (m)
3 × F 3	3 layers hot bitumen	0.009	216
2 × F 3 + 1 × J 2	3 layers hot bitumen	0.010	240
2 × F 3 + 1 × V 60	3 layers hot bitumen	0.009	213
2 × V 60 + 1 × J 2	3 layers hot bitumen	0.010	234
2 × F 60 + 1 × V 60	welded	0.007	158
2 × F 50 + 1 × A 60	welded	0.007	153
3 × F 60	welded	0.0075	158

(3) Plastic foil watertight membranes

Material[a]	Applied with	Thickness, s (m)	μs (m)
PVC	laid loose	0.0008	11
PVC	1 layer hot bitumen	0.0023	43
PVC + 1 × V 60	2 layers hot bitumen	0.0058	123
PIB	laid loose	0.0012	300
PIB + 1 × V 60	1 layer hot bitumen	0.0047	380
HP	1 layer hot bitumen	0.0027	70
HP + 1 × A 60	1 layer hot bitumen	0.0047	120
HP + 1 × V 60	2 layers hot bitumen	0.0062	150
HP	laid loose	0.0008	40
Butyl	laid loose	0.0015	450

[a]PVC = polyvinylchloride; PIB = polyisobutylene; HP = hypalon.

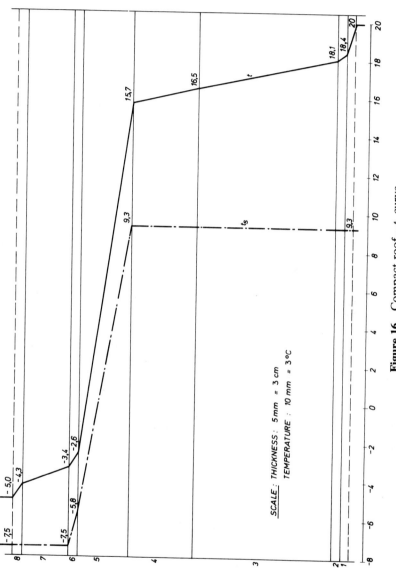

Figure 16 Compact roof—t_s curve

(i) As the vapour flow (from equation (8)) equals $\Delta p/\rho$, and we know that the same flow passes through all layers, we see that

$$\Delta p_n = \Delta p/\rho \times \rho_n$$

where n stands for the nth layer. Thus

$$\frac{\Delta p}{\rho} = \frac{847.8}{6113.971 \times 10^6}$$

is a constant factor that only has to be multiplied by the ρ value of each layer, and leads to the Δp values of Table 8 (column 14); by subtracting the values found from the actual vapour pressure on the inside, we obtain the variation of the actual vapour pressure across the construction.

(j) A dewpoint temperature corresponds to each actual vapour pressure, i.e. a temperature that has that vapour pressure as saturation vapour pressure.

For these temperatures we again use Table 3, but now in the opposite way. We now know the pressure and look for the corresponding temperature. We always take the temperature corresponding to the next highest pressure value.

(k) Now we plot the t_s curve on Figure 15 as shown on Figure 16 for clarity.

(l) As t and t_s curves do not cross each other, there is no condensation.

IV. PRACTICAL EXAMPLE 2: Classical Built-up roof

The reader is referred to Table 11. As the solution is identical to the previous example, we shall only repeat briefly the items that have changed.

(i) $L = 0.034(1.667 - 0.451) = 0.041$ m, and thus we say that the applied thickness must be 0.045 m in order to satisfy both conditions put forward.

(ii) The t and t_s plots across the construction are given on Figure 17.

Conclusion

Condensation occurs although a vapour barrier of good quality with 0.1 mm aluminium foil was applied. Now we have to check the outside temperature at which condensation will occur under the watertight membrane. For this we use a simplified calculation method whereby we do not take the resistance of the various layers against water vapour diffusion into con-

Table 11

1 No. of layer	2 Material	3 L (m)	4 k (W/m°C)	5 L/k (m²°C/W)	6 Δt (°C)	7 t (°C)	8 t_m (°C)	9 μ	10 s (m)	11 μs (m)	12 N (kJ/kg m)	13 $\rho = \mu s N$ (N h/kg)	14 Δp (Pa)	15 P_r (Pa)	17 t_s (°C)	17 Remarks
1	Inside surface resistance	—	—	0.106	1.5	20	—	—	—	—	—	—	0.11	1169.20	9.3	
2	Plaster coat	0.01	0.46	0.022	0.3	18.5	18.35	6	0.01	0.06	1.441×10^6	0.086×10^6		1169.09	9.3	
3	Reinforced concr. slab	0.15	1.40	0.107	1.5	18.2	17.45	34	0.15	5.10	1.444×10^6	7.364×10^6	9.74			
4	Screed to falls	0.07	1.28	0.055	0.8	16.7	16.30	20	0.07	1.40	1.449×10^6	2.029×10^6	2.68	1159.35	9.2	P_s for 15.9°C
5	Vapour barrier or vap. check	—	—	—	—	15.9	15.90	—	—	180	1.450×10^6	261.0×10^6	345.11	1156.67	9.1	= 1806.6 Pa
6	Thermal insulation	0.045	0.034	1.324	18.7	15.9	6.55	40	0.045	1.80	1.490×10^6	2.682×10^6	3.55	811.56	4.0	P_s for 20°C
7	BU roofing	0.01	0.19	0.053	0.7	−2.8	−3.15	—	—	240	1.533×10^6	367.92×10^6	486.52	808.01	4.0	= 2338.4 Pa
8	Gravel	0.05	0.80	0.063	0.9	−3.5	−3.95	1.2	0.05	0.06	1.538×10^6	0.092×10^6	0.12	321.52	−7.5	Allowable ϕ_i $\dfrac{1806.6}{2338.4}$
9	Outside surface resistance	—	—	0.045	0.6	−4.4	—	—	—	—	—	—		321.40	−7.5	$= \dfrac{1806.6}{2338.4}$ = 0.77
						−5.0										
10	Total			1.775	25.0							641.173×10^6	847.80			

$P_r = \phi \times P_s$

Inside air temperature $t_i = 20$°C

Outside air temperature $t_o = -5$°C

Temperature difference inside–outside $\Delta t = 25$°C

$P_r = 0.50 \times 2338.4 = 1169.2$ Pa

$P_{r\,o} = 0.80 \times 401.73 = 321.4$ Pa

Relative humidity $\phi_i = 0.50$

Relative humidity $\phi_o = 0.80$

$$\frac{\Delta p}{\rho} = \frac{847.8}{641.173 \times 10^6}$$

Thermal transmittance $U = \dfrac{1}{1.775} = 0.563$ W/m²°C

Actual heat flow $Q = U\Delta t = 14.1$ W/m²

Allowable $\phi_i = 0.77$ (to keep t_s on cold side of vapour barrier)

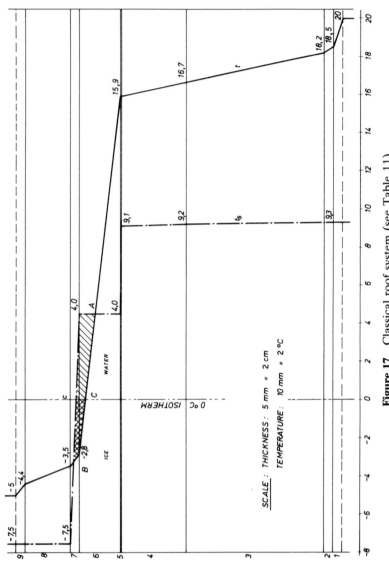

Figure 17 Classical roof system (see Table 11)

sideration. This means that

$$\Delta t_1 = \frac{\Delta t}{1.775} \times 0.106$$

$$\Delta t_2 = \frac{\Delta t}{1.775} \times 0.022$$

$$\Delta t_3 = \frac{\Delta t}{1.775} \times 0.107$$

$$\Delta t_4 = \frac{\Delta t}{1.775} \times 0.055$$

$$\Delta t_5 = \frac{\Delta t}{1.775} \times —$$

$$\Delta t_6 = \frac{\Delta t}{1.775} \times 1.324$$

$$\overline{20 - 9.3} = \frac{\Delta t}{1.775} \times \overline{1.614}$$

or

$$\Delta t = \frac{20 - 9.3}{1.614} \times 1.775 = 11.8°C$$

as

$$\Delta t = t_i - t_o$$

or

$$11.8 = 20 - t_o$$

so

$$t_o = 20 - 11.8 = 8.2°C$$

The period for which we have outside temperatures below 8.2°C is relatively long. So it is not surprising that interstitial condensation is a problem even in those houses and buildings where relative humidity of the air at 20°C is low, as the 50% in our example.

We shall now apply the Glaser method to check the quantity of water vapour (see Figure 18). We choose the scales and plot the graph. As we plot pressure against resistance, we obtain

$$G = \Delta p/\rho \quad (N \ kg/m^2 \ N \ h)$$

which is a flow. When no condensation occurs, we have the same flow of water vapour passing through each layer and of course through the construction. This is given on the graph by the straight line that links up P_{ri}

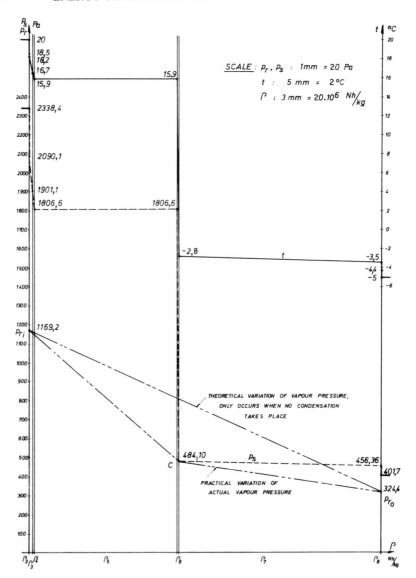

Figure 18   Quantity of water vapour

with P_{ro}. When we have condensation, it is a broken line linking up P_{ri} with C and P_{ro}.

The resistance of the thermal insulation applied (P_6) is so small that the points M and N on the graph of Figure 7 coincide at C. The line $P_{ri} - C$ gives us the flow of water vapour migrating in, and the line $C - P_{ro}$ the flow of water vapour out. Thus we have

$$g_E = \frac{\Delta p}{\rho} = \frac{1169.2 - 484.1}{\rho_2 + \rho_3 + \rho_4 + \rho_5 + \rho_6} = \frac{685.1}{273.161 \times 10^6} \text{ kg/m}^2 \text{ h}$$

or

$$g_E = \frac{685.1}{273.161} = 2.5 \text{ mg/m}^2 \text{ h}$$

and

$$g_o = \frac{\Delta p}{\rho} = \frac{484.10 - 324.4}{\rho_7 + \rho_8} = \frac{159.7}{367.92 + 0.092} = 0.4 \text{ mg/m}^2 \text{ h}$$

and the amount that condenses is

$$2.5 - 0.4 = 2.1 \text{ mg/m}^2 \text{ h}$$

So over a period of two months, we have the amount condensing given by

$$2.1 \times 24 \times 30 \times 2 = 3.024 \text{ g/m}^2$$

Now we have to check summer conditions: $t_i = t_o = 12°C$ and $\phi_i = \phi_o = 0.70$. This leads to the following ρ values:

$$\rho_2 = 0.06 \times (1.467 \times 10^6) = 0.088 \times 10^6$$
$$\rho_3 = 5.10 \times (1.467 \times 10^6) = 7.482 \times 10^6$$
$$\rho_4 = 1.40 \times (1.467 \times 10^6) = 2.054 \times 10^6$$
$$\rho_5 = 180 \times (1.467 \times 10^6) = 264.060 \times 10^6$$
$$\rho_6 = 1.80 \times (1.467 \times 10^6) = 2.641 \times 10^6$$
$$\rho_7 = 240 \times (1.467 \times 10^6) = 352.080 \times 10^6$$
$$\rho_8 = 0.06 \times (1.467 \times 10^6) = 0.088 \times 10^6$$
$$\overline{628.493 \times 10^6 \text{ N h/kg}}$$

Again we plot a Glaser graph, which is now far simpler as only one temperature occurs (see Figure 19). In this case $P_s = 1402.4$ Pa and $P_r = 0.7 \times 1402.4 = 981.68$ Pa.

By diffusion, part of the humidity (condensation) present at C will leave towards the inside of the building (path I) and towards the outside (path II) so,

$$g_I = \frac{1402.4 - 981.68}{\rho_2 + \rho_3 + \rho_4 + \rho_5 + \rho_6} = \frac{420.72}{276.325} = 1.5$$

$$g_{II} = \frac{420.72}{\rho_7 + \rho_8} = \frac{420.72}{352.168} = 1.2$$

$$\text{Total} = 2.7 \text{ mg/m}^2 \text{ h}$$

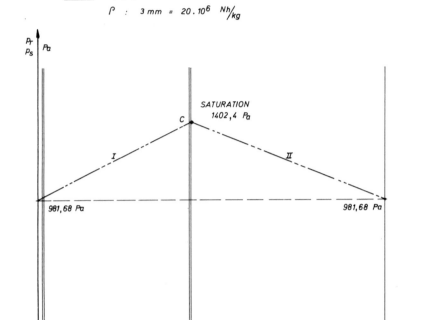

Figure 19   Glaser diagram

In three months we evaluate

$$2.7 \times 24 \times 30 \times 3 = 5.8 \ \text{g/m}^2$$

In theory there is no year-by-year accumulation to be feared in this case. But these conditions can change completely when

(a) a cheaper vapour check is chosen and applied, offering less resistance;
(b) materials with a lower K value than concrete are used on the warm side of the vapour check; and
(c) humidity conditions become more severe on the inside.

On the other hand, we should not forget that condensation occurs in the colder period, the period during which we most need our thermal insulation; if it becomes wet due to condensation, it no longer gives the same efficiency.

V. PRACTICAL COMPARISON

A. Summer–Winter conditions

We shall consider the same inside conditions $t_i = 20°C$ and $\phi_i = 0.50$, but on the outside we propose a practical temperature on top of the watertight membrane of 50°C (which can be reached during sunny periods).

1. Compact roof system

(a) Movement

We shall consider relative movement in joints between Foamglas blocks. Knowing that the thermal coefficient of linear expansion for concrete is $10 \times 10^{-6}\,°C^{-1}$ and for Foamglas $8.5 \times 10^{-6}\,°C^{-1}$ (see Table 12), we shall

Table 12 Coefficient of thermal linear expansion

Steel	11 to 12	$\cdot\,10^{-6}/°C$
Concrete	10	$\cdot\,10^{-6}/°C$
Glass	8 to 9	$\cdot\,10^{-6}/°C$
cellular glass	8.5	$\cdot\,10^{-6}/°C$
Copper	16	$\cdot\,10^{-6}/°C$
Stainless steel	about 16	$\cdot\,10^{-6}/°C$
Aluminium	24	$\cdot\,10^{-6}/°C$
Zinc	29	$\cdot\,10^{-6}/°C$
Plastic foams	60 to 70	$\cdot\,10^{-6}/°C$
Perlite board	20 to 25	$\cdot\,10^{-6}/°C$

determine the relative movement in the joints between Foamglas blocks of 60 cm = 45 cm. For this we calculate the variation in length of the screed and the Foamglas panel, over a winter–summer period (see Figure 20).

$$\Delta L_{Foamglas} = 0.60 \times (37.2 - 6.55) \times 8.5 \times 10^{-6}\,m$$
$$= 0.156\ mm$$
$$\Delta L_{concrete} = 0.60 \times (24.9 - 16.1) \times 10 \times 10^{-6}\,m$$
$$= 0.053\ mm$$

Relative movement in joint = 0.103 mm

This movement is easily taken up by the bitumen in the joint, as well as by the bitumen layer under the Foamglas where the temperature only varies between 15.7 and 25.4°C.

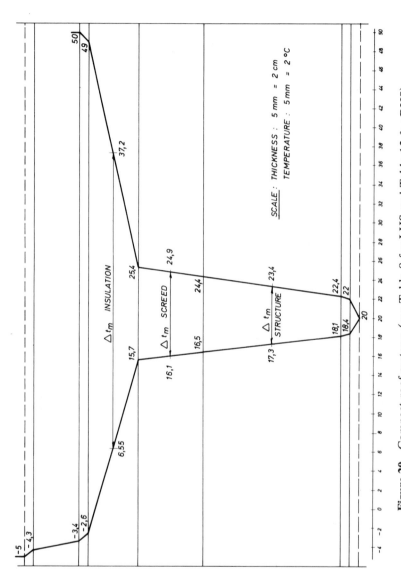

Figure 20 Compact roof systems (see Table 8 for LHS and Table 13 for RHS)

Table 13

1 No. of layer	2 Material	3 L (m)	4 k (W/ m °C)	5 L/k (m² °C/ W)	6 Δt (°C)	7 t (°C)	8 t_m (°C)	9 μ	10 s (m)	11 μs (m)	12 N (kJ h/ kg m)	13 $\rho = \mu s N$ (N h/kg)	14 Δp (Pa)	15 P_r (Pa)	16 t_s (°C)	17 Remarks
1	Inside surface resistance	—	—	0.106	2.0	20										
2	Plaster coat	0.01	0.46	0.022	0.4	22	22.2									
3	Reinforced concr. slab	0.15	1.40	0.107	2.0	22.4	23.4									
4	Screed to falls	0.07	1.28	0.055	1.0	24.4	24.9									
5	Foamglas® T2	0.055	0.044	1.250	23.6	25.4	37.2									
6	BU roofing	0.01	0.19	0.053	1.0	49.0	59.5									
7						50.0										
8																
9																
10	Total			1.593	30.0								847.8			

$P_r = \phi \times P_s$

$P_{r\,i} = 0.50 \times 2338.4 = 1169.2$ Pa
$P_{r\,o} = 0.80 \times 401.73 = 321.4$ Pa
Relative humidity $\phi_i = 0.50$

Relative humidity $\phi_o = 0.80$

Inside air temperature $t_i = 20°C$
Surface temperature on top of watertight membrane $t_o = +50°C$
Temperature difference $\Delta t = 30°C$

Thermal transmittance $U = \dfrac{}{} = $ ———— W/m² °C

Actual heat flow $= U \Delta t = 30$ W/1.593/m²
Allowable $\phi_i =$

(b) Resistance against water vapour diffusion

Although the μ value for Foamglas is infinite, we based our calculations on a μ value of 70 000, found in European literature for Foamglas applied in bitumen. This value corresponds to the μ value of bitumen itself, and means that we only considered the weakest point in the composition of that layer. Still, the μs value for a 4 cm thick Foamglas layer (which is μs = 70 000 × 0.04 = 2800 m) is far higher than the μs value of any vapour barrier (vapour check).

(c) The influence of the joints between blocks

The following two points are relevant:

(i) the ratio of joint area to Foamglas area (see Figure 21); and

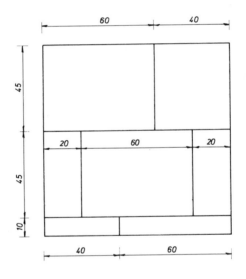

A – METAL DECK ROOF
  WIDTH OF JOINTS   2mm
  FILLED WITH BITUMEN

B – CONCRETE ROOF
  WIDTH OF JOINTS   3 mm
  FILLED WITH BITUMEN

JOINT AREA :   A : 0,2 × 750 = 150 cm²
               B : 0,3 × 750 = 225 cm²

OR : A : 1,5 % BITUMEN   98,5 % FOAMGLAS ®
      B : 2,25 % BITUMEN   97,75 % FOAMGLAS ®

Figure 21   Ratio of joints to Foamglas®

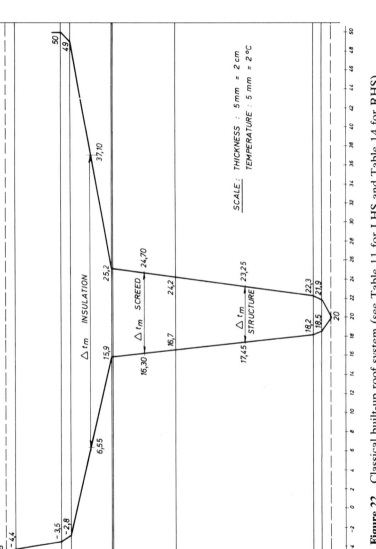

Figure 22 Classical built-up roof system (see Table 11 for LHS and Table 14 for RHS)

(ii) the k value of bitumen (0.16 W/m °C) compared with the k value of wood (0.15 W/m °C).

Both facts explain why the influence of the joints with respect to heat flow and vapour flow can be neglected.

2. Classical Built-up roof

The reader is referred to Figure 22.

(a) Movement

Movement in joint, calculated on 1 m panel length, is

$$\Delta L_{\text{thermal insulation}} = (37.10 - 6.55) \times 60 \times 10^{-6} = 1.833$$
$$\Delta L_{\text{concrete}} = (24.70 - 16.3) \times 10 \times 10^{-6} = \underline{0.084}$$
$$\text{Relative movement in joint (mean value)} = \overline{1.749 \text{ mm}}$$

This movement is far more serious, taking into consideration that an extra movement due to the change of relative humidity takes place as well.

The nature of cellular plastics allows deformation. These movements cause the joint to open in V form, as the thermal changes are even greater at the top side (outside) than on the bottom side (inside). The presence of a slope (falls) makes these materials move to the lowest point, whereby wide open joints occur in the upper parts.

These top parts are the origin of many problems:

(i) extra heat loss due to higher convection;
(ii) more condensation due to underpressure in winter;
(iii) the watertight membrane is no longer supported in this area; and
(iv) because cycling temperatures and important stresses appear in the watertight membrane owing to the constant movement of the support (thermal insulation), signs of fatigue occur over the joint area. The characteristic sign of this is a reduction of the membrane's thickness in this area.

(b) Resistance against water vapour diffusion

As can be seen from the practical example 2, the resistance against water vapour diffusion depends entirely upon the quality, the type, and the μs value of the vapour check, as well as on the care with which it has been put down.

When the joints are too wide, and due to movements as explained in (a), the influence of the joints is more significant. These joint problems are most severe in winter (maximum contraction of the thermal insulation causes the widest gaps in the joints).

Table 14

1 No. of layer	2 Material	3 L (m)	4 k (W/m°C)	5 L/k (m²°C/W)	6 Δt (°C)	7 t (°C)	8 t_m (°C)	9 μ	10 s (m)	11 μs (m)	12 N (kJ h/kg m)	13 $\rho = \mu s N$ (N h/kg)	14 Δp (Pa)	15 P_r (Pa)	16 t_s (°C)	17 Remarks
1	Inside surface resistance	—	—	0.106	1.9	20	—									
2	Plaster coat	0.01	0.46	0.022	0.4	21.9	22.10									
3	Reinforced concr. slab	0.15	1.40	0.107	1.9	22.3	23.25									
4	Screed to falls	0.07	1.28	0.055	1.0	24.2	24.70									
5	Vapour barrier or vap. check	—	—	—	—	25.2	25.20									
6	Thermal insulation	0.045	0.034	1.324	23.8	25.2	37.10									
7	BU roofing	0.01	0.19	0.053	1.0	49.0	49.50									
8						50.0										
9																
10	Total			1.667	30.0											

$P_r = \phi \times P_s$

Inside air temperature $t_i = 20°C$
Surface temperature on top of watertight membrane $t_o = +50°C$
Temperature difference $\Delta t = 30°C$

$P_{ri} = 0.50 \times 2338.4 = 1169.2$ Pa
$P_{ro} = 0.80 \times 401.73 = 321.4$ Pa
Relative humidity $\phi_i = 0.50$
Relative humidity $\phi_o = 0.80$

$\dfrac{\Delta p}{\rho} =$

Thermal resistance $U = \dfrac{\quad}{\quad} = \quad$ W/m² °C
Actual heat flow $Q = U \Delta t =$ W/m²
Allowable $\phi_i =$

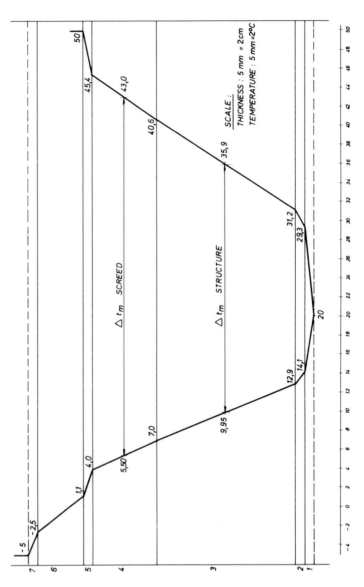

Figure 23 Noninsulated roof (see Table 15 for LHS and Table 16 for RHS)

3. Noninsulated roof

(a) Movement

Movement of structure per 1 m length is given by

$$\Delta L = (35.9 - 9.95) \times 10 \times 10^{-6} = 0.260 \text{ mm/m}$$

which reduces to

$$\Delta L = (23.4 - 17.3) \times 10 \times 10^{-6} = 0.061 \text{ mm/m}$$

for a thermally insulated roof. Thus the movement is reduced to a quarter with respect to the noninsulated roof (see Figure 23).

(b) Condensation

The condensation area is also greater (see Figure 24). The quantity of water vapour migrating in (see Table 17) and Figure 25 is

$$g_1 = \frac{1169.2 - 813.4}{\rho_2 + \rho_3 + \rho_4} = \frac{355.8}{9.704 \times 10^{-6}} = 36.7 \text{ mg/m}^2 \text{ h}$$

The quantity of water vapour migrating out is

$$g_2 = \frac{813.4 - 321.4}{\rho_5 + \rho_6} = \frac{492}{362.011} = 1.4 \text{ mg/m}^2 \text{ h}$$

and the amount that condenses is

$$36.7 - 1.4 = 35.3 \text{ mg/m}^2 \text{ h}$$

(c) Summer conditions

When we consider summer conditions $t_i = t_o = 12°C$ and $\phi_i = \phi_o = 0.70$ as in practical example 2, we can determine the quantity of humidity evacuated by diffusion.

As the ρ values in this case do not alter very much, we can simplify by using the ρ values obtained for the winter period. Taking the vapour pressures as mentioned in example 2 (Figure 19), this leads to:

$$g_I = \frac{1402.4 - 981.68}{9.704} = 43.40 \text{ mg/m}^2 \text{ h}$$

$$g_{II} = \frac{1402.4 - 981.68}{362.011} = 1.16 \text{ mg/m}^2 \text{ h}$$

$$\text{Total} = 44.56 \text{ mg/m}^2 \text{ h}$$

We can already conclude that there is no big danger of a yearly accumulation of humidity, unless circumstances change as mentioned in example 2.

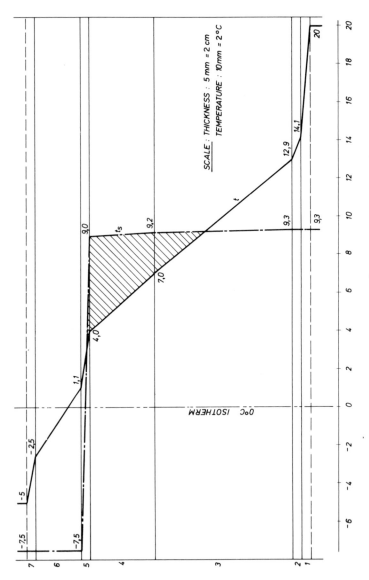

Figure 24 Noninsulated roof (see Table 17 for RHS)

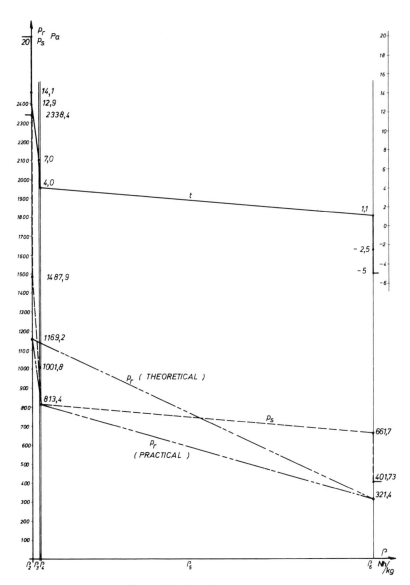

Figure 25 Noninsulated roof

Table 15

1 No. of layer	2 Material	3 L (m)	4 k (W/m °C)	5 L/k (m² °C/W)	6 Δt (°C)	7 t (°C)	8 t_m (°C)	9 μ	10 s (m)	11 μs (m)	12 N (kJ h/kg m)	13 $\rho = \mu s N$ (N h/kg)	14 Δp (Pa)	15 P_r (Pa)	16 t_s (°C)	17 Remarks
1	Inside surface resistance	—	—	0.106	5.9	20										
2	Plaster coat	0.01	0.46	0.022	1.2	14.1										
3	Reinforced concr. slab	0.15	1.40	0.107	5.9	12.9	9.95									
4	Screed to falls	0.07	1.28	0.055	3.0	7.0	5.50									
5	BU roofing	0.01	0.19	0.053	2.9	4.0										
6	Gravel	0.05	0.80	0.063	3.6	1.1										
7	Outside surface resistance	—	—	0.045	2.5	−2.5										
8						−5.0										
9																
10	Total			0.451	25.0											

$P_r = \phi \times P_s$

Inside air temperature $t_i = 20°C$

Outside air temperature $t_o = -5°C$

Temperature difference inside–outside $\Delta t = 25°C$

$P_r = 0.50 \times 2338.4 = 1169.2$ Pa

$P_{r\,o} = 0.80 \times 401.73 = 321.4$ Pa

Relative humidity $\phi_i = 0.50$

Relative humidity $\phi_o = 0.80$

$\dfrac{\Delta p}{\rho} =$

Thermal transmittance $U = \dfrac{1}{0.451} = 2.22$ W/m² °C

Actual heat flow $Q = U\Delta t = 55.4$ W/m²

Allowable $\phi_i =$

Table 16

No. of layer	Material	L (m)	k (W/m °C)	L/k (m² °C/W)	Δt (°C)	t (°C)	t_m (°C)	μ	s (m)	μs (m)	N/N (kJ h/kg m)	ρ = μsN (N h/kg)	Δp (Pa)	P_r (Pa)	t_s (°C)	Remarks
1	Inside surface resistance	—	—	0.106	9.3	20										
2	Plaster coat	0.01	0.46	0.022	1.9	29.3										
3	Reinforced concr. slab	0.15	1.40	0.107	9.4	31.2	35.90									
4	Screed to falls	0.07	1.28	0.055	4.8	40.6	43.00									
5	BU roofing	0.01	0.19	0.053	4.6	45.4										
						50.0										
6																
7																
8																
9																
10	Total			0.343	30.0											

$P_r = \phi \times P_s$

Inside air temperature $t_i = 20°C$ $\qquad$ $P_{ri} =$

Surface temperature on top of watertight membrane $t_o = +50°C$ $\qquad$ $P_{ro} =$

Relative humidity $\phi_i = 0.50$

Temperature difference $\Delta t = 30°C$ $\qquad$ Relative humidity $\phi_o =$

$\dfrac{\Delta p}{\rho} =$

Thermal transmittance $U = \dfrac{}{\rule{1cm}{0.4pt}} = \rule{1cm}{0.4pt}$ W/m² °C

Actual heat flow $Q = U\Delta t = \rule{1cm}{0.4pt}$ W/m²

Allowable $\phi_i =$

Table 17

No. of layer	Material	L (m)	k (W/m°C)	L/k (m²°C/W)	Δt (°C)	t (°C)	tm (°C)	μ	s (m)	μs (m)	N (kJ h/kg m)	ρ = μsN (N h/kg)	Δp (Pa)	P_I (Pa)	t_s (°C)	Remarks
1	Inside surface resistance	—	—	0.106	5.9	20	—	—	—	—	—	—				
2	Plaster coat	0.01	0.46	0.022	1.2	14.1	13.50	6	0.01	0.06	1.461×10^6	0.088×10^6	0.20	1169.20	9.3	
3	Reinforced concr. slab	0.15	1.40	0.107	5.9	12.9	9.95	34	0.15	5.10	1.475×10^6	7.523×10^6	17.16	1169.00	9.3	
4	Screed to falls	0.07	1.28	0.055	3.0	7.0	5.50	20	0.07	1.40	1.495×10^6	2.093×10^6	4.77	1151.84	9.2	
5	BU roofing	0.01	0.19	0.053	2.9	4.0	2.55	240	—	240	1.508×10^6	361.92×10^6	825.46	1147.07	9.0	
6	Gravel	0.05	0.80	0.063	3.6	1.1	-0.70	1.2	0.05	0.06	1.522×10^6	0.091×10^6	0.21	321.61	-7.5	
7	Outside surface resistance	—	—	0.045	2.5	-2.5	—	—	—	—	—	—		321.40	-7.5	
8						-5.0										
9																
10	Total			0.451	25.0							371.715×10^6	847.8			

$P_r = \phi \times P_s$
Inside air temperature $t_i = 20°C$
Outside air temperature $t_o = -5°C$
Temperature difference inside–outside $\Delta t = 25°C$

$P_{ri} = 0.50 \times 2338.4 = 1169.2$ Pa
$P_{ro} = 0.80 \times 401.73 = 321.4$ Pa
Relative humidity $\phi_i = 0.50$
Relative humidity $\phi_o = 0.80$

$$\frac{\Delta p}{\rho} = \frac{847.8}{371.715 \times 10^6}$$

Thermal transmittance $U = \dfrac{1}{0.451} = 2.2$ W/m²°C

Actual heat flow $Q = U\Delta t = 55.4$ W/m²

Allowable $\phi_i =$

4. Insulated roof

An insulated roof will be considered with the thermal insulation applied on the inside of the supporting structure (see Tables 18 and 19, together with Figures 26, 27 and 28).

(a) Movement

Movement of supporting structure is given by $\Delta L = (47.05 + 1.25) \times 10 \times 10^{-6}$ m $= 0.483$ mm/m which is 86% more than the movement of a noninsulated structure.

(b) Condensation

The condensation area has increased compared with the classical insulated roof with the insulation applied on the outside of the supporting structure. Even the $0°C$ isotherm has moved towards the inside of the structure and reaches to the thermal insulation.

The quantity of water vapour condensing in the structure reaches

$$g_1 = \frac{1169.2 - 484.10}{\rho_2 + \rho_3 + \rho_4 - \rho_5 + \rho_6} = \frac{685.10}{272.05} = 2.5 \text{ mg/m}^2 \text{ h}$$

$$g_2 = \frac{484.10 - 321.40}{\rho_7 + \rho_8} = \frac{162.7}{368.252} = 0.4 \text{ mg/m}^2 \text{ h}$$

and the total that condenses is 2.1 mg/m^2 h. As determined in example 2, condensation below the watertight membrane already starts when the outside air temperature reaches in this case $8.2°C$. How many nights a year do we have temperatures equal to, or below, $8.2°C$?

Condensation may seriously increase when inside or outside conditions of temperature and relative humidity change or for the reasons given in example 2.

(c) Construction

This construction does not allow us to use the thermal inertia of the structure. Thermal insulation applied on the inside of the structure should remain an exception and only be used for those buildings that are heated for a few hours per day or per week.

Allowable ϕ_i increased from 0.77 to 0.89. In this case the degree of comfort is reached more rapidly as the mass of the building does not have to be warmed up, due to the fact that the thermal insulation reduces heat flow between inside ambient air and the building structure.

Table 18

1 No. of layer	2 Material	3 L (m)	4 k (W/m °C)	5 L/k (m²°C/W)	6 Δt (°C)	7 t (°C)	8 t_m (°C)	9 μ	10 s (m)	11 μs (m)	12 N (kJ h/kg m)	13 ρ = μsN (N h/kg)	14 Δp (Pa)	15 P_T (Pa)	16 t_s (°C)	17 Remarks
1	Inside surface resistance	—	—	0.106	1.5	20	—	—	—	—	—	—		1169.20	9.3	
2	Plaster coat	0.01	0.46	0.022	0.3	18.5	18.35	6	0.01	0.06	1.441×10^6	0.086×10^6	0.11	1169.09	9.3	
3	Vapour barrier or vap. check	—	—	—		18.2	18.20	—	—	180	1.441×10^6	259.38×10^6	343.44			P_s for 18.2°C
4	Thermal insulation	0.045	0.034	1.324	18.7	18.2	8.85	40	0.045	1.80	1.480×10^6	2.664×10^6	3.53	825.65	4.3	= 2090.1 Pa
5	Reinforced concr. slab	0.15	1.40	0.107	1.5	-0.5	-1.25	34	0.15	5.10	1.525×10^6	7.778×10^6	10.30	822.12	4.2	P_s for 20°C
6	Screed to falls	0.07	1.28	0.055	0.8	-2.0	-2.40	20	0.07	1.40	1.530×10^6	2.142×10^6	2.84	811.82	4.0	= 2338.4 Pa
7	BU roofing	0.01	0.19	0.053	0.7	-2.8	-3.15	—	—	240	1.534×10^6	368.16×10^6	487.46	808.98	4.0	Allowable
8	Gravel	0.05	0.80	0.063	0.9	-3.5	-3.95	1.2	0.05	0.06	1.538×10^6	0.092×10^6	0.12	321.52	-7.5	$\phi_i = \dfrac{2090.1}{2338.4}$
9	Outside surface resistance	—	—	0.045	0.6	-4.4	—	—	—	—	—	—		321.40	-7.5	= 0.89
						-5.0										
10	Total			1.775	25.0							640.302×10^6	847.8			

$P_r = \phi \times P_s$

Inside air temperature $t_i = 20°C$
Outside air temperature $t_o = -5°C$
Temperature difference inside–outside $\Delta t = 25°C$

$P_{ri} = 0.50 \times 2338.4 = 1169.2$ Pa
$P_{ro} = 0.80 \times 401.73 = 321.4$ Pa
Relative humidity $\phi_i = 0.50$
Relative humidity $\phi_o = 0.80$

$\dfrac{\Delta p}{\rho} = \dfrac{847.8}{640.302 \times 10^6}$

Thermal transmittance $U = \dfrac{1}{1.775} = 0.56 \ \text{W/m}^2 \, °C$

Actual heat flow $Q = U\Delta t = 14.1 \ \text{W/m}^2$
Allowable $\phi_i = 0.89$ (to keep t_s on cold side of vapour barrier)

Table 19

No. of layer	Material	L (m)	k (W/m °C)	L/k (m²°C/W)	Δt (°C)	t (°C)	t_m (°C)	μ	s (m)	μs (m)	$\frac{N}{N}$ (kJ h/kg m)	$\rho = \mu s N$ (N h/kg)	Δp (Pa)	P_r (Pa)	t_s (°C)	Remarks
1	Inside surface resistance	—	—	0.106	1.9	20	—									
2	Plaster coat	0.01	0.46	0.022	0.4	21.9	22.10									
3	Vapour barrier or vap. check	—	—	—	—	22.3	22.30									
4	Thermal insulation	0.045	0.034	1.324	23.8	22.3	34.20									
5	Reinforced concr. slab	0.15	1.40	0.107	1.9	46.1	47.05									
6	Screed to falls	0.07	1.28	0.055	1.0	48.0	48.50									
7	BU roofing	0.01	0.19	0.053	1.0	49.0	9.50									
8						50.0										
9																
10	Total			1.667	30.0											

$P_r = \phi \times P_s$

Inside air temperature $t_i = 20°C$

Surface temperature on top of watertight membrane $t_o = +50°C$

Temperature difference $\Delta t = 30°C$

$P_{ri} = 0.50 \times 2338.4 = 1169.2$ Pa
$P_{ro} = 0.80 \times 401.73 = 321.4$ Pa
Relative humidity $\phi_i = 0.50$
Relative humidity $\phi_o = 0.80$

$\dfrac{\Delta p}{\rho} =$

Thermal transmittance $U = \dfrac{\quad}{\quad} = \qquad$ W/m²°C

Actual heat flow $Q = U\Delta t = \qquad$ W/m²

Allowable $\phi_i =$

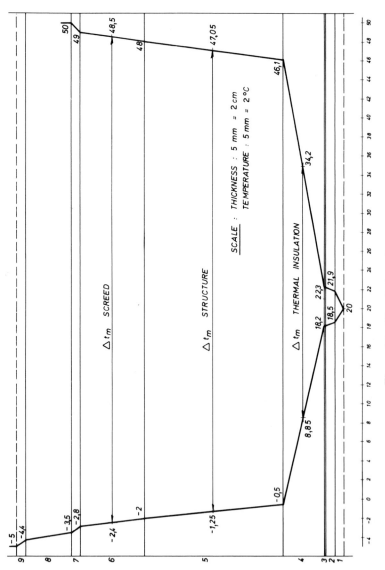

Figure 26 Classical roof system

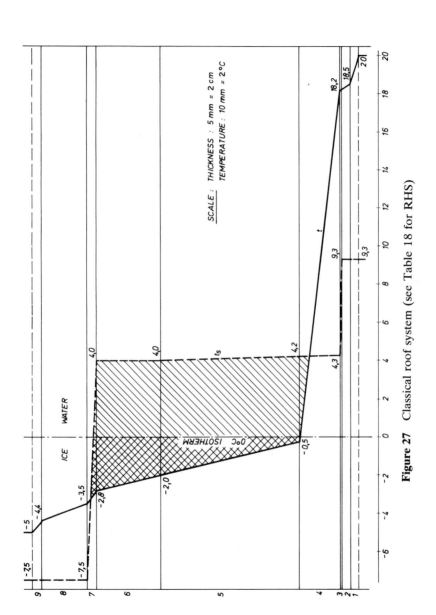

Figure 27 Classical roof system (see Table 18 for RHS)

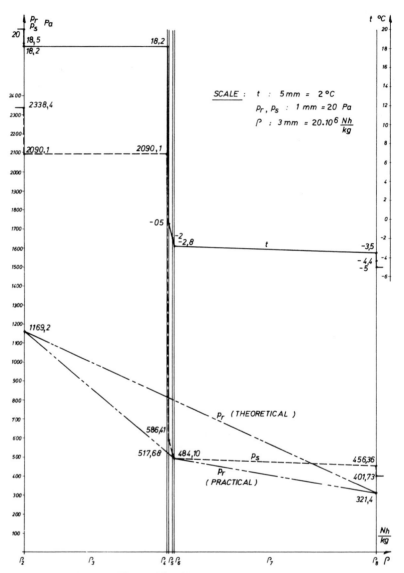

Figure 28 Classical roof system

VI. WHAT DID WE ACHIEVE BY APPLYING THERMAL INSULATION?

A. Energy savings

These are given by considering

Non insulated roof, $U = 2.22$ W/m^2 °C
Insulated roof, $U = 0.59$ W/m^2 °C
Savings $= 1.63$ W/m^2 °C or 1.63 W h/m^2 h°C

The energy consumption over the winter period is determined on the basis of the number of degree days, as given by the Department of Energy (1977).

This number varies with the region, whereby the climatic conditions over the last 20 years are taken into consideration. For example: Thames Valley $= 2022$°D.

Supposing an efficiency of our central heating system of $\eta = 0.80$, we save (only considering heat flow through the roof), as 1° day $= 1$°C $\times$ 24 h, the amount of energy:

$$\frac{1}{0.80} \times 1.63 \left(\frac{W\,h}{h\,m^2\,°C} \right) \times 2022 \times 24(°C\,h) = 98\,876 \left(\frac{W\,h}{m^2} \right)$$

$$= 100 \text{ kWh/m}^2$$

B. Protection

Protection of the supporting structure against temperature variations that occur on the outside are achieved: movement of noninsulated structure is 0.260 mm/m; movement of insulated structure (insulation on the outside of the structure) is 0.061 mm/m; and movement of structure when insulation is applied on the inside is 0.483 mm/m.

C. Improvement of comfort

(a) *Winter Period:* temperature difference between surface temperature and ambient air temperature on inside of building is:

noninsulated roof $= 5.9$°C
insulated roof $= 1.6$°C

(b) *Summer period*: temperature difference is:

noninsulated roof $= 9.3$°C
insulated roof $= 2.0$°C

It is clear that a barracks climate is avoided in summer, as no heat is radiated on the head of the occupants of the top floor.

(c) As less water vapour is lost through the structure when applying gas-tight Foamglas, relative humidity conditions on the inside will improve, whereby fewer air humidifiers are needed.

D. Control of condensation

It is clear that this is a very important item which is not often considered. When we control condensation, we increase the efficiency of the thermal insulation and avoid problems and degradation of building materials, furniture, and expensive machinery and apparatus. We also increase safety, because we have cut the risks of corrosion of electric contacts, which means reduced fire risks due to electrical short circuits.

In the case of a classical built-up roof, we should consider that when humidity is lost the same way as heat, it has to be replaced in order to maintain comfort. When installing air humidifiers, we consume several litres of water per day and per room. Where does all the water go to? A great deal condenses in the structure, as the flow is continuous.

E. Thermal inertia

Use to a maximum the thermal inertia of the mass of the building. This will reduce the influence of temperature variations that occur on the outside on the ambient inside air conditions. This makes the central heating or air conditioning control systems run more smoothly, which means less energy consumption and less wearing out of the installation.

VII. WHAT ARE THE ADVANTAGES OF THE COMPACT ROOF SYSTEM WITH FOAMGLAS CELLULAR GLASS THERMAL INSULATION?

(1) Energy savings keep constant in time. They do not increase due to loss of efficiency of the thermal insulation. The k value of this material is not influenced by condensation nor by the formation of ice crystals.

(2) Glass and also Foamglas have practically the same thermal coefficient of linear expansion as concrete, masonry, and steel. This avoids movement between layers which means that slip layers, where water can flow in a noncontrollable manner, are no longer necessary. This allows us to save material and labour.

(3) It is a dimensionally stable thermal insulation that does not warp, swell, move nor shrink. It keeps its shape and withstands load with-

out a reduction in thickness. It is chemically inert and has a very good chemical resistance. No aging takes place.

(4) As proven before, no vapour barrier is required because bitumen is also water tight, and the movement in the joints is so small that it is taken up by the elasticity of the bitumen without increasing water vapour diffusion.

(5) As only about 2% of the area is joints area against 98% of cellular glass area, the influence of the bitumen-filled joints can be neglected. Bitumen is not considered to form a thermal short circuit, as it has practically the same k value as wood.

(6) For big roof areas, the compact system offers a big advantage in safety. At the end of the day's work, the area covered with Foamglas is dry and stays dry. There are no independent layers where rain water, such as overnight rain showers, may penetrate, wetting the thermal insulation from the very beginning and spoiling the complete roof structure. We should keep in mind that it is very easy to get water in, but it is very difficult to get water out.

(7) Once the roof is finished, an accidental puncture of the watertight membrane does not affect the roof structure. This local damage should be repaired, the cost of such a repair being negligible in comparison with the classical built-up roof, where such damage as accidental punctures cause serious and often irreparable damage.

(8) In the case of fire, even during the application of the watertight membrane with hot bitumen, the supporting structure is and remains protected against high temperature. Foamglas, being inorganic, and thus not combustible, resists fire. There is no admission of oxygen into the joints, and therefore the bitumen in the joints cannot burn.

(9) The compact roof system is the result of the logical analysis of flat roof problems: a rational built-up roof where each layer has its function to play. Such a rational solution to the problems is also an economical solution that pays off in time.

REFERENCES

Department of Energy, 1977, *Fuel Efficiency Booklet* No. 7, HMSO, London.

BIBLIOGRAPHY

Glaser, H. (1958). Temperatur und Dampfdruckverlauf in einer homogenen Wand bei Feuchtigkeitsausscheidung, Kältetechnik 10
Glaser, H. (1959). Graphisches Verfahren zur Untersuchung von Diffusionsvorgängen, Kältetechnik 11
Seiffert, K. (1974). Wasserdampfdiffusion im Bauwesen, Wiesbaden-Berlin Bauverlag G.m.b.H.

Energy Conservation and Thermal Insulation
Edited by R. Derricott and S. S. Chissick
© 1981 John Wiley & Sons Ltd.

CHAPTER 22

The thermal insulation of roofs

L. M. Hohmann

LMH Design, London

1. WHY WE NEED TO GIVE SPECIAL CONSIDERATION TO THE THERMAL INSULATION OF ROOFS

The short answer is that if you can design a roof properly, any other element of building enclosure is a comparatively easy subset of the problems encountered in roofs.

A roof is the most complex element of a building enclosure. Some definitions will immediately illustrate this. A roof is the construction enclosing a building or structure from above. This in turn consists of:

(i) *roofing*, which is the outer weatherproof covering of a roof;
(ii) *ceiling*, which is the surface or covering of the underside of a roof forming the upper enclosing surface of the rooms immediately below; and
(iii) a *structure* which is always necessary between roofing and ceiling to support them and itself.

Already we see that the roof over our head is not a simple issue. Even the simplest form of roof—the tent—where roofing and ceiling are one material, already needs a separate structure (poles, pegs, and guy ropes) to keep it up.

A roof must withstand the highest environmental loadings of any part of the building enclosure both from the outside as well as from within:

from the outside: rain, snow, frost, wind, sunshine, and traffic from maintenance to terraces, roof gardens, and parking decks; look at the temperature range alone—any roofing surface in Britain is subject to a 100°C change in temperature when you consider evaporation and radiation cooling in winter nights (−20°C) and solar gain in summer (+80°C);

from the inside: buildings need to be heated for comfort, and most heat

559

is lost, or could be saved, in the roof; all human activity in buildings generates water vapour, be it in households, offices or through industrial processes; water vapour, one of the gases which make up 'air', is lighter than the common gases (oxygen, nitrogen, carbon dioxide) and will tend to rise to the ceiling.

Both heat and water vapour will naturally tend to escape through roofs in preference to any other direction. All water vapour, without exception, will sooner or later end up again as condensation—let us remember our schoolboy essays on the water cycle. It is the function of design, then, to be able to ensure that this condensation takes place only as 'rain on the plains of Spain', say, and not as condensation in the building fabric.

If design does not achieve this much, then all attempts at conserving energy through insulation will be futile, because nothing converts an insulator into a conductor of heat quicker than its getting wet through condensation, assuming, of course, that we manage to make the roofing leak-proof.

We all know of the effects of not designing specifically for the safe conduct of water vapour from our buildings: the words *condensation* and *mould growth* still make all connected with housing shudder; swimming pool roofs and their problems are just beginning to hit us.

A checklist of the performance requirements for roofs may show how this limitation of my topic relates to all other aspects of roof design: the list is given as a health and safety aspect followed by a list of the requirements to be considered.

A. Structural safety

(1) Strength to sustain and transmit deadweight, imposed, wind, and operational loadings;
(2) deformation, thermal, and humidity effects;
(3) deformation, deflection;
(4) deformation, impact;
(5) corrosion resistance; and
(6) resistance to biological attack.

B. Absence of dampness

(1) Conduct of precipitation, rain;
(2) support of precipitation, snow;
(3) shape, in the light of B(1) and (2) and A(1)–(4);
(4) water permeability;
(5) thermal effects—temperature, radiation;
(6) impact strength;

(7) adhesion—wind uplift control;
(8) corrosion resistance;
(9) thermal movement;
(10) humidity effects;
(11) biological effects;
(12) airflow effects, and
(13) vapour-flow effects.

C. Fire safety

(1) Fire resistance;
(2) ignitability, combustibility;
(3) surface spread of flame;
(4) fire propagation; and
(5) combustion products (smoke, toxic gases)

D. Comfort

(1) Thermal insulation;
(2) thermal capacity;
(3) surface temperatures;
(4) relative humidity;
(5) heat sources;
(6) ventilation;
(7) air permeability;
(8) vapour conduct/condensation risks; and
(9) acoustic properties—airborne and impact sound transmission/absorption.

E. Secondary functions

(1) Surface reflectivity;
(2) transmission of light;
(3) rainwater discharge;
(4) penetrations and fixtures; and
(5) roof terraces, decks, gardens.

I have said that I shall limit myself here to roof insulation free from condensation risks. In the checklist, my point of entry is therefore at D(8), vapour conduct/condensation risks. But not a single one of these checklist headings can be considered in isolation: the performance of a roof is the conjunct of all items in the list. I shall refer to the relevant other points necessary to discuss even just this one aspect as I go along.

One aspect that I shall not discuss further in detail, although it falls clearly within the orbit of the title which, after all, is 'The Thermal Insulation of Roofs', is found under fire safety C(5), combustion products. However, not all insulation materials are combustible, or contain combustible binders. I am sure however that most are aware that the smoke emission, the melting of thermoplastics, and possible toxic gases produced by some plastic foam insulants are causing great concern worldwide, and are receiving the attention of legislatures to tighten the safety regulations in this respect.

2. THE DESIGN PRINCIPLES

Next to the demand for structural stability, the absence of dampness is the most important requirement in the design of buildings. Rain, rising damp, and condensation on surfaces known to be cold—like window panes—are usually catered for in design. These sources of dampness are, however, always localized in time and space. Of greater fundamental importance is the less well understood source of possible dampness in the form of water vapour which migrates continually through all external floors, walls, and roofs of a building.

The most forceful reminder of the existence of this source of dampness was delivered by the number of roofs and walls suffering more from the effects of internal water vapour than from external rain. It is as well to remember that—amongst all other important points of view regarding a home–a house is also a machine for living in: not a very good one if it gets damp.

There are only three basic reasons to which any fault in building is attributable:

(1) faulty *design*;
(2) faulty *workmanship*; or
(3) faulty *use*.

Under *design* is to be understood the disposition of material things in three dimensions, as well as the specification of what these things are and how they are to be fitted together. Under *workmanship* is to be understood the consistent quality of materials as specified, as well as the skills, equipment, and operational organization of assembly, test, and commission. Under *use* it is as well to remember that any design can be made to fail if it is used in a way for which it was never designed. On the other hand, preknowledge of the range of uses is essential design information, as is preknowledge of materials and workmanship possibilities. Of 500 defects examined by BRE, 42% were due to design, 47% were due to workmanship, and 11% were due to unexpected use. Dampness was involved in 50% of all defects. (Building Design, 1974).

The three causes of faults are inseparably linked in two further ways:

(1) in their *operational sequences*, where design extends to include specifi-

cation of materials and workmanship as descriptions, followed by construction which physically assembles what it is intended by design;

(2) in their *logical sequence*, which clarifies the obligation of the designer, because: faultless use of faultless materials applied with faultless workmanship will not correct a faulty design.

Now to the design principles for the thermal insulation of roofs predictably free from condensation risks, pointing to the necessary design tools which exist:

(1) the governing physical factors in enclosure design, especially in housing and specifically for roofs;
(2) the design principles that have evolved through the costly experience of others, along the lines of Oscar Wilde's definition: 'Experience is the name everyone gives to their mistakes';
(3) the use of *temperature diagrams* as a design tool for assessing condensation risks or safety:
  (a) steady state method, and
  (b) dynamic method;
(4) theory in practice:
  (a) comparative long-term flat roof experiments, and
  (b) the cure of condensation problems in pitched roofs.

The avoidance of dampness in buildings—to come back to our common business—requires that all external sources of water are stopped from entering, and that all internal production is conducted away without wetting the fabric.

The internal production of water consists not only of soil or waste water (with which we are not concerned here) but also of water vapour whose safe discharge is equally important—literally: typical household activity produces some 7 kg of water in the form of vapour each day; if drying clothes is taken into account, this figure is around 15–20 kg/day. This is the equivalent of two buckets of water being emptied inside a dwelling every day—but not into any sink or basin! Instead, this internal production of water as vapour forms part of the mixture of gases called air, to which it is added continually within a building, from where it will discharge itself—to the outside air where conditions make it possible, or back into liquid water through condensation if its continued gaseous existence is prevented, for there are limits to the amount of water vapour that can be held by air.

Reference to Figure 1 will give an understanding of the interplay between air temperature, the amount of water that air can carry, and the resultant characteristic known as partial vapour pressure. Condensation is, of course, a ubiquitous fact of life—rain, snow, fog, dew, and frost are all forms of condensation. In that sense, all water vapour will eventually end up as condensation.

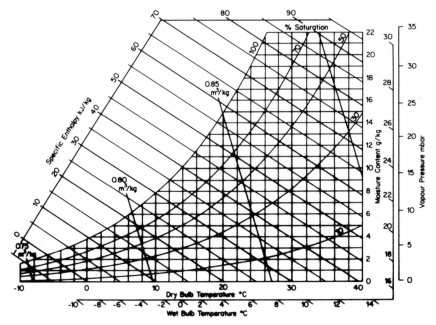

Figure 1 'Know your air: the interrelationships of temperature, relative humidity, water content, partial vapour pressures, and dewpoint temperatures[1].

Building design has to ensure that all the water vapour produced inside condenses only outside. To fulfil this function, three requirements need always be considered:

(1) ventilation for the gaseous transport of water vapour;

(2) heating, which will enable air to carry more vapour; and

(3) building enclosures specifically designed to be free from condensation risks since gases, including water vapour, migrate through practically every known building material other than glass or metal.

The energy which propels this migration comes from the pressure difference between the partial vapour pressures on the two sides of a roof, wall or floor.

Ventilation, heating, and enclosure design are about as interdependent as the legs of a three-legged table in the design against dampness from condensation. This must be remembered, for here I discuss only the principles of enclosure design (exemplified by roofs); the other two 'legs'—heating and ventilation—are referred to where necessary but their design belongs elsewhere.

Roofs are part of the enclosure of a building by which its primary function is established, namely that of shelter of an 'inside' condition to be stabilized within relatively narrow limits for comfort, against 'outside' conditions which vary beyond bearable limits. To illustrate the design principles, factors affecting housing in south-east England are chosen; principles and design method are otherwise valid for other combinations of conditions.

Inside conditions: These are desirable to be held within an air temperature range of 18–23°C at relative humidities of 40–60%. Unless anything else is known to be reliably maintained, inside air at 20°C and 65% RH may be taken for calculation purposes and these are assumed to cover localized temporary excesses of humidities in kitchens, bathrooms or similar rooms.

Outside conditions: Design temperatures of −1°C in winter are assumed for outside air, and summer air at 25°C. But air temperatures alone are misleading; radiation losses and gains are of overriding importance in roof design. In winter, surface temperature may fall to 10°C below air temperature on clear nights, while in summer surface temperatures of 80°C can be reached. Outside air is assumed for design purposes to be at −1°C and 80% RH.

Apart from considerations of heat loss or gain and vapour migration produced by these differences in humidities and temperatures, the effects of thermal movements produced by the variation in surface temperatures must not be overlooked.

In flat roofs, the stability of the base on which the waterproofing rests and to which it is usually bonded is particularly important. In the illustration (Figure 2) the difference that insulation can make to stabilize the base against temperature movements is highlighted.

Basic design principles can now be stated for the more critical winter conditions:

(1) where the production of water vapour is continuous, ventilation and heating must be continuous for its safe conduct to the outside;
(2) to keep the cost of heating tolerable, insulation must be provided;
(3) insulation should be provided on the outside of the structure to reduce its thermal movement and to make use of its thermal capacity as heat store;
(4) vapour migration affects all parts of the enclosure–its safe conduct must be ensured to avoid condensation problems; and
(5) as the thermal transmittance (*U* value) of an enclusure is not related to its vapour migration characteristics, it is meaningless as an indicator of condensation risks.

With these basic considerations in mind, there are only three known principles of roof design that can be regarded as safe:

(1) the 'cold roof' principle;
(2) the 'warm roof' principle; and
(3) the 'inverted roof' also known as the 'protected membrane roof' principle.

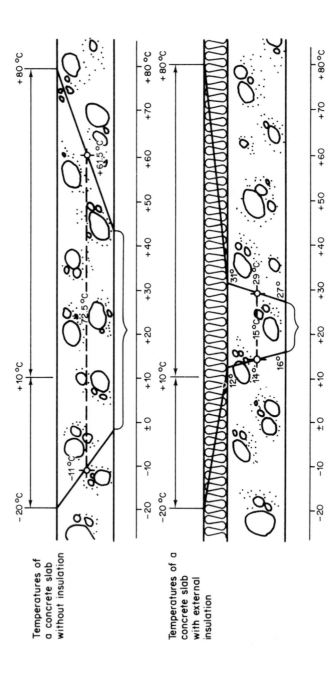

Figure 2 The influence of insulation on the thermal movement of structures. (Reproduced by permission of Esser, K. G.)

As design principles they are equally valid for any part of an enclosure—roof, wall or floor. Definitions of the three principles are given now, together with brief descriptions of how they satisfy safe conduct of water vapour migration. A worked example of a temperature diagram for each of these principles will later both explain the working of each principle more fully together with an indication of how they are translated into materials and detail design.

A. 'Cold Roof' design principle

A cold roof is a roof that has an outer skin providing weatherproofing functions separated from an inner skin which provides for insulation and structural functions by an airspace that is effictively ventilated by outside air. All insulation calculably required to keep structural movement within tolerable limits and to provide any desired standard of insulation must be provided on the outer side of the inner skin over which outside air conditions must be made to prevail. The inner skin must be known to be free from any condensation risk as may be indicated by a diagram of actual and dewpoint temperatures.

The most common form of the 'cold roof' is, of course, the traditional tiled roof which guides the rain down the slope while outside air blows relatively freely through the many joints in the outer, tile skin, with the roof space practically always at outside air condition. Felts or even plastic foils have been introduced under roofing tiles as a second defence against windblown rain and snow. Wherever this is the case, the cold roof principle is broken and the roof becomes a certain condensation trap unless air circulation in the roof space is specifically provided by other means (see Figures 3–8).

A warning, therefore: not every roof with an airspace is a cold roof. The criterion is the effective ventilation of the airspace (see Seiffert, 1970).

The openings provided for ventilation should be as large as possible; they can never be too large. There is not a single case on record where a

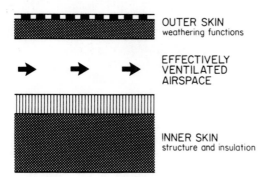

OUTER SKIN
weathering functions

EFFECTIVELY
VENTILATED
AIRSPACE

INNER SKIN
structure and insulation

Figure 3 The cold roof design principle

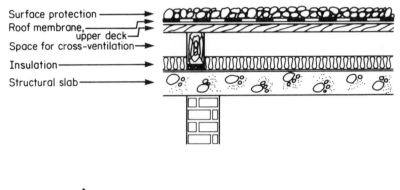

Surface protection
Roof membrane,
          upper deck
Space for cross-ventilation

Insulation

Structural slab

Figure 4   Cold roofs can be of any shape. (Reproduced by permission of Esser, K. G.)

Figure 5   Cold roof—extreme solution for an exhibition stand. (Reproduced by permission of Friedr. Vieweg & Sohn)

Figure 6 Platform building—Waterloo Station: easily a perfect cold roof

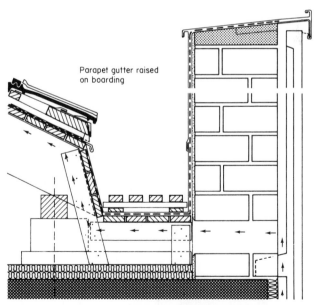

Parapet gutter raised
on boarding

Figure 7 Design detail must incorporate design principle: gutter detail for a cold roof with specific provision for ventilation. Note also the 'cold wall'. (Architect: Klans Göbel)

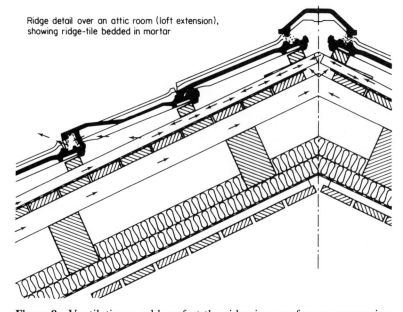

Ridge detail over an attic room (loft extension),
showing ridge-tile bedded in mortar

Figure 8   Ventilating a cold roof at the ridge in a roof space conversion. Note that internally, the void above the suspended ceiling—the internal boarding—is part of the internal air through keeping all joints open. The membrane under the insulation is the *wind control* and need not act as a vapour barrier; in fact its vapour resistance should be as *low* as possible while separating internal from external air successfully. Special provision is made to ventilate the cold roof void at its highest point with discharge through a row of special vent tiles either side of ridge. (Architect: Klans Göbel)

cold roof has suffered from too much ventilation. As a rule, 17 cm²/m₂ of roof area in plan should be provided along two opposite sides of the roof and in addition, along the ridge 20 cm²/m² of roof area should be provided, all as evenly spaced as possible over the whole length of the roof.

B. 'Warm Roof' design principle

A warm roof is a roof with an outer weatherproofing layer which is not separated from other parts of the enclosure sandwich by an effectively ven-

WEATHERING
INSULATION
VAPOUR BARRIER

STRUCTURE

Figure 9   The warm roof design principle

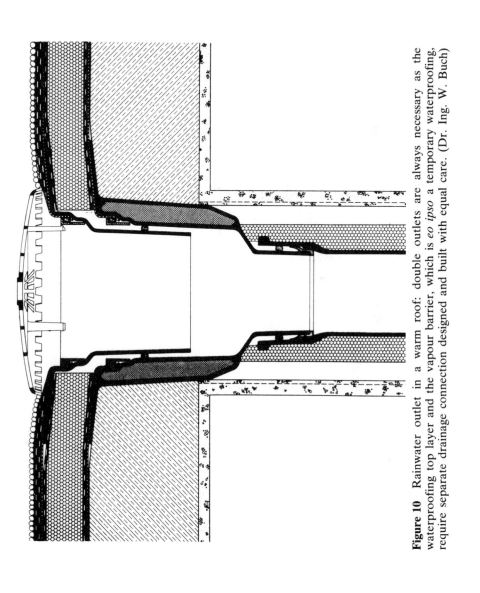

Figure 10 Rainwater outlet in a warm roof: double outlets are always necessary as the waterproofing top layer and the vapour barrier, which is *eo ipso* a temporary waterproofing, require separate drainage connection designed and built with equal care. (Dr. Ing. W. Buch)

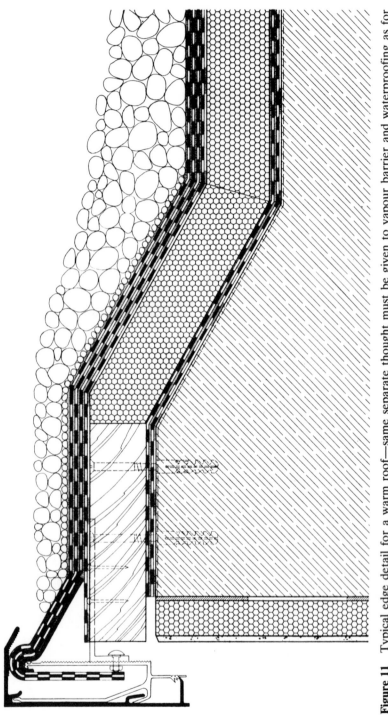

Figure 11 Typical edge detail for a warm roof—same separate thought must be given to vapour barrier and waterproofing as for outlets. (Dr. Ing. W. Buch)

tilated airspace. Extremely dry and nonabsorbent insulation must be used, and kept dry during installation, which must be protected against vapour ingress on the warm side by a vapour barrier. The amount of insulation required is determined by the acceptable thermal movement of the structure, by the desired degree of thermal insulation, both of which may be overriden by the necessity to keep the whole of the enclosure sandwich free from condensation risk as may be determined by a diagram of actual and dewpoint temperatures (see Figures 9–11).

In a warm roof there is no ventilation to give safe conduct to any water vapour that penetrates through the structure and insulation. Waterproofing layers used in flat roofs are highly vapour-resistant outer skins allowing only extremely small quantities of vapour to escape. The only way to interpret the safe conduct rule here is to ensure that even less water vapour can every migrate towards this outer skin than it is capable of letting through. This is the specific role of vapour barriers. As a rule of thumb, the vapour barrier layer on the warm side of the insulation should provide three to four times as much theoretical resistance to vapour migration as the outer skin, to allow for inevitable flaws in the best of workmanship.

The warm roof design principle makes the highest demands on design knowledge and detailing skill, as well as demanding meticulous workmanship in dry conditions, plus exceptional supervision. Every sin will come home to roost sooner or later (typically between five and eleven years) and this design prinicple has no built-in recovery mechanism.

With the recent arrival of dynamic calculation methods by easily performed graphic means the warm roof principle is currently undergoing reappraisal. The climatic data both for internal and external conditions throughout the year need to be known reliably in any case before a dynamic assessment can be made. Therefore, if the summer recovery cycle can be relied on to remove with a considerable margin of safety all of the winter condensate, then vapour barriers may become extinct. But the effect of condensation on the insulant, mainly the lowering of the U value, also needs to be allowed for.

C. The 'Protected membrane roof' design principle (inverted roof)

The protected membrane roof consists, from inside to out, of ceiling, structure, weatherproofing, insulation, and wind-uplift control. The insulation has to remain effective as insulant under the effects of weather. The amount of insulation required is determined by the tolerable thermal movement of the structure, by the desired degree of insulation, and by the necessity to keep the structure free from condensation risks as may be determined by a diagram of actual and dewpoint temperatures.

The protected membrane roof is the most promising development in flat

INSULATION
WEATHERING

STRUCTURE

Figure 12 The inverted roof design
principle

roof design and construction. There is now several years' experience of these roofs in Hungary, Germany, Switzerland, Canada, and the USA. Several systems are undergoing trials with the Greater London Council, not only in regard to new installations but also as a means of correcting faulty flat roofs of the pseudo-dry-screed variety (Figures 12–18).

The advantages of this flat roof design principle appear very convincing:

(1) Not only the structure but also the waterproof membrane is insulated against temperature fluctuations.
(2) The membrane is protected against thermal shock, solar degradation, and mechanical damage.

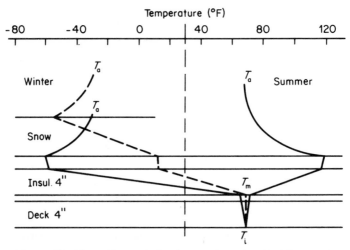

Figure 13 The effect of the inverted roof principle on the temperature fluctuations of the roof membrane—the results of the Alaska experiments of the US Army Corps of Engineers: unprotected membranes were previously exposed to temperature fluctuations between −60°F and +120°F (a range of 180°F or 100°C). This is the same temperature difference as here, but in a far lower temperature band (mean 0°C; UK mean +30°C). Before adopting the inverted roof principle, the US Army experienced something like 75% roof membrane failures after two years of installation

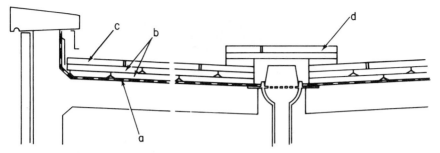

Figure 14 General arrangement of a protected membrane roof as practised by CRREL (a) Waterproof membrane; (b) Insulation boards; (c) Pavers for protection; (d) Drain cover

(3) The membrane nevertheless remains easily accessible for inspection.

(4) The amount of insulation may be increased at any time without disturbance of function or use.

(5) The membrane need not be bonded to the deck except at flashings and is therefore free from stresses by substructure movements.

(6) The wind-uplift defence consists of weights to hold the roof components down, either in the form of gravel or of concrete slabs whose thickness can be chosen to suit local wind conditions, easily increased around the perimeter.

(7) Construction is less dependent on favourable weather conditions. There is no danger of accumulating moisture in the roof that becomes sealed-in.

(8) The roof is self-drying both upwards and downwards.

(9) The protective upper surface—no longer a fragile waterproof skin—offers greater flexibility in the design and opportunity for use.

(10) It provides the only safe means by which the insulation of existing flat roofs may be increased.

Mention must briefly be made again of heating and ventilation, for even with enclosures known to be free from condensation risks within their own terms of reference, ventilation and heating must be permanent throughout the enclosure of a building at calculable minimum levels to avoid condensation. Personal choice of comfort level by the user can only start above this minimum level determinable by the design requirement to avoid dampness from condensation. The necessary level of this background heating requires that increasing attention is paid to the economics of insulation.

But increased standards of insulation will remain economical only if vapour migration takes place in accordance with the rules of safe conduct out-

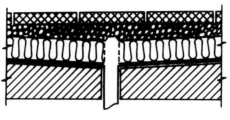

(a) Gravel drainage and levelling layer

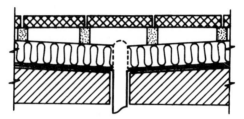

(b) Traffic surface on supports

Open Joints of surface Paving

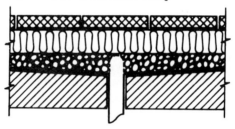

Drainage and Levelling Layer of Gravel
under the Insulation

Figure 15   Inverted roofs: general arrangements of falls, insulation, and wind uplift control. (Reproduced by permission of Division of Building Research, National Research Council of Canada)

lined here. Nothing will convert an insulator into a conductor of heat quicker than dampness; the first step in its prevention is preventive design.

3. THE USE OF STEADY-STATE TEMPERATURE DIAGRAMS IN DESIGN

All design amounts to a prediction that whatever is constructed in accordance with it will withstand all use loadings. The existence of condensation problems is reason enough to call for these predictions in enclosure design

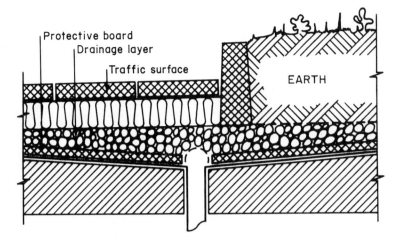

Roof Terraces with Traffic Surfacing or Landscaping

Figure 16 Inverted roofs: general arrangement of roof gardens and terraces. (Reproduced by permission of Division of Building Research, National Research Council of Canada)

Figure 17 Place Bonaventure, Montreal: a general view. This is a hotel with internal roof gardens and terraces over exhibition centre and shops. Inverted roofs were used

Figure 18   As Figure 17, showing view of roof gardens (following the general arrangement as shown in Figure 16)

to be easily calculable. Temperature diagrams are a form of such calculations which can be done quickly by a simple graphical method.

Temperature diagrams assume 'steady-state' conditions, i.e. an unvaried duration of the chosen design assumptions of environmental conditions lasting long enough for all parts of the enclosure to reach that balanced state shown by any steady-state calculation. This is a considerable simplification of the variations of real conditions. But what is it we ask of a design tool? No more than that it should be as quick and simple to use as possible (or it will not be used at all—which is, of course, why there are so many problems with roofs) while still giving safe but not wasteful predictions of design safety within known limits.

In the design of building enclosures, steady-state conditions can quite realistically be assumed especially with respect to housing, since during the critical heating season household activities will always produce internal temperatures and vapour pressures substantially higher than outside air.

Apart from heating and ventilation to ensure comfort, health, and the removal of water vapour, enclosures which usually consist of layers of differing materials in form of a sandwich must in themselves be free from condensation risks, not only on the surface but throughout the entire thickness of the sandwich. Diagrams of actual and dewpoint temperatures through such sandwiches are a means of making the existence or absence of condensation risks visible at the design stage.

For calculation purposes, the winter design conditions are taken as:

outside air, $-1°C$ at 80% RH
inside air, 20°C at 65% RH

In summer, outside surface temperatures on roofs may rise to 70°C or more, and in winter they may fall to $-12°C$ or more through radiation losses on clear nights. We are all aware that such variations in temperature are the main source of stress on any external roof membrane; effects on thermal movements have already been mentioned.

In regard to vapour migration through enclosures, the design assumptions chosen are assumed to cover temporary lower surface temperatures; considering night radiation losses, however, highlights the false economy of not heating at night if the structure is allowed to cool below dewpoint.

The principles governing the movement of water vapour through a material are the same as those that govern the movement of heat through a material; the calculation methods in both cases therefore follow the same pattern. Anyone able to do a U value calculation for the transmittance of heat can do a U_v value calculation for the transmittance of water vapour in exactly the same way, as the comparison of definitions, symbols, and basic equations in Table 1 shows.

In order to produce a temperature diagram with the relations given in Table 1, we need facts and figures as to the characteristics of moist air (see Figure 1) as well as to the characteristics of the materials proposed for use, namely their thermal resistivities, readily available for example in IHVE (1970), and their vapour resistivities (see Table 2).

Instead of using the actual resistivities to vapour migration of the various materials or of air (with its cumbersome units of GN s/kg), it is simpler to say that the resistivity of air is the standard unit of measurement, give it the dimensionless symbol μ, and conduct all calculations in air-equivalent vapour resistances R_{va} with the simple unit of m for metres.

The 'air-equivalent vapour resistivity' or μ value of a material states how many times more resistant it is than a layer of air of the same thickness (... and at the same temperature and barometric pressure, although these can be ignored for most practical purposes; material properties and workmanship are too variable in building to call for too much precision).

The best way to explain the usefulness of diagrams of actual and dewpoint temperatures (t and t_d diagrams) as a design aid is perhaps through worked examples, taking the three design principles—cold, warm, and inverted—in turn.

Table 1 Definitions, symbols, and basic equations used for calculating U_v values in the same way as U values, and relationships for sandwich constructions

	Symbol	Units	Relationship
(a) HEAT			
Heat		J	$1\text{ J} = 1\text{ N m}$ ($1\text{ N} = 1\text{ kg m/s}^2$)
Heat flow		W	$1\text{ W} = 1\text{ J/s}$
Temperature	t	°C	
Dewpoint temperature	t_d	°C	
Outside temperature	t_o	°C	
Inside temperature	t_i	°C	
Temperature difference	Δt	°C	$t = t_o - t_i$
Thermal conductivity	k	W/m °C	
Thermal resistivity .	r	m °C/W	$r = 1/k$
Material thickness	L	m	
Thermal resistance	R	m² °C/W	$R = rL = L/k$
Surface resistances			
outer	R_{so}	m² °C/W	
inner	R_{si}	m² °C/W	
Thermal conductance	C	W/m² °C	$C = 1/R$
Thermal transmittance	U	W/m² °C	$U = 1/R + 1/R_{si} + 1/R_{so}$
(b) VAPOUR			
Vapour		kg	(of water)
Vapour flow		kg/s	
Vapour pressure	p	N/m²	
Outside vapour pressure	p_o	N/m²	
Inside vapour pressure	p_i	N/m²	
Vapour pressure difference	Δp	N/m²	$p = p_o - p_i$
Vapour conductivity	k_v	kg m/N s	
Vapour resistivity	r_v	N s/kg m	$r_v = 1/k_v$
Material thickness	L	m	
Vapour resistance (1)	R_v	N s/kg	$R_v = r_v L = L/k_v$
Surface effects	—	—	(negligible)
Air-equivalent vapour resistivity	μ		
Vapour resistivity of air	$r_{v(air)}$	GN s/kg m	$r_{v(air)} \simeq 5$
Vapour resistance (2)	R_v	GN s/kg	$R_v = r_{v(air)} \mu L$
Air-equivalent vapour resistance	R_{va}	m	$R_{va} = \mu L$
Vapour resistance (3)	R_v	GN s/kg	$R_v = r_{v(air)} R_{va}$
Vapour transmittance	U_v	kg/GN s	$U_v = 1/R_v$

Sandwich constructions (n-layer sandwich)

(a) HEAT

Resistance
$$R = R_{si} + L_1/k_1 + L_2/k_2 + \cdots + L_n/k_n + R_{so}$$
$$R = R_{si} + R_1 + \cdots + R_n + R_{so}$$

Transmittance
$$U = 1/R$$

(b) VAPOUR

Resistance
$$R_{va} = \mu_1 L_1 + \mu_2 L_2 + \cdots + \mu_n L_n$$
$$R_v = r_{v(air)} R_{va}$$

Transmittance
$$U_v = 1/R_v$$

Table 2 Some μ values

Material	Density (kg/m³)	μ value	Thickness, L (m)	R_{va} value (m)
Motionless air	~1.19	1[a]	1.000[a]	1.000[a]
Gypsum plaster	1600	5		
Aerated concrete	800	5		
Brickwork	1700	8	(any)	(μL)
Dense concrete	2300	28		
Asbestos cement	1800	60		
Granite marble	3000	∞		
Resinbonded plywood:				
8 mm	650	73	0.008	0.574
20 mm	650	128	0.020	2.560
Glass wool, mineral wool	200	1.2	0.050	0.060
Wood-wool slabs	390	5	0.050	0.250
Cork board	100	5	0.050	0.250
Exp. polystyrene	16	25	0.050	1.250
	25	40	0.050	2.000
Foamed polyurethane	30	60	0.050	3.000
50 mm cellular glass slabs bedded and jointed in hot bitumen		70000	0.050	3500.000
Bitumen priming		800	0.001	0.800
Roofing felt (good quality)		3000	0.0012	3.600
Asphalt, 20 mm		1000	0.020	20.000
1 × felt with 2 × hot bit.		10000	0.002	20.000
2 × felt with 3 × hot bit.		10000	0.004	40.000
3 × felt with 3 × hot bit.		10000	0.008	80.000
0.2 mm aluminium foil		700000	0.0002	140.000
Polyisobutylene, 2.5 mm		360000	0.0025	900.000
Glass, metal (in sheet form)		∞		∞

[a]The definition of μ.

A. Cold roof

With outside air conditions operant in the airspace, insulation in the outer skin is unnecessary. Conversely, the inner skin must contain all the required (or desired) insulation and must in itself be free from condensation risk. Only the inner skin needs to be considered as to condensation safety.

(i) Design assumptions

The inside conditions are:

$$t_i = 20°C \text{ at } 65\% \text{ RH has}$$
$$p_i = 1500 \text{ N/m}^2$$
$$t_{di} = 13.5°C$$

The conditions in the airspace and outside are:

$$t_o = -1°C \text{ at } 80\% \text{ RH has}$$
$$p_o = 450 \text{ N/m}^2$$
$$t_{do} = -3.7°C$$

(ii) Material assumptions

We have 25 mm expanded polystyrene (subscript 2) on top of 150 mm dense concrete (subscript 1) and the following values apply:

$$k_1 = 2.05 \text{ W/m °C} \quad k_2 = 0.041 \text{ W/m °C} \quad R_{si} = 0.106 \text{ m}^2 °C/W$$
$$L_1 = 0.150 \text{ m} \quad L_2 = 0.025 \text{ m} \quad R_{so} = 0.045 \text{ m}^2 °C/W$$
$$\mu_1 = 28 \quad \mu_2 = 25$$

(iii) Temperature distribution

From Table 1 we have

$$R = R_{si} + R_1 + R_2 + R_{so}$$

where

$$R_{si} = 0.106 \text{ m}^2 °C/W$$
$$R_1 = L_1/k_1 = 0.150/2.05 = 0.073$$
$$R_2 = L_2/k_2 = 0.025/0.041 = 0.610$$
$$R_{so} = 0.045$$

and so we obtain

$$R = 0.834 \text{ m}^2 °C/W$$

(iv) Vapour pressure distribution

We have

$$R_v = R_{v1} + R_{v2}$$

and

$$R_v = r_{v(air)} \mu L$$

so we obtain

$$R_v = r_{v(air)}\mu_1 L_1 + r_{v(air)}\mu_2 L_2 \text{ (GN s/kg)}$$

$r_{v(air)}$ is the same constant in every layer. We know both ends of the dewpoint temperature curve in advance: t_{di} and t_{do} are determined by the initial assumptions of design temperatures and humidities. The distribution between the two endpoints is therefore determined only in proportion to the various products μL of the sandwich; multiplying each product by a constant factor will not alter these proportions. (Where the amount of water vapour transported through the enclosure needs to be known, multiplying the sum of all μL values by the $r_{v(air)}$ constant will easily give it.) Usually, the products μL are all we need to know, i.e. we can work in air-equivalent layer thicknesses measured in metres rather than use the complex vapour resistance measured in giganewton-seconds per kilogram. The resistance we deal in then becomes the air-equivalent vapour resistance R_{va}:

$$R_{va} = \mu L \text{ (m)}$$

This convention makes calculation of the vapour pressure distribution even simpler than that of heat distribution:

$$R_{va} = R_{va1} + R_{va2}$$

Thus

$$R_{va1} = \mu_1 L_1 = 28 \times 0.150 = 4.200 \text{ m}$$

$$R_{va2} = \mu_2 L_2 = 25 \times 0.025 = 0.625$$

give us

$$R_{va} \qquad\qquad = 4.825 \text{ m}$$

With these thermal and vapour resistances, the 't and t_d diagram' can be drawn as shown in Figure 19 (using Figure 1 to translate vapour pressures into dewpoint temperatures after their proportional distribution has been established by help of the air-equivalent vapour resistances R_{va}).

Figure 19 shows the distribution of acutal temperature—line (t)—at all points throughout the sandwich higher than the dewpoint temperature—line (t_d)—which indicates that, given the airspace conditions are provided as assumed by 'effective ventilation', no condensation can occur when the inside conditions are held reasonably stable at 20°C and 65% RH. When altering any design assumption one at a time, a considerable

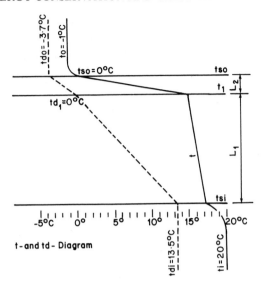

t- and td- Diagram

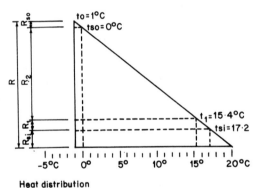

Heat distribution

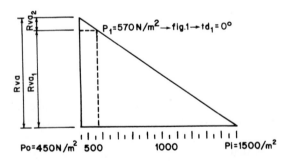

Vapour pressure distribution

Figure 19 Diagram of actual and dewpoint temperatures for the inner skin of a cold roof

range of changing situations can be explored: At what outside temperature will the roof show condensation, i.e. will the actual temperature curve (t) fall below the dewpoint temperature curve (t_d), and where? At the initial outside assumptions of temperature and humidity, what is the highest internal vapour pressure than can be accommodated without condensation risk?

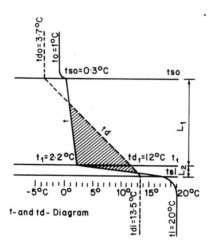

t- and td- Diagram

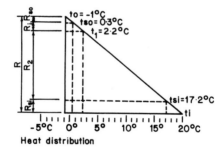

Heat distribution

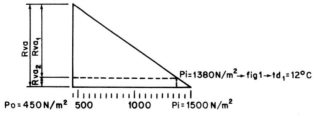

Vapour pressure distribution

Figure 20 As Figure 19, showing the effect of incorrect location of the insulation: interstitial condensation is predictable

Can we omit the insulation? This would leave only the two surface resistances and the concrete resistance $(R_{si} + R_1 + R_{so})$ in the picture, and the actual temperature distribution will show an internal surface temperature $t_u = 11.4°C$ lower than the internal dewpoint temperature $t_{si} = 13.5°C$. That means that surface condensation will result at this point (point in the diagram, surface in reality).

Why not try and correct a faulty situation by adding insulation on the inside? Another diagram will show the reason quickly enough. We make the same assumptions do the same calculations, and obtain results as before, but have the reverse order for concrete and EPS in all diagrams; the result is shown in Figure 20. Here we see the inner surface free from condensation risk, but within the sandwich the actual temperature does fall below the dewpoint temperature, indicating that condensation must be expected to occur; in this case we have interstitial condensation.

Notice that in both cases a pure U value calculation would give identical results, $U = 1/R = 1/0.834 = 1.200$ W/m² °C.

B. Warm roof

(i) Design assumptions and material properties:

▬▬▬▬▬▬▬▬▬▬▬▬▬▬▬▬▬▬▬	(5) 20 mm asphalt
xxxxxxxxxxxxxxxxxxxxxxxxxxxxxxxxxxxxxxxxxxxxxxxxxxxx	(4) 50 mm corkboard
— — — — — — — — — — — — — —	(3) 2 mm felt in hot bit. vapour barrier
	(2) 150 mm dense concrete
▬▬▬▬▬▬▬▬▬▬▬▬▬▬▬▬▬▬▬	(1) 4 mm gypsum plaster

The inside air conditions are:

$t_i = 20°C$ at $RH_i = 65\%$
$t_{di} = 13.5°C$
$p_i = 1550$ N/m²

The outside air conditions are:

$t_o = 1°C$ at $RH_o = 80\%$
$t_{do} = -3.7°C$
$p_o = 450$ N/m²

The thermal surface resistances are:

$R_{si} = 0.106$ m² °C/W
$R_{so} = 0.045$ m² °C/W

The following values apply for the various layers:
(1) Plaster
$L_1 = 0.004$ m
$k_1 = 0.460$ W/m °C
$\mu_1 = 5$
(2) Concrete
$L_2 = 0.150$ m
$k_2 = 2.050$ W/m °C
$\mu_2 = 28$
(3) Vapour barrier
$L_3 = 0.002$ m
$k_3 = 0.200$ W/m °C
$\mu_3 = 10\ 000$
(4) Cork
$L_4 = 0.050$ m
$k_4 = 0.050$ W/m °C
$\mu_4 = 5$
(5) Asphalt
$L_5 = 0.020$ m
$k_5 = 1.150$ W/m °C
$\mu_5 = 1000$

(ii) Temperature distribution

We have

$R_{si} = 0.106$ m² °C/W
$R_1 = L_1/k_1 = 0.004/0.460 = 0.009$
$R_2 = L_2/k_2 = 0.150/2.050 = 0.073$
$R_3 = L_3/k_3 = 0.002/0.200 = 0.010$
$R_4 = L_4/k_4 = 0.050/0.050 = 1.000$
$R_5 = L_5/k_5 = 0.020/1.150 = 0.017$

and so

$R' = 1.215$ m² °C/W

For the design assumption that outside surface resistance is balanced by radiation loss, we thus obtain

$R = R' + R_{so}$
$R = 1.215 + 0.045 = 1.260$ m² °C/W

and since $U = 1/R$, we obtain the U value as

$U = 1/1.260 = 0.793$ W/m² °C

(iii) Vapour pressure distribution

We have

$$R_{va1} = \mu_1 L_1 = 5 \times 0.004 \qquad = 0.020 \text{ m}$$
$$R_{va2} = \mu_2 L_2 = 28 \times 0.150 \qquad = 4.200$$
$$R_{va3} = \mu_3 L_3 = 10\,000 \times 0.002 = 20.000$$
$$R_{va4} = \mu_4 L_4 = 5 \times 0.050 \qquad = 0.250$$
$$R_{va5} = \mu_5 L_5 = 1000 \times 0.020 \quad = 20.000$$

and so

$$R_{va} = 44.470 \text{ m}$$

This figure of $R_{va} = 44.470$ m means that this entire warm roof sandwich has a resistance to vapour migration equal to a layer of air some 44 m thick.

(iv) Vapour transmittance

This calculation is, for our present purpose of checking condensation safety through t and t_d diagrams, as little required as the thermal transmittance, and is included here only to show the close resemblance of the method, result, and dimension in both cases. We use

$$R_v = r_{v(air)} R_{va}$$

where $r_{v(air)} = 5.0$ GN s/kg m and R_{va} is given in (iii) So

$$R_v = 5 \times 44.470 = 222.350 \text{ GN s/kg}$$

Now, since $U_v = 1/R_v$, we get

$$U_v = 1/222.350 = 0.0045 \text{ kg/GN s} = 0.0045 \times 10^{-9} \text{ kg/N s}$$

The seeming strangeness of the dimension for vapour transmittance (kg/N s) will now also disappear when we remember that it is a short-ened version of $kg/(m^2 \text{ s } N/m^2)$, less strange thus:

$$U_v: \frac{kg}{m^2 \text{ s } N/m^2} \qquad \text{compared to} \qquad U: \frac{J}{m^2 \text{ s } °C} \quad (J/s = W)$$

With the results of the heat and vapour resistances R and R_{va} we can again arrive at a diagram of actual temperatures and dewpoint temperatures (using Figure 1 for the latter to translate pressures into the corresponding dewpoint temperatures), as shown in Figure 21.

The first thing that will be noticeable is that the actual temperature will *always* be lower than the dewpoint temperature t_d directly underneath the weatherproofing in winter conditions. Which means exactly what it says.

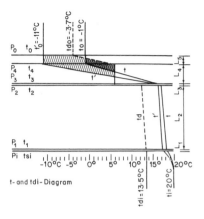

t- and tdi- Diagram

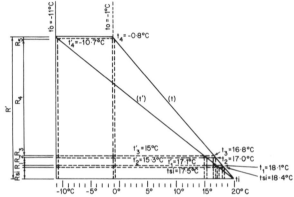

Temperature distribution diagram

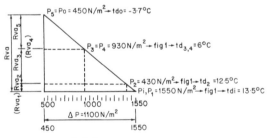

Vapour pressure distibution

Figure 21 Diagram of actual and dewpoint temperatures for a warm roof. Note that no matter what you try, there will always be indication of a condensation risk under the waterproofing. Actual condensation can only be avoided if no vapour gets there that could condense and if the insulation is laid absolutely dry and kept dry before being sealed in. Pressure equalization layers under the waterproofing *and* under the vapour barriers are the minimum recommended precautions in addition. Dynamic assessment methods should be used

Condensation there can only be avoided if there is almost no vapour that could condense, giving rise to the following requirements specific to warm roofs:

(1) Insulation materials for use in warm roofs must be really dry, and be kept dry until sealed-in. This is one of the operations on site where design is most easily defeated if it is not matched by supervision of work on site.

(2) $RH_i = 65\%$ at $20°C$ has been assumed for calculation purposes. Where this fairly high value, or even higher values, of vapour pressure are reached only for short periods of time, absorbent surfaces in the rooms such as plasters may act as buffers—absorbing a certain amount of moisture which will evaporate back into the room again later; provided, of course, that the room in question remains to be heated and ventilated. Intermittent heating can be a form of very false economy as it does away with all such recovery cycles, in which case very definite design measures to cope with the resultant migration problems have to be taken.

(3) In the example illustrated, the vapour barrier provided (or assumed and here checked) would be regarded inadequate: as a rule '. . . the barrier layer on the warm side of the insulation should provide three to four times as much resistance as that on the cold side' (Seiffert, 1970), p. 129.

Vapour barriers–general remarks

The practical solutions adopted when barrier layers are introduced into the construction of walls or roofs usually fall into one of the following three groups:

(1) Single layers of roofing felt or plastic foils for which the value of the air-equivalent vapour resistance $R_{va} = \mu L$ (the R_{va} value) is of the order of 10 m or less: for a descriptive term of what they do, one could call this order of barrier perhaps 'vapour migration retarders'.

(2) Metal foils embedded in protective layers of bitumen felt or plastic, with R_{va} values between 10 and 100 m: this range is what is usually meant or implied by a 'vapour barrier' in buildings. However, point (3) must be considered.

(3) Enclosures that include sheet materials with sealed joints (curtain walls, sheet metal roofing, certain plastic roofing, etc.) possess R_{va} values of thousands of metres right up to infinity. This is the region of 'vapourtight barriers'. Where the outer layers of enclosures are of this order of gastightness any design approach along the 'warm'

principle demands vapourtight barriers and insulation materials (apart from equally demanding H + V considerations).

Table 2 includes some of the more common R_{va} values of building materials.

C. Inverted roof

(i) Design assumptions and material properties:

────────────────	(4) 40 mm paving slabs, open joints, laid loose on plastic spacers
────────────────	
XXXXXXXXXXXXXXXXXXXXXXXXXXXXXXXXXXXXXXXXXX	(3) 80 mm (2 × 40 mm) insulation
XXXXXXXXXXXXXXXXXXXXXXXXXXXXXXXXXXXXXXXXXX	slabs, loose, staggered jts
= =	(2) 8 mm 3-ply felt roofing in hot bit.
────────────────	
────────────────	(1) 200 mm dense concrete

The inside air conditions are:

$t_i = 20°C$ at $RH_i = 65\%$

$t_{di} = 13.5°C$

$p_i = 1550 \text{ N/m}^2$

The outside air conditions are:

$t_o = 1°C$ at $RH_o = 80\%$

$t_{do} = 3.7°C$

$p_o = 450 \text{ N/m}^2$

$t_{so} = t_o$

The surface resistances are:

$R_{si} = 0.106 \text{ m}^2 °C/W$

$R_{so} = 0.045 \text{ m}^2 °C/W$

The following values apply for the various layers:

(1) Concrete

$L_1 = 0.200 \text{ m}$

$k_1 = 0.050 \text{ W/m °C}$

$\mu_1 = 28$

(2) Three-ply felt

$L_2 = 0.008 \text{ m}$

$k_2 = 0.200 \text{ W/m °C}$

$\mu_2 = 10\,000$

(3) Insulation
   $L_3 = 0.080$ m
   $k_3 = 0.063$ W/m °C
   $\mu_3 = 1^*$
(4) Paving slabs
   $L_4 = 0.040$ m
   $k_4 = 2.050$ W/m °C
   $\mu_4 = 1^*$

(ii) Temperature distribution

we have

$R_{si} = 0.106$ m^2 °C/W
$R_1 = L_1/k_1 + 0.200/2.050 = 0.098$
$R_2 = L_2/k_2 + 0.008/0.200 = 0.040$
$R_3 = L_3/k_3 + 0.080/0.063 = 1.260$
$R_4 = L_4/k_4 + 0.040/2.050 = 0.019$

and so

$R = 1.533$ m^2 °C/W

(iii) U-value (thermal transmittance)

$R = R + R_{so}$

and

$U + 1/R$

so

$U = 1/(1.533 + 0.045) = 0.635$ W/m^2 °C

(iv) Vapour pressure distribution

We have

$R_{va1} = \mu_1 L_1 = 28 \times 0.200 \quad = 5.600$ m
$R_{va2} = \mu_2 L_2 = 10\,000 \times 0.008 = 80.000$
$R_{va3} = \mu_3 L_3 = 1 \times 0.080 \quad = 0.080$
$R_{va4} = \mu_4 L_4 = 1 \times 0.040 \quad = 0.040$

and so

$R_{va} = 85.720$ m

$^*\mu_3$ and μ_4 are assumed equal to $\mu_{air} = 1$, being loose laid and open-jointed.

Figure 22 shows the resultant t and t_d diagram. The system is free from any condensation risk inside the waterproofing. The upper part of the insulation shows a similar result as in Figure 4, but here there is nothing that could go wrong; condensation above the waterproofing is simply drained away like rain or dew, or is left to evaporate again; the choice of insulation material is, of course, governed by the specification condition, see p. 573, that the insulant remains unaffected in its insulating properties by weather, i.e. by water, frost, and humidity.

4. ASSESSMENT OF CONDENSATION RISKS THROUGH GRAPHIC METHODS NOT BASED ON STEADY-STATE CONDITIONS

These methods have been reduced from mathematical models as first described for a wider audience by Seiffert (1970) to fairly simple graphical methods which pass my previous test for such methods of design mentioned earlier; it must be simple enough to be used by designers.

As one would expect, however, it is more complex than a steady-state approach simply because a whole-year cycle has to be considered, requiring the tabulation of environmental data throughout the year for all likely geographical areas to be readily available.

In some parts of the world these facts already reach architects free of cost as part of trade literature, mainly because building legislation demands proof of condensation safety enclosures in the same way as structural safety calculations as part of the building approval system.

BASF (1975) gives description of this calculation in English. With these methods of prediction we can now remove that part of the faults that have previously led to condensation failures: faulty design. Faulty workmanship and misuse are in the court of the building industry and building users.

Again we must remember that enclosure design is only one of three 'legs of the table': heating and ventilation being the other two. If a high degree of insulation, now safely designable, reduces the fabric losses considerably, ventilation heat losses become paramount, the thermal capacity of building structures requires increased attention, and heating systems will have to be rethought from scratch. In turn this not only means that the house will have become a 'fully engineered machine for living in', if energy, money, and the balance of payments are to be improved through better insulation of buildings, but also that the users of these machines must be educated to use them properly—at least until micro-miniaturization becomes cheap enough for buildings to run themselves through sensor-coupled automation.

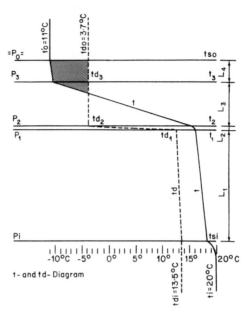

t - and td - Diagram

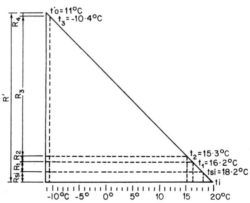

Temperature distribution

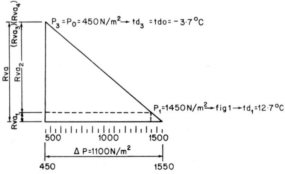

Vapour pressure distribution

Figure 22 Diagram of actual and dewpoint temperatures for an inverted roof (see text)

5. SUMMARY OF REQUIRMENTS FOR HIGHLY INSULATED ROOFS

Requirements

(1) Structural integrity, including the prevention of deterioration over the design life through thermal and moisture effects.
(2) Maintenance of the conditions of space use within the building, unaffected by moisture ingress or interstitial condensation, and maintaining internal surface temperatures at or above levels required for comfort.
(3) Maintaining throughout the design life the economic balance between initial and running costs.

It should be noted that highly insulated buildings require separate consideration of internal surface temperatures for comfort balance. It is for this reason that double-glazing will have to be used, as a single-glazed window will act as a radiation heat sink—or 'cold' radiator—producing discomfort leading to energy waste because it can only be counteracted by much increased air temperatures not otherwise necessary. Again, economic cost–benefit exercises based on heat-loss calculations alone are entirely misleading.

Design steps

(1) Establish design parameters.
  (i) Position and use of the intended building or room.
  (ii) Determine materials, their properties, size, sequence, and thicknesses.
  (iii) Compare with checklist under section 1 above.
(2) Calculation and assessment of thermal behaviour.
  (i) Check compliance with legal minimum requirements.
  (ii) Calculate the thermal transmittance (U value)
  (iii) Determine the surface temperatures and interface temperatures throughout the enclosure element.
  (iv) Determine the economic insulation standard considering energy costs.
  (v) Determine thermal movements.
(3) Establish satisfactory behaviour against precipitation and wind.
  (i) Separately.
  (ii) In combination.
  (iii) Check that combined effects are considered in the assessment of thermal behaviour.
(4) Calculate and assess condensation risks—steady state.
  (i) Check for surface condensation.
  (ii) Check for interstitial condensation.

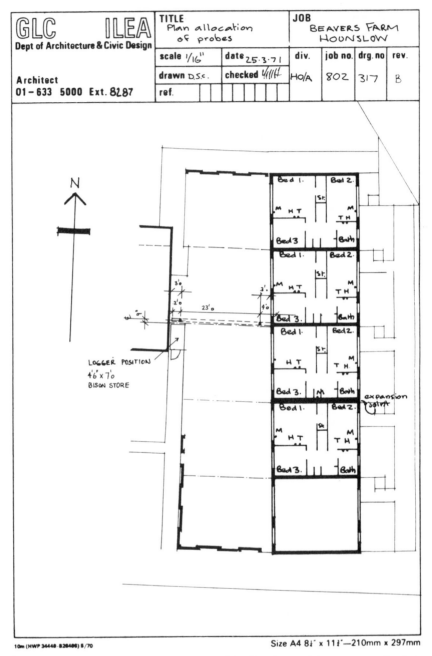

Figure 23 Beavers Farm flat roof trials (GLC): plan of house terrace. Probes were installed in and over a bedroom and bathroom in each of the four roof types. (Reproduced by permission of Greater London Council)

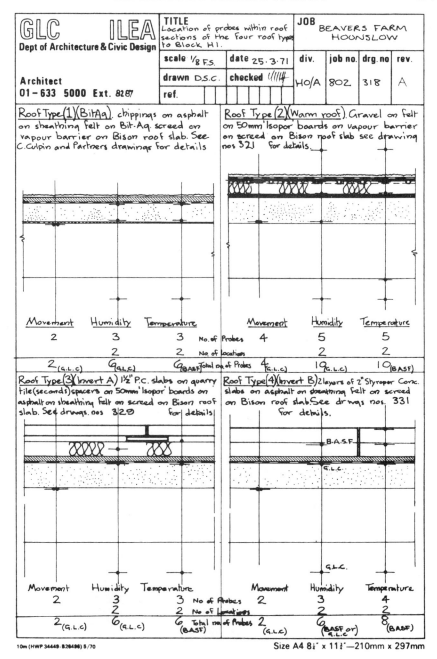

Figure 24 Beavers Farm flat roof trials (GLC): section through the four types showing the location of movement, temperature, and humidity/moisture probes installed—a total of 68 probes. (Reproduced by permission of Greater London Council)

(5) Steady-state check indicates condensation risks.
   (i) Consider changing design principle.
   (ii) Consider change in materials and sequence.
   (iii) Employ dynamic calculation method and determine the amount of condensation, its effect on insulation properties, and the effect of annual cycling.
(6) Repeat until satisfactory.

6. TWO CASE STUDIES OF HIGHLY INSULATED ROOFS

The field study reported by the Greater London Council (1974, 1975), is a direct comparison of four different flat roof systems (two warm roofs, two

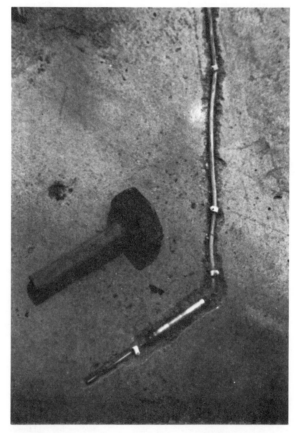

Figure 25 Beavers Farm flat roof trials (GLC): a temperature probe in position at surface of dense screed to falls under the vapour barrier of a warm roof. (Reproduced by permission of Greater London Council)

inverted roofs) in a single terrace of houses of identical construction in all other respects.

This field study is also the most satisfactory example of how to conduct a practical evaluation of theory in a real situation: starting with an architect with a bee in his bonnet, a real building contract, an insulation manufacturer who supplied calibrated humidity and temperature sensors and a data logger to collect the information, main contractor and roofer willing to cooperate within an existing contract, a building system manufacturer willing to indemnify (and paid) extra costs over contract for a building owner who is not permitted to spend a penny for such exercises (though must spend many pounds for any failures), and a competent Building Sciences group to evaluate the data—and it was all done without delay to contract. Rare indeed! The trial roofs were laid in the summer of 1971. Figures 23–30 show the installation. Detailed results were reported by the Greater London Council (1975) (see Figures 31 and 32).

Condensation problems have, however, nothing to do with roof shape as mono-pitched asbestos cement roofs over houses on a second GLC estate, which led to reports of severe rain penetration, demonstrate. With the best will in the world no leaks could be found by inspecting the roofs on site, but water was seen to seep from under the asbestos cement sheets (Figure 33). So they were opened up (see Figure 34). At first no one would believe that such

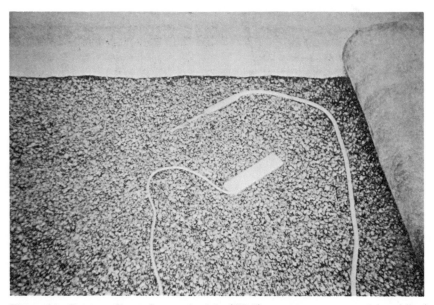

Figure 26 Beavers Farm flat roof trials (GLC): a temperature and humidity/moisture probe installed on the surface of an insulating screed in a warm roof; sheathing felt being laid ready for asphalting. (Reproduced by permission of Greater London Council)

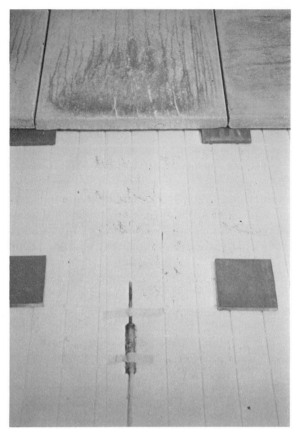

Fig. 27

Fig. 28

Figure 29 Beavers Farm flat roof trials (GLC): the gravel-protected felt warm roof completed. (Reproduced by permission of Greater London Council)

Figure 27 Beavers Farm flat roof trials (GLC): a temperature probe installed on the surface of the insulation in an inverted roof. Note the quarry tile spacers to give an airspace under paving slabs which provide the wind uplift control. These quarry tiles proved the only 'failure' after seven years: they crumble from frost attack. Purposemade plastic spacers should be used. (Reproduced by permission of Greater London Council)

Figure 28 Beavers Farm flat roof trials (GLC): felt warm roof. Note the surface of insulation boards grooved to provide pressure equalization. EPS boards melt at 80°C and the first waterproofing felt layer had to be brushed to allow bitumen to cool sufficiently to avoid melting the EPS before folding over onto the insulation. Subsequent layers were 'poured-and-rolled' in the normal way. (Reproduced by permission of Greater London Council)

Figure 30   Beavers Farm flat roof trials (GLC): the four trial roofs completed. In the foreground is an inverted roof using two layers of cement-bound polystyrene bead boards, followed by an inverted roof using EPS moulded boards protected by paving slabs on spacers; thirdly a warm roof employing an insulating screed. All these three roofs employed asphalt for the waterproofing membrane. The fourth trial roof over two houses in the background is a warm roof using EPS moulded boards for insulation and three layers of felt in hot bitumen as waterproofing; protected by a reflective gravel layer. The cables from the 68 probes pass over a gantry to the data logger housing visible on the right. Installation of the roofs and probes took place in August 1971, data logging commenced full-time in February 1972 after the houses were occupied, and was continued until September 1974. Since then, only the insulation in the two inverted roof types is monitored each summer and winter by visual inspection and moisture measurements from samples removed. But the experiment is a long-term study, and the probes and connecting wires are left in place should future behaviour of the roofs warrant a detailed logging exercise again. (Reproduced by permission of Greater London Council)

Figure 31   Beavers Farm flat roof trials (GLC): inspection of inverted roof type 4 after three years. The polystyrene–cement slabs have not deteriorated but moisture content is higher than hoped. Note the asphalt forming ridges through cold-flow where drainage grooves had been provided on the under- side of insulation slabs. (Reproduced by permission of Greater London Council)

Figure 32   Beavers Farm flat roof trials (GLC): inspection of inverted roof type 3 after three years. Note the more pronounced ridge pattern formed by asphalt cold-flow up into the EPS board grooves. The asphalt temperatures in winter were consistently higher than in roof type 4, which means the insulation was better, but that also made the asphalt 'flow' more. (Reproduced by permission of Greater London Council)

Fig. 31

Fig. 32

Figure 33 Harold Hill (GLC): inspection of roofs reported to have 'severe leaks' showed water running onto roof from underside of upper sheets, pointing to condensation as the problem source. (Reproduced by permission of Greater London Council)

Figure 34 Harold Hill (GLC): roofs opened for inspection showed all internal surfaces dripping with moisture, timbers wet, with stains and mould growth. (Reproduced by permission of Greater London Council)

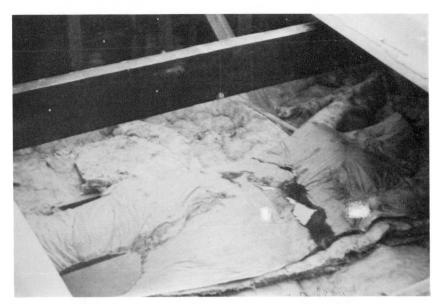

Figure 35 Harold Hill (GLC): pools of water collected from dripping roof beams—all due to condensation in the unventilated roof void; no leaks in the weathering details were ever found. The lesson to be learnt is that no airspace in a roof must *ever* be allowed to figure as contributory to its insulation or *U* value, or the results will be similar to this. (Reproduced by permission of Greater London Council)

Figure 36 Harold Hill (GLC): plywood beams under heavy attack from condensation and moulds. (Reproduced by permission of Greater London Council)

Figure 37   Harold Hill (GLC): condensation water running down on building behind weatherboarding. (Reproduced by permission of Greater London Council)

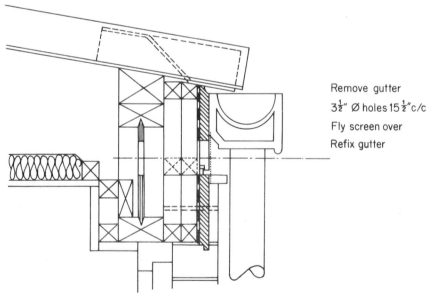

Remove gutter

$3\frac{1}{2}''$ Ø holes $15\frac{1}{2}''$c/c

Fly screen over

Refix gutter

Figure 38   Harold Hill (GLC): proposed remedy—convert roof space to cold roof and provide ventilation as detail at gutter. (Reproduced by permission of Greater London Council)

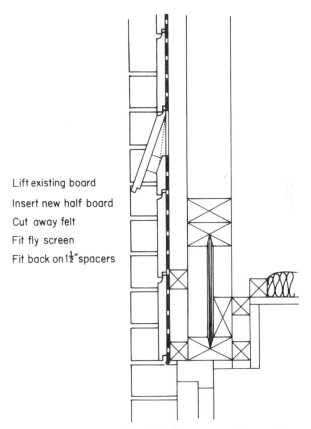

Lift existing board
Insert new half board
Cut away felt
Fit fly screen
Fit back on 1½″ spacers

Figure 39 Harold Hill (GLC): proposed remedy—cold roof ventilation at lower region of weatherboarding [Figure 40]. (Reproduced by permission of Greater London Council)

quantities of water (see Figure 35) could accumulate in a roof from condensation (Figures 36 and 37), and the search was intensified 'to find those leaks'. None were found.

Until someone remembered my 'buckets of water' explanation (see Section 2 above) and under heavy suspicion of everyone else involved, I was allowed to test the cold roof theory in practice on the worst cases found on site.

A look at the detailed drawings showed that the roof design was neither a 'warm' nor 'inverted' roof; it was also not a 'cold' roof because the effective ventilation of the airspace was missing. Token vent grilles had been installed in the weatherboarded area, but their effect was insignificant in relation to the minimum rules I set out above (in Section 2A). So I proposed conversion to 'cold roofs' by providing effective ventilation at no less than the minima

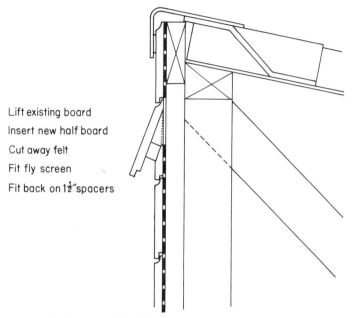

Lift existing board
Insert new half board
Cut away felt
Fit fly screen
Fit back on 1½"spacers

Figure 40   Harold Hill (GLC): [Figure 39] and close to ridge; fly screens behind all vent slots. (Reproduced by permission of Greater London Council)

Figure 41   Harold Hill (GLC): the remedy carried out—appearance at gutter; vent holes covered by fly screens. (Reproduced by permission of Greater London Council)

Figure 42 Harold Hill (GLC): Overall view of modified terrace—gutter side. (Reproduced by permission of Greater London Council)

Figure 43 Harold Hill (GLC): the remedy carried out—detail appearance of vent slots in weatherboarding. Note vent grilles originally installed: they are far too small to provide any 'effective ventilation' as required for a cold roof. (Reproduced by permission of Greater London Council)

Figure 44   Harold Hill (GLC): overall view of modified terrace—ridge side. (Reproduced by permission of Greater London Council)

quoted. The modifications were carried out in October 1977 (see Figures 38–44).

The results are monitored by the Scientific Adviser, taking eight modified houses, and eight unmodified houses of identical construction with similar problems as controls. The results shown in Tables 3 and 4 demonstrate that the cold roof principle works:

(a) the moisture content of roof timbers after only two months has risen $2\frac{1}{4}$ times as much in the unmodified controls as in the converted houses;
(b) the bedroom–loft partial vapour pressure difference in the modified houses is $2\frac{1}{2}$ times higher than in the controls; and
(c) the loft–external air partial vapour pressure difference in the modified houses is only one-third of that in the controls.

Table 3   Moisture contents (%) of ply-beams

	Modified		Controls	
	4 Nov. 77	10 Jan. 78	4 Nov. 77	10 Jan. 78
Top rail	15.3	16.8	15.8	19.0
Web	17.6	18.6	20.3	24.3
Bottom rail	13.5	15.6	14.9	17.9
Average rise		1.5		3.4

Table 4   Partial vapour pressure differences (mbar)

	Modified		Controls	
	4 Nov. 77	10 Jan. 77	4 Nov. 77	10 Jan. 77
Bedroom–loft	3.25	1.47	0.48	0.58
Loft–ext. air	1.8	0.47	3.4	1.59

In addition, houses in the control group report 'water through the ceiling' again, as in the previous year. The modified houses have reported no such complaints.

BIBLIOGRAPHY

Aamot, H. W. C., and Schaefer, O. 1976. *Protected Membrane Roofs in Cold Regions*, CRREL Report 76–2, Corps of Engineers, US Army, Cold Regions Research and Engineering Laboratory, Hanover, N.H., USA.

BASF, 1975, *Diffusion and Condensation of Water Vapour: Terms and Concepts*, Ti-1031e, January; *Diffusion and Condensation of Water Vapour*: Design Data and Calculations, Ti-1032e, December, BASF United Kingdom Ltd (Plastics Division), Cheadle Hulme, Cheadle, Cheshire (ref. BASF AG/D-67).

Bobran, H. W., 1967, *Handbuch der Bauphysik* (*Handbood of Building Physics*), Ullstein GmbH, Frankfurt/M and Berlin.

British Standards Institution, 1975, *Code of Basic Data for the Design of Buildings*: *the Control of Condensation in Dwellings*, BS 5250:1975.

Building Design, 1974, BD, 6 December 1974, Morgan Grampian, London, p. 7.

Building Research Establishment, 1972, Condensation, *BRE Digest*, 110, HMSO, London.

Building Research Establishment, 1974, *A New Approach to Predicting the Thermal Environment in Buildings at the Early Design Stage*, BRE Current Paper, CP 2/74.

Building Research Establishment, 1975, *Energy Conservation: a Study of Energy Consumption in Buildings and Possible Means of Saving Energy in Housing*, BRE Current paper, CP 56/75.

Construction Development (Co-ordination) Committee, 1976, *Performance Requirements for Roofs*, Technical Note no. 6, CDCC (DES, DHSS, SDD, DOE), HMSO, London, November.

Greater London Council, 1972, *The Diffusion of Water Vapour into Roof Structures* (incorporating notes by H. W. McInnes), Development and Materials Bulletin, No. 60 (second series), December, Department of Architecture and Civic Design, Greater London Council.

Greater London Council, 1974, *Monitored Field Study of Flat Roof Systems*, Development and Materials Bulletin, No. 75 (second series), May, Department of Architecture and Civic Design, Greater London Council.

Greater London Council, 1975, *Monitored Field Study of Flat Roof Systems—Final Report*, Development and Materials Bulletin, No. 90 (second series), December, Department of Architecture and Civic Design, Greater London Council.

HMSO, 1970, *Condensation in Dwellings*, Pt. 1, *A Design Guide*, HMSO, London.

IHVE, 1970, *IHVE Guide*, The Institution of Heating and Ventilating Engineers, London, 1970, 1971, and 1972.

Seiffert, K., 1970, *Damp Diffusion and Buildings (Prevention of Damp Diffusion Damage in Building Design)*, translated from the German by A. B. Phillips and F. H. Turner, Elsevier, Amsterdam, London, and New York.

Timmerberg, C. H. (ed.), 1974, *Details . . . Details*, Klaus Esser KG, Düsseldorf.

Energy Conservation and Thermal Insulation
Edited by R. Derricott and S. S. Chissick
© 1981 John Wiley & Sons Ltd.

CHAPTER 23

Case study: energy conservation and solar energy developments in Australia

S. V. Szokolay

University of Queensland

INTRODUCTION

Australia is a 'lucky country'. It is large, rich in mineral resources, and small in population. Its most populated areas have an amicable climate. The attitude of many (if not most) people to the energy question would be expressed as 'we have got no energy problem'. Figure 1 illustrates the energy flow pattern through the Australian economy in broad terms. There is a large export of high-quality coal. The only energy import is petroleum. In 1974–75, the import was 36% of the total use, but since then this has increased to over 50% and it is predicted to increase to 70% by 1985. Responsible people admit that 'there is a liquid fuel problem'.

In January 1977 the federal government convened a National Energy Advisory Committee and in June 1978 the National Energy Research, Development and Demonstration Council (NERDDC) was established. After some six months of deliberation, NERDDC has allocated over $15 million for various energy projects. Their assessment of the relative importance of various energy industries is indicated by the magnitude of grants: coal research got the largest slice, almost half of the total amount. Liquid fuel production from coal and solar/biomass conversion received the second largest share. Solar energy research and development projects received less than $1 million.

An energy consciousness is however being created by individuals and small groups working at many levels. A Thermal Insulation Institute was established in 1975 and their journal *Thermal Insulation* has been published since 1976. The Institution of Engineers set up a Task Force on

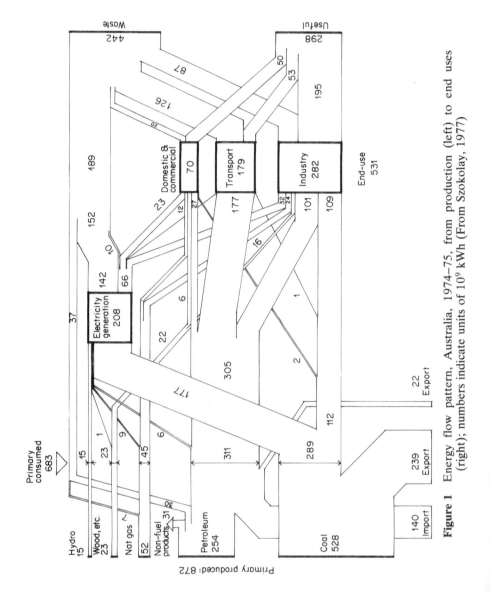

Figure 1 Energy flow pattern, Australia, 1974–75, from production (left) to end uses (right); numbers indicate units of 10^9 kWh (From Szokolay, 1977)

Energy in 1975, and their recommendations were published two years later (Institution of Engineers, 1977). In 1978 the Institution of Engineers, in cooperation with the Royal Australian Institute of Architects (and other bodies), organized a competition and a symposium on Energy Conservation in Buildings.

Newspapers and popular magazines are full of energy-related items. Energy symposia, conferences, and meetings at all sorts of levels are so numerous that one has already lost count. Although short-sighted pragmatic attitudes still prevail, a new consciousness began to emerge, believing that there are values other than monetary, that the wastage of energy is inherently wrong, and realizing that our resources are finite.

CLIMATE

The fraction of the national total primary energy consumption used for the heating and cooling of buildings is (approximately)

  (i) 22% in the UK,
 (ii) 21% in the USA, and
(iii) 12% in Australia.

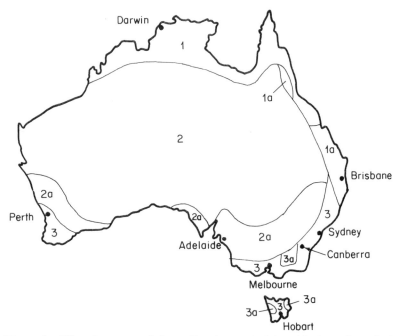

Figure 2 Climatic zones of Australia: 1, tropical humid; 1a, subtropical humid; 2, hot dry; 2a, dry warm temperate; 3, temperate; 3a, cool temperate

The main reason for this much lower fraction in Australia is the climate. The continent extends from 11°S (Cape York) to 43°S latitude (Tasmania), from hot–humid to cool–moderate climates. Figure 2 gives a broad definition of climatic zones in Australia. Most of the population is contained between latitudes 27°S and 38°S (40% of the continent's population lives in two cities: Sydney (34°S) and Melbourne (38°S)). This zone is characterized by hot summers and mild winters. The mean minimum temperature of the coldest month (July) is 8°C in Sydney and 6°C in Melbourne. Even in Hobart (43°S) the July mean minimum is only 5°C.

The heating requirement is not very large. An indication of this is given by the annual number of degree-days (i.e. the cumulative temperature deficit from 18°C):

Brisbane	440°C-days	Adelaide	1270°C-days
Sydney	880	Melbourne	1870
Perth	1040	Hobart	2200

For comparison, the degree-day values are 2610 for London, 3130 for Edinburgh, and 3300 for Boston (Massachusetts, USA). The coldest of Australian climates is comparable to the south of England.

BUILDING DESIGN

Work carried out at the CEBS (Commonwealth Experimental Building Station) in the late 1940s led to the design recommendations first published by Drysdale (1952) and summarized in Table 1. The rather cautious recommendation of 25 mm insulation was then a step forward.

Subsequently the CSIRO (Commonwealth Scientific and Industrial Research Organisation) carried out a study of desirable insulation in houses (Muncey, 1955). His recommendations are summarized in Table 2. This was a further improvement, as in most instances the thicknesses recommended are greater than those in Table 1, and the method has a rational basis: cost of insulation versus the benefit of heating cost savings. However, the basis of the above recommendations was a heating cost of 0.57¢/kWh and an insulation cost of about 31¢/m² for 25 mm of glass wool. Since then the cost of insulation increased by a factor of three, but the heating costs escalated by a factor of six. The main criticism is, however, that the cost of winter heating alone was taken into account and the cost of summer cooling or the value of improved indoor thermal conditions in the summer were disregarded.

There is no legal requirement for the thermal insulation of buildings, except in an indirect form. Prior to 1960 the minimum height for habitable rooms was 2.7 m. Since then 2.4 m is allowable where 50 mm insulation is provided in the roof/ceiling. In New Zealand minimum insulation stan-

Table 1 Design recommendations for climatic zones in Australia

	Hot–humid zone	Hot-arid zone	Temperate zone
Walls	If sunlit: heavyweight or insulated frame; if shaded: frame construction	Heavyweight for day rooms, frame construction for bedrooms	Mass unimportant, but for inland areas the hot–arid zone recommendations apply
Roofs	Pitch and finish selected for weather resistance; construction may be light or heavy	As for hot–humid	As for hot–humid
Floors	Suspended timber or concrete on ground	As for hot–humid, but any under-floor space to be enclosed	As for hot–humid
Insulation	25 mm glass wool or reflective foil above ceiling and in framed walls not shaded	25 mm glass wool or reflective foil above ceiling and in all framed walls	25 mm glass wool or reflective foil is beneficial above the ceiling or heated spaces
Shading	Walls: by eaves or trellis with vegetation; windows with external devices	As for hot–humid	Walls: by eaves; windows: internal blinds adequate
Surfaces	Preferable to have light colours for noninsulated walls or roofs	As for hot–humid	Unimportant, but light colours preferable for noninsulated walls or roofs

Table 2   Recommended thickness (mm) of insulation (glass wool or equivalent)

	Cavity brick house	Timber-framed house	
	Roof	Roof	Wall
Brisbane	0	0	0
Sydney	30	15	8
Perth	15	11	8
Adelaide	55	33	26
Melbourne	60	40	30
Hobart	74	55	44

dards were established in 1978 for all new buildings and interest-free loans are available for the insulation of existing houses.

Both of the above two sets of recommendations ignore or underestimate the role of thermal capacity or mass. Recent work has shown that in a well designed massive house there would be no heating or cooling required at any time of the year in Brisbane and in large areas of inland (hot–arid) Australia. Even in the hot–humid zones (e.g. Darwin) the massive construction shows distinct advantages over the traditional lightweight and fully cross-ventilated buildings, for all rooms used during the day. For air-conditioned buildings, massive construction is preferable even for bedrooms (Szokolay, 1977).

A standard brick-veneer house in the Melbourne winter may have a daily heating requirement of 150 kWh. A thickness of 50 mm ceiling insulation alone could reduce this by 25% (to 118 kWh). Rearrangement of fenestration, use of concrete slab floor and cavity brick walls can reduce it to 70 kWh. It is readily demonstrable that good design can reduce the annual domestic heating and cooling energy requirements by 50%.

SOLAR WATER HEATERS

In Brisbane, where there is very little space heating or cooling demand, the energy used for domestic water heating can be as much as 50% of the energy consumed by the household. In national terms, water heating is about 25% of the domestic energy consumption.

Serious research into solar water heaters was pioneered by the CSIRO Division of Mechanical Engineering in the 1950s. By the early 1960s definitive recommendations were published (CSIRO, 1964) and several manufacturers were licensed to use the patented designs. Figure 3 shows the increase of solar water heater production in Australia. The annual growth rate exceeds 60% in recent years. The interesting point to emphasize here is that this industry has developed on its own, without any form of

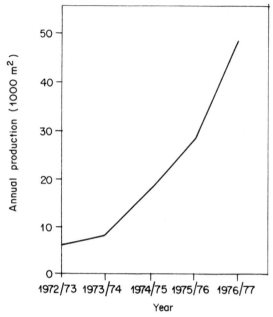

Figure 3   The growth of solar collector production
in Australia

government assistance, in spite of the assistance given to other, competing
energy industries. Much more spectacular growth (300% per annum) was
achieved in the USA, but this was brought about partly by direct governmental funding and partly by indirect support: tax rebates and other incentives.

   In northern parts of Australia, roughly north of the Tropic of Capricorn,
the standard water heater types can operate without any auxiliary energy
source, giving a solar fraction of 1. In the southern parts it would be
impracticable and uneconomic (although technically possible) to build such
systems. It is usual to aim at a solar fraction of 0.75 and obtain the remaining 25% from an electrical element or a gas heater. In this case the utilization rate of the collectors will be more satisfactory.

   Thermosiphon systems are the most popular, where the tank is located
above the level of the solar collector and the water heated in the collector
panel rises automatically to the tank. Figure 4 shows a late 1960s water
heater, which has a separate tank. Later installations normally have the
tank under the ridge of the roof, with the collectors laid on the roof, near
the eaves. The integral tank system (Figure 5) is today the best-selling
model. Its performance is 2–3% less, but it is one compact unit, which can
be sold 'over the counter' and it is easy to install.

Figure 4   An early thermosiphon solar water heater

Figure 5   A solar water heater with integral tank

For installations in existing buildings ('retrofitting', to use the American term), pumped systems may have a certain advantage. An existing hot water cylinder, often at floor level, can be coupled to a new set of collectors on the roof by a pipe loop, using a small, 18–20 W circulating pump. In larger installations, pumped systems are used without exception.

A report by the CSIRO Solar Energy Studies Unit (Proctor and Morse, 1977) showed that in the food industry 89% of the energy is used for process heating and that 46% of this (i.e. 41% of the total) is used at temperatures between 60 and 100°C. In absolute terms this is a much greater quantity than the energy used in domestic water heating. Industrial solar installations could produce a greater contribution to the national energy supply.

Since then two large pilot plants have been completed, as cooperative projects between CSIRO and private industry:

(1) a Coca Cola bottling plant, near Canberra, providing hot water for a can warmer, using 94 m^2 collectors and a 21 m^3 storage tank; and

(2) a brewery, near Adelaide, supplying hot water for the pasteurization of beer in sealed bottles at 60°C, using 178 m^2 collectors and two storage tanks totalling 52 m^3.

Both are performing satisfactorily. A completely privately funded project has been completed recently in Perth, using 75 m^2 collectors and a 12 m^3 storage tank, and supplying warm water for a Kodak photoprocessing laboratory.

SOLAR HEATING SYSTEMS

The collection and heat transfer fluid may be air or water and, accordingly, we can distinguish air and water systems (the latter is referred to in the USA as 'hydronic systems'). As the space heating requirement is small in Australia, there is insufficient incentive for the development of such systems. Some research and development work has been carried out for the last 15 years, or so, and the pace of such work is accelerating.

Water systems

Water systems are scaled-up versions of the pumped domestic hot water system. Whilst for the latter 3–5 m^2 collectors are used with 210–350 litre tanks, for space heating both the collector area and the tank size are increased by a factor of about 10. The hot water collected is used for space heating, most often through fan–coil units or in an embedded coil floor-warming system.

Figure 6 SOLARCH house, at Fowler's Gap, near Broken Hill, New South Wales

A notable example of such a water system is the 'Solarch' house, designed by John. A. Ballinger, of the University of New South Wales (Figure 6). It was built at Fowler's Gap (near Broken Hill) in 1977–78. It is a single-storey, three-bedroom house of 143 m² floor area. The north-facing roof incorporates 47 m² collectors of the 'trickle-plate' type. Water is pumped from a 4 m³ underground tank up to the ridge of the roof, distributed by a perforated pipe and it trickles down the black surface of the metal trough roof, under a transparent cover. The heated water is drained back into the tank and can be circulated through a coil of 18 mm polyethylene pipes, embedded in the floor screed at about 250 mm centres.

The house also includes various energy conservation measures, e.g. wall insulation to a *U* value of 0.36 W/m² °C and roof insulation to 0.28 W/m² °C (100 mm glass wool), as well as a 'direct gain' system, i.e. solar radiation entering through windows absorbed by and stored in the concrete floor. Rain water from the roof is collected in a large tank. A small wind generator provides electricity for lighting.

Air systems

Air systems are very popular in the USA, especially in areas where freezing can occur, as they do not require any frost protection measures. The CSIRO has developed a very efficient air heater collector, using a 'V'-corrugated black copper foil as the absorber. Storage is normally provided in a 'rock bin', i.e. an insulated container filled with crushed rock or pebbles. Such a system has been in use at the CSIRO laboratories in Melbourne for over 14 years. It uses 28 m^2 collectors and a 9.5 m^3 rock bin storage.

A low-energy house was built on the same site in 1978. It is a modified version of a popular design by a local building firm: a single-storey, three-bedroom house of 112 m^2 floor area (Figure 7). Air heater collectors of 19 m^2 area are built in with the roof, at a tilt of 58° (latitude +20°). Two ribbed aluminium roofing sheets are rivetted together, face-to-face, to form the absorber, with airways running horizontally. The cover is a corrugated acrylic sheet, to match the corrugated steel roofing sheet profile. The underfloor rock bin storage is 96 m^2 in area and 0.5 m deep. There is no warm air withdrawal facility: the store itself provides a direct floor-warming effect. The system has been designed for a solar fraction of 0.65.

Another interesting house was completed in Hobart in January 1979, designed by David M. Button, a final-year architecture student, for a private client. It is a two-storey, two-bedroom house of 140 m^2 floor area. Besides various conservation measures it includes 15 m^2 air heater collectors, of the CSIRO 'V'-corrugated type. A rock-bin storage of 9 m^3 is located under the living-room floor. It is expected that this system will provide the total heating requirement left after the various conservation measures, without any auxiliary heat input, although this is probably too optimistic (Figures 8a and b).

Figure 7   CSIRO low-energy house, at Highett, Melbourne, Victoria

Figure 8   Two views of Button house, at Mount Nelson, Hobart, Tasmania

Many solar space heating systems (particularly water systems) installed in Germany, Japan or the USA are quite complicated and sophisticated. Practically all systems installed in Australia are simple and inexpensive. The main aim seems to be cost-effectiveness, rather than reaching the ultimate in efficiency.

SOLAR COOLING SYSTEMS

The many available systems can be grouped into two categories: open- and closed-cycle systems. The open-cycle systems are not really cyclic in operation: outdoor air is used, which is treated, supplied to the house and/or discharged to the atmosphere. In the closed-cycle systems, a fluid is circulated in a closed loop, transferring heat from a source to a sink.

Open-cycle systems

The simplest *open-cycle system* is the *direct evaporative cooler*, such as that shown in Figure 9. Water is evaporated into the supply airstream, which is thus cooled and humidified. Such coolers have a long tradition in Australia. They are simple, inexpensive, and very effective in hot–dry inland areas and also in the dry season of the hot–humid zones.

Where the high atmospheric humidity limits the evaporation (thus the cooling) potential, and makes the addition of further moisture undesirable, an indirect evaporative cooler can be used. The CSIRO has developed an inexpensive plate heat exchanger. An airstream driven through every second space between the plates is evaporatively cooled and then discharged. The supply air passed through the alternate spaces transfers some of its

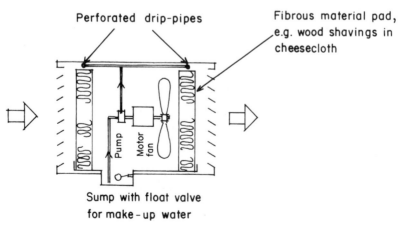

Figure 9 A direct evaporative cooler

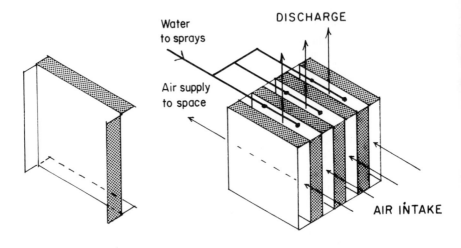

Figure 10   Plate heat exchanger, an indirect evaporative cooler

heat through the plates to the evaporatively cooled air, before entering the room (Figure 10).

The rock-bed regenerative cooler (RBR cooler) is an invention patented by the CSIRO (Robeson, 1970). Two bins of crushed rock are used in a 10 min cycle intermittent operation. The exhaust air is led through one rock bed, which is sprayed with water, and thus evaporatively cooled. The air intake passes through the other rock bed, which has been cooled previously. In every 10 min the flow is reversed. The system works well with very small energy consumption, but its capital cost is high and it does not offer significant advantages over the indirect evaporative cooler.

Dehumidification can be achieved by liquid or solid chemical absorbents, such as silica gel, activated alumina, calcium chloride, lithium chloride, lithium bromide or ethylene glycol. The saturated absorbent can be reactivated by solar heated desorption. Many theoretical studies and some experimental work has been carried out and various system configurations have been proposed, but there is no conclusive result.

Closed-cycle systems

Closed-cycle cooling systems are of two types: compression or absorption cycles. The absorption cycles are adaptable for solar operation. These are binary systems, using a refrigerant (e.g. ammonia or water) and an absorbent (e.g. water, H_2O, or lithium bromide, LiBr). Figure 11 shows the principles of a LiBr/H_2O absorption cooler. As solar heat of at least 76°C is applied at the generator, the water is expelled from the LiBr solution.

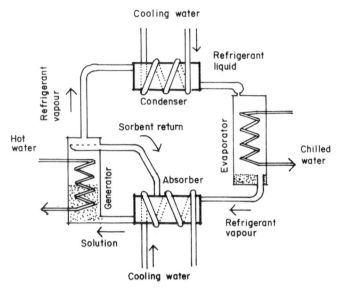

Figure 11   Principles of absorption chillers

The water vapour is condensed in a water-cooled vessel, admitted through a small orifice into the evaporator, where it picks up heat from its environment (producing chilled water), and returned into the absorber, where it is dissolved in the LiBr again. The circulation is driven by a vapour lift pump.

Such an absorption cooler was first applied to solar operation at the University of Queensland in 1965 (Sheridan and Carr, 1967). The experiment was a success, but further development was prevented by lack of funds. Since then several American and Japanese firms have developed and started marketing such systems. A number of projects have been completed in the last two years using a Japanese cooler. The first one was a solar air-conditioned house at Mount Cotton, near Brisbane (Figure 12). This is a 150 m², three-bedroom residence, built as a demonstration house. The peak cooling load of a conventional house of comparable size would be some 11 kW. Correct orientation, fenestration (no windows on east and west sides), and full shading for the summer six months would reduce this to 8.5 kW. By replacing the generally used brick-veneer construction with cavity brick walls (on east and west walls the cavity is filled with urea formaldehyde foam) and the timber floor with a concrete slab, the cooling load is reduced to 5.2 kW. The house and the system were designed by the author and completed in July 1978.

A 4.6 kW absorption chiller is used, in conjunction with a 1.3 m³ chilled water tank, which acts as a buffer. When the cooling load is less than

Figure 12   Mount Cotton solar house, near Brisbane, Queensland

Figure 13   Eldridge Clinic, Bankstown, near Sydney, New South Wales

4.6 kW, chilled water is supplied to the tank. At peak load times chilled water is also taken from the tank to the fan–coil unit. The purpose of this system is to avoid intermittent operation of the chiller, which has been shown to be rather inefficient. Hot water to the chiller generator is supplied by a system consisting of 40 m^2 selective surfaced collectors and a 2 m^3 hot water tank.

A much larger solar air-conditioning system has been installed in the same year at Bankstown (a suburb of Sydney) on a medical building (Figure 13). The collectors are mounted on a saw-tooth roof in seven arrays. A close-up picture of one of the arrays is shown in Figure 14.

The most ambitious project so far has been designed by the Federal Department of Construction and is now being built in Townsville. It is a two-storey office block of 1080 m^2 floor area. It is to have two absorption chiller units of 27 kW cooling capacity each, served by 320 m^2 water heater collector panels. Hot water is stored in a 21 m^3 tank and chilled water in a 14 m^3 tank. A similar capacity conventional chiller will also be installed to allow comparative study of the two systems.

Figure 14   Close-up of a collector array, Eldridge Clinic

PASSIVE SOLAR SYSTEMS

The generic term for all the systems discussed so far is 'active systems'. Active, in the sense that some heat transfer fluid (air or liquid) is circulated in pipes or ducts by motor-driven pumps or fans. As distinct from these, we can talk about 'passive systems', where the heat transfer and heat transport take place spontaneously, e.g. relying on fluid flow induced by the temperature effects themselves. Some authors consider the thermosiphon water heater as a passive system.

For space heating purposes the best known passive system is the Michel–Trombe glazed mass wall, which acts as both collector and thermal storage. This system has been used in a house built in Hobart in 1978 (Figure 15). It was designed by two final-year architecture students, Michael Leach and Greg Strickland. The north elevation includes an 18 m² solar wall. This consists of 300 mm hollow concrete blocks, the hollows filled with a weak concrete mix. The outer face is painted black. Single glazing is supported on pressed galvanized steel sheet mullions, 75 mm in front of this surface. Vents are provided through the wall at floor level and near the ceiling. When the wall is heated by solar radiation, the warm air rises and enters the room at ceiling level, whilst cooler air is drawn from the room at floor level. This thermosiphon air circulation provides about 30% of the heating. The massive wall stores a large amount of heat, which will be released to the room from the internal surface by radiation and convection, with a time delay of up to 10 h. This provides the remainder of the

Figure 15   Ferguson house, at Mount Nelson, Hobart, Tasmania

heating. The only auxiliary heat source is a wood-burning fireplace, which has a steel fire-box, including several air passages.

An interesting experiment is being carried out on the last panel at the western end of this wall. The surface of the concrete block wall is sheeted with a copper foil, applied by an adhesive, and has a selective black copper oxide finish. Its absorptance for solar wavelengths is about 0.95 and its emittance at normal operating temperature is less than 0.1. The radiant heat loss will be reduced quite substantially and more heat will be stored in the mass wall. The magnitude of this benefit is currently being evaluated, in comparison with the ordinary black painted walls.

Domestic hot water is provided by four flat plate collector panels of $0.8 \, \text{m}^2$ each, located on the greenhouse roof. The tank is housed in a chimney-like structure, at a level higher than the collectors, permitting a thermosiphon circulation. This system is designed to give 60% of the water heating requirement, the remaining 40% supplied by an electric immersion heater.

An earlier house built in Perth by R. Lawrance (in 1975) includes a variant of the above system, which is completely original. A $21 \, \text{m}^2$ section of the north wall has a 75 mm brick-on-edge outer skin, a 65 mm cavity and a 110 mm brick inner skin. The cavity side of this inner skin is lined with aluminium-foil-faced building paper. As the cavity is heated, a thermosiphon airflow is generated to and from the room. The outside face of the brick-on-edge wall is painted black and it is covered by a transparent plastic sheet mounted on 25 mm battens. The system provides some 75% of the house's heating requirement. In summer the cavity is vented to the outside air.

NEW COLLECTOR DEVELOPMENTS

Commercially available flat plate collectors at present are suitable for low-temperature operation, up to 80°C. CSIRO research workers believe that advanced flat plate collectors should be suitable for medium-temperature operation, up to 150°C. Such a collector is currently being developed. It would use a pressed sheet metal absorber, with a selective chromium black surface ($\alpha/\varepsilon > 8$), a single sheet of low-iron-content toughened glass, with an antireflective (etching) treatment of the outside surface, and a convection suppression device between the glass and the absorber surface.

For higher-temperature collection, evacuated tubular glass collectors have been developed by Philips in Germany, by Corning Glass, Owens-Illinois, and GEC in the USA, and by Sanyo in Japan. The vacuum between the inner absorber tube and the outer glass eliminates convection losses, and thus collection efficiency is greatly improved, especially at higher temperatures. In some, the inner tube is copper. This can have a selective surface, and thus the radiant losses are also reduced. There are, however, difficulties in producing a metal-to-glass bond which would retain a permanent vacuum. Others are of an all-glass construction, where there is no problem with the seal, but—until recently—no selective surfacing existed which could be satisfactorily applied to glass.

The Sydney University Solar Energy Group (notably Brian Window) produced (in 1977) a selective surface which can be applied to glass and retains its characteristics at high operating temperatures. An iron carbide compound is deposited on the glass absorber tube over a copper base by a vacuum sputtering technique. With such tubes, using a back reflector, stagnation temperatures up to 350°C can be achieved and a normal operating temperature of 200°C is sustainable under average conditions. The work now continues for the optimization of array configuration and manifolding arrangements.

An alternative for higher-temperature collection is the use of concentrating collectors. These must have a tracking mechanism, which is expensive; they only collect beam radiation. The University of New South Wales team, lead by John Giutronich developed a nontracking (stationary) asymmetrical, nonimaging concentrating collector, which produces a concentration rate up to 5, collects a large part of the diffuse radiation, and achieves temperatures up to 200°C.

There are some new ideas also in flat plate collector technology. The BHP (Broken Hill Proprietary Ltd) Laboratories in Melbourne, in con-

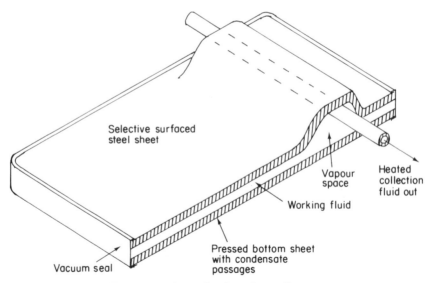

Figure 16   A condensing plate collector

junction with AMIRA (Australian Mineral Industries Research Association), developed a flat plate collector which works on the heat pipe principle. A sealed evacuated space is created between two steel sheets. The plate thus formed is installed at a tilt, with a horizontal heat removal tube penetrating it across the top (Figure 16). A working fluid evaporates in the plate when there is a solar input, the vapour rises to the top, condenses on the heat removal tube surface, transmitting its latent heat, and the fluid trickles back to the plate. This gives a simple, quick-response, low-thermal-capacity collector which is easy to manufacture, free of frost risk, and has very little resistance to the heat removal fluid flow.

A heat pipe is also used along the focal length of a parabolic trough concentrating collector developed by the Little brothers at Mount Isa. The top (condensing) end is immersed in the water tank. The long axis of the trough is tilted at the latitude angle and allows an east-to-west single-axis tracking movement. Their invention is a very simple and inexpensive tracking device which operates without any motor-driven mechanism by differential heating of a fluid in two cylinders on either side of a control diaphragm.

OTHER WORK

The above gives a review of work in the area of thermal applications of solar energy. No less significant is the work in other areas. At least three centres are engaged in photovoltaic device development. The CSIRO is doing fundamental studies on CdS and Cd_2SnO_4 films. Lyons at the University of Queensland is investigating CdTe and GaAs films. Godfrey and Green at the University of NSW are using silicon, not in the p–n junction configuration but in MIS (metal–insulation–semiconductor) cells. The University of Queensland also has a project using optical concentrators in conjunction with water-cooled photoelectric devices.

Several centres are involved in solar/biomass/ethanol fuel production research. The University of Queensland team is primarily concerned with the agronomy of potential fuel crops, notably cassava (tapioca), whilst the Sydney University group concentrates on conversion technology.

Wind is in fact a secondary form of solar energy, and thus it may not be inappropriate to include some notes on wind energy developments. Both the high solidity pumping windmills and the low solidity (propellor) type aerogenerators have a long history in Australia. They were widely used up to the early 1950s. The established manufacturers went into obscurity during the two decades of the cheap fuel era and most farms have substituted diesel generators for their windmills. These firms are now re-asserting themselves. Dunlite Pty Ltd is marketing the well tested 2 kW unit, but also developed a new 5 kW aerogenerator which can operate with only moderate windspeeds. Several projects are aimed at producing hybrid generating systems, e.g. a wind + solar–thermal–electric generation.

Much of this work is very promising. The greatest need seems to exist in bridging the gap between research and the industrial or commercial reality. This is a slow process and has a long lead time. It has been suggested recently by the Premier of Victoria that the various forms of solar energy utilization could contribute as much as 10% of the national energy supply by the year 2000. This would, however, require a solar energy industry comparable in size, capital investment, and labour force to our present motor car industry.

REFERENCES

CSIRO, 1964, *Solar Water Heaters*, Circular No. 2, CSIRO, Division of Mechanical Engineering, Melbourne.

Drysdale, J. W., 1952, *Designing Houses for Australian Climates*, Commonwealth Experimental Building Station, Bulletin No. 6, Sydney.

Institution of Engineers (Australia), 1977, *Recommendations for an Energy Policy for Australia* (incl. working papers), Canberra.

Muncey, R. W., 1955, Optimum thickness of insulation for Australian houses, *Aus. J. Appl. Sci.*, **6** (4).

Proctor, D. and Morse, R. N., 1977, Solar energy for the Australian food processing industry, *Solar Energy*, **19** (1), 63–72.

Robeson, K. A., 1970, Evaluation of evaporative rock bed regenerative coolers, *Aus. Refrig. Air Cond. Heatg*, **24** (2, 3), 21–45, 48–51.

Sheridan, N. R. and Carr, W. H., 1967, *A Solar Air Conditioned House in Brisbane*, Univ. of Queensland, Solar Research Note, No. 2.

Szokolay, S. V., 1977, *Air Conditioning in Tropical Australia and the Role of Solar Powered Methods*, PhD Thesis, Univ. of Queensland.

Energy Conservation and Thermal Insulation
Edited by R. Derricott and S. S. Chissick
© 1981 John Wiley & Sons Ltd.

CHAPTER 24

Case study: houses for people to build

Cedric Green

University of Sheffield

In this chapter some experimental approaches to the design of low-energy, low-cost housing are described and illustrated, and their implications discussed. The houses concerned were all designed to be partially or wholly built by relatively unskilled persons, partly from necessity but also from a desire to explore the problems of designing for owner-builders. The cost of new housing now is forcing increasing numbers of people to try to make their own houses, either collectively in self-build groups or individually, and too few architects have so far addressed themselves to the problem, with the result that the standard of design is deplorably low and many possibilities that are opened up by the elimination of the standard contracting procedure are not being realized. Acknowledgement must be made, though, to the architect who is an exception to this state of affairs, Walter Segal, who has evolved a type of design and approach to construction that has produced a number of very successful houses, and some of his precepts and ideas have influenced this project.

The first house described here is 'Delta', a detached timber-framed, four-bedroom house (Figure 1), designed by the writer for his own family, intended originally to be partially self-built, but in the end constructed by Mr A. T. Saunders and his son Stanley. It was also an experiment in passive solar heating using a south-facing lean-to conservatory and was evaluated for one heating season. As a result of the lessons learned from the computer analysis of the data from Delta, a project was begun at Sheffield University to build a solar house incorporating everything that had proved to be successful, and improving on those aspects which were not. The Sheffield project (entitled SHED—Solar Housing Experimental Design) was conceived more clearly for construction by self-builders, and the first phase was built by students without any experience of building, under the supervision of the writer. The project is in two parts, the first being the construction of a part of the dwelling as a self-contained unit to test various

635

Figure 1   Delta: photograph from south-east

methods of construction, insulation, ventilation, solar collection, and thermal storage, and the second, completion of the whole house incorporating any lessons learned from the first.

Delta was built to try to prove a point or, to put it more scientifically, to test a hypothesis. It was suspected that the renewed interest in the use of solar energy in the early 1970s and the emphasis on complex technological subsystems for solar collection and distribution obscured the simple fact that the most efficient and economical system was the whole building, designed in an integrated way to act as collector and store without the need for separate subsystems. The thermal properties of greenhouses and conservatories have been familiar since glass became an abundant building material in the nineteenth century and solar overheating of overglazed south-facing walls is a phenomenon of which no architect could fail to be aware in the twentieth century. Many architects have commented on the surprisingly low heating costs of well insulated houses with south-facing glazed walls, and often unwittingly have produced highly efficient 'solar collectors' for space heating in winter.

The scientific justification for this statement should be briefly introduced at this point. The 'efficiency' of a solar collector is the percentage of the radiation falling on it that is absorbed as useful heat to the interior of the building, and all other things being equal, is inversely proportional to the difference in temperature between the air inside the glass (or other transparent plane) and that outside. The greater the temperature difference, the more the heat loss through the front of the collector, and thus the less heat left for use inside. Leaving aside such specialized types as vacuum tubes with water pipes in the middle, most systems that use water or similar fluids in a collector under glass rely on achieving temperatures over 40°C, the minimum temperature of the return water in a central heating system. The temperature of the air trapped in a sealed collector rises much higher than this, and on a sunny day in winter the temperature difference between the air inside and outside the collector can be as high as 45°C, with radiation on the outside of the glass over a 4 h period of 1200 w h/m². Thus the heat loss through the front glass of the collector will be 1008 W h/m², leaving 192 W h/m² to be absorbed by the water. This is a collection efficiency of 16%. Under the same conditions of sunlight and outside temperature, a window to a space in which the temperature does not rise above 20°C and where the difference between inside and outside temperature is 15°C loses 336 W h/m², leaving 864 W h/m² available for heating the interior—a collection efficiency of 72%. Thus it is clear that an ordinary window to a building that has enough thermal capacity to absorb the solar gain without overheating is about the most efficient solar collector that has been invented.

Delta was designed with this knowledge, but also with an awareness that if the area of glass was increased greatly in a well insulated building, the heat gain could be greater than the thermal mass of the building was able to absorb. The temperature would then rise to reduce the efficiency, and, worse than that, produce discomfort. But for other reasons, to do with ease of construction by the owner, the house was to be of timber construction, and would therefore be low in thermal mass. There seems to be an insoluble contradiction between the requirements for self-build and the needs of an efficient solar heating system. The solution was to provide thermal storage in a heavy ground floor raft with black tiled finish (containing the underfloor back-up heating system), to provide the south-facing glazing as

Figure 2 Delta: the conservatory from the entrance showing shading blinds drawn over the upper half of the glazing

a separate conservatory from which warm air could be drawn into the house as required (or excluded to prevent overheating in summer) and to separate the living accommodation from conservatory by glazing and heavy curtains (Figure 2) which could be used for retaining stored warmth at night. The 30 m^2 of conservatory glazing become the outer plane of a double-glazed south wall, but which provided useful accommodation in the 'cavity'. The space in the conservatory could be ventilated and blinds under the outer glass sloping at 53° were adequate to control heat gain to the interior in summer. The building was totally passive in its use of solar heat and was therefore not expensive to build. Figures 3 and 4 show plan and section.

There was additional thermal storage in the concrete blockwork walls of the 'core' consisting of kitchen, cloakroom, and shower, containing all the drainage and electrical services. Originally this was to be done by a building contractor, and left with a temporary PVC roof membrane, providing a fully serviced site hut from which owner and friends could work, putting up the rest of the timber-framed and clad structure. Although it was not carried out like this, the concept of constructing all the services first and concentrating them in a core, perhaps even prefabricated, has a lot of attraction in owner-built housing. In the case of Delta, the idea imposed a discipline on the whole design that resulted in a relatively low cost—£11 000 in 1974.

In addition to a higher than usual standard of insulation ($U = 0.3$ for walls and roof), openings were weatherstripped, and access into and out of the house was, in winter, through the conservatory which provided an air-lock, and so air change was kept to the minimum that was comfortable. It was intended from the start to evaluate the precise thermal performance, and so the lack of gas service in the village made the choice of off-peak electric storage heating inevitable, being the only source that could be precisely metered. The ground floor slab was the basic heat store while upstairs block storage units were placed with their backs to the inner conservatory glazing to absorb direct radiation from the sun. Evaluation consisted of recording temperatures and humidity in various zones of the house and outside, off- and on-peak electricity inputs, recording activities, numbers of people, and weather, all of which were put together with records from a nearby meteorological station and analysed with a computer program at Sheffield University.

The results were encouraging if not as good as originally hoped. Over a four-month period from the last week in December to April 1975, the amount of radiation on the various glazed planes of the house that would provide any heat gain was 7951 kWh—a figure derived by putting the meteorological data provided by Dr F. Cope of the Levington Research Station into a computer program developed by Professor J. K. Page at

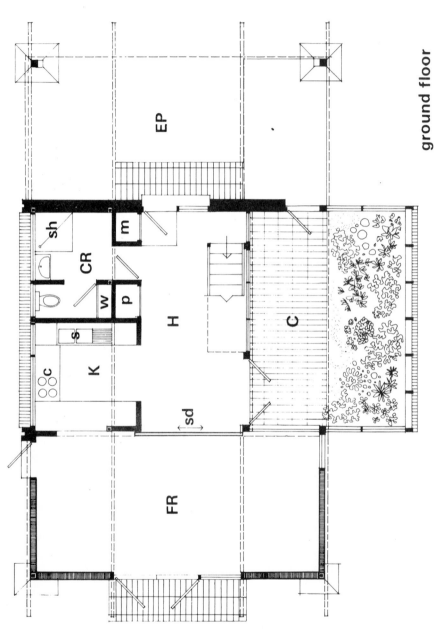

ground floor

Figure 3 Delta; ground floor plan. FR—family room; H—hall; K—kitchen; CR—cloakroom with shower and WC and all services in core with kitchen; C—conservatory; EP—entry porch and temporary car shelter; sd—sliding door; w—hot water tank; m—meters; p—cupboard; s—sink

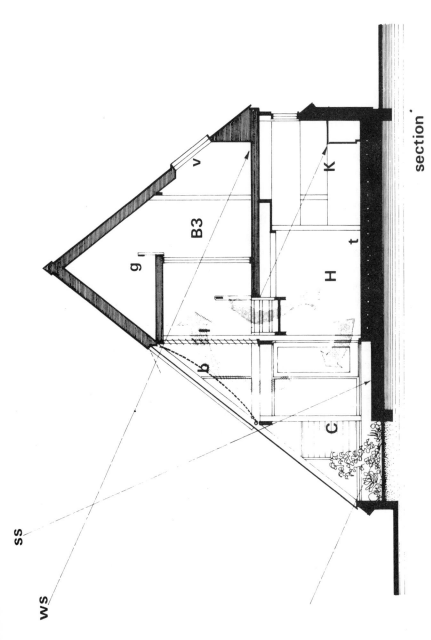

Figure 4 Delta: section. B3—bedroom 3; g—gallery; v—roof window; b—shading blinds; t—black tiled concrete floor raft for thermal storage; ss—summer sun angle; ws—winter sun angle; l—louvres

Sheffield University to predict quite accurately the amount of radiation falling on surfaces at given orientation, slope, position, and altitude. The total amount of energy used by the building for heating during the period was 12673 kWh of which 5480 kWh was provided by the Sun. This figure included the heat required to warm the conservatory itself, and represents a gross percentage of available energy of 69%. But if the heating of the conservatory itself were ignored, and one only took the heat gained to the interior living areas, the amount was 2919 kWh which gave a nett efficiency of 37%. Over a third of the total available radiation on the glazing was collected as useful heat to the interior, and the proportion of interior heating needed which was provided by the Sun was 23%. But because the conservatory is more than just a solar collector, and is an integral part of the living space of the house, the solar energy that is used to keep it warm during the day should be counted as useful gain, and the ratio of solar energy to total energy required increases to 43%.

But there were ways in which the system could be improved both to gain efficiency and to increase the total gain to an amount that would provide a much larger percentage of the heating needed. In Delta, the temperature rose much higher than necessary in the conservatory, with a consequent lowering of efficiency; there was not enough thermal storage for solar gain, or rather, the gain did not go into storage because of rugs, furniture, and a high polish effectively preventing sunshine from warming the black floor tiles. If there had been a much larger area of conservatory glazing, the overheating would have been uncontrollable. The experiment showed that there is a limit to the area of conservatory glazing in a totally passive system that is related to the amount of thermal storage that can be provided. But it is also very critical how the mass of the thermal storage is placed to receive direct sunshine. Delta would have worked much better if there had been a black painted mass concrete wall 150 mm thick for the full two storeys beside the staircase inside the inner plane of glass. Another important lesson from Delta was that warm air from the conservatory came into the upper floor and kept the bedrooms warm all day while they were not being used, and the underfloor heating on the ground floor kept the living areas warm all night while they were not being used. The whole house was unnecessarily warm for 24 h and it is easy to see that a system that allows a zone to cool down a little while it is not being used will make a saving of about 10% in the total energy required.

As a result of this experience, and a study of the economics of various systems for solar collection and energy conservation in housing, a new prototype house was designed for construction by students at the Department of Architecture at Sheffield University where the author was working. Whereas Delta was a cautious and extremely restrained approach to the problem of energy conservation, the SHED was conceived as a much more ambitious projection into a hypothetical future. Delta had to satisfy all the conservative criteria of a Building Society to obtain a mortgage (which was why it could not in the end be owner-built), meet the existing Building Regulations, and satisfy a Planning Authority in a rural county where anything that looked slightly strange would be refused planning permission. SHED, because it was a purely experimental building, financed without Building Society help, on a temporary site, could disregard some of these constraints if they conflicted with the main aims of the project. The future to which the project looked was one in which the costs of fossil fuels had risen so much that makers of houses would be looking for ways of reducing domestic energy consumption to negligible amounts. But at the same time it was felt that a serious energy crisis would have such an effect on the overall economy that constraints on housing construction costs would be just as severe as they are now. So whatever the costs of fuel, it would be the lowest-cost methods of energy conservation that would succeed, provided that they did not give lower standards of amenity than we expect now. The architectural objective became an extremely important one—to produce an inexpensive but attractive house which was demonstrably successful in both conserving energy and in providing the required standards of comfort and shelter.

The design also projects an ecologically sensitive way of life in an urban framework. Detached houses in rural areas will never form more than a very small proportion of all housing, and the real problems for the future will be the conservation of resources of all kinds: energy, water, mineral, land, existing building stock, food, trees, vegetation, and landscape, in existing and new cities and towns. It is assumed that pressures of population will put greater pressure on the land available, require more intensive cultivation of food-producing crops, and protection of rural land from suburban sprawl, and require better use of urban gardens and open spaces for the production of food and useful plants.

Thus many features of Delta, despite its conception for the rural context, would be transferable to the urban house: the conservatory as a basis for the solar collection system and for plant growing; the high standard of insulation and ventilation control; timber construction for ease of owner-building and use of a renewable organic material; and a direct and semi-passive airborne solar heating collection and heat transfer system for efficiency and economy. The significant changes were: a more compact plan that could be used in terraced or semidetached form at a medium density; much more thermal storage in the form of stacked solid bricks or blocks through which warm air from the collector would be circulated; and a much larger area of south-facing glazing to conservatory or roof to increase

Figure 5 The SHED project: completed first-phase building from the south-west

the percentage of solar energy collected. The precise relationship between these last two features was calculated to minimize the chances of uncontrollable overheating at any time of year. In addition it was calculated that if it were necessary to aid the heat transfer from air collector and conservatory to heat store by means of a fan, the energy consumed would be negligible and the additional capital cost could be kept low if the ducts were integrated into the construction rather than treated as a separate subsystem; the potential increase in effectiveness and efficiency would quickly pay for the extra cost.

The first phase of the project (Figures 5 and 6) was the construction of

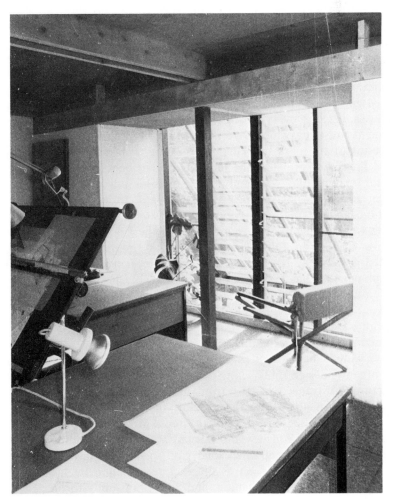

Figure 6   SHED: interior of first phase being used as project office

40 m^2 of the house with a conservatory, heat stores, and enough features of the tentative final house design to test them before building the whole thing (Figures 7 and 8). Three students, doing their year in practical training, were employed under a Job Creation Scheme to work on the project, from doing basic working drawings, obtaining consents, ordering materials, digging foundations, fabricating the frame and panels right through to glazing and finishing the first stage. In the course of the project, other students worked on it in voluntary work parties. Throughout this process, the system of construction devised for use by relatively unskilled users was being put to the test. An added complication, but as it turned out in the end, a benefit, is that the building has to be removed from the site when the experiment is over, and so in order not to waste all the material and work in fabricating the components, it was designed to be completely demountable. The benefits are that it is extremely easy to change things and experiment with different arrangements of heat stores, solid and glazed panels, and to add on the second phase of the design—the rest of the three-bedroom house (Figures 8–10). The discipline that this requirement imposed gave the building an aesthetic quality similar to traditional Japanese houses, because every dimension was determined by the proportions of standard board sizes, which were used and cut to avoid any waste, and also as a result of the dark-stained timber frame, light infill panels, and suspended floor. But probably the most significant benefit is that the method used for making the wall and roof panels—sandwiches of marine ply, 150 mm extruded polystyrene, and chipboard—provides structure, high insulation, and internal and external finish simultaneously. Adhesives and jointing were problems that were solved in the first phase, and so should not give trouble in future. As there is no interstitial cavity and the extruded polystyrene has a closed-cell structure, condensation has not been the problem that has been found in other highly insulated buildings with low ventilation rates. Finally if this method of construction can be accepted, it should be of great benefit to owner-builders because of the ease with which they can extend or change their houses, as well as making construction much easier and faster.

There were a number of other important features that arose from the basic organic design philosophy—an approach that has been labelled 'ecotecture'. A frame over the front of the conservatory, in addition to providing tension support for the structure of the glazing to keep it as light as possible, was also there to support rapid-growing climbing plants like vines, hops or beans which would put out a canopy of shading leaves over the glazing in the summer when solar gain was not required. Plants inside the conservatory in winter act as filters for ventilation by carbon dioxide conversion and oxygen production, and add significantly to the amount of solar collection in two ways: energy from the sun is absorbed in transpiration and deposited in the heat store when the water vapour condenses, and

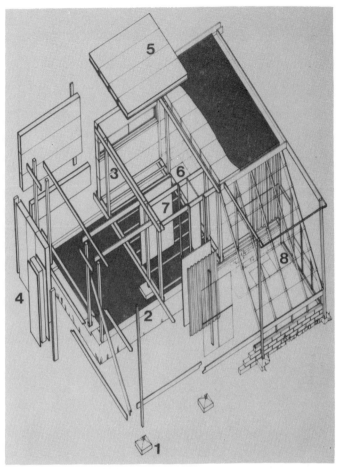

Figure 7 SHED: exploded component drawing of first-phase prototype. *Key:* 1, Foundations—galvanized steel shoe anchored into concrete pad foundation; strip foundation for low-level block wall. 2, Subfloor construction—fireclay heat storage blocks stacked to allow warm air circulation around them on 100 mm extruded polystyrene insulating slabs; damp-proof membrane below. 3, Basic timber frame—structural timber posts and beams bolted together with galvanized steel bolts and timber connectors. 4, Wall panels—150 mm extruded polystyrene slabs sandwiched by plywood on the external face and chipboard on the internal face; panel-sized horizontal or vertical double-glazed timber window. 5, Roof panels—150 mm extruded polystyrene slabs sandwiched by two sheets of plywood; membrane covering panels protected by camomile turf roof finish. 6, Floor construction—chipboard panels supported by softwood timber joists spanning between main beams. 7, Heat store construction—insulation board on timber framing surrounding heat storage blocks; black-painted profiled metal sheet behind float glass on timber bars. 8, Conservatory construction—light timber glazing bars supporting 'horticultural' glass bedded on PVC-covered foam strips held down with cadmium-plated clips, with acrylic sheet hinged ventilators; timber frame supporting ties to purlin and wires for external plants

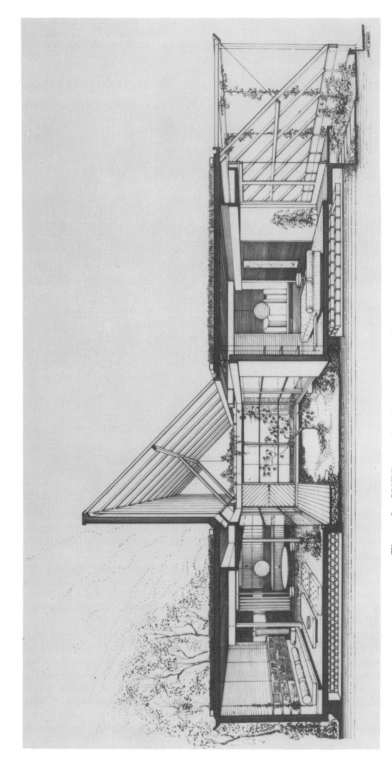

Figure 8   SHED: sectional perspective of the final house design

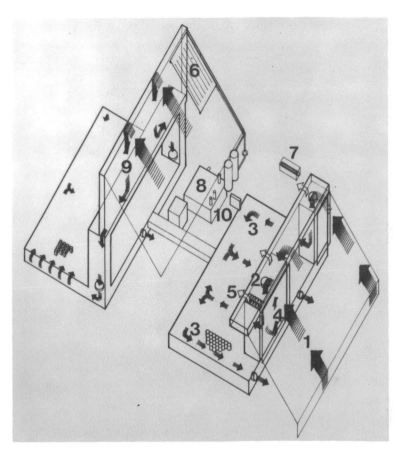

Figure 9 SHED: diagram of the thermal system of second-phase
house. *Key*: 1, Hot air in the conservatory rises and enters the
horizontal duct at the top of the heat store. 2, When the temperature of this air is hotter than the bottom of the heat store, a fan
pushes it down the vertical heat store; as it passes through the heat
storage material the air gives up its heat. 3, Another fan assists the
flow of air through the horizontal heat store back into the conservatory after giving up its heat. 4, At the bottom of the heat store the
air is free to enter the airspace between the vertical glazing and the
black metal sheet; the hot air rises into the top of the heat store,
sucking into the space more air from the bottom of the heat
store by thermosiphonic action. 5, Flaps at the top of the heat store
may be opened by occupants to release hot air to the living space;
this may be fan assisted at night. 6, Low-cost flexible solar collector
under conservatory glazing connected to first hot water tank; the
system may be extended to cover the length of the conservatory
during summer, removing excess space heat to heat domestic water.
7, Cooker fumes recirculation unit. 8, Waste hot water holding tank
under floor of services unit. 9, Warm air from top of court conservatory is drawn down with the hot air produced in vertical high-level
collector and pushed into the heat store under the floor. 10,
Electronic 'black box' automatically switches fans and flap mechanisms according to temperature differences in spaces and stores

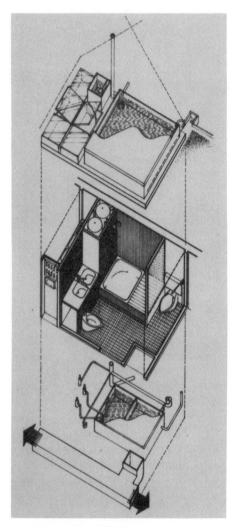

Figure 10 SHED project: exploded drawing of low-energy and water-conserving bathroom/services core unit

the added area of leaves increases the heat transfer to air. As a part of the resource conservation effort, the use of water is minimized (Figure 10), basically by collection of rain water from the roof for washing purposes, and so the flat part of the roof is covered with a camomile herb turf, which acts as an organic filter at the same time as it protects the roof finish and provides a little more insulation. Thus plants are used as a functioning part

Figure 11 SHED project: perspective and block layout of medium-density low-rise housing

of the natural conditioning of the building which is made as much as possible from organic products or recycled materials.

The first phase was completed in time for evaluation over the winter of 1978–79 and, although at the time of writing, this has not been completed, enough has been done to know that the performance is up to the expectations shown on the graph of predicted performance (Figure 12). The conclusions at this stage have led to the final design of the complete house to be built as the second phase in 1979, illustrated here (Figures 8–10). But in addition to the single house, designs were produced for groups incorporating different types and sizes of house, and layouts exploring the problems of overshadowing, density, and open space distribution and use. A number of private commissions for small low-energy houses or extensions to be built partially by their owners have given an opportunity for applying the experimental system, and a few of them are illustrated here (Figures 13–15). All these designs formed an entry to the Sixth Misawa Homes International Design Competition in Japan, in which they were awarded third prize.

The conclusions that can be drawn from this project are interesting and valid even at the present level of energy and resource costs. The implications go beyond housing, and the system for solar collection, storage, and conservation can be applied to many other types of building. The self-build aspects are valid only for the housing field, but are applicable as much to the rehabilitation and thermal improvement of existing building stock as to brand new housing, and the final report will include sections on conservatory extensions with methods for integrating new heat stores and controls in existing cellars, subfloor spaces, and underground. But the basic thinking behind the thermal system lends itself to much wider application—in fact to any building where low-cost means are sought for drastically reducing continuous heating energy consumption, where siting and orentation are suitable for autumn and spring solar collection, and where sufficient inclined glazing may be provided as conservatories, arcades, glazed courts, corridors, or any type of sheltered semi-external space. The kind of design that would result has been anticipated in many schemes that have used patent glazing in large areas, and the means to make that kind of design extremely effective in energy terms are largely invisible. There would be a much stronger orientation bias with the only noticeable feature being vertical collector glazing on the north side of glazed spaces, and the gentle hum of fans putting the excess heat into underground heat stores.

The suggestion of an ecologically sensitive way of life implied by the prototype house may not be accepted very widely for some time in all its details, but the scheme does not depend entirely upon it. It depends on an integrated combination of methods of energy conservation, construction,

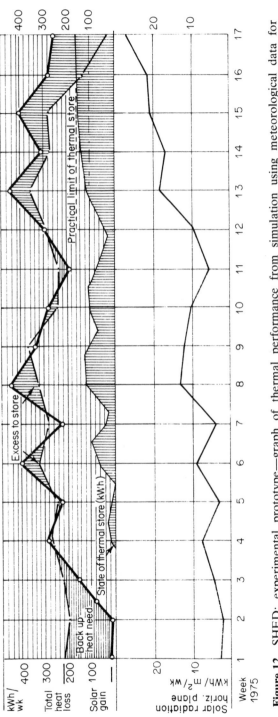

Figure 12 SHED: experimental prototype—graph of thermal performance from simulation using meteorological data for January to April 1975. Note that no backup heating is required after fourth week in January; estimated total heating requirement for shortened 6–8 week season is 1200 kWh

Figure 13 Proposed low-cost autonomous house on small rural site with water-powered electrical supply from stream at end of garden

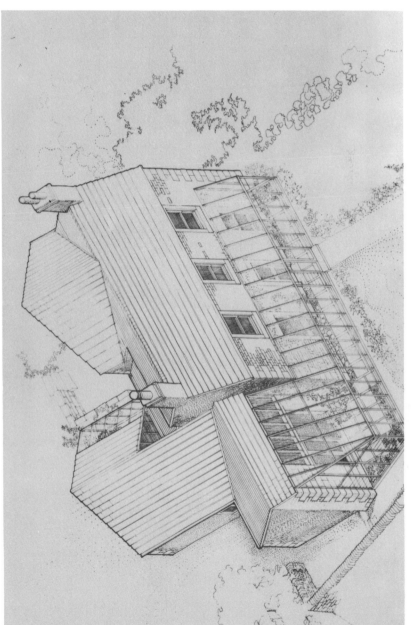

Figure 14 Extension to thick-walled stone cottage to improve thermal performance

Figure 15 Housing design for hypothetical site

planning, simple technology, and intelligent use, none of which are particularly radical, and none of which is developed to optimum levels of performance. Used together, they produce an effect that is greater than the sum of the parts and, because so many elements are used to perform more than one function and support each other, the result is more economical than an engineering approach which isolates the subsystems and develops each one in parallel with all the others. This approach developed as a response, not just to the energy problem, but to a whole complex interlocking nexus of design challenges that centred on the design of an urban house in the context of changes to the way that we are able to use resources in the very near future.

Energy Conservation and Thermal Insulation
Edited by R. Derricott and S. S. Chissick
© 1981 David Croghan
Published 1981 John Wiley & Sons Ltd

CHAPTER 25

The architect in energy conservation: a case study of a Cambridge College

David Croghan

Anglian Architects, Cambridge

1. INTRODUCTION

A full energy conservation programme for an existing building will usually commence with the collection and assessment of information on the building fabric, the type and performance of services installations, the pattern of use, fuel accounting methods, and general management procedures. For the consultant, this initial phase will be often the most demanding and time-consuming.

The second phase of his work concerns the identification of any areas that are deficient in relation to the current state of the art, and the consideration of how these may be improved economically. This usually will involve the continuation of the monitoring of fuel use and comfort conditions that will have been established during the earlier phase, detailed design of any fabric and services modifications and maintenance work, obtaining estimates, and cost–benefit analyses.

The third phase is the implementation of beneficial works, including letting contracts, site supervision, and general administration such as checking accounts. Finally, the building owner should be advised on any desirable variation of management procedures and energy accounting methods. The preparation and supply of a compendium of fully updated energy-related information on the subject building may also be useful, together with the recommendation that all future 'energy events' should be logged.

Effective energy conservation consultancy may thus involve a good deal more than the adjustment of a boiler, the casual prescription of loft insulation, or the appearance of a solar panel. It may involve taking a building to pieces—usually only on paper, but sometimes literally—and putting it back together again in a way that corrects defects and introduces modern techniques. Also, as above, it may involve the close study of how a building is used and managed. Frequently it will be found that no plans or other physical records exist, and these require to be prepared.

Much of this work is within the normal scope of an architect, though—where he is not already experienced in renovation work—he may have to adjust to operating a pattern almost the reverse of that normally involved in the production of new buildings.

The architect is used to working with other specialists, both fellow consultants and manufacturers and contractors, welding them together to form a team that achieves the result required by his client. This is the same in energy conservation work as in 'new build'. The architect customarily takes overall responsibility for any project, and his training has embraced aspects of the work involved in most of the required specialities. He is the only member of the team usually in direct contact with the building owner, and is practised in taking a comprehensive and impartial view of the problem. He has no financial connections with those trading in building supplies, and he will not attempt to sell double-glazing to answer a problem best solved by a thermostat.

Most architects spend most of their time on technical design and technical administration rather than on aesthetic design. It is from this often hidden but firm base that architects frequently have taken the technical initiative: not perhaps in basic physics and engineering, but certainly in application. Architects have been consciously insulating walls and roofs, unasked, since the emergence of the profession. Some have had heat pumps and solar panel installations for 25 years or more, and a number of buildings exploiting passive solar energy were built in the 1930s.

The author of this paper is an architect in general practice and also now much involved in the development of consultants' fieldwork aids. These include a solid-state memory-store temperature recording 'bug' and associated desk-top microprocessor that should greatly ease speed and accuracy in monitoring the performance of heating and air-conditioning systems.

2. SYNOPSIS

An energy conservation programme in one of the old Cambridge Colleges produced immediate savings of 16%, balancing fuel price inflation. Over a three-year period, the college will reduce energy consumption by 45% and costs by 54%, with full payback of the investment in adopted measures and consultancy.

The study first established the function and operational efficiency of the existing system, and a proposal for transferring the bulk of the heating load from four main boilers onto a single modern unit was validated and implemented. At the same time the buildings were closely surveyed physically and in terms of pattern of use, and various insulation standards and management measures were optimized using the British Gas Corporation's THERM computer program.

3. BACKGROUND

Pembroke College, Cambridge, was founded in 1347 and appears today as a complex of mainly mediaeval and Victorian buildings. These are generally three stories high and built around courtyards (Figures 1 and 2). The total floor area is 12 400 m^2 (133 500 ft^2) spread over a site area of about 2 hectares (5 acres).

Between 1970 and 1976 the College experienced price rises of 560% for oil and 410% for steam coal, while electricity and gas prices increased some 250%. Overall energy costs for similar consumption doubled every three years to reach nearly £25 000, and expenditure approaching £39 000

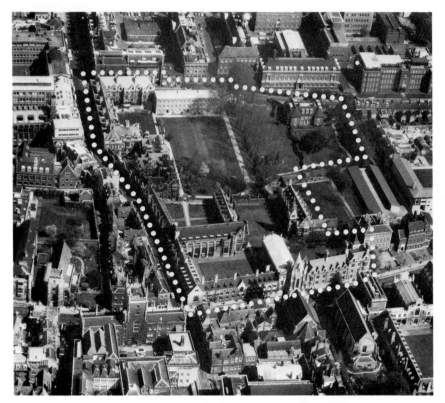

Figure 1 Pembroke College, Cambridge, from the air (Photo: Cambridge University Collection: copyright reserved)

Figure 2   Pembroke College from Trumpington Street

was in view for 1978–79. By the same year the costs of maintaining, stoking, and insuring aging boiler plant were expected to have risen from a nominal amount to some £10 000 per annum.

The College, whose income does not respond to inflation of this order, decided that a full utilities conservation study was required and commissioned the present author, who for some years had acted as consulting architect in the College's buildings renovation programme.

4. INITIAL INVESTIGATIONS

As frequently is the case with such institutions, mechanical services had been installed piecemeal over many years. The consultant's first and often lengthy task is to unravel and record them.

In this case, 'as fixed' drawings were available for nine building modernization contracts that had been carried out over the last 16 years. These covered the provisions for space heating and domestic hot water in individual blocks comprising about half the College. The remainder required to be surveyed and equipment tested before an overall picture could be developed.

At inception, there were four oil-fired boilers, one solid-fuel stem boiler, and two small gas-fired units. Incoming utilities passed through 12 main electricity meters, 9 main gas meters, 13 water meters, and three oil storage tanks. In addition there were some 300 gas and electricity bye-meters.

5. ENERGY AUDIT

Detailed information on energy purchases, applications, cost distribution, and annual variation is central to any utilities conservation study.

The College provided detailed fuel accounts for the year preceding commencement of the study, 1975–76, and supply tariffs were checked. The further breakdown of these accounts led to the development of an annual audit system based on the completion of two *pro formas*. These were designed to present summary information of immediate practical use. They measure performance over a given accounting period, aid budgeting, and act as a watchdog in the maintenance of a conservation regimen. Energy Audit Form A gives amounts and costs of purchased energy and a series of indices useful to annual comparisons. Figure 3 shows it completed for the base year 1975–76. A similar compilation for 1976–77 indicated the 'heating energy index' reduced by 15% and the 'total energy index' by 13%. Though net energy cost had increased by 5.5%, extension of 1975–76 consumption at 1976–77 prices showed that without conservation activity energy cost would have advanced by 16%. In actuality overall energy system expenditure reduced by 0.7%, due in part to the reduction of maintenance expenditure on equipment being phased out.

The same form was completed for a notional 1978–79 in which consumption was maintained at 1975–76 levels but costs advanced in accordance with current experience. This showed gross energy system cost increased 62% and left no doubt that conservation action was necessary.

Figure 4 predicts the actual 1978–79 situation: energy consumption reduced by 44.6% and costs reduced by 53.5%. This is offered with confidence as predictions for 1977–78—the year in which system modifications

PEMBROKE COLLEGE CAMBRIDGE

SUMMARY OF ENERGY AND WATER CONSUMPTION AND EXPENDITURE FOR 1975/76 (August to July financial year)
(Main Site, including Master's Lodge - total floor area[a] 12 402 m²)

A. ENERGY

Type of Energy	Quantity Purchased				average price/unit[b]	cost £ [b]	Energy -Imperial units equivalent therms[c]	cost per therm	Energy - S.I.units[d] equivalent gigajoules	cost per gigajoule
	tons	litres	therms	kWh						
steam coal	175				£30.10	5 259	48 650	10.8p	5 121	£1.03
house coal	1.25				£45.30	57	331	17.2p	35	£1.63
oil		171 007			5.35p	9 153	62 256	14.7p	6 553	£1.40
gas			18 051		14.0p	2 527	18 051	14.0p	1 900	£1.33
electricity				426 262	1.85p	7 865	14 536	54.1p	1 530	£5.14

NET ENERGY COST [e]   £ 24 861
add: maintenance   3 508
    stoker's wages   2 118
GROSS HEAT & LIGHT COST £ 30 487

TOTAL ENERGY CONSUMPTION   143 824 THERMS   15 139 GIGAJOULES

Number of 'Degree Days' in period (below 15.5°C base)[f]   2 233

B. WATER CONSUMPTION in period [1 322] x 1000 gals.   TOTAL WATER COST £ [451] [g]   Average cost per 1000 gals [34.1]p.[g]

C. CONSUMPTION INDICES [h] :

① Water Consumption Index
(gals/m²/pa)

$$106.1$$

② Electrical Energy Index
(therms/m²/pa)

$$1.17$$

③ Heating Energy Index [i]
(therms/m²/pa x 0.63 x 1.12)

$$7.36$$

④ Total Energy Consumption Index
(Index 2 + Index 3)

$$8.53$$

NOTES

[a] *floor area measured inside external walls excluding unheated basements, boiler rooms, open porches and cloisters, but including internal walls, cupboards, etc.*

[b] *including fixed charges where applicable.*

[c] *measure of gross heat input by direct fuel to heat conversion (i.e. not adjusted 'useful therms' taking into account system efficiencies as commonly used for comparisons of heat outputs).*
Conversion factors used: Steam Coal – 278 therms/ton; House Coal – 265 therms/ton; Gas – unity;
Oil (35 secs.) – 0.364 therms/litre; Electricity – 0.0341 therms/kWh

[d] *Gigajoules (10⁹ joules) equivalent to 9.5 therms or 278 kWh*

[e] *includes space heating, domestic hot water, cooking and lighting.*

[f] *as published by Department of Energy for period September to May. Annual comparisons give indication of climatic factor affecting energy consumption ('Degree Days', Fuel Efficiency booklet No.7, HMSO 1977 and D of E monthly tables).*

[g] *excluding local water rate.*

[h] *as proposed by University of Cambridge Estate Management & Building Service for annual internal comparisons and for external comparisons with other institutions ('UK Universities Utilities Consumption Survey 1974/75', E.M.B.S. Sep 1976).*

[i] *index ③ excludes all electricity, but includes water heating and cooking with space heating where other fuels used. The numerical factors adjust for building type and climate on 'degree day' basis, both as proposed in op.cit.(h) above. The College is taken as wholly residential (i.e., no separate consideration of its communal and teaching areas) and the occupation factor of 0.63 is an annual constant. The Degree day factor is derived by dividing the Sep-May total for East Anglia in the given year into a constant of 2500 that gives a base factor of 1.0 for areas in Scotland.*

CROGHAN + LEWIS   © 1977
associated architects and environmental consultants
39 NEWTON ROAD CAMBRIDGE ENGLAND telephone 0223-63364

Figure 3 Energy Audit Form A completed for 1975–76 (base year of study)

PEMBROKE COLLEGE CAMBRIDGE

SUMMARY OF ENERGY AND WATER CONSUMPTION AND EXPENDITURE FOR 19 78/79 (August to July financial year)
(Main Site, including Master's Lodge - total floor area[a] 12 402 m²)

A. ENERGY

Type of Energy	Quantity Purchased				average price/unit[b]	cost £ [b]	Energy -Imperial units		Energy - S.I.units[d]	
	tons	litres	therms	kWh			equivalent therms[c]	cost per therm	equivalent gigajoules	cost per gigajoule
steam coal	– –									
house coal	2.25				£84.32	190	596	31.9p	63	£3.02
oil		– –								
gas			67 768		19.9 p	13 486	67 768	19.9p	7 135	£1.89
electricity				336 641	2.72p	9 157	11 479	79.8p	1 208	£7.58

NET ENERGY COST [e] £ 22 833
add: maintenance 200
    stoker's wages - - -
GROSS HEAT & LIGHT COST £ 23 033

TOTAL ENERGY CONSUMPTION   79 843 THERMS   8 406 GIGAJOULES

Number of 'Degree Days' in period (below 15.5°C base)[f]   2 273   (20-year average)

B. WATER CONSUMPTION in period ☐ x 1000 gals.   TOTAL WATER COST £ ☐ [g]   Average cost per 1000 gals ☐ p.[g]

C. CONSUMPTION INDICES [h] :

(1) Water Consumption Index
(gals/m²/pa)

$$\boxed{\text{N/A}}$$

(2) Electrical Energy Index
(therms/m²/pa)

$$\boxed{0.93}$$

(3) Heating Energy Index [i]
(therms/m²/pa x 0.63 x 1.1)

$$\boxed{3.82}$$

(4) Total Energy Consumption Index
(Index 2 + Index 3)

$$\boxed{4.75}$$

NOTES

[a] *floor area measured inside external walls excluding unheated basements, boiler rooms, open porches and cloisters, but including internal walls, cupboards, etc.*

[b] *including fixed charges where applicable.*

[c] *measure of gross heat input by direct fuel to heat conversion (i.e., not adjusted 'useful therms' taking into account system efficiencies as commonly used for comparisons of heat outputs).*
Conversion factors used: Steam Coal - 278 therms/ton; House Coal - 265 therms/ton; Gas - unity;
Oil (35 secs.) - 0.364 therms/litre; Electricity - 0.0341 therms/kWh

[d] *Gigajoules (10⁹ joules) equivalent to 9.5 therms or 278 kWh*

[e] *includes space heating, domestic hot water, cooking and lighting.*

[f] *as published by Department of Energy for period September to May. Annual comparisons give indication of climatic factor affecting energy consumption ('Degree Days', Fuel Efficiency booklet No.7, HMSO 1977 and D of E monthly tables).*

[g] *excluding local water rate.*

[h] *as proposed by University of Cambridge Estate Management & Building Service for annual internal comparisons and for external comparisons with other institutions ('UK Universities Utilities Consumption Survey 1974/75', E.M.B.S. Sep 1976).*

[i] *index (3) excludes all electricity, but includes water heating and cooking with space heating where other fuels used. The numerical factors adjust for building type and climate on 'degree day' basis, both as proposed in op.cit.[h] above. The College is taken as wholly residential (i.e., no separate consideration of its communal and teaching areas) and the occupation factor of 0.63 is an annual constant. The 'degree day' factor is derived by dividing the Sep-May total for East Anglia in the given year into a constant of 2500 that gives a base factor of 1.0 for areas in Scotland.*

CROGHAN + LEWIS   © 1977
associated architects and environmental consultants
39 NEWTON ROAD CAMBRIDGE ENGLAND telephone 0223-63364

Figure 4   Energy Audit Form A completed for 1978–79 (prediction)

PEMBROKE COLLEGE CAMBRIDGE

ANALYSIS OF ENERGY APPLICATIONS, COST DISTRIBUTION, AND ANNUAL VARIATIONS FOR 1976/77

Each cell is shown as: value · (proportion %) · ±increase/decrease on previous year.

FUEL	TOTAL ENERGY consumption and cost [GJ / £ (a)]	CENTRAL HEATING [GJ / £ (b)]	DOMESTIC HOT WATER (excluding Kitchen) [GJ / £ (c)]	SUPPLEMENTARY HEATING [GJ / £ (d)]	KITCHEN & BUTTERY (cooking & hot water) [GJ / £ (e)]	LIGHTING [GJ / £ (f)]	PUMPS and FANS [GJ / £ (g)]	GARDEN (glasshouse, etc.) [GJ / £ (h)]	MISCELLANEOUS (eg gym hobs, laundry) [GJ / £]
STEAM COAL (GJ)	4 068 · 31 · −21	1 597 · +0.1	619 · −30		1 852 · −30				
STEAM COAL (£)	4 897 · 19 · −7	1 923 · +17	745 · −18		2 229 · 46 · −18				
HOUSE COAL (GJ)	63 · 0.5 · +80			63 · 100 · +80					
HOUSE COAL (£)	125 · 0.5 · +119			125 · 100 · +119					
OIL (GJ)	5 621 · 42 · −14	3 627 · 65 · −18	1 994 · 35 · −6						
OIL (£)	9 139 · 35 · −0.2	5 897 · −5	3 242 · +9						
GAS (GJ)	2 033 · 15 · +7	159 · [i] · −30	12 · E · 0	623 · 31 · +32	956 · 47 · +3			208 · 10 · +9 E	75 · 4 · +4
GAS (£)	3 233 · 12 · +28	253 · −17	19 · +19	991 · +59	1 520 · +23			331 · +30 E	119 · +24
ELECTRICITY (GJ)	1 504 · 11 · −2		425 · 28 · −20	218 · 15 · +54	132 · 9	450 · 30 · +1 E	183 · 12 · 0	50 · 3 · 0 E	46 · 3 · 0
ELECTRICITY (£)	8 837 · 34 · +12		497 · −8	1 281 · +75	776 · +15	2 644 · +30 E	1 075 · +14	294 · +14 E	270 · +14
TOTALS: ENERGY (GJ)	13 289 · −12	5 383 · 41 · −14	3 050 · 23 · −14	904 · 7 · +40	2 940 · 22 · −21	450 · 3 · 0	183 · 1 · 0	258 · 2 · +7	121 · 1 · +3
EXPENDITURE (£)	26 231 · +5.5	8 073 · 31 · −1	6 503 · 25 · −1	2 397 · 9 · +70	4 525 · 17 · −2	2 644 · 10 · +14	1 075 · 4 · +14	625 · 2 · +22	389 · 2 · +17
av. cost/GJ (£)	1.97 · +20	1.50 · +15	2.13 · +15	2.65 · +22	1.54 · +22	5.87 · +14	5.87 · +14	2.42 · +14	3.21 · +13

NOTES

(a) Values as invoiced in year. No allowance made for opening and closing balances.

(b) Values are balances of fuel totals after deduction of other known or estimated amounts.

(c) Values for Coal and Oil based on 2 X deliveries in summer (Apr - Sep). Value for Gas as metered or estimated (E). Value for Electricity based on meter readings in pilot study of electric water heater consumption.

(d) Values for Gas and Electricity are totals of metered supplies to individual rooms. In case of Electricity common consumption of storage heaters is added as metered centrally. Value for Coal as invoiced. Value for Electricity also includes some use for minor domestic appliances where socket outlets on metered room circuit.

(e) Values for Gas and Electricity as metered. Value for Steam Coal based on 1.5 X deliveries in summer (Apr - Sep).

(f) Value estimated largely on basis of residual balance.

(g) Value based on installed load with allowance for continuous or intermittent operation.

(h) Value for Gas metered. Value for Electricity estimated from installed load.

E An estimated value. Remains a constant as at base year (1975/76) unless known alteration in consumption or means become available for more accurate assessment.

ADDITIONAL NOTE

[i] Consumption of Old Library gas furnace

CROGHAN + LEWIS   © 1977
associated architects and environmental consultants
39 NEWTON ROAD CAMBRIDGE ENGLAND telephone 0273-63364

Figure 5 Energy Audit Form B completed for 1976-77

were partly complete—seem likely to be realized (though final figures are not available at the time of writing).

Energy Audit Form B provides the analysis of energy applications, cost distribution, and annual variations. This is the main predictive and 'watchdog' tool. The example in Figure 5 is completed for 1976–77. The main items of consumption are derived from meter readings. Some of the smaller items are rough estimates, but others (such as 'pump and fans') are by load calculation or are balances after known attributions.

Given this degree of breakdown, predictions may be made fairly readily by substituting the calculated effects of any system or management variations.

6. HEATING PLANT ALTERATIONS

A master plan for reorganizing the College's heat generation and distribution system had been prepared in 1969—well before the 'energy crisis'—with a view to implementation in stages related to the general renovation of the residential accommodation.

This recommended the development of a single main boilerhouse—one of those already existing—with constant-temperature underground mains serving a number of 'services rooms': control, calorifier, and distribution points within the various College buildings.

Some authorities prefer multiple boiler installations to 'district heating', but with well insulated mains (here running about 220 m) heat losses in a centralized system can be a mere fraction of those arising in boiler rooms, and capital cost, maintenance, and space demands are also less.

A 950 kW (3.25 million Btu/h) Allen Ygnis fire-tube, oil-fired boiler was installed in the Orchard Building in 1970 to serve nearby blocks, and prior to completion of some landscaping work in 1973 a pair of 150 mm heating mains was run about half the length of the College (Figure 6). However, this was not brought into use and at the inception of the present study three other main boilers remained on stream: a pair of oil-fired units each of about 250 kW (850 000 Btu/h) rated output, and an old solid-fuel vertical steam boiler rated at 380 kW (1.3 million Btu/h).

The latter unit served 'live steam' to the main kitchen, but the bulk of its output was immediately cooled down to 65°C (150°F) in two adjacent steam-to-water calorifiers supplying domestic hot water and central heating. Investigation found the actual requirement for steam nominal. It would seem that over the years alterations in the kitchen had relieved the boiler of much of its original load, and the residue was readily replaced by four new independent gas-fired kitchen appliances (two hot cupboards and two self-contained steaming ovens). When subsequently this boiler was removed and its load transferred to units that could be metered, it was

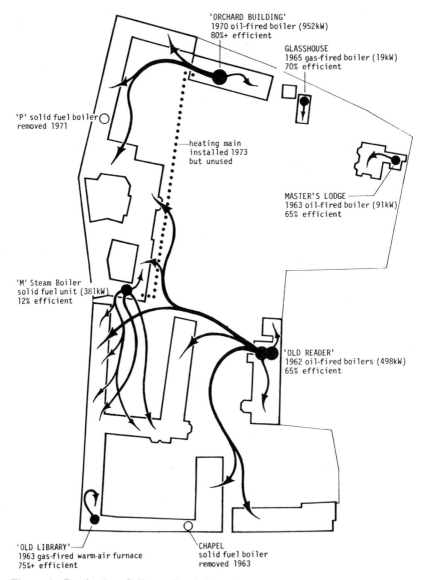

Figure 6 Pembroke College Cambridge, heat generation and distribution system before July 1977

found that energy input had been eight times more than useful energy output, a thermal efficiency of 12%. Taking into account stoking and maintenance, the cost of the duty originally performed by this boiler was reduced by 93%.

This alteration had been anticipated in the heating master plan, and an important part of the study now reported was the substantial implementation of the 1969 proposals. At that time the installation consisted of eight boilers spread over the College site. Prior to the major alterations of mid-1977, seven boilers remained (Figure 6). These had a total rated output of nearly 2000 kW (6.8 million Btu/h) with a low maximum utilization of about 42%, resulting in under-running and consequent low efficiency. The plant as modified (Figure 7) manages with half this rated output at a comfortable utilization of 75%.

The bulk of this output is from the Allen Ygnis unit, which has a measured combustion efficiency of 80%. This was converted from oil to gas fuel: though the fuel price advantage is not large, the benefits in controllability and to monitoring give significant overall advantage. As above, the old solid-fuel boiler was removed, but three other boilers—of domestic size and gas-fired—remain in use to serve remote or special-purpose areas.

The pair of oil-fired boilers from the original system remain 'in mothball': in an emergency they are able to heat half the College and supply hot water to the whole. The effects of any main boiler or distribution breakdown are limited by the provision of supplementary gas or electric fires in most rooms. The central kitchen—crucial to the functioning of a college—has been provided with a wall-hung Potterton 'Netaheat' 22 kW (75 000 Btu/h). This is a domestic central heating boiler here used to boost the temperature of a 450 litre (100 gallon) domestic hot water cylinder that normally has a hot 'cold feed' from central boiler supplies. In an emergency this booster can satisfy the whole of the kitchen's hot water needs.

All heating and hot water circuits are pumped, and all pumps are clock-controlled. There are some 20 'spring reserve' type clocks regulating an installed electrical load of some 12 kW. The careful programming of these together with reduction in heat losses from distribution pipework accounts for nearly 3% of total energy savings.

Space heating subcircuits are controlled by Satchwell 'Climatronic' weather-sensitive compensators operating three-port valves supplied from the constant-temperature boiler main. These controls were modified by adding a further time clock that enables the standard 'night setback' adjustment to be used to provide an alternative, preset daytime temperature. The system is not run at night.

Domestic hot water is produced in quick-recovery calorifiers sited as near as possible to the point of use. These are supplied from the boiler mains and controlled by thermostats operating motorized valves.

The whole forms a basically simple concept with largely automatic plant, and this seems the basic prerequisite for operational efficiency. The pre-existing system was complex and labour-intensive, operated largely by tradition, and difficult to monitor.

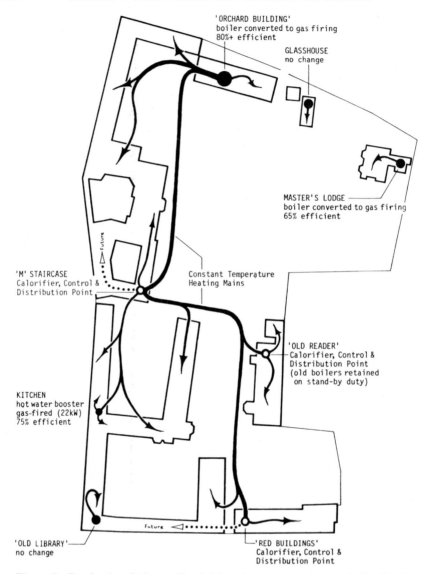

Figure 7 Pembroke College, Cambridge, heat generation and distribution system after September 1977

7. COMPUTER ANALYSIS

The validation of the above heating plant strategy depended on the detailed assessment of heating loads and their relationship to the potential of existing plant. It is unlikely that the College could have accepted the cost and disturbance of complete reservicing.

Traditional methods for calculating building heat load assume a steady thermal state and, though often laborious, cannot take full account of all factors and give only very limited information. For most practical purposes, this amounts to providing little more than an indication of the peak heat load of the subject building in relation to an assumed constant minimum outside temperature and a selected constant internal 'comfort level' at a single moment in time. This is a restricted and sometimes misleading concept: heating plant based on such calculations is typically oversized, with boilers underutilized and thus inefficient. There is the corollary that expectations of the advantage of insulation and control measures may be inflated.

At inception of the present study, the College was found to have appreciable excess boiler capacity, some 50%, as above. This was due to the partial implementation of the heating master plan of 1969 that left three old boilers in daily use that cost much to run and contributed little. Though empirical indications were strong that the one modern boiler already installed would in all probability suffice, a calculation method was sought to provide reliable validation. The method was also required to provide a reliable basis for the appraisal of other fuel conservation measures.

A number of computer programs for predicting the thermal performance of buildings are commercially available, but the basic assumptions and approximations built into them are often obscure and the indications are of differences in the sophistication of their basic models. In selecting the British Gas Corporation's THERM program, advantage was taken of a new program developed from first principles, and also the direct assistance of its originators was available.

THERM uses a mathematical model that describes the many complex, interactive, and essentially transient heat transfer processes that take place inside and outside a building to give a full indication of the dynamic performance of the building in use. Among factors that cause transient variations are: swings in external temperature, sun and shade, thermal inertia of the building fabric, intermittency in heating, delay in control, reponse, ventilation, and adventitious gains (as from lighting and occupancy). The effect of these can be adequately accounted only by performing calculations through time.

The THERM output provides a complete record of the subject building's behaviour at each point in time over a specified simulation period (such as a college term) using a 20-year climatic database. A typical summary output gives average and peak loads, maximum and minimum internal temperatures, and a load frequency spectrum.

In the present case, application of THERM was primarily the study of the interactive effects of a series of 13 specified conservation measures and related conditions. These are defined in terms of 'simulations' and were considered in three groups, as in Table 1. Group A takes the building as existing and varies the comfort level. Group B defines various proposals for additional insulation. Group C proposes optional diurnal heating programmes.

Table 1   Schedule of THERM simulations

Subject building: Pembroke College, Cambridge, P-Q-R staircases

Simulation	Building details	Comfort level (°C)	(°F)
Group A			
1	Building as existing: 50 mm vermiculite insulation in horizontal ceilings; no insulation in sloping ceilings or dormers; average hourly air change rate 2.5 (2 ach in studies); assumed room occupancy: 33% 9 a.m.–5 p.m. 66% 5 p.m.–10 p.m. 100% 10 p.m.– 9 a.m. continuous central heating with no supplementary heating	15.5	60
2	As Sim. 1 but comfort level varied	18.3	65
3	As Sim. 1 but comfort level varied	20.0	68
4	As Sim. 1 but comfort level varied	21.0	70
Group B			
5	25 mm vermiculite insulation added to horizontal ceilings (to give 75 mm total); otherwise as Sim. 1 but comfort level varied	18.3	65
6	50 mm polystyrene insulation added to sloping ceilings; 75 mm vermiculite insulation added to dormer ceilings; otherwise as Sim. 5	18.3	65
7	Insulated lining added to inner face of external wall (9 mm foil-backed plasterboard over 30 mm mineral wool); otherwise as Sim. 6	18.3	65
8	Draught sealing (as by new windows) to reduce average hourly air change rate to 1.9 (1 ach in studies); otherwise as Sim. 6	18.3	65
9	Double glazing by provision of secondary windows fitted inside stone mullions giving approx. 75 mm airspace; otherwise as Sim. 6	18.3	65
10	No ceiling insulation; otherwise as Sim. 1 but comfort level varied	18.3	65

Table 1   Schedule of THERM simulations (continued)

		Comfort level	
Simulation	Building details	(°C)	(°F)
Group C			
11	As Sim. 6 but continuous heating replaced by intermittent programme: 'on' at 06.30, 'off' at 22.00	18.3	65
12	As Sim. 6 but continuous heating replaced by intermittent programme with diurnal variation:		
	'background' 07.30–12.00	15.5	60
	'comfort' 13.00–22.00	18.3	65
13	As Sim. 6 but continuous heating replaced by intermittent programme with diurnal variation:		
	'background' 07.30–12.00	15.5	60
	'comfort' 13.00–22.00	20.0	68
14	As Sim. 6 but considers measure of condensation protection introduced in Christmas Vacation	10.0	50

The computer study considered only one of the College's buildings over one term in detail, but this was such that results could be extrapolated to cover other blocks and the full heating season with reasonable accuracy. A summary of results in terms of varying heat loads, and an outline of the study development, are shown in Table 2. The proposals for alternative diurnal heating programmes were further studied for sample days. Graphs of temperature and heat load against time for Simulation 12 (the programme subsequently adopted) are shown in Figure 8. Taken with other examples, these indicated that the initial morning warm-up to 15.5°C (60°F) and required afternoon boost to 18.3°C (65°F) were achievable and that the thermal inertia of the heavyweight building fabric allowed heating to be turned off at night (as was already the pattern) without room temperatures falling excessively or increasing total fuel consumption.

Overall the THERM study confirmed that the College's heating load could all be transferred to the single modern boiler, a measure that reduced total energy consumption 23% and accounted for just over half the total saving. The THERM results also formed the basis for estimating the energy saving potential, and ultimately the cost–benefit, of a series of further management and insulation measures.

8. MANAGEMENT MEASURES

The proposal for diurnal variation of the space heating level was suggested by a pattern of use in which most rooms are unoccupied in mornings while students attend lectures. The lower morning comfort levels could be topped-up if required by remaining occupants using their individually metered supplementary heaters, but in practice this seems not to be general. Taken together with the essential prerequisite of adjusting pre-existing arbitrary heating levels, this reduced total energy consumption by nearly 5% and accounted for over 10% of total saving.

The accurate determination of heating levels where some 250 rooms are involved is a daunting procedure. In the absence of a large number of individual chart recorders, and as wiring to multichannel recorders was not considered feasible, only a small sample could be considered. Seven rooms were selected, spread over different buildings and different floors, with different orientations and served by different heating controls. Each was fitted with a simple 'maximum–minimum' thermometer from a matched set, and these were visited twice each week (once in the morning, once after midday) for 16 weeks. On each occasion 'peak', 'low', and 'present' temperatures were logged. Subsequently an assessment was made of the deviation from the target norm of 18.3°C (65°F). Using the THERM results, an approximation could then be made of the excess heat load and fuel consumption. Variation in average heating levels between rooms was found to be as much as 7°C (13°F), with over one-third of the rooms having an excess of 3°C, resulting in an increase in heat load of 24%.

This method of survey is crude and time-consuming and cannot be recommended for the periodic fine 'tuning' of large buildings that can be a fruitful source for fuel economy. The author has proposed an outline for a relatively inexpensive, but accurate and robust, solid-state memory temperature recorder and transcription unit, and this is now being developed.

The room heating survey included observation of the frequency with which windows of study bedrooms were left open during the heating season, and this was found to be appreciable. No economic heating system can sustain the losses consequent upon the much increased rate of room air change that results from open windows. The THERM study indicated that an increase of 1 ach (air change per hour)—the difference between modern draughtstripped windows and older, leaky units (both closed)—added 22% to room heat load.

Measures to encourage the closure of windows have been taken, and on the assumption that without onerous discipline these can be only partially effective, their value has been assessed modestly as equivalent to a reduction of 0.5 ach, which represents an overall energy saving of nearly 2%.

The development of a ready method for the site measurement of adventitious air movement through windows and doors (perhaps employing ultrasonics) would be valuable to the practice of energy conservation.

The largest individual saving under the heading of management measures—nearly 11% of total energy savings—resulted from reducing the

Table 2 Results and analysis of THERM simulations

Subject building: P-Q-R staircases (36 SBUs)[a]
Heating period: Lent term (13 Jan.–12 March)[b]
Climatic data: 19-year average (1952–1971)

Outline of study development and application action path

Building as existing, Sept. 1977 ——→ A. Optional comfort levels compared
→ heat load increasing markedly with rising comfort level: 18.3°C norm adopted

——→ B. Insulation measures compared
→ preferred roof insulation adopted in subsequent simulations
→ marked effect of air change rate noted

——→ C. Heating programmes compared
→ advantage of diurnal variation noted and response rate studied
→ optional measure examined

D. Determination of energy costs and possible economies
→ capital costs of measures estimated

| | A | B | C | Heat load variation (%) increase(+)/decrease(−) | |
Simulation (Table 1)	Average heat load (kW/h)	Peak load (kW/h)	95% load[c] (kW/h)	Average load	Peak load
				(a) Comparison with Sim. 1	
1	43.7	76.6	—	—	—
2	58.3	91.2	83.1	+33.4	+19.1
3	67.3	100.1	—	+54.0	+30.7
4	72.5	105.4	—	+66.0	+37.6
				(b) Comparison with Sim. 2	
5	58.1	90.9	82.8	−0.3	−0.3
6	51.9	81.1	75.3	−11.0	−11.1
7	48.4	77.1	72.0	−17.0	−15.5
8	42.3	65.7	62.4	−27.4	−28.0
9	38.9	60.9	58.7	−33.3	−33.2
10	60.7	94.8	87.3	+4.1	+4.0
				(c) Comparison with Sim. 6	
11	46.1	99.1[d]	96.1	−11.2	+22.2
12	42.8	99.1	95.4	−17.5	+22.2
13	46.3	99.1	97.9	−10.8	+22.2
14[b]	27.5	49.0	49.0	n.a.	n.a.

[a]'Study bedroom units' include share of circulation and ancillaries.
[b]Simulation 14 relates to Christmas Vacation period (4 Dec.–12 Jan.).
[c]Approx. peak load if min. exterior base temperature of 0°C adopted (as in traditional heat load calculations). Col. B results reflect actual exterior temperatures that at times drop below 0°C.
[d]Specified maximum designed output of heating installation.

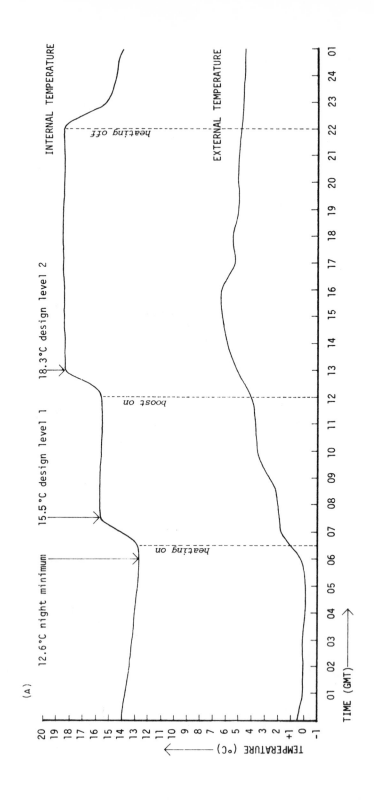

(A)

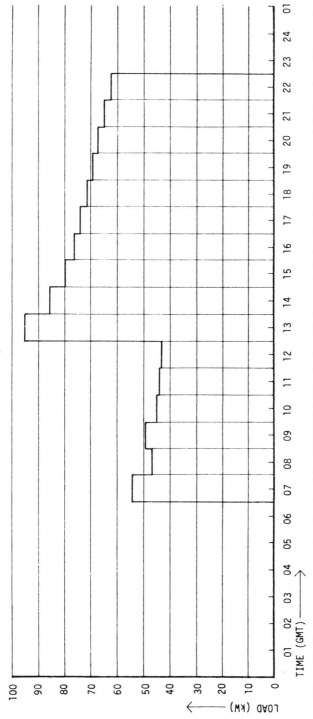

Figure 8 Temperature and heating load by 'THERM' Simulation 12 for typical day in the Lent term. (A) Graphs of internal and external temperatures. (B) Load frequency spectrum

use of electric immersion heaters (a total installed load of 62 kW) where there were alternative supplies of hot water from calorifiers heated by the main boiler, and limiting tap water temperatures (from both the remaining electric heaters and central supplies) to 50°C (120°F). Operating for only six months in 1976–77 these measures reduced energy consumed in the provision of domestic hot water by 21%.

The high proportion of energy and cost consumed in the provision of domestic hot water was a significant finding of the energy audit and may indicate a usual, but perhaps not widely recognized, situation in comparable institutions. Taking kitchen hot water together with general supplies, the energy consumed in year-round production was seen virtually to equate with that for space heating in 1975–76 and 1976–77, each accounting for some 40% of total energy use.

Though in the fully modified scheme the provision of domestic hot water consumes appreciatively less energy than originally, its demands remain close to those of space heating. It accounts for 31% of total energy consumption (compared with 33% for space heating) and 27% of expenditure (compared with 23% for space heating).

9. INSULATION MEASURES

The THERM study—as above and Table 1—included a series of options for thermal insulation measures. These were specified with the variety of existing provisions in mind with a view to examining the value of upgrading.

All roof spaces in the College were inspected to assess the need for, and feasibility of, increased insulation. Appropriate measures were tabulated and priced.

Simulations for wall lining, replacement windows, and double glazing were included in the THERM study for general interest: it was not thought likely that these would be cost-effective nor, in many cases, feasible.

The upgrading of roof insulation to a general level of 75 mm over horizontal ceilings, and the introduction of 50 mm rigid insulation between rafters above the sloping ceilings of attic rooms—with care taken not to produce conditions that might encourage interstitial condensation—is computed to reduce overall energy consumption by 2.3%.

This saving is relatively small, partly because about one-third of the roof area involved covers rooms with little or no central heating in which notional savings are likely to be absorbed by increased comfort, but mainly because of the inherently good thermal qualities of the College's old buildings and provisions previously made during a modernization programme over the last 16 years.

In many places the level of roof and wall insulation prior to the study was already equal to, or better than, the standards of the pre-1975 Building Regulations. The further measures aimed to bring roofs up to the current new building standard (U value 0.6 W/m^2°C) as far as possible without recourse to major work such as the removal of slates and tiles. Thick brick and masonry walls with U values of about 1.2–1.5 were not considered to call for improvement (the current standard being 1.8). This would be difficult and expensive to achieve, and it was also suspected that treatment such as low-thermal-capacity lining might well reduce the thermal inertia and be counterproductive, particularly in the maintenance of room temperatures at night when the heating system is off.

In the majority of the modernized rooms, heating is by hot water convectors inserted into existing stone fire surrounds, and heat loss to the fabric is minimal. Elsewhere some 116 radiators are located against external walls, and these were insulated with adhesive aluminium foil placed on the inner surface of the walls. AGA 'Thermovision' infrared photography taken prior to the work showed a distinct pattern of heat loss associated with these situations. The insulation was undertaken for a modest charge by the Cambridge Friends of the Earth and is estimated to have reduced total energy consumption by 0.5%.

10. COST-BENEFIT ANALYSES

The conservation measures at Pembroke College were considered on the basis of full cost–benefit analyses. In each case four methods were employed:

(1) Comparison of annual saving with a 12½% simple interest return on capital outlay in alternative investment.
(2) Capital payback period related to the anticipated life of the measure.
(3) Annual saving related to total 'annual cost' based on interest in alternative investment at 12½% and linear depreciation.
(4) Consideration of the 'equivalent present value' of future savings over life of measure under six scenarios of interest and inflation rates.

Most measures satisfied all test methods, often to considerable financial advantage.

11. SOLAR ENERGY

The energy conservation programme at Pembroke College has so far involved little of technical novelty. Though many of the calculations were computer assisted, the only 'new technology' hardware is to be found in the solid-state heating compensator controls, and these have been a standard item for some years. It is arguably the case that the study has taken existing practice to its practical limit, and further economies may be made only by harnessing ambient energy.

The use of solar energy to supplement water heating was considered in some detail. A number of small sections of the College rely on electricity for water heating, and a detailed study of one of these indicated that a domestic-type solar energy system was likely to be marginally economic, but it was felt that the quality and performance of hardware options were not yet convincing. However it is contemplated that advances in the state of the art—and in fuel prices—could make such installations attractive before long.

A large system with a solar collector panel area of $100 \, \text{m}^2$ has been examined in outline and its energy contribution assessed at 180 GJ per annum (1710 therms per annum) of useful heat. Assuming that the summer production of domestic hot water by the main boiler is some 40% efficient, the solar contribution would be the equivalent of 450 GJ of purchased energy. This is about one-third of the summer period consumption, and capital costs would be recovered within 10 years.

Two such systems designed to provide the bulk of the College's summer hot water would contribute the balance required to reduce total annual energy consumption to half the 1975–76 level.

12. RESULTS OF THE STUDY

An outline of the conservation measures, costs, and resulting energy use reductions is tabulated in Table 3, and the savings are illustrated graphically in Figure 9. At year 3 (1978–79) the overall energy saving is 45% and the cost saving is 54%.

The main conservation activity has concentrated on central heating, domestic hot water provision, and central cooking which together accounted for 88% of total energy use in the base year 1975–76. Isolating these, the energy saving at 1978–79 is 51% and the cost saving is 69%.

Considering the full picture, heating plant alterations costing about £21 500 produce 56% of total energy savings. Management measures involving little cost contribute 28%, and insulation measures account for the balance of 6%.

The overall capital payback period based solely on fuel cost reductions is two years, but in fact all costs (including fees for 40 weeks of consultant's work) will have been reimbursed by savings in total energy system operating costs accruing during the $2\frac{1}{2}$ years of the study and associated implementation work: the project has in effect been self-financing and leaves the College with a long-term advantage.

Table 3 Results of the conservation study

Conservation measure	Energy reduction (GJ)	(%)	Proportion of total saving (%)	Capital cost (£)	(%)	Cost–benefit tests Saving per £1000 cap. cost (GJ/a)	Payback at energy cost £2.4/GJ (years)	'Present value' of fuel cost savings over 20 years (£) 10% interest, 16% inflation	10% interest, 14% inflation	Interest = inflation
1. Plant alterations										
(a) Implementation of single boiler and distribution main system; gas-firing	3491	23.1	52	20900	67.5	167	2.6			
(b) Additional time clocks on pumps	193	1.3	2.9	378	1.2	511	0.8			
(c) Fully closing air damper on boiler	84	0.6	1.2	294	1.0	286	1.5			
Section results	3768	25	56	21572	70	174	3	370771	285756	180864
2. Management measures										
(a) Reduce central heating season by 4 weeks	172	1.1	2.6	nil						
(b) Limit domestic hot water to 50°C (120°F) restrict use of elec. water heaters	721	4.8	10.7	nil						
(c) Adjust arbitrary space temperatures	413	2.7	6.1	nil						

Table 3 Results of the conservation study (continued)

Conservation measure	Energy reduction (GJ)	(%)	Proportion of total saving (%)	Capital cost (£)	(%)	Cost–benefit tests Saving per £1000 cap. cost (GJ/a)	Payback at energy cost £2.4/GJ (years)	'Present value' of fuel cost savings over 20 years (£) 10% interest, 16% inflation	10% interest, 14% inflation	Interest = inflation
(d) Adopt heating prog: a.m. 15.5°C (60°F) p.m. 18.3°C (65°F)	318	2.1	4.7	594	2.0	535	0.8			
(e) Encourage closure of windows	259	1.7	3.9	nil						
(f) Improved boiler maintenance	136	0.9	2.0	nil						
(g) Misc. consultant activity	532	3.5	7.9	by fee						
Section results	2551	17	38	594	2	4294	0.1	251018	193468	122448
3. Insulation measures										
(a) Upgrade roof insul. to approx. 75 mm	343	2.3	5.0	8634	27.9	40	10.5			
(b) Foil backing to 116 radiators	71	0.5	1.0	167	0.5	425	1			
Section results	414	3	6	8801	28	47	9	40738	31398	19872
Overall results	6733	45	100	30967	100	217	2	662527	510631	323184

In 1978–79 the annual reduction in fuel cost is about £16 000. Applying the principles of compound interest in reverse to discount future sayings back to the present, and taking an interest rate of 10% and a fuel cost inflation rate of 16% (as in the last two years), the 'present value' of savings over 20 years is some £650 000. Even if it is considered that interest rates and fuel cost inflation will equate, the present capital sum equivalent to the future savings is not inconsiderable at over £320 000.

13. COMMENT

Energy savings in the range 30 to 40% are reported not infrequently, but the Pembroke College expectation is perhaps an unusually good result for a large complex of old buildings in continuous residential occupation and where a sizable proportion of ultimate consumption is in the hands of several hundred student occupants enjoying a high degree of autonomy. This is not an industrial process under total management control, nor an intermittently used office block where simple central measures can reduce overheating.

Though all savings must to some extent reflect initial deficiencies, it is doubtful that it is the case that Pembroke College started from any outstandingly bad position. Its fabric was found to have inherently good thermal qualities, and the heating systems in its constituent blocks—if not all the central heat production plant—are mainly of relatively recent installation, were intensively designed, and remain in advance of the present norm.

In the year prior to any significant conservation activity, Pembroke's heating energy index was slightly less than one of its entirely modern collegiate neighbours and its total energy bill 40% less than an old public school of similar floor area in the same region.

It seems possible, therefore, that the Pembroke College case study may contain useful indications for other similar institutions, even where they may consider themselves fairly efficient by existing standards. Given a body of similar case studies, it should be possible to begin the definition of 'energy targets'. At present the energy manager cannot easily know whether his institution is economical or otherwise.

ACKNOWLEDGMENTS

The material in this study appears by kind permission of The Master and Fellows of Pembroke College, Cambridge, and is based upon information presented to the National Energy Management Conference organised by the Department of Energy in October 1978.

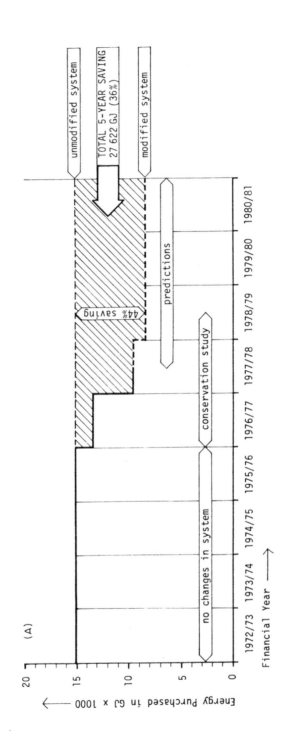

(A)

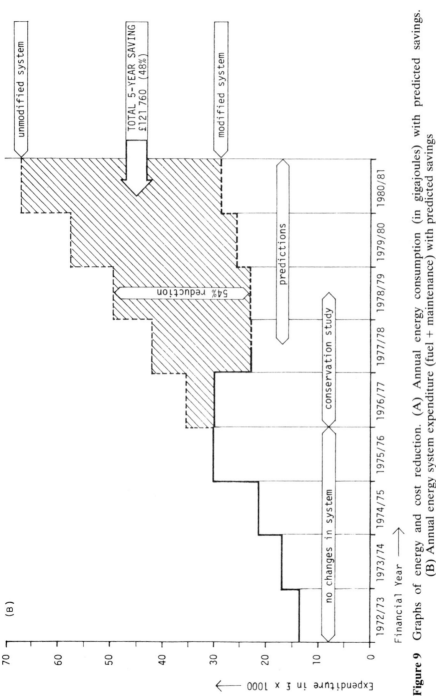

Figure 9 Graphs of energy and cost reduction. (A) Annual energy consumption (in gigajoules) with predicted savings. (B) Annual energy system expenditure (fuel + maintenance) with predicted savings

Mr A. R. Robinson of Messrs ESG Sales Ltd of Cambridge and London developed the strategy for the modification of the heating plant in consultation with the author, and his firm subsequently undertook the bulk of the capital works. Mr S. F. S. Whitehouse, Cost Reduction Engineer to the University of Cambridge, gave much helpful advice during implementation.

The preparation of this paper was made possible by the generosity of the British Gas Corporation, whost participation in the application of the THERM computer program is also gratefully acknowledged.

Energy Conservation and Thermal Insulation
Edited by R. Derricott and S. S. Chissick

CHAPTER 26

New perspectives for kinetic energy storage

Jean-Pierre Barthelemy

Aerospatiale, Les Mureaux

ABSTRACT

Research into satellite flywheels has identified products with two possible fields of application:

(a) providing important angular momentum for low mass, and
(b) storing energy in kinetic form, by utilization of two counter-rotating wheels.

Such developments have been made possible by designing and proving well adapted suspension systems, motors, and rotors. The present contribution describes several of those designs, and discusses the possibility of their extension to large industrial applications on the ground, such as back-up energy supply for telecommunication centres.

I. INTRODUCTION

Such wheels, utilized since the early times of human acitvity, still present much interest in many areas. Their present and foreseen possibilities cover a wide range of applications as much for space as for ground activities.

The necessity for satellite flywheels that achieved very stringent characteristics gave birth to strenuous development efforts mainly oriented in two directions.

(i) Magnetic bearings which eliminate the wear problems and most of the power dissipation occurring with other types of bearing: the effort described here was concentrated on 'the one active axis' type of magnetic bearing with radial centring performed passively, a single axial servoloop maintaining the rotor in the axial direction. This concept is the simplest and the most reliable.

(ii) Rotors which utilize fibre composite materials in different configurations: the result was to obtain interesting levels of energy per unit mass involving sufficiently high peripheral speeds. The main problems were to provide compatibility of the rotor with the stresses induced by centrifugal forces and simultaneously to stabilize the initial balancing in spite of rim elongation.

However, for applications on the ground, rotor efficiency (stored energy per unit mass) is not a dictating factor towards delicate technologies, and simpler designs are possible.

The result of this work was to obtain completely integrated systems including the rotor, the magnetic suspension, the motor–generator, and associated electronics. It enables us to introduce energy in electric form, to store it in kinetic form, and to recover it in electric form, the wheel being placed in a vacuum container for ground operations. Satellite flywheel systems have been produced, and are the subject of further development by Aerospatiale, France, for INTELSAT, ESA, and CNES.

A system of kinetic energy storage using the same design principle is presently being developed for CNET (French Telecommunication Agency), which will be capable of a power of 3 kW. This unit constitutes a first step in the production of the kinetic wheels for use on the ground.

II. DESCRIPTION OF MAGNETIC BEARINGS AND ASSOCIATED SUBSYSTEMS

The dominant and most important feature is the essentially passive nature of the magnetic suspension systems employed. Passive magnetic suspension, i.e. that based on maximal use of permanent magnets, is used in preference to alternative methods based entirely on electronically controlled electromagnets, in view of the primary objective to maximize reliability. The passive suspension configuration adopted employs permanent magnets operating in the attraction mode for radial centring of the flywheel hub, and a single servo-driven electromagnet actuator for axial position control.

In this way, four out of the six degrees of freedom of the suspended rotor are constrained passively, and only one actively. The sixth degree of freedom is the desired rotational motion of the wheel which, of course, requires no constraint. The realization of the various suspensions and other elements involved are described separately below.

1. Passive permanent-magnet radial bearings

Radial centring of the rotor is effected by means of two permanent-magnet radial bearings, the dimensions, geometry, and design of which are chosen according to the specific requirements of each wheel. One possible con-

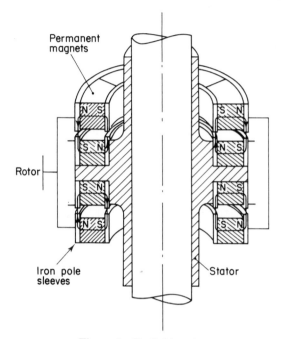

Figure 1 Radial bearing

figuration is shown in Figure 1. Each bearing comprises four radially magnetized rings of segmented construction, fitted with soft-iron pole sleeves on their inner and outer curved surfaces.

Oppositely magnetized rings are attached to the flywheel rotor and stator respectively. The concentrated axial field set up in the gaps between adjacent pole ring and faces gives appreciable radial centring forces for quite small radial rotor displacements with respect to the stator. Radial stiffnesses of the order of 1.5×10^5 to 5×10^5 N/m per bearing are easily achieved for quite small bearing dimensions, whilst ratios of radial stabilizing/axial destabilizing stiffnesses of the order of 0.5 to 0.25 are typical.

A number of constructional variations are possible. One possibility with some important constructional advantages is to employ radially instead of axially stacked ring pairs. The latter configuration has been adopted in more recently developed wheels.

2. Active electromagnetic axial bearing

The attractive forces between the iron pole sleeves in the radial bearings give not only a radial restoring effect but also tend to decentre the rotor in an axial direction.

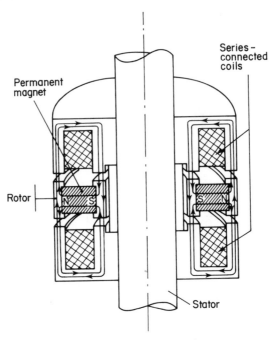

Figure 2   Axial bearing

In order to counteract these axial decentring forces, an electromagnetic axial bearing is employed. This bearing forms the power element of a servoloop which automatically controls the axial position of the rotor. Figure 2 shows the configuration adopted. A pair of series-connected iron-clad coils attached to the stator modulates the field produced by a radially magnetized Samarium–Cobalt permanent–magnet ring attached to the rotor. The direction of the resulting force depends on the direction of current flow in the coils. Compared with a straightforward electromagnet without permanent-magnet bias, this arrangement produces much larger axial forces per ampere-turn and exhibits a nearly linear force versus current relationship.

3. Axial servoloop

A block diagram of the electronic servoloop used to control the electromagnet axial bearing is shown in Figure 3. The system employed is essentially a position control scheme which keeps the rotor at an axial position where applied forces (in 1g) are exactly compensated by lifting forces due to the permanent magnets in the radial and axial bearings. This system has the advantage of very low suspension power requirements in both 1g

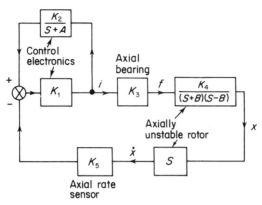

Figure 3 Axial position control system

and $0g$ operating conditions as well as simplifying the electronic and mechanical realization. The input signals used are the axial rate of the suspended rotor as measured by a small pick-up coil mounted in the wheel stator and the current flowing in the axial bearing coils. A separate logic is used for initial lift-off of the rotor.

The stability of the system is evident from the root locus plot shown in Figure 4. By appropriate choice of the electronic gain factors K_1 and K_2, all closed-loop poles are brought in the left half-plane.

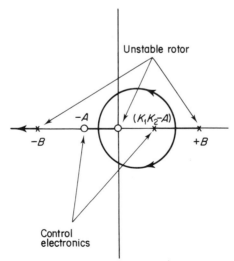

Figure 4 Root locus of axial control system

4. Radial dampers

The passive radial magnetic bearings described above may be likened to almost perfect lossless springs as far as their radial restoring characteristics are concerned. Together with the rotor mass they form an undamped second-order vibratory system which is excited by unbalanced forces during rotor spin-up as well as precessional torques at normal operating speed. In order to prevent excessive amplitudes of rotor motion and the possibility of mechanical contact occurring between rotor and stator parts, radial damping of the rotor motion must be introduced. The radial damping devices employed are shown in Figure 5. Each wheel contains two dampers, each damper being made up of four permanent-magnet rings attached to the rotor and a copper disc attached to the stator. Radial motion of the magnet rings with respect to the copper disc induces eddy currents in the latter which interact with the permanent-magnet fields and yield the required damping force. Rotational drag is small due to the azimuthal continuity of the fields.

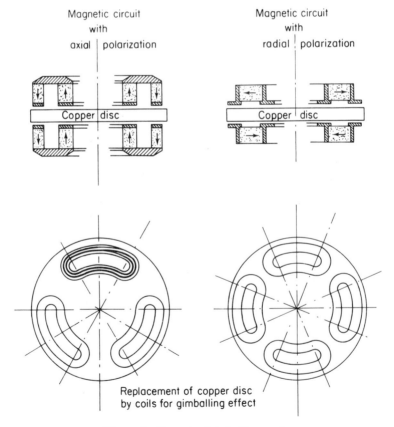

Figure 5   Damping/gimballing system

This type of damper was selected after extensive research involving a wide variety of alternative damping methods. Its chief advantages are the very effective damping characteristics, the nonreliance on structural deformation for its operation, and the easy possible adaptation at a combined active damper/vernier alignment device by substituting the copper disc by a system of coils, if particular applications require such a vernier gimballing effect.

5. Motor

The motor is of the brushless DC type with electronic commutation based on high-frequency sensing devices. The construction is shown in Figure 6. The rotating-field structure consists of an iron ring fitted with a number of rare-earth permanent-magnet poles. The stationary armature comprises several coils of multistrand insulated conductors encapsulated in a cylindrical epoxy-resin support. The use of ferrous materials in the stator armature is specifically avoided in order that no radial decentring forces due to attraction between the armature and field parts are introduced. The motors are characterized by high efficiency, high delivered power-to-mass ratio, and very smooth rational torque free of magnetic indent effects.

6. Emergency bearings

In order to avoid damage to the magnetic bearings in the event of a suspension failure at high wheel speeds or excessive slewing rates being applied, an emergency bearings system consisting of a dual pair of angular contact dry-lubricated ball-bearings is provided. These bearings act as both axial and radial excursion stops and only come into operation when either

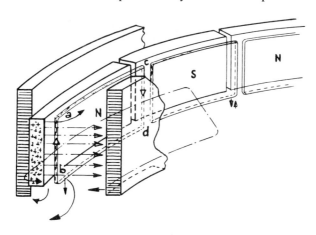

Figure 6 Motor construction

axial or radial excursions become too great. The emergency bearings are attached to the rotor of the wheel and are fitted with contact touchdown pads. This configuration ensures consistently smooth touchdown behaviour even at very high rotor speeds.

7. Electronics

The electronic circuits control the current in the axial servoloop actuator and in the motor coils. Three types of electronics have been developed:

(a) utilization of operational amplifiers,
(b) utilization of discrete components, and
(c) utilization of thick-film hybrid electronics.

8. Adaption of the designs to a ground system

A potential cost-saving area is obviously the electronics in the case of a broad production line. Another simplification can be made in the shape of the coils of the motor static armature, a larger available volume making it possible to adopt a simpler shape.

The study of the magnet's implementation and of the central axis, which must achieve high stiffness and precision positioning, indicates that requested technologies are not fundamentally different from currently applied ones in the field of motors, or even, in some mechanical aspects, the car industry. It has to be noted that the motor concept has been presently validated by successful tests up to 3 kW.

III. DESCRIPTION OF THE ROTORS

1. Overall concepts

The rotor is constituted of several functional parts:

(a) the central part, which contains the magnetic bearings elements and the motor magnetic circuit;
(b) the rim, which produces the main part of the inertia; and
(c) the connection between the central part of the rotor and the rim.

Two main problems are in relation to the rotor rim and its connection to the centring rings:

(i) to accept the stresses due to centrifugal forces; and

(ii) to keep an accurate balancing, i.e. to maintain the coincidence between the axis of the rotor centring rings and the main axis of inertia, with a high level of accuracy. This accuracy will have to remain stable in spite of elongation due to centrifugal forces, of temperature variations, and of time effects associated with the above parameters.

Several types of rotor were developed. For medium speeds up to 15 000 rpm and peripherical speeds of 250 m/s, light alloy rotors were developed. They are of interest for producing angular momentum (satellite applications) but they do not allow important levels of stored energy for a reasonable mass. For high-speed rotors, two main techniques were utilized: the subcircular rotor' and the 'cycloprofile rotor'.

An interesting order of magnitude is given by the comparison of the energy stored in a flywheel with another type of energy: for example let us suppose that a flywheel with a 10 kg thin rim on a diameter of 62 cm rotates at 30 000 rpm and is utilized on a giant hoist operating without losses. The energy of the 10 kg rim rotating at a peripheral speed of 1000 m/s would be able to take a mass of 100 kg up to a height of 5000 m. Obviously this example is just given to compare the two different physical aspects of this energy but it clearly indicates the wide amount of energy which can be stored in a small mass rotating at high speed: even with energy conversion efficiencies far less than unity, many possibilities of application are apparent.

Several conclusions can be drawn from these orders of magnitude:

(a) the concept of the rotor has to accommodate the important elongations which are involved, particularly at the balancing point; and
(b) the long-term operation which will be generally requested for energy storage systems will necessitate a relatively high safety margin between the working stresses and the ultimate tensile strength.

2. Ground applications

The design complexity of rotors for ground application is defined by their maximum energy storage capacity. Below approximately 1 kWh deliverable, a very simple rotor, in steel or light alloy, with cylindrical outer shape, can be foreseen. With less than 0.4 m diameter and a rotating speed of 12 000 rpm, the storage of 1 kWh can be ensured without exceeding forbidden stress levels in the peripheral zones.

Beyond 1 kWh, the external part must be constrained by wound filament material, and concepts like 'subcircular' or 'cycloprofile' are to be applied. In addition, the dynamic stability throughout the range of rotational speeds calls for servocontrol electronics.

IV. CONCLUSION

If the main functional hardware is considered, it can be seen that it is adaptable to ground applications:

(a) without significant new developments for a range below 3 kW in power and 1 kWh in deliverable energy, and
(b) with additional adaptations for larger storage capacities.

The cost studies performed show that kinetic storage is a competitor to batteries and associated electronics. No special rooms or large surfaces are required. Electronic racks are limited in volume. Owing to the inherent reliability of magnetic suspension, no maintenance is necessary except 'on request', in case of alarms. Contrary to batteries, ability for performance is always guaranteed. Large flexibility for the user's applications is also provided, such as direct supply of alternating current. Moreover, if the level of output is adjusted accordingly, the wheel may then serve as a level stabilizer. Another range of applications, of increasing interest with the scope of energy savings, is the use of a wheel (especially a larger wheel) as a complementary energy source, allowing us to avoid overdimensioning of nominal systems to face consumption peaks.

Energy Conservation and Thermal Insulation
Edited by R. Derricott and S. S. Chissick
© 1981 John Wiley & Sons Ltd.

CHAPTER 27

Index of manufacturers and suppliers of heating, cooling and thermal insulation products in the UK

S. S. Chissick and R. Derricott

The following information has been compiled by the authors and is believed to be correct at the date of preparation (early 1979). It is intended to be of value for the benefit of readers who may need to seek further information on the great range and variety of products and services available. The details refer to a limited number of manufacturers and suppliers of products and services and the list of organizations is not meant to be exhaustive. Many other manufacturers and suppliers exist and the noninclusion of these does not imply that they are regarded as unsatisfactory in any way. The inclusion of a particular organization is not meant to be an endorsement of that manufacturer or supplier, and none of the organizations' products and services have been tested in any formal sense.

For further information on these manufacturers and suppliers and for information concerning their products and services, readers are directed to the following sources:

Barbour, 1979, Barbour Compendium of Building Products, Barbour Index Ltd, Windsor
Greater London Council, 1979, Description of Materials and Workmanship, Preambles to Bills of Quantities, SMM6 Edition, GLC, London.
National Building Commodity Centre, 1979, Commodity File, NBCC Ltd., London.
Standard Catalogue Information Services, 1979, Architects Standard Catalogue, SCIS Ltd., Tonbridge, Kent.

AIR CONDITIONING UNITS

AAF Ltd,
  Basington Lane, CRAMLINGTON, Northumberland, NE23 8AF
  Tel: Cramlington (0670) 713477
Airdale Air-Conditioning Ltd,
  Clayton Wood Rise, West Park, LEEDS, West Yorkshire, LS16 6RF
  Tel: Leeds (0532) 787191
Aquafan Cooling Towers Ltd,
  183 Hampton Road, TWICKENHAM, Middlesex, TW2 5NG
  Tel: 01-977 0231
Beta-Plus Ltd,
  177 Haydons Road, LONDON, SW19 8TB
  Tel: 01-543 1142
F. H. Biddle Ltd,
  Newton Road, NUNEATON, Warwickshire, CV11 4HP
  Tel: Nuneaton (0682) 384233
Canadian Chromalox Co. Ltd,
  Grangefield Industrial Estate, Richardshaw Road, Pudsey, LEEDS, West
  Yorkshire, LS28 6QN
  Tel: Leeds (0532) 577911
Carter Cooling Towers,
  Redhill Road, Hay Mills, BIRMINGHAM, B25 8EY
  Tel: 021-772 4300
Celsius Refrigeration & Air-Conditioning Equipment Ltd,
  Wiggenhall Industrial Estate, WATFORD, Hertfordshire, WD1 8AW
  Tel: Watford 45021
Climate Equipment Ltd,
  Highlands Road, Shirley, SOLIHILL, West Midlands, B90 4NL
  Tel: 021-705 7601
Clipper Air Handling Units Ltd,
  Raans Road, AMERSHAM, Buckinghamshire, HP6 6HY
  Tel: Amersham 21212
L. A. Cotter & Co. Ltd,
  Fletchers Lane, Walsall Road, WILLENHALL, West Midlands,
  WV13 2QW
  Tel: Willenhall (0902) 67231
Delta RA Ltd,
  Hollands Road, HAVERHILL, Suffolk, CB9 8PT
  Tel: Haverhill (0440) 2653/7
Denco Miller Ltd,
  P.O. Box 11, Holmer Road, HEREFORD, HR4 9SJ
  Tel: Hereford (0432) 68151

Ductwork Engineering Systems Ltd,
2 Commerce Way, CROYDON, Surrey, CR9 4PH
Tel: 01-681 0771
Dunham Bush Ltd,
Fitzherbert Road, Farlington, PORTSMOUTH, Hampshire, PO6 1RR
Tel: Cosham (0705) 370961
EER Air Conditioning Products Ltd,
Britannia House, Birch Street, ASHTON UNDER LYNE, Lancashire,
O17 0PP
Tel: 061-308 2137
Electra Air Conditioning Services Ltd,
66–68 George Street, LONDON, W1H 5RG
Tel: 01-487 5606
Enviro-Aesthetic Engineering Co. Ltd,
9 Albion Place, MAIDSTONE, Kent, ME14 5DY
Tel: Maidstone (0622) 50157
Fair-Air Ltd,
Sybron Way, Millbrook Industrial Estate, CROWBOROUGH, East
Sussex, TN6 3JW
Tel: Crowborough 273/8
Film Cooling Towers (1925) Ltd,
Chancery House, Parkshot, RICHMOND, Surrey, TW9 2RH
Tel: 01-940 6494
Heenan Coolers Ltd,
P.O. Box 20, Shrub Hill Road, WORCESTER, WR4 9EX
Tel: Worcester (0905) 23461
IMI Paxman Ltd,
Bradford Road, BRIGHOUSE, West Yorkshire, HD6 4AL
Tel: Brighouse (0484) 714361/714584
Lennonx Industries Ltd,
P.O. Box 43 Lister Road, BASINGSTOKE, Hampshire, RG22 4AR
Tel: Basingstoke (0256) 61261
Lesney Environmental Controls Ltd,
Worcester Trading Estate, Blackpole Road, WORCESTER, WR3 8HR
Tel: Worcester (0905) 52561
Luwa (UK) Ltd,
Luwa House, 36–38 Goldsworth Road, WOKING, Surrey, GU21 1JT
Tel: Woking 21441/6
Microflow Pathfinder Ltd,
Fleet Mill, Minley Road, FLEET, Hampshire, GU13 8RD
Tel: Fleet 28441

Midlands Efficiency Services Ltd,
    Industrial Estate, Muckows Hill, HALESOWEN, West Midlands,
    B62 8DR
    Tel: 021-501 1211
Multivent Ltd,
    Bridport Road, Edmonton, LONDON, N18 1SL
    Tel: 01-803 3366
Myson Copperad Ltd,
    Old Wolverton, MILTON KEYNES, Buckinghamshire, MK12 5PT
    Tel: Milton Keynes (0908) 312641
Novenco Ltd,
    Tundry Way, Chainbridge Road, BLAYDON ON TYNE, Tyne & Wear,
    NE21 5SN
    Tel: Blaydon (0632) 444511
Ozonair Engineering Co. Ltd,
    Quarry Wood Industrial Estate, Aylesford, MAIDSTONE, Kent,
    ME20 7NB
    Tel: Maidstone (0622) 77861
Packaged Chillers (Winson Group),
    St Annes House, 45 Park Street, LUTON, Bedfordshire, LU1 3JX
    Tel: Luton (0582) 421861
Qualitair (Air-Conditioning) Ltd,
    Castle Road, Eurolink, SITTINGBOURNE, Kent ME10 3RH
    Tel: Sittingbourne (0795) 75461
Searle Manufacturing Co Ltd,
    Newgate Lane, FAREHAM, Hampshire, PO14 1AR
    Tel: Fareham 236151
Sinclaire Air Conditioning Ltd,
    22 Queen Annes Gate, Westminster, LONDON, SW1H 9AH
    Tel: 01-222 9200
South London Electrical Equipment Co. Ltd,
    Lanier Works, Hither Green Lane, LONDON, SE13 6QD
    Tel: 01-318 3021
Standard & Pochin Ltd,
    Evington Valley Road, LEICESTER, LE5 5LS
    Tel: Leicester (0533) 736114
Temperature Ltd,
    Newport Road, SANDOWN, Isle of Wight
    Tel: Sandown (0983) 402221
Thermoscreens Ltd,
    Avenger Close, Chandlers Ford Industrial Estate, EASTLEIGH,
    Hampshire, SO5 3DQ
    Tel: Eastleigh 4731

TI Creda Ltd,
  Creda Works, P.O. Box 5, Blythe Bridge, STOKE ON TRENT,
  Staffordshire, ST11 9LJ
  Tel: Blythe Bridge 2281
Toshiba (UK),
  Toshiba House, Frimley Road, CAMBERLEY, Surrey
  Tel: Camberley (0276) 62222
Trane Ltd,
  Donibristle Industrial Estate, DUNFERMLINE, Fife, KY11 5JP
  Tel: Dalgety Bay (0383) 823551
Trendpam Engineering Ltd,
  29 Nork Way, BANSTEAD, Surrey, SM7 1PB
  Tel: Burgh Heath 59325
Venduct Ltd,
  141 Barkby Road, LEICESTER, LE4 7LW
  Tel: Leicester (0533) 766636
Ventilation Equipment & Conditioning Ltd,
  Vequip Works, 320 Latimer Road, LONDON, W10 6QR
  Tel: 01-969 7553
Walker Air Conditioning Ltd,
  Dublin Industrial Estate, Finglas Road, DUBLIN 11, Eire
  Tel: Dublin 300844
Weathermaker Equipment Ltd,
  268 Abbeydale Road, WEMBLEY, Middlesex, HA0 1PR
  Tel: 01-998 7721
West Benyon Ltd,
  242 High Street, BROMLEY, Kent, BR1 1PQ
  Tel: 01-460 0081
Westinghouse Electric SA,
  1 The Curfew Yard, Thames Street, WINDSOR, Berkshire
  Tel: Windsor 54711
Woods of Colchester Airpac Division,
  Tufnell Way, COLCHESTER, Essex, CO4 5AR
  Tel: Colchester (0206) 44122
Wright Air Conditioning Ltd,
  Sampson Road North, Camp Hill, BIRMINGHAM, B11 1BL
  Tel: 021-773 8421

BLOCKS: CONCRETE

Aglite (Midlands) Ltd,
  Wellington Street, RIPLEY, Derbyshire, DE5 3DZ
  Tel: Ripley (0773) 3881

ARC Concrete Ltd,
  Elms Court, West Way, Botley, OXFORD, OX2 9LR
  Tel: Oxford (0865) 46351
Barton Block Co. Ltd,
  Bedford Road, Barton, BEDFORD, MK45 4PP
  Tel: Luton (0582) 881267
Beachcroft Concrete Partitions Ltd,
  80 Capworth Street, Leyton, LONDON, E10 7JA
  Tel: 01-539 7768
Border Concrete Products Ltd,
  Jedburgh Road, KELSO, Borders, TD5 8JG
  Tel: Kelso (05732) 2393
E. G. Bradley Building Products Ltd,
  Okus, SWINDON, Wiltshire, SN1 4JJ
  Tel: Swindon (0793) 28131
Brand & Rae Ltd,
  Russell Mill, Springfield, CUPAR, Fife, KY15 5QX
  Tel: Cupar (0334) 2828
Celcon Ltd,
  129 Kingsway, LONDON, WC2B 6NU
  Tel: 01-405 6267
Clondalkin Concrete Ltd,
  CLONDALKIN, Co. Dublin, Eire
  Tel: Dublin 512222
Colinwell Concrete Ltd,
  37 Colinglen Road, Dunmurray, BELFAST
  Tel: Belfast (0232) 618145
Concrete Masonry Ltd,
  Llay, WREXHAM, Clwyd, LL12 0TL
  Tel: Gresford (097883) 2101
P. & B. Connoly Ltd,
  Redford, DUNGANNON, Co. Tyrone
  Tel: Moy (076278) 208
H. Dean Ltd,
  Weybread, DISS, Norfolk, IP21 5UI
  Tel: Harleston (0379) 852201
Devon Concrete Works Ltd,
  Taw Wharf, BARNSTAPLE, Devon, EX31 2AB
  Tel: Barnstaple (0271) 3021
Dunbrik (Ulster) Ltd,
  Hopefield Avenue, BELFAST, BT15 5AQ
  Tel: Belfast (0232) 745401

Durox Building Units Ltd,
  Linford, STANFORD LE HOPE, Essex, SS17 0PY
  Tel: Stanford le Hope (03756) 3344
ECC Quarries Ltd,
  Northernhay House East, Northernhay Place, EXETER, Devon
  EX4 4QP
  Tel: Exeter (0392) 52231
Edenhall Concrete Products Ltd,
  Barbary Plains, Edenhall, PENRITH, Cumbria, CA11 8SP
  Tel: Langwathby (076881) 366
A. E. Evans, 'Evalast' Works, Llanymynech, OSWESTRY, Salop.
  Tel: Oswestry (0691) 830571
Forticrete Ltd,
  Park Lane West, BOOTLE, Merseyside, L30 6UJ
  Tel: 051-521 3545
John Fyfe Ltd,
  Whitemyres Avenue, Mastrick, ABERDEEN, AB9 1PX
  Tel: Aberdeen (0224) 691333
Samuel Gilchrist (Lenaderg) Ltd,
  Lenaderg, BAINBRIDGE, North Yorkshire
  Tel: Bainbridge (08206) 24271
Gillmans Masonry Ltd,
  Gillmans Industrial Estate, BILLINGSHURST, West Sussex
  Tel: Billingshurst (040381) 3696
Harrison's Limeworks,
  Flusco, PENRITH, Cumbria
  Tel: Shap 241
Hemel Hempstead Lightweight Concrete Ltd,
  Redbourn Road, HEMEL HEMPSTEAD, Hertfordshire
  Tel: Hemel Hempstead (0442) 54321
Hilcrete Ltd,
  Hilton, DERBY, DE6 5FQ
  Tel: Etwall (028373) 2277
C.A.E.C. Howard Ltd,
  St John's Works, BEDFORD, MK42 0DR
  Tel: Bedford (0234) 63171
Hunnable Concrete Products Ltd,
  Rayne Road, BRAINTREE, Essex, CM7 6RY
  Tel: Braintree (0376) 26126
Hydroprest Concrete Ltd,
  P.O. Box 8, St Vincent House, Normanby Road, SCUNTHORPE, South
  Humberside
  Tel: Scunthorpe (0724) 3217

Interfuse Ltd,
  4/10 Brook Street, Syston, LEICESTER, LE7 8GS
  Tel: Leicester (0533) 609666
Inverness Precast Concrete Ltd,
  Longman Road, INVERNESS, IV1 1SD
  Tel: Inverness (0463) 35174
Kingston Minerals Ltd,
  Perryfield House, Wakeham, PORTLAND, Dorset
  Tel: Portland (0305) 820207
Landbeach Concrete Co. Ltd,
  Ely Road, Waterbeach, CAMBRIDGE
  Tel: Cambridge (022) 860315
Lightweight Aggregates Ltd,
  Snowdown, DOVER, Kent, CT15 4LT
  Tel: Nonington (0304) 840894
Lignacite Products Ltd,
  51–53 Victoria Street, LONDON, SW1H 0EZ
  Tel: 01-222 1442
Lytag Ltd,
  Hamilton House, 111 Marlowes, HEMEL HEMPSTEAD, Hertfordshire,
  HP1 1BB
  Tel: Hemel Hempstead (0442) 69331
Mendip Stone and Concrete Co. Ltd,
  Flax Bourton, BRISTOL
  Tel: Flax Bourton (027583) 3361
Mixconcrete Masonry Ltd,
  Southcrete Works, Ashford Road, Chartham, CANTERBURY, Kent
  Tel: Chartham (022773) 551
North Down Quarries Ltd,
  Ballybarnes Quarry, NEWTOWNARDS, Co. Down
  Tel: Belfast (0232) 622211
Pascall, Thomas & Sons Ltd,
  Frittenden, CRANBROOK, Kent, TN17 2AZ
  Tel: Frittenden (058080) 216
Penryn Granite Ltd,
  Mabe, PENRYN, Cornwall
  Tel: Penryn (0326) 73851
Plastmor Ltd,
  P.O. Box 44, KNOTTINGLEY, West Yorkshire, WF11 0DN
  Tel: Knottingley (0977) 83221
Rainford, William (Holdings) Ltd,
  Heysham Road, LIVERPOOL, L30 6UQ
  Tel: 051-525 5991

Russlite (Scotland) Ltd,
  Gartshore Twechar, Kilsyth, GLASGOW, G65 9TW
  Tel: Kilsyth (0236) 822461
Sabey Kirby & Sons Ltd,
  Langholm, 16 High Bondgate, BISHOP AUCKLAND, County Durham,
  DL14 7PH
  Tel: Bishop Auckland (0388) 603228
Sage & Down Ltd,
  Kewstoke Road, WESTON SUPER MARE, Avon, BS22 9LE
  Tel: Weston Super Mare (0934) 24181
Saunders (Ipswich) Ltd, Sproughton Road, IPSWICH, Suffolk, IP1 5AN
  Tel: Ipswich (0473) 44611
Scott (Toomebridge) Ltd,
  TOOMEBRIDGE, Co. Antrim
  Tel: Toomebridge (064883) 272
Siporex Ltd,
  Carfin Road, Newarthill, MOTHERWELL, Strathclyde, ML1 5AA
  Tel: Motherwell (0698) 64111
Skegness Cast Stone Ltd,
  40 Old Wainfleet Road, SKEGNESS, Lincolnshire
  Tel: Skegness (0754) 2431
Stowell Concrete Ltd,
  Kenn, CLEVEDON, Avon, BS21 6TL
  Tel: Clevedon (0272) 876001
TAC Construction Materials Ltd,
  Brick and Block Division, Gorsey Lane, Halton, WIDNES, Cheshire,
  WA8 0RL
  Tel: 051-423 1161
A. & E. Tate Ltd,
  Coleford Road, SHEFFIELD, S9 5NN
  Tel: Sheffield (0742) 441081
Thakeham Tiles Ltd,
  Heath Common, Storrington, PULBOROUGH, West Sussex,
  RH20 3AD
  Tel: Storrington (09066) 2381
Thermalite Ltd,
  Station Road, Coleshill, BIRMINGHAM, B46 1HP
  Tel: Coleshill (0675) 62081
Tiling Construction Services Ltd,
  Quarries Division, Northern Region, Parliament Street, HARROGATE,
  North Yorkshire, HG1 2RF
  Tel: Harrogate (0423) 68092

Warecrete Concrete Products Ltd,
  Marsh Lane, London Road, WARE, Hertfordshire, SE12 9QD
  Tel: Ware (0920) 2468
Wath Concrete Products Ltd,
  Wath, RIPON, North Yorkshire, HG4 5EP
  Tel: Melmerby (076584) 308
Western Blocks Ltd,
  42 Upton Towans, Loggans, HAYLE, Cornwall, TR27 5BL
  Tel: Hayle (0736) 753128
Wilson Precast Concrete,
  Mains Road, BEITH, Strathclyde, KA15 2HR
  Tel: Beith (05055) 2711
Wotton Concrete Products Ltd,
  London Road, Wick, BRISTOL, BS15 5SJ
  Tel: Abson (027582) 3371

BOILERS

Acoustics & Envirometrics Ltd,
  Winchester Road, WALTON ON THAMES, Surrey, KT12 2RP
  Tel: Walton on Thames 47644/6
Afos Ltd,
  Manor Estate, Anlaby, HULL, North Humberside, HU10 6RL
  Tel: Hull (0482) 52152
Agaheat Appliances Glynwed Domestic & Heating Appliances Ltd,
  P.O. Box 37, Foundry Lane, LARBERT, Central, FK5 4PL
  Tel: Larbert (03245) 2233
Agaheat Appliances Glynwed Domestic & Heating Appliances Ltd,
  P.O. Box 30, Ketley, TELFORD, Salop, TF1 1BR
  Tel: Telford (0952) 3973
Baxi Heating,
  P.O. Box 52, Bamber Bridge, PRESTON, Lancashire, PR5 6SN
  Tel: Preston (0772) 36201
Beeston Boiler Co. (Successors) Ltd,
  Beeston, NOTTINGHAM, NG9 2DN
  Tel: Nottingham (0602) 254271
Boulter Boilers Ltd,
  Meteor Close, Norwich Airport Industrial Estate, NORWICH,
  Norfolk, NR6 6HG
  Tel: Norwich (0603) 411817
Brinton Bros (Boilers) Ltd,
  Ashley Street, Rowley Regis, WARLEY, West Midlands, B65 0JJ
  Tel: 021-599 1026
Broag Ltd,
  Thyssen House, Molly Millars Lane, WOKINGHAM, Berkshire,
  RG11 2PY
  Tel: Wokingham (0734) 783434

Chaffoteaux Ltd,
Concord House, Brighton Road, Salfords, REDHILL, Surrey, RH1 5DX
Tel: Horley (02934) 72744

Cosybug Ltd,
2 Napier Road, BROMLEY, Kent, BR2 9JA
Tel: 01-464 3263

Curwen & Newbury Ltd,
Westcroft Works, Alfred Street, WESTBURY, Wiltshire, BA13 3DZ
Tel: Westbury (0373) 823646

W. H. Dean & Son Ltd,
Accrington Road, BURNLEY, Lancashire, BB11 5DS
Tel: Burnley (0282) 25901

Delheat Ltd,
91 Spinney Hill, Addlestone, WEYBRIDGE, Surrey, KT15 1AZ
Tel: Ottershaw (093287) 3456

Doulton Heating Systems Ltd,
School Lane, Knowsley, PRESCOT, Merseyside, L34 9HJ
Tel: 051-546 8225

Dunsley Heating Appliance Co. Ltd,
Fearnought, Holmfirth, HUDDERSFIELD, West Yorkshire, HD7 2TU
Tel: Holmfirth (048489) 2635

Glow-Worm Ltd,
Nottingham Road, BELPER, Derbyshire, DE5 1JT
Tel: Belper (077382) 4141

Harvey Habridge Ltd,
Haldane, Halesfield 11, TELFORD, Salop., TF7 4PJ
Tel: Telford (0952) 585545

Heatpak 77 Ltd,
Raynham Road Industrial Estate, BISHOP'S STORTFORD, Hertfordshire, CM23 5PN
Tel: Bishop's Stortford (0279) 55031

Heatrae-Sadia Heating Ltd,
Hurricane Way, Norwich Airport, NORWICH, NR6 6EA
Tel: Norwich (0603) 44144

Hill-Foster Ltd,
262 Uxbridge Road, Hatch End, PINNER, Middlesex, HA5 4HS
Tel: 01-428 0265

IMI Santon Ltd,
Somerton Works, NEWPORT, Gwent, NPT 0XU
Tel: Newport (0633) 277711

Interoven Ltd,
70 Fearnley Street, WATFORD, Hertfordhire, WD1 7DE
Tel: Watford 46761

Kayanson Engineers Ltd,
  1–3 Market Square, CHESHAM, Buckinghamshire,
  HP5 1BH
  Tel: Chesham (02405) 2128
S. Keeping (Developments) Ltd,
  Little Always, CULLOMPTON, Devon, EX15 1RA
  Tel: Craddock (0884) 40846/7
McFarlane Bros (Heat) Ltd,
  Glasgow Road, Rutherglen, GLASGOW, G73 1SW
  Tel: 041-647 4485
Multitherm Ltd,
  TURVEY, Bedfordshire, MK43 8ET
  Tel: Turvey (023064) 360
Myson Domestic Products Ltd,
  Residential Heating Division, ONGAR, Essex, CM5 9RE
  Tel: Ongar (0277) 362222
Nor-Cal Engineering (GB) Ltd,
  Nor-Cal House, 106 Stafford Road, WALLINGTON, Surrey,
  SM6 8PX
  Tel: 01-647 6644
OGB Products Ltd,
  Churchbridge Industrial Estate, Oldbury, WARLEY, West Midlands,
  B69 4LH
  Tel: 021-544 5720
Perkins Boilers Ltd,
  Mansfield Road, DERBY, DE2 4BA
  Tel: Derby (0332) 48235
Potterton International Ltd,
  Brooks House, Coventry Road, WARWICK, CV34 4LL
  Tel: Warwick (0926) 43471
Radiation-Ascot Ltd,
  Nottingham Road, BELPER, Derbyshire, DE5 1JT
  Tel: Belper (077382) 4141
Albert Roberts,
  Wollescote Foundry & Engineering Works, STOURBRIDGE, West
  Midlands, DY9 8SL
  Tel: Lye (038482) 2161
Robinson Willey Ltd,
  Mill Lane, LIVERPOOL, L13 4AJ
  Tel: 051-228 9111
Rossfor Associates Ltd,
  296b Station Road, HARROW, Middlesex, HA1 2DX
  Tel: 01-427 8454

Smith & Wellstood Ltd,
  BONNYBRIDGE, Central, FK4 2AP
  Tel: Bonnybridge (032481) 2171
Stelrad Group Ltd,
  P.O.Box 103, National Avenue, HULL, North Humberside, HU5 4JN
  Tel: Hull (0482) 492251
Taylor & Portway Ltd,
  Rosemary Lane, HALSTEAD, Essex, CO9 1HR
  Tel: Halstead (07874) 2551
R. Tomlinson (Boilers) Ltd,
  Lotherton Way, Aberforth Road Trading Estate, Garforth, LEEDS
  Tel: Leeds (0532) 861122
Trianco Redfyre Ltd,
  Stewart House, Brook Way, LEATHERHEAD, Surrey, KT22 7LY
  Tel: Leatherhead (03723) 76453
UA Engineering Ltd,
  Canal Street, SHEFFIELD, South Yorkshire, S4 7ZE
  Tel: Sheffield (0742) 21167
Warmback Ltd,
  Manor Works, Short Road, Leytonstone, LONDON, E11 4RH
  Tel: 01-539 6601
Worcester Engineering Co. Ltd,
  Diglis, WORCESTER, WR5 3DG
  Tel: Worcester (0905) 356224

CAVITY INSULATION

Advance Insulation Ltd,
  10 Sefton Drive, WILMSLOW, Cheshire, SK9 4EL
  Tel: Wilmslow (09964) 26188.
B. R. Ainsworth & Co. Ltd,
  Bracol House, 6 Thames Road, BARKING, Essex, IG11 0HZ
  Tel: 01-594 7277
Alexander Insulation Consultants Ltd,
  Anaconda Works, Springfield Lane, SALFORD, M3 7LR
  Tel: 061-832 2121
John Baker (Insulation) Ltd,
  Superfoam House, High Street, HENFIELD, Sussex, BN5 8HN
  Tel: Henfield (079155) 3561
British Industrial Plastics Ltd,
  Chemicals Division, P.O. Box 6, Popes Lane, Oldbury, WARLEY,
  West Midlands, B69 4PD
  Tel: 021-552 1551

Cape Insulation Services Ltd,
  Rosanne House, Bridge Road, WELWYN GARDEN CITY,
  Hertfordshire, AL8 6UE
  Tel: Welwyn Garden 31155
Cavity Foam Insulation Ltd,
  Milton, GILLINGHAM, Dorset, SP8 5PY
  Tel: Gillingham (07476) 2777
Corrutex Ltd,
  Spurlings Yard, Wallington, FAREHAM, Hampshire, PO17 6AD
  Tel: Fareham 5292
Cosywall Insulation Ltd,
  60 Buxton Road, Hazel Grove, STOCKPORT, Cheshire
  Tel: 061-487 1833
Energstor Ltd,
  47a Mosside Road, Derriaghy, BELFAST, BT17 9HH
  Tel: Belfast (0232) 619006
Energy Conservation Consultants,
  15 Bluebell Wood, BILLERICAY, Essex, CM12 0ES
  Tel: Billericay (02774) 52887
Fitwarmth Ltd,
  Homefield House, Langley, HITCHIN, Hertfordshire
  Tel: Stevenage (0438) 820391
GC Insulation Ltd,
  Wilmington House, Wilmington, DARTFORD, Kent, DA2 7EF
  Tel: Dartford 70022
Homefoam Cavity Wall Insulation Ltd,
  7 Mallard Avenue, Sandal, WAKEFIELD, West Yorkshire, WF2 6SL
  Tel: Bradford (0274) 496152
Impact Insulation (Leeds) Ltd
  Building 6, Whitehall Estate, Whitehall Road, LEEDS, LS12 5JB
  Tel: Leeds (0532) 793222
Indefoam Ltd,
  Unit P I, Kingsditch Trading Estate, Kingsditch Lane
  CHELTENHAM, Gloucestershire
  Tel: Cheltenham (0242) 30221
Insuwall Ltd,
  Bankfield Industrial Estate, South Reddish, STOCKPORT, Cheshire,
  SJ5 7SE
  Tel: 061-480 3444
Lynham Insulations Ltd,
  6 Highbury Buildings, Portsmouth Road, Cosham, PORTSMOUTH,
  Hampshire
  Tel: Cosham (07018) 72467

Maxi-Foam Ltd,
  3 Grange Road, MIDDLESBROUGH, Cleveland
  Tel: Middlesbrough (0642) 210100
MBS Insulation Ltd,
  Corringham Road Industrial Estate, GAINSBOROUGH,
  Lincolnshire, DN21 1QB
  Tel: Gainsborough (0472) 5050
MDC Group Services,
  77 Wyle Cop, SHREWSBURY, Salop.
  Tel: Shrewsbury (0743) 52584
Megafoam Ltd,
  Morley Road, TONBRIDGE, Kent, TN9 1RA
  Tel: Tonbridge (0732) 357022
Modern Plan Insulation (Airedale) Ltd,
  Summerville Road, BRADFORD, West Yorkshire, BD7 1NS
  Tel: Bradford (0274) 23089
Modern Plan Insulation (Cardiff) Ltd,
  12 Splott Road, CARDIFF, CF2 2BZ
  Tel: Cardiff (0222) 26220
Modern Plan Insulation (Godiva) Ltd,
  Dubarry Lodge, Somers Road, RUGBY, Warwickshire
  Tel: Rugby (0788) 74967
Modern Plan Insulation (Holdings) Ltd,
  Willowcraft Works, Broad Lane, Cottenham, CAMBRIDGE
  Tel: Willingham (0954) 50155
Modern Plan Insulation (NI) Ltd,
  75 Palmerston Road, BELFAST, BT4 1QD
  Tel: Belfast (0232) 651010
Modern Plan Insulation (North West) Ltd,
  Insulation House, Coniston Street, LEIGH, Lancashire, WN7 1XH
  Tel: Leigh (0942) 604121
Alvin J. Patterson Ltd,
  50 Holywood Road, BELFAST, BT4 1NT
  Tel: Belfast (0232) 652101
Payne's Insulation Co. Ltd,
  Ynysfor Moelfre, ANGLESEY, Gwynedd, LL72 8LB
  Tel: Moelfre (024888) 598
Pheonix Insulation Ltd,
  Whewell Street, Birstall, BATLEY, West Yorkshire, WF17 9PQ
  Tel: Batley (0924) 478261
Propotec (D. W. & I.) Ltd,
  Whitehall, Chapel Street, Exning, NEWMARKET, Suffolk
  Tel: Exning (063877) 543

Rentokil Ltd,
  Insulation Division, Felcourt, EAST GRINSTEAD, West Sussex, RH19 2JY
  Tel: East Grinstead (0342) 23661
Rockwool Company (UK) Ltd,
  St Agnes House, Cresswell Park, Blackheath, LONDON, SE3 9RD
  Tel: 01-318 3915
RR Home Insulation (NI) Ltd,
  62 Antrim Street, LISBURN, Northern Ireland, BT28 1AY
  Tel: Lisburn (08462) 5505
Sankey Building Supplies Ltd,
  Station House, Harrow Road, WEMBLEY, Middlesex
  Tel: 01-903 3811
Saveheat (Cavity Wall Insulation) Ltd,
  34 Back Sneddon Street, PAISLEY, Strathclyde
  Tel: 041-887 0927
Structurecare Ltd,
  433 Street Lane, LEEDS, West Yorkshire, LS17 6HQ
  Tel: Leeds (0532) 685543
Tropic Foam Ltd,
  Cocklebury Road, CHIPPENHAM, Wiltshire
  Tel: Chippenham (0249) 51161
Veejay Foam Insulation Co.,
  24 West Park Road, CORSHAM, Wiltshire, SN13 9LN
  Tel: Corsham (0249) 713693
Warmawall Insulation Services Ltd,
  221 Old Christchurch Road, BOURNEMOUTH, Dorset, BH8 8MF
  Tel: Bournemouth (0202) 293477
Westholme Insulations Ltd,
  Pottery Close, Winterstoke Road, WESTON SUPER MARE, Avon, BS23 3YH
  Tel: Weston Super Mare (0934) 29121

CEILING AND UNDERFLOOR HEATERS

Beau Design Services Ltd,
  Pipeline House, 26–30 Theobald Street, BOREHAMWOOD, Hertfordshire, WD6 4SG
  Tel: 01-953 4065
F. H. Biddle Ltd,
  Newton Road, NUNEATON, Warwickshire, CV11 4HP
  Tel: Nuneaton (0682) 384233

Comyn Ching & Co (Solray) Ltd,
  110 Golden Lane, LONDON, EC1Y 0SS
  Tel: 01-253 8414
Dunham-Bush Ltd,
  Fitzherbert Road, Farlington, PORTSMOUTH, Hampshire, PO6 1RR
  Tel: Cosham (0705) 370961
Enviro Aesthetic Engineering Co. Ltd,
  9 Albion Place, MAIDSTONE, Kent, ME14 5DY
  Tel: Maidstone (0622) 50157
ESWA Ltd,
  203 Blackfriars Road, LONDON, SE1 8NS
  Tel: 01-928 7356
HVE Thermoduct Ltd,
  Greenholme Mills, Burley in Wharfedale, ILKLEY, West Yorkshire,
  LS29 7EA
  Tel: Burley in Wharfedale (09435) 3587
Multibeton (UK) Ltd,
  2 Bamborough Gardens, LONDON, W12
  Tel: 01-749 3197
Phoenix Burners Ltd,
  34–44 Tunstall Road, LONDON, SW9 8DA
  Tel: 01-733 1181
Rapaway Ltd,
  Eurohurst House, Oakenshaw Road, Shirley, SOLIHULL, West
  Midlands, B90 4PE
  Tel: 021-745 3144
Sulzer Bros. (UK) Ltd,
  FARNBOROUGH, Hampshire
  Tel: Farnborough (0252) 44311
TA Controls Ltd,
  Lea Industrial Estate, Lower Luton Road, HARPENDEN, Hertfordshire,
  AL5 5EQ
  Tel: Harpenden (05827) 67991
Thermaflex Ltd,
  Queensway Industrial Estate, GLENROTHES, Fife, KY7 5PZ
  Tel: Glenrothes (0592) 752212

CONTROLS

AMF Venner,
  Electrical Products Division (UK), AMF International Ltd, AMF House,
  Whitby Road, BRISTOL, BS4 4AZ
  Tel: Bristol (0272) 778381

Andrews-Weatherfoil Ltd,
  185 Bath Road, SLOUGH, Berkshire, SL1 4AP
  Tel: Slough (0753) 23871
Anglo-Nordic Thermal Holdings Ltd,
  74 London Road, KINGSTON UPON THAMES, Surrey, KT2 5BR
  Tel: 01-549 0901
Appliance Components Ltd,
  Cordwallis Street, MAIDENHEAD, Berkshire, SL6 7BQ
  Tel: Maidenhead (0628) 72121
Bailey & Mackey Ltd,
  Baltimore Road, BIRMINGHAM, B42 1DE
  Tel: 021-357 4204
Barflo Ltd,
  56 Cavendish Place, EASTBOURNE, East Sussex, BN21 3RN
  Tel: Eastbourne (0323) 27877
BM Control Valves Ltd,
  Cogmore Lane CHERTSEY, Surrey, KT16 9AR
  Tel: Chertsey (09328) 60124
Brannan Thermometers,
  CLEATOR MOOR, Cumbria, CA25 5QE
  Tel: Cleator Moor (0946) 810413
The Bucks Diecasting Co. Ltd,
  Britannia Foundry, BURNHAM, Buckinghamshire, SL1 7JN
  Tel: Burnham (06286) 4001
BVMI Ltd,
  MIL Division, Shaw Road, Bushbury, WOLVERHAMPTON, West
  Midlands, WV10 9NN
  Tel: Wolverhampton (0902) 20496
Compression Joints Ltd,
  Oldmixton, WESTON SUPER MARE, Avon, BS24 9BA
  Tel: Weston super Mare (0934) 24171
Conex–Sanbra Ltd,
  Whitehall Road, TIPTON, West Midlands, DY4 7JU
  Tel: 021-557 2831
Crane Ltd,
  11 Bouverie Street, LONDON, EC4Y 8AH
  Tel: 01-353 6511
Crater Controls Ltd,
  Lower Guildford Road, Knaphill, WOKING, Surrey, GU21 2EP
  Tel: Brookwood (04867) 2571
Danfoss Ltd,
  Horsenden Lane South, GREENFORD, Middlesex, UB6 7QE
  Tel: 01-998 5040

Daxima Ltd,
  Reynard Mills Trading Estate, Windmill Road, BRENTFORD,
  Middlesex, TW8 9NG
  Tel: 01-568 4621
Drayton Controls (Engineering)Ltd,
  WEST DRAYTON, Middlesex, UB7 8JW
  Tel: West Drayton 44021
Ekco Heating & Appliances,
  Drury Lane, HASTINGS, East Sussex, TN34 1XN
  Tel: Hastings (0424) 42914
Ellis Miller Ltd,
  Market House, St Judes Road, Englefield Green, EGHAM, Surrey,
  TW20 0BU
  Tel: Egham (07843) 5302/5387
Eltron (London) Ltd,
  Accrington Works, Strathmore Road, CROYDON, Surrey, CR9 2NA
  Tel: 01-689 4341
Findley Irvine Ltd,
  Bog Road, PENICUIK, Lothian, EH26 8BR
  Tel: Penicuik (0968) 72111
The Finnish Valve Co. Ltd,
  138 Kenley Road, LONDON, SW19 3HT
  Tel: 01-540 4581
Haylake Ltd,
  Holtspur Parade, Holtspur, BEACONSFIELD, Buckinghamshire,
  HP9 1DX
  Tel: Beaconsfield (04946) 71819
Honeywell Ltd,
  Charles Square, BRACKNELL, Berkshire, RG12 1EB
  Tel: Bracknell (0344) 24555
The Horne Engineering Co. Ltd,
  P.O. Box 7, Rankine Street, JOHNSTONE, Strathclyde, PA5 8BD
  Tel: Johnstone (0505) 21455
The Horstmann Gear Group Ltd,
  Newbridge Works, Newbridge Road, BATH, Avon, BA1 3EF
  Tel: Bath (0225) 21141
IEM Supplies Co. Ltd,
  50–52 Boscombe Road, LONDON, W12 9HU
  Tel: 01-749 2512
ISS Clorius Ltd,
  Redwood House, Bristol Road, Keynsham, BRISTOL, BS18 2BB
  Tel: Keynsham (02756) 61166

ITT Controls,
  Maclaren Division, 333 West Street, GLASGOW, G5 8JE
  Tel: 041-429 2191
ITT Reznor,
  Park Farm Road, FOLKESTONE, Kent, CT19 5DR
  Tel: Folkestone (0303) 59141
Jackson Instruments & Controls Ltd,
  122a Western Road, LEICESTER, LE3 0GB
  Tel: Leicester (0533) 546649
Kane-May Ltd,
  Burrowfield, WELWYN GARDEN CITY, Hertfordshire, AL7 4SR
  Tel: Welwyn Garden (07073) 31051
Kent-Moore UK Ltd,
  Robinair Division, 19–21 Stockfield Road, Acocks Green,
  BIRMINGHAM, B27 6AJ
  Tel: 021-707 6955
Landis & Gyr Ltd,
  Victoria Road, North Acton, LONDON, W3 6XS
  Tel: 01-992 5311
Marflow Engineering Ltd,
  Oak Road, WEST BROMWICH, West Midlands, B70 8HR
  Tel: 021-553 6361
Myson Domestic Products Ltd,
  ONGAR, Essex, CM5 9RE
  Tel: Ongar (02776) 4311
Negretti & Zambra Ltd,
  Stocklate, AYLESBURY, Buckinghamshire, HP20 1DR
  Tel: Aylesbury (0296) 5931
Nettle Accessories Ltd,
  Whitegate Broadway, Chadderton, OLDHAM, Lancashire, OL9 9QG
  Tel: 061-652 1111
Pactrol Controls Ltd,
  P.O. Box 123, 46 Greenhey Place, SKELMERSDALE, Lancashire,
  WN8 9SA
  Tel: Skelmersdale (0695) 22191
K. W. Paulus & Co. Ltd,
  NMT House, Heston Aerodrome, HOUNSLOW, Middlesex TW5 9ND
  Tel: 01-897 6037
Peglers Ltd,
  Belmont Works, St Catherine's Avenue, DONCASTER, South
  Yorkshire, DN4 8DF
  Tel: Doncaster (0302) 68581

Photain Controls Ltd,
Unit 18, Hangar No. 3, The Aerodrome, Ford, ARUNDEL, West Sussex, BN18 0BE
Tel: Littlehampton (09064) 21531/4

Paul Poddy Ltd,
16 Minerva Road, LONDON, NW10 6HJ
Tel: 01-965 3462

Potterton International Ltd,
Brooks House, Coventry Road, WARWICK, CV34 4LL
Tel: Warwick (0926) 43471

PP Controls Ltd,
Cross Lances Road, HOUNSLOW, Middlesex, TW3 2AD
Tel: 01-572 3331

Randall Electronics Ltd,
Ampthill Road, BEDFORD, MK42 9ER
Tel: Bedford (0234) 64621

Samson Controls (London) Ltd,
Holmethorpe Avenue, REDHILL, Surrey, RH1 2NU
Tel: Redhill 66391

Sangamo Time Controls,
Industrial Estate, PORT GLASGOW, Strathclyde, PA14 5XG
Tel: Port Glasgow (0475) 45131

Satchwell Sunvic Ltd,
Watling Street, MOTHERWELL, Strathclyde, ML1 3SA
Tel: Motherwell (0698) 66277

Sauter Automation Ltd,
165 Bath Road, SLOUGH, Berkshire, S11 4AA
Tel: Slouth (0753) 39221

Skil Controls Ltd,
Greenhey Place, East Gillibrands, SKELMERSDALE, Lancashire, WN8 9SB
Tel: Skelmersdale (0695) 23671

Slaven Enterprises Ltd,
Mere Farm House, Hannington, NORTHAMPTON, NN6 9SU
Tel: Walgrave St Peters (060126) 579

Smith Meters Ltd,
170 Rowan Road, Streatham Vale, LONDON, SW16 5JE
Tel: 01-764 5011

Smiths Industries Ltd,
Waterloo Road, Cricklewood, LONDON, NW2 7UR
Tel: 01-452 3333

Sopac Regulation (UK) Ltd,
  Oaklands House, Solartron Road, FARNBOROUGH, Hampshire,
  GU14 7Q1
  Tel: Farnborough (0252) 514329
Starkstrom (London) Ltd,
  258 Field End Road, Eastcote, RUISLIP, Middlesex , HA4 9UW
  Tel: 01-868 3732
C. H. Stewart (Components) Ltd,
  P.O. Box 76, GREAT DUNMOW, Essex, CM6 3NR
  Tel: Great Dunmow (0371) 820868
Sulzer Bros (UK) Ltd,
  FARNBOROUGH, Hampshire, GU14 7LP
  Tel: Farnborough (0252) 44311
Superswitch Electric Appliances Ltd,
  7 Station Trading Estate, Blackwater, CAMBERLEY, Surrey,
  GU17 9AH
  Tel: Camberley (0276) 34556
TA Controls Ltd,
  Lea Industrial Estate, Lower Luton Road, HARPENDEN,
  Hertfordshire, AL5 5EQ
  Tel: Harpenden (05827) 67991
Tacotherm Ltd,
  Morison House, Rankine Road, Daneshill Estate, BASINGSTOKE,
  Hampshire, RG24 0PH
  Tel: Basingstoke (0256) 61181
Temfix Engineering Co. Ltd,
  FARNBOROUGH, Hampshire, GU14 7LP
  Tel: Farnborough (0252) 515151
H. Warner & Son Ltd,
  3 Foundation Street, IPSWICH, Suffolk, IP4 1DT
  Tel: Ipswich (0473) 53702
J. Williams (Energy Services) Ltd,
  The Furlong, Berry Hill Industrial Estate, DROITWICH,
  Worcestershire, WR9 9AJ
  Tel: Droitwich (09057) 0905
Yorkshire Imperial Fittings,
  P.O. Box 166, LEEDS, West Yorkshire, LS1 1RD
  Tel: Leeds (0532) 701107
G. H. Zeal Ltd,
  Lombard Road, Merton, LONDON, SW19 3UU
  Tel: 01-542 2283

ELECTRIC HEATERS AND APPLIANCES

Allan Haigh & Co. Ltd,
  Savile Park Works, Moorfield Street, HALIFAX, West Yorkshire, HX1 3AX
  Tel: Halifax (0422) 53952
Belling & Co. Ltd,
  Bridge Works, Southbury Road, ENFIELD, Middlesex
  Tel: 01-804 1212
Boscombe Engineering Ltd,
  145 Sterte Road, POOLE, Dorset, BH15 2AF
  Tel: Poole (02013) 5141
E. Chidlow & Co. Ltd,
  Castle Court, Castle Street, SHREWSBURY, Salop., SY1 2BG
  Tel: Shrewsbury (0743) 55776
Chieftain Industries Ltd,
  Grange Road, Houstoun Industrial Estate, LIVINGSTON, Lothian, EH54 5DD
  Tel: Livingston (0589) 32223
Claudgen Ltd,
  Wembley Hill Estate, South Way, WEMBLEY, Middlesex, HA9 0DF
  Tel: 01-902 3682
Reginald Clayton,
  19 Seaside Road, ST LEONARDS ON SEA, East Sussex, TN38 0AL
  Tel: Hastings (0424) 427437
Diffusion Radiator Company Ltd,
  Lyon Road, Hersham Trading Estate, WALTON ON THAMES, Surrey, KT12 3QA
  Tel: Walton on Thames 40197
Dimplex Heating Ltd,
  Millbrook, SOUTHAMPTON, SO9 2DP
  Tel: Southampton (0703) 777117
Ekco Heating & Appliances,
  Drury Lane, HASTINGS, East Sussex, TN34 1XN
  Tel: Hastings (0424) 429141
Eltron (London) Ltd,
  Accrington Works, Strathmore Road, CROYDON, Surrey, CR9 2NA
  Tel: 01-689 4341
Erskine Westayr (Engineering) Ltd,
  P.O. Box 16, IRVINE, Strathclyde, KA12 8JL
  Tel: Irvine (0294) 75211

Eurohurst Services Ltd,
  Eurohurst House, Oakenshaw Road, Shirley, SOLIHULL, West
  Midlands, B90 4PE
  Tel: 021-745 3141
GEC-Xpelair Ltd,
  P.O. Box 220, Deykin Avenue, Witton, BIRMINGHAM, B6 7JH
  Tel: 021-327 1984
Hanovia Lamps Ltd,
  480 Bath Road, SLOUGH, Berkshire SL1 6BL
  Tel: Burnham (06286) 4041
Hartington Engineering Co. Ltd,
  117–119 Hartington Road, LONDON, SW8 2HD
  Tel: 01-720 7301
Heatovent Electric Ltd,
  Lomond Street, GLASGOW, G22 6JQ
  Tel: 041-336 8321
Isopad Ltd,
  Stirling Way, BOREHAMWOOD, Hertfordshire, WD6 2AF
  Tel: 01-953 6242
Kaloric Heater Co. Ltd,
  31 Beethoven Street, LONDON, W10 4LJ
  Tel: 01-969 1367/8/9
La Belle Cheminee Ltd,
  85 Wigmore Street, LONDON, W1H 9FB
  Tel: 01-486 7486/7
David C. Lesser & Co. Ltd,
  10 Balaclava Road, Kings Heath, BIRMINGHAM 14
  Tel: 021-444 5435
Metal Pressings Ltd,
  Care Road, FAVERSHAM, Kent, ME13 7TN
  Tel: Faversham (079582) 3225
Metway Electrical Industries Ltd,
  Metway Works, Canning Street, BRIGHTON, East Sussex,
  BN2 2ES
  Tel: Brighton (0273) 606433
Power-lectric Ltd,
  Church Street, STROUD, Gloucestershire, GL5 1JL
  Tel: Stroud (04536) 77951
Redring Electric Ltd,
  Redring Works, PETERBOROUGH, PE2 9JJ
  Tel: Peterborough (0733) 60431
Spredaire Ltd,
  Baker Street, HIGH WYCOMBE, Buckinghamshire, HP11 2RX
  Tel: High Wycombe (0494) 33134

Stabilag (ESH) Ltd,
  34 Mark Road, HEMEL HEMPSTEAD, Hertfordshire, HP2 7DD
  Tel: Hemel Hempstead (0442) 64481
TI Creda Ltd,
  Creda Works, Blythe Bridge, STOKE ON TRENT, Staffordshire,
  ST11 9LJ
  Tel: Blythe Bridge (07818) 2281
TI Sunhouse Ltd,
  P.O. Box 70, WALSALL, West Midlands, WS1 3DL
  Tel: Walsall (0922) 25551
Unidare Engineering Ltd,
  Surrey Street, GLOSSOP, Derbyshire, SK13 9BX
  Tel: Glossop 4815
Wellco Holdings Ltd,
  9 Lower Grosvenor Place, LONDON, SE1N 0EN
  Tel: 01-828 0104/5
Wynbourne-Satoba Ltd,
  82–86 City Road, LONDON, EC1Y 2BJ
  Tel: 01-251 4442
Wyndawaye Systems Ltd,
  London Road, BRANDON, Suffolk
  Tel: Thetford (0842) 810415

GRILLES AND DIFFUSERS

Air Diffusion Devices Ltd,
  Birchwood Trading Estate, 144–146 Great North Road, HATFIELD,
  Hertfordshire
  Tel: Hatfield 72601
Air Flo Heating Supplies Ltd,
  Bruce Grove, WICKFORD, Essex, SS11 8DA
  Tel: Wickford 2231
Airflow Air Bricks Ltd,
  Oxhey Lane, WATFORD, Hertfordshire, WD1 4RQ
  Tel: Watford (0923) 21579
Airflow (Nicoll Ventilators) Ltd,
  45 High Ridge Crescent, NEW MILTON, Hampshire, BH25 5BT
  Tel: New Milton (0425) 611547
Argosy Fenton Ltd,
  Hertford Road, BARKING, Essex, IG11 8BT
  Tel: 01-594 1081
Barber & Colman Ltd,
  Marsland Road, Brooklands, SALE, Cheshire, M33 1UI
  Tel: 061-973 2277

Bat Building & Engineering Products Ltd,
  Halesfield 9, TELFORD, Salop, TF7 4LD
  Tel: Telford (0952) 586193
J. D. Beardmore & Co. Ltd,
  Field End Road, RUISLIP, Middlesex, HA4 0QG
  Tel: 01-864 6811
F. H. Biddle Ltd,
  Newtown Road, NUNEATON, Warwickshire, CV11 4HP
  Tel: Nuneaton (0682) 384233
Brooke Air Sales,
  Arterial Road, RAYLEIGH, Essex
  Tel: Rayleigh (0268) 772266
BTR Permali RP-Ltd,
  Hydroglas Works, Bristol Road, GLOUCESTER, GL1 5TT
  Tel: Gloucester (0452) 28671
John Caddick & Sons Ltd,
  Spoutfield Tilieries, STOKE ON TRENT, Staffordshire, ST4 7BX
  Tel: Newcastle (Staffs) (0782) 616413
Colt International Ltd,
  New Lane, HAVANT, Hampshire, PO9 2LY
  Tel: Havant (0705) 451111
Consoles Ltd,
  29 Lyon Road, Hersham Trading Estate, WALTON ON THAMES,
  Surrey, KT12 3QF
  Tel: Walton on Thames 24046/8
H. W. Cooper & Co. Ltd,
  Page House, 33 Pages Walk, LONDON, SE1 4SF
  Tel: 01-237 1767/9
Crittall Construction Ltd,
  Ventura Division, Halford Lane, Smethwick, WARLEY, West Midlands,
  B66 1BJ
  Tel: 021-558 2191
Dewey Waters & Co. Ltd,
  Cox's Green, Wrington, BRISTOL, BS18 7QS
  Tel: Wrington (0934) 862601
Drawn Metal Ltd,
  Swinnow Lane, LEEDS, West Yorkshire, LS13 4NE
  Tel: Pudsey (0532) 565661
Ellis Miller Ltd,
  Market House, St Judes Road, Englefield Green, EGHAM, Surrey,
  TW20 0BU
  Tel: Egham 5302/5387

The Expanded Metal Co. (Mfg) Ltd,
  1 Butler Place, LONDON, SW1H 0PS
  Tel: 01-222 7766
John Fyfe Ltd,
  Whitemyres Avenue, ABERDEEN, AB9 1XP
  Tel: Aberdeen (0224) 691333
GGS,
  Nutwood Close, Brockham, BETCHWORTH, Surrey
  Tel: Betchworth 2813
Greenwood Airvac Ventilation Ltd,
  P.O. Box 3, Brookside Industrial Estate, Rustington,
  LITTLEHAMPTON, West Sussex, BN16 3LH
  Tel: Rustington 71021
Hart & Cooley, Tuttle & Bailey,
  7 Chesham Place, LONDON, SE1X 5HN
  Tel: 01-235 9754
Hartington Engineering Co. Ltd,
  117–119 Hartington Road, LONDON, SW8 2HD
  Tel: 01-720 7301
Henderson Safety Tank Co. Ltd,
  1 Queensway, CROYDON, Surrey, CR0 4RH
  Tel: 01-688 7777
James Hill & Co. Ltd,
  Sea Street, HERNE BAY, Kent, CT6 8LB
  Tel: Herne Bay 63666
Industrial Acoustics Co. Ltd,
  Walton House, Central Trading Estate, STAINES, Middlesex,
  TW18 4XB
  Tel: Staines 56251
Interlite Linear Controls Ltd,
  Interlite House, Eskdale Road, Uxbridge Industrial Estate,
  UXBRIDGE, Middlesex, UB8 2RT
  Tel: Uxbridge (0895) 56331
George Kraemer & Co. Ltd,
  Upper Evingar Road, WHITCHURCH, Hampshire, RG28 7EU
  Tel: Whitchurch 2162
Luwa (UK) Ltd,
  Luwa House, 36–38 Goldsworth Road, WOKING, Surrey, GU21 1JT
  Tel: Woking 21441/6
Marley Extrusions Ltd,
  Lenham, MAIDSTONE, Kent, ME17 2DE
  Tel: Maidstone (0622) 54366

Martingale Technical Services Ltd,
St Johns Industrial Estate, PENN, Buckinghamshire, HP10 8HR
Tel: Penn 5158/9

Maxi-Flow Ltd,
2 Kingswood Road, Penge, LONDON, SE20 7BN
Tel: 01-778 8575

Newridge Engineering Co. Ltd,
Belville 2 Works, Patterson Street, BLAYDON ON TYNE,
Tyne & Wear, NE21 5RT
Tel: Blaydon (0632) 446000

Ozonair Engineering Co. Ltd,
Quarry Wood Industrial Estate, Aylesford, MAIDSTONE, Kent,
ME20 7NB
Tel: Maidstone (0622) 77861

Don V. Powell (Air Diffusion) Ltd,
P.O. Box 49, 2 Manor Road, ALTRINCHAM, Cheshire, WA15 9RS
Tel: 061-941 2161

H. H. Robertson (UK) Ltd,
Cromwell Road, ELLESMERE PORT, South Wirral, Cheshire,
L65 4DS
Tel: 051-355 3622

Royalair Ltd,
697 Stirling Road Trading Estate, SLOUGH, Berkshire, SL1 4ST
Tel: Slough 30133

Silavent Ltd,
32 Blyth Road, HAYES, Middlesex, UB3 1DG
Tel: 01-573 2822

Stadium Ltd,
Queensway, ENFIELD, Middlesex, EN3 4SD
Tel: 01-804 4343/5131

Thermotank Div., Hall-Thermotank Products Ltd,
Helen Street, GLASGOW, G51 3HH
Tel: 041-445 2444

Warm Air Components Ltd,
Fryers Road, BLOXWICH, West Midlands, WS3 2XL
Tel: Bloxwich (0922) 75766/7

Wavin Plastics Ltd,
P.O. Box 12, HAYES, Middlesex, UB3 1EY
Tel: 01-573 7799

Thos K. Webster (UK) Ltd,
Boughton Industrial Estate, NEWARK, Nottinghamshire, NG22 9LF
Tel: Mansfield (0623) 860577

Woods of Colchester Ltd,
  Tufnell Way, COLCHESTER, Essex, CO4 5AR
  Tel: Colchester (0206) 44122
Zest Equipment Co. Ltd,
  Hendor House, Cray Road, SIDCUP, Kent, DA14 5DX
  Tel: 01-302 0131

HEAT PUMPS

Andrews Industrial Equipment Ltd,
  36 Lewis Road, MITCHAM, Surrey, CR4 3XQ
  Tel: 01-648 6174
Dunham-Bush Ltd,
  Fitzherbert Road, Farlington, PORTSMOUTH, Hampshire, PO6 1RR
  Tel: Cosham (0705) 370961
EER Air Conditioning Services Ltd,
  66–68 George Street, LONDON, W1 5RG
  Tel: 01-487 5606
Lennox Industries Ltd,
  P.O. Box 43, Lister Road, BASINGSTOKE, Hampshire, RG22 4AR
  Tel: Basingstoke (0256) 61261
Myson Copperad Ltd,
  Air Conditioning Division, Old Wolverton, MILTON KEYNES,
  Buckinghamshire, MK12 5PT
  Tel: Milton Keynes (0908) 312642
Packaged Chillers (Winson Group),
  St Annes House, 45 Park Street, LUTON, Bedfordshire, LU1 3JX
  Tel: Luton (0582) 421861
Stiebel Eltron Ltd,
  25–26 Lyvedone Road, Brackmills, NORTHAMPTON, NN4 0ED
  Tel: Northampton (0604) 66421
WR Heat Pumps Ltd,
  Unit 2, The Causeway, MALDON, Essex, CM9 7PU
  Tel: Maldon (0621) 56611
Westair Dynamics Ltd,
  Central Avenue, EAST MOLESEY, Surrey, KT8 0QZ
  Tel: 01-979 9031

INSULATION: QUILTS AND MATS

B. R. Ainsworth & Co. Ltd,
  20–26 Bignold Road, Forest Gate, LONDON, E7
  Tel: 01-534 7580

D. Anderson & Son Ltd,
  Stretford, MANCHESTER, M32 0YL
  Tel: 061-865 4444
Bradford Insulation Industries Pty Ltd (UK Agents), Parbury Henty & Co. Pty Ltd,
  44–45 Chancery Lane, LONDON, WC2A 1JB
  Tel: 01-405 5261
Cape Insulation Ltd,
  Kerse Road, STIRLING, Central, FK7 7RW
  Tel: Stirling (0786) 3100
Carborundum Co. Ltd,
  The Mill Lane, Rainford ST HELENS, Merseyside, WA11 8LP
  Tel: Rainford (074488) 2941
Fibreglass Ltd,
  ST HELENS, Merseyside, WA10 3TR
  Tel: St Helens (0744) 24022
Hodgson and Hodgson Ltd,
  Crown Industrial Estate, Anglesey Road, BURTON ON TRENT, DE14 3PA
  Tel: Burton on Trent (0283) 64772/3/4
Inswool Insulation Company (Falcon) Ltd,
  34 Hailsham Industrial Estate, Station Road, HAILSHAM, East Sussex, BN27 2EP
  Tel: Hailsham (0323) 841934
Johns-Manville (GB) Ltd,
  Parkbridge House, The Little Green, RICHMOND, Surrey, TW9 1QU
  Tel: 01-940 4181
Kay Metzeler Ltd,
  New Mill Park Road, DUKINFIELD, Cheshire SK16 5LL
  Tel: 061-330 7311
Kitson's Insulation Products Ltd,
  Kitson House, P.O. Box 4, London Road, BARKING, Essex, IG11 8DA
  Tel: 01-594 5544
A. Latter & Co. Ltd,
  43 South End, CROYDON, Surrey CR9 1AN
  Tel: 01-688-9335/8
F. McNeill & Co. Ltd,
  10 Lower Grosvenor Place, LONDON, SW1W 0ER
  Tel: 01-834 6022
Joseph Nadin Ltd,
  1st Floor, 74 King Street, MANCHESTER, M2 4NJ
  Tel: 061-834 5627

Newalls Insulation Co. Ltd,
   WASHINGTON, Tyne & Wear, NE38 8JL
   Tel: Washington (0632) 461111
PH Thermal Products Ltd,
   Fairfield Works, Glenview Road, BINGLEY, West Yorkshire
   Tel: Bingley (09766) 7931
Rockwool Company (UK) Ltd,
   St Agnes House, Cresswell Park, Blackheath, LONDON, SE3 9RD
   Tel: 01-318 3915
Roclaine Ltd,
   Fir Tree Place, Church Road, ASHFORD, Middlesex, TW15 2PH
   Tel: (07842) 59139
Ruberoid Building Products Ltd,
   1 New Oxford Street, LONDON, WC1A 1PE
   Tel: 01-405 8797
Vencel Resil Ltd,
   Ocean Works, West Street, ERITH, Kent, DA8 1DD
   Tel: Erith 36922

INSULATION SLABS

B. R. Ainsworth & Co. Ltd,
   20–26 Bignold Road, Forest Gate, LONDON, E7
   Tel: 01-534 7580
Baxenden Chemical Co. Ltd,
   The Paragon Works, Baxenden, ACCRINGTON, Lancashire, BB5 2SL
   Tel: Accrington (0254) 381631
British Gypsum Ltd,
   Ferguson House, 15 Marylebone Road, LONDON, NW1 5JE
   Tel: 01-486 1282
BTR Silvertown Ltd,
   Factory Road, Thameside Industrial Estate, Silvertown, LONDON, E16
   Tel: 01-476 3200
Bulstrode Plastics & Chemical Co. Ltd,
   Bulstrode House, Bowater, Woolwich, LONDON, SE18 5TH
   Tel: 01-855 6806
Cape Boards & Panels Ltd,
   Iver Lane, UXBRIDGE, Middlesex, UB8 2JQ
   Tel: Uxbridge (0895) 37111
Cape Insulation Ltd,
   Kerse Road, STIRLING, Central FK7 7RW
   Tel: Stirling 3100

Cemoss Equipment Ltd,
  Radford Way, BILLERICAY, Essex, CM12 0DD
  Tel: Billericay (02774) 51105
Chemicals Trading Co. Ltd,
  25 Berkeley Square, LONDON, W1X 6DH
  Tel: 01-499 1246
E. J. Clay Ltd,
  Claygate House, Albert Road North, Reigate, Surrey
  Tel: Reigate 45741
Concargo Ltd,
  Winterstoke Road, WESTON SUPER MARE, Avon, BS24 9AH
  Tel: Weston super Mare (0934) 28221
Coolag Ltd,
  P.O. Box 3, Charlestown, GLOSSOP, Derbyshire, SK13 8LE
  Tel: Glossop (04574) 3227
Cork Gowers Ltd,
  Vulcan Street, BOOTLE, Merseyside, L20 4HL
  Tel: 051-922 1917
Dow Chemical Co. Ltd,
  Heathrow House, Bath Road, HOUNSLOW, Middlesex, TW5 9QY
  Tel: 01-759 2600
Dunlop Semtex Ltd,
  BRYNMAWR, Gwent, MP3 4XN
  Tel: Brynmawr (0495) 310000
Falcon Group of Companies,
  Unit A3, Halesfield 11, TELFORD, Salop., TF7 4PQ
  Tel: Telford (0952) 586692
Fibreglass Ltd,
  ST HELENS, Merseyside, WA10 3TR
  Tel: St Helens (0744) 24022
Harrison & Jones (Eurotherm) Ltd,
  Chaul End Lane, LUTON, Bedfordshire
  Tel: Luton (0582) 597261
Hodgson & Hodgson Ltd,
  Crown Industrial Estate, Anglesey Road, BURTON ON TRENT,
  Staffordshire, DE14 3PA
  Tel: Burton on Trent (0283) 64772
Imperial Chemical Industries Ltd,
  Purlboard Insulation Products, P.O. Box 6, BILLINGHAM, Cleveland,
  TS23 1LD
  Tel: Stockton on Tees (0642) 553601
Johns-Manville (Great Britain), Ltd,
  Parkbridge House, The Little Green, RICHMOND, Surrey, TW9 1QU
  Tel: 01-948 4181

William Kenyon & Sons (Vicuclad) Ltd,
  Albert Works, Dukinfield, Cheshire, SK16 4UP
  Tel: 061-330 5651
Micropore International Ltd,
  Hadzor Hall, Hadzor, DROITWICH, Worcestershire, WR9 7DJ
  Tel: Droitwich (09057) 4211
Mundet Cork & Plastics Ltd,
  Vicarage Road, CROYDON, Surrey, CR9 4AR
  Tel: 01-688 4142
Newalls Insulation Co. Ltd,
  WASHINGTON, Tyne & Wear, NE38 8JL
  Tel: Washington (0632) 461111
W. H. O'Gorman Manufacturing Ltd,
  School Lane, Chandlers Ford, EASTLEIGH, Hampshire, SO5 3DE
  Tel: Chandlers Ford (04215) 67316
Pittsburgh Corning (United Kingdom) Ltd,
  4 St James Chambers, North Mall, The Green, LONDON, N9
  Tel: 01-807 5454
Plaschem Ltd,
  Morris Street, Dumers Lane, Radcliffe, MANCHESTER, M26 9GF
  Tel: 061-766 9711
Rockwool Co. (UK) Ltd,
  St Agnes House, Cresswell Park, Blackheath, LONDON, SE3 9RD
  Tel: 01-318 3915
Ross Warmafoam Ltd,
  Poron Works, TORPOINT, Cornwall, PL11 3AX
  Tel: Millbrook (07552) 551
Vencel Resil Ltd,
  Ocean Works, West Street, ERITH, Kent, DA8 1DD
  Tel: Erith 36922
Western Cork Co. Ltd,
  Trade Street, CARDIFF, CF1 5RQ
  Tel: Cardiff (0222) 33926

MECHANICAL

Aidelle Products Ltd,
  Div. of Airflow Developments Ltd, Lancaster Road, HIGH
  WYCOMBE, Buckinghamshire, HP12 3QP
  Tel: High Wycombe (0494) 25252
Air Technology Ltd,
  28 Birmingham Street, Oldbury, BIRMINGHAM, B69 4DS
  Tel: 021-544 6707

Airscrew Howden Ltd,
  WEYBRIDGE, Surrey, KT15 2QR
  Tel: Weybridge 45511
AK Fans Ltd,
  20 Upper Park Road, LONDON, NW3 2UR
  Tel: 01-586 0266
Anglo Nordic Ltd,
  74 London Road, KINGSTON UPON THAMES, Surrey, KT2 6PZ
  Tel: 01-549 0901
British Organ Blowing Co. Ltd,
  Coleman Street, DERBY, DE2 8NN
  Tel: Derby (0332) 74112/3
Clipper Air Handling Units Ltd,
  Raans Road, AMERSHAM, Buckinghamshire, HP6 6HY
  Tel: Amersham 21212
Colchester Fan Marketing Co. Ltd,
  Hillbottom Road, Sands Industrial Estate, HIGH WYCOMBE,
  Buckinghamshire, HP12 4HR
  Tel: High Wycombe (0494) 28905
  Egerton Works, Egerton Street, Farnworth, BOLTON, Lancashire.
Colt International Ltd,
  New Lane, HAVANT, Hampshire, PO9 2LY
  Tel: Havant (0705) 451111
Coolvent Ltd,
  Hillbottom Road, Sands Industrial Estate, HIGH WYCOMBE,
  Buckinghamshire, HP12 4HR
  Tel: High Wycombe (0494) 28905
Dynamic Plastics Ltd,
  Egerton Works, Egerton Street, Farnworth, BOLTON, Lancashire.
  BL4 7ER
  Tel: Bolton (0204) 75154
Electric Fans & Controls Ltd,
  Hillbottom Road, Sands Industrial Estate, HIGH WYCOMBE,
  Buckinghamshire, HP12 4HR
  Tel: High Wycombe (0904) 28905
GEC-Xpelair Ltd,
  P.O. Box 220, Deykin Avenue, Witton, BIRMINGHAM, B6 7JH
  Tel: 021-327 1984
Greenwood Airvac Ventilation Ltd,
  P.O. Box 3, Brookside Industrial Estate, Rustington,
  LITTLEHAMPTON, Sussex, BN16 3LH
  Tel: Rustington 71021
Holyhead Engineering Ltd,
  Meadow Lane, BILSTON, West Midlands, WV14 9NJ
  Tel: Sedgley 4477

Hydor Co. Ltd,
  Salisbury Road, Downton, SALISBURY, Wiltshire, SP5 3JJ
  Tel: Downton (0725) 21422
London Fan Co. Ltd,
  75–81 Stirling Road, LONDON, W3 8DJ
  Tel: 01-992 6923
Midland Fan Co. Ltd,
  212 Aston Road, BIRMINGHAM, B6 4LQ
  Tel: 021-359 1588/9
Novenco Ltd,
  Tundry Way, Chainbridge Road, BLAYDON ON TYNE, Tyne & Wear,
  NE21 5SN
  Tel: Blaydon (0632) 444511
NuAire Ltd,
  Western Industrial Estate, CAERPHILLY, Mid Glamorgan
  Tel: Caerphilly (0222) 885911
Ozonair Engineering Co. Ltd,
  Quarry Wood Industrial Estate, London Road, Aylesford,
  MAIDSTONE, Kent, ME20 7NB
  Tel: Maidstone (0622) 77861
Plastic Constructions Ltd,
  Midland Fabrications Division, Evelyn Road, Sparkhill,
  BIRMINGHAM, B11 3JJ
  Tel: 021-773 4951
Plastics Design & Fabrications (Glasgow) Ltd,
  Plade Works, Gladstone Avenue, Barrhead, GLASGOW,
  G78 1QT
  Tel: 041-881 8241
Shipley Fan Co. Ltd,
  106 Dockfield Road, SHIPLEY, West Yorkshire
  Tel: Bradford (0274) 581337
Silavent Ltd,
  32 Blyth Road, HAYES, Middlesex, UB3 1DG
  Tel: 01-573 2822
R. W. Simon Ltd,
  System Works, Hatchmoor Industrial Estate, GREAT TORRINGTON,
  Devon, EX38 7HP
  Tel: Torrington 3721
Smiths Industries Precision Fan Co.,
  Burford Road, WITNEY, Oxfordshire, OX8 5EE
  Tel: Witney (0993) 2929
Standard & Pochin Ltd,
  Evington Valley Road, LEICESTER, LE5 5LS
  Tel: Leicester (0533) 735114

Thermor Electrical Appliances Ltd,
  Madison House, Molesey Avenue, EAST MOLESEY, Surrey, KT8 0SA
  Tel: 01-979 4461/4
J. B. Thorne & Son Ltd,
  Marsden Avenue, Queniborough, LEICESTER, LE8 8FL
  Tel: Leicester, (0533) 605757
Vent-Axia Ltd,
  Fleming Way, CRAWLEY, West Sussex, RH10 2NN
  Tel: Crawley (0293) 26062
Ventilating Equipment Supply Co,
  West Way, Walworth Industrial Estate, ANDOVER, Hampshire,
  SP10 5AR
  Tel: Andover (0264) 66325
Ventilation Equipment & Conditioning Ltd,
  Vequip Works, 320–322 Latimer Road, LONDON, W10 6QR
  Tel: 01-969 7553
Warm Air Components Ltd,
  Fryers Road, BLOXWICH, West Midlands, WS3 2XL
  Tel: Bloxwich (0922) 75766/7
Watkins & Watson Ltd,
  Westminster Road, WAREHAM, Dorset, BH20 4SP
  Tel: Wareham 6311
Woods of Colchester Ltd,
  Tufnell Way, COLCHESTER, Essex, CO4 5AR
  Tel: Colchester (0206) 44122

MULTIPLE GLAZING UNITS

Thomas Bennett Ltd,
  Goodman Street, LEEDS, West Yorkshire, LS10 1QN
  Tel: Leeds (0532) 702121
Bomert, Teves & Blankley Ltd,
  Pembroke House, 44 Wellesley Road, CROYDON, Surrey,
  CR9 3PD
Custom-Made Double Glazing Ltd,
  729 Tudor Estate, Abbey Road, Park Royal, LONDON, NW10 7XY
  Tel: 01-965 0126
Martin Dunn Ltd,
  142 High Street, WEST BROMWICH, West Midlands, B70 6JH
  Tel: 021-553 0551
Glas-Seal of Ulster Ltd,
  Belfast Road, BALLYNAHINCH, Co. Down
  Tel: Ballynahinch (023856) 2932

Multiglas Ltd,
  Arklow Road, LONDON, SE14 6EB
  Tel: 01-692 8282
Parka Double Glazing Ltd,
  44 Milton Road, College Milton North, East Kilbride, GLASGOW,
  G74 5BU
  Tel: East Kilbride (03552) 38527
Pilkington Brothers Ltd,
  ST HELENS, Merseyside, WA10 3TT
  Tel: St Helens (0744) 28882
Plyglass Ltd,
  Cotes Park, Somercotes, DERBY, DE5 4PL
  Tel: Alfreton (077383) 3321
Regniers, Ekile & Co. (London) Ltd,
  450 High Road, ILFORD, Essex, IG1 1UN
  Tel: 01-478 8201
Thermovitrine Ltd,
  P.O. Box 7, Broadway, HYDE, Cheshire, SK14 4QW
  Tel: 061-368 5711
Twinwindow (UK) Ltd,
  160 St Anne's Road, Denton, MANCHESTER, M34 3DY
  Tel: 061-336 9611
UBM Glass Ltd,
  Techno Trading Estate, Bramble Road, SWINDON, Wiltshire, SN2 6EX
  Tel: Swindon (0793) 24592

PUMPS

Anglo-Nordic Thermal Holdings Ltd,
  74 London Road, KINGSTON UPON THAMES, Surrey, KT2 6PZ
  Tel: 01-549 0901
Armstrong Pumps Ltd,
  Peartree Road, Stanway, COLCHESTER, Essex, CO3 5JX
  Tel: Colchester (0206) 79491
John Cherry & Sons Ltd,
  Beckside, BEVERLEY, North Humberside, HU17 0PS
  Tel: Hull (0482) 881436
Graham Precision Pumps Ltd,
  The Forge, CONGLETON, Cheshire
  Tel: Congleton (02602) 4721
Grundfos Pumps Ltd,
  Grovebury Road, LEIGHTON BUZZARD, Bedfordshire,
  LU7 8TL
  Tel: Leighton Buzzard (0525) 374876

Hill-Foster Ltd,
262 Uxbridge Road, Hatch End, PINNER, Middlesex, HA5 4HS
Tel: 01-428 0266

Holden & Brooke Ltd,
Sirius Works, West Gorton, MANCHESTER, M12 5JL
Tel: 061-273 8262

Myson Circulators Ltd,
Old Meadow Road, Hardwick Industrial Estate, KING'S LYNN,
Norfolk, PE30 4PP
Tel: King's Lynn (0553) 64821

Myson Pumps Ltd,
Phoenix Works, Wakefield Road, BRIGHOUSE, West Yorkshire,
HD6 1PE
Tel: Brighouse (0484) 718531

Pullen Pumps Ltd,
58 Beddington Lane, CROYDON, Surrey, CR9 4PT
Tel: 01-684 9521

Sihi Ryaland Pumps,
Bridgewater Road, Broadheath, ALTRINCHAM, Cheshire, WA14 1NB
Tel: 061-928 6371

Thermofire Engineering Co. Ltd,
Nelson Works, STROUD, Gloucestershire, GL5 2HW
Tel: Stroud (04536) 4348

Turney Turbines Ltd,
Highmead Works, 73 Station Road, HARROW, Middlesex, HA1 2TZ
Tel: 01-427 3449

RADIATORS

Acoustics & Envirometrics Ltd,
Winchester Road, WALTON ON THAMES, Surrey, KT12 2RP
Tel: Walton on Thames 47644/6

Anglo-Nordic Thermal Sales Ltd,
74 London Road, KINGSTON UPON THAMES, Surrey, KT2 5BR
Tel: 01-549 0901

Barlo Radiators Ltd,
Foundry Lane, HORSHAM, West Sussex, RH13 5TQ
Tel: Horsham (0403) 61653

F. H. Biddle Ltd,
Newtown Road, NUNEATON, Warwickshire, CV11 4HP
Tel: Nuneaton (0682) 384233

Boscombe Engineering Ltd,
145 Sterte Road, POOLE, Dorset, BH15 2AF
Tel: Poole (02013) 5141

Broag Ltd,
   Thyssen House, Molly Millars Lane, WOKINGHAM, Berkshire,
   RG11 2PY
   Tel: Wokingham (0734) 783434
Carron Co.,
   FALKIRK, Central, FK2 8DW
   Tel: Falkirk (0324) 24999
Celsius Air Conditioning & Refrigeration Equipment Ltd,
   Wiggenhall Industrial Estate, WATFORD, Hertfordshire, WD1 8AW
   Tel: Watford 45021
Comyn Ching & Co. (Solray) Ltd,
   110 Golden Lane LONDON, EC1Y 0SS
   Tel: 01-253 8414
Diffusion Radiator Co. Ltd,
   Lyon Road, Hersham Trading Estate, WALTON ON THAMES, Surrey,
   KT12 3QA
   Tel: Walton on Thames 40197/8
Doulton Heating Systems Ltd,
   School Lane, Knowsley, PRESCOTT, Merseyside, L34 9HJ
   Tel: 051-546 8225/7
Dunham-Bush Ltd,
   Fitzherbert Road, Farlington, PORTSMOUTH, Hampshire, PO6 1RR
   Tel: Cosham (0705) 370961
Euro-Heat Ltd,
   70–72 Fearnley Street, WATFORD, Hertfordshire, WD1 7DE
   Tel: Watford 46762
Faral Tropical UK,
   72 Hoblands, HAYWARDS HEATH, West Sussex, RH16 3NB
   Tel: Haywards Heath (0444) 57524
Finrad Ltd,
   62 Norwood High Street, LONDON, SE27 9NP
   Tel: 01-670 6987
Finstyle Ltd,
   Firth Mill, Firth Street, SKIPTON, North Yorkshire, BD23 2PS
   Tel: Skipton (0756) 2739
Gold Radiators GB,
   12 Anerley Station Road, LONDON, SE20 8QE
   Tel: 01-659 5216/7
John Henning (Engineering) Ltd,
   Unicorn Works, Waringstown, CRAIGAVON, Co. Armagh, BT66 8QB
   Tel: Waringstown (076288) 346
Hudevad Britain,
   262 Hook Road, CHESSINGTON, Surrey, KT9 1PF
   Tel: 01-391 1327

Myson Domestic Products Ltd,
  ONGAR, Essex, CM5 9RE
  Tel: Ongar (02776) 4311
OGB Products Ltd,
  Churchbridge Industrial Estate, Oldbury, WARLEY, West Midlands,
  B69 4LH
  Tel: 021-544 5720
Property Mechanical Products Ltd,
  133 Rushey Green, LONDON, SE6 4BD
  Tel: 01-697 8618
Runtalrad (1970) Ltd,
  Ridgway Industrial Estate, Ridgway Road, IVER, Buckinghamshire
  Tel: Iver (0753) 654142
S & P Coil Products Ltd,
  Evington Valley Road, LEICESTER, LE5 5LU
  Tel: Leicester (0533) 730771
Spur Engineering Ltd,
  138 Kenley Road, LONDON, SW19 3HT
  Tel: 01-540 4581
Stelgrad Group Ltd,
  P.O. Box 103, National Avenue, HULL, North Humberside, HU5 4JN
  Tel: Hull (0482) 492251
Temfix Engineering Co. Ltd,
  FARNBOROUGH, Hampshire, GU14 7LP
  Tel: Farnborough (0252) 515151
UBM (Mechanical Services) Ltd,
  145 Larkhall Lane, LONDON, SW4 6RG
  Tel: 01-720 8115
Veha (UK) Ltd,
  Unit D, St David's Trading Estate, SALTNEY, Clwyd
  Tel: Chester (0244) 673822
Warmastyle Ltd,
  Firth Mill, First Street, SKIPTON, North Yorkshire, BD23 2PS
  Tel: Skipton (0756) 60921

SOLAR

Antarim Ltd,
  Solar Works, New Street, PETWORTH, West Sussex, GU28 0AS
  Tel: Petworth (0798) 43005
Bexley Glass & Glazing Contractors Ltd,
  37 High Street, BEXLEY, Kent, DA5 1AB
  Tel: Crayford 53311/4

Calorsol Ltd,
  Lancaster Road, SHREWSBURY, Salop., SY1 3NG
  Tel: Shrewsbury (0743) 51578
Don Engineering (South West) Ltd,
  Wellington Trading Estate, WELLINGTON, Somerset, TA21 8SS
  Tel: Wellington (Som) (082347) 3181
Doulton Heating Systems Ltd,
  School Lane, Knowsley, PRESCOT, Merseyside, L34 9HJ
  Tel: 051-546 8225
Drake & Fletcher Ltd,
  Parkwood, Sutton Road, MAIDSTONE, Kent, ME15 9NW
  Tel: Maidstone (0622) 55531
Electra Air Conditioning Services Ltd,
  66–68 George Street, LONDON, W1H 5RG
  Tel: 01-487 5606
Industrial (Anti-Corrosion) Services Ltd,
  Britannica House, 214–224 High Street, WALTHAM CROSS,
  Hertfordshire, EN8 7DU
  Tel: Waltham Cross (0992) 22368
Insolar Technology,
  Shaftesbury Crusade, Kingsland Road, St Philips, BRISTOL, Avon
  Tel: Bristol (0272) 552834
Lucas Electrical Ltd,
  Parts & Service Division, Great Hampton Street, BIRMINGHAM,
  B18 6AU
  Tel: 021-236 5050
McKee Solaronics Ltd,
  12 Queenborough Road, SOUTHMINSTER, Essex, CM0 7AB
  Tel: Maldon (0621) 772477
Packaged Chillers (Winson Group),
  St Annes House, 45 Park Street, LUTON, Bedfordshire, LU1 3JX
  Tel: Luton (0582) 421861
Rapaway Ltd,
  Durohurst House, Oakenshaw Road, Shirley, SOLIHULL, West
  Midlands, B90 4PE
  Tel: 021-745 3141
Redpoint Associates Ltd,
  Cheyney Manor, SWINDON, Wiltshire, SN2 2PA
  Tel: Swindon (0793) 284440
Robinsons Developments Ltd,
  Robinson House, Winnall Industrial Estate, WINCHESTER,
  Hampshire, SO23 8LH
  Tel: Winchester (0962) 61777

Rohm & Haas Ltd,
  Lennig House, 2 Masons Avenue, CROYDON, Surrey, CR9 3NB
  Tel: 01-686 8844
Ruberoid Contracts Ltd,
  22 Thomas Street, Cirencester, Gloucester GL1 2EW
  Tel: (0285) 61281
Satchwell Control Systems Ltd,
  P.O. Box 57, Farnham Road, SLOUGH, Berkshire, SL1 4UH
  Tel: Slough (0753) 23961
Skinner Heating Ltd,
  Estate Road, NEWHAVEN, East Sussex, BH9 0AL
  Tel: Newhaven (07912) 5121
Solar Collector Designs Ltd,
  11 Boyn Hill Avenue, MAIDENHEAD, Berkshire
  Tel: Maidenhead (0628) 24909
Solar Water Heaters Ltd,
  153 Sunbridge Road, BRADFORD, West Yorkshire, BD1 2PA
  Tel: Bradford (0274) 24664
Solarsense Ltd,
  48–52 Goldsworth Road, WOKING, Surrey, GU21 1LE
  Tel: Woking (04862) 65620
Solartherm International/Eumed Ltd,
  Westering House, Bel-Royal St Lawrence, JERSEY, Channel Islands
  Tel: Jersey (0534) 23872
Spencer Solarise Ltd,
  Units 3 & 4, Queens Way, ANDOVER, Hampshire
  Tel: Andover (0264) 51625
Stiebel Eltron Ltd,
  25 Lyveden Road, Brackmills, NORTHAMPTON, NN4 0ED
  Tel: Northampton (0604) 66421
Sunpower Ltd,
  Coombe Park, Chillington, KINGSBRIDGE, Devon, TQ7 2JD
  Tel: Frogmore (054853) 347
Sunsense Ltd,
  1 Lincoln Road, Northborough, PETERBOROUGH, PE6 9BL
  Tel: Peterborough (0733) 252672
TA Controls Ltd,
  Lea Industrial Estate, Lower Luton Road, HARPENDEN,
  Hertfordshire, AL5 5EQ
  Tel: Harpenden (05827) 67991
Thermoray Ltd,
  33 High Street, COWBRIDGE, South Glamorgan, CF7 7AE
  Tel: Cowbridge (04463) 4639

Uniflo Water Treatment International Ltd,
    Unit 2, Haywood Way, Iveyhouse Lane, HASTINGS, East Sussex,
    TN5 4PL
    Tel: Hastings (0424) 429682
W. B. Solar Economy Ltd,
    Balksbury Hill, Upper Clatford, ANDOVER, Hampshire, SP11 7LW
    Tel: Andover (0264) 51522
Washington Engineering Ltd,
    Industrial Road, Hertburn Industrial Estate, WASHINGTON, Tyne &
    Wear, NE37 2SB
    Tel: Washington (0632) 463001
J. Williams (Energy Services) Ltd,
    The Furlong, Berryhill Industrial Estate, DROITWICH, Worcestershire,
    WR9 9AJ
    Tel: Droitwich (09057) 3701

SOLID FUEL HEATERS

Agaheat Appliances Glynwed Domestic & Heating Appliances Ltd,
    P.O. Box 37, Foundry Lane, LARBERT, Central, FK5 4PL
    Tel: Larbert (03245) 2233
Associated Builders Merchants Ltd,
    182 Cranbrook Road, ILFORD, Essex, IG1 4LT
    Tel: 01-554 1015
Baxi Heating Ltd,
    P.O. Box 52, Bamber Bridge, PRESTON, Lancashire, PR5 6SN
    Tel: Preston (0772) 36201
A. Bell & Co. Ltd,
    Kingsthorpe, NORTHAMPTON, NN2 6LT
    Tel: Northampton (0604) 712505
Classic Garden Furniture Ltd,
    Audley Avenue, NEWPORT, Salop., TF10 7DS
    Tel: Newport (0952) 813311
Dunsley Heating Appliances Co. Ltd,
    Fearnought, Huddersfield Road, Holmfirth, HUDDERSFIELD, West
    Yorkshire, HD7 2TU
    Tel: Holmfirth (048489) 2635
Ellis Sykes & Sons Ltd,
    Victoria Works, Howard Street, STOCKPORT, Cheshire
    Tel: 061-477 5626
Elstead Forge Ltd,
    Mill Lane, ALTON, Hampshire, GU34 2QG
    Tel: Alton (0420) 82377

Fosse Warmair,
  Old Farm, Norton Road, Iverley, STOURBRIDGE, West Midlands,
  DY8 2RU
  Tel: Hagley (0562) 885898
Grahamston Iron Co.,
  P.O. Box 5, Gowan Avenue, FALKIRK, Central, FK2 7HH
  Tel: Falkirk (0324) 22661
Home Stoves Ltd,
  113 Warwick Avenue, LONDON, W9 2PP
  Tel: 01-289 1667
Interoven Ltd,
  70–72 Fearnley Street, WATFORD, Hertfordshire, WD1 7DE
  Tel: Watford 46761
S. Keeping (Developments) Ltd,
  Little Always, CULLOMPTON, Devon EX15 1RA
  Tel: Craddock (0884) 40846/7
La Belle Cheminee Ltd,
  85 Wigmore Street, LONDON, W1H 9FB
  Tel: 01-486 7486/7
Logfires (Woodstoves) Ltd,
  24 Harrow Farm Estate, Froxfield, MARLBOROUGH, Wiltshire
  Tel: Great Bedwyn (06727) 682
Marrs Enfield Design Ltd,
  7 Fleetside, EAST MOLESEY, Surrey, KT8 0NF
  Tel: 01-979 0671
Ocees Components & Structures Ltd,
  49–50 Knightsbridge Court, 13 Cloan Street, LONDON, SW1X 9LL
  Tel: 01-235 1453/6
Ouzledale Foundry Co. Ltd,
  P.O. Box 4, Long Ing, Barnoldswich, COLNE, Lancashire BB8 6BN
  Tel: Barnoldswick (0282) 813235
Richard Quinell Ltd,
  Rowhurst Forge, Oxshot Road, LEATHERHEAD, Surrey, KT22 0EN
  Tel: Leatherhead 75148
Seaboard International Ltd,
  2 Regent Street, LONDON, SW1Y 4NY
  Tel: 01-839 4971
James Smellie Ltd,
  Ivanhoe Works, Stafford Street, DUDLEY, West Midlands DY1 2AD
  Tel: Dudley (0384) 52320
Smith & Wellstood Ltd,
  BONNYBRIDGE, Central FK4 2AP
  Tel: Bonnybridge (032481) 2171

Taylor & Portway Ltd,
  Rosemary Lane, HALSTEAD, Essex, CO9 1HR
  Tel: Halstead (07874) 2551/2960
TI Parkray Ltd,
  Park Foundry, BELPER, Derbyshire, DE5 1WE
  Tel: Belper (077382) 3741
Tinderpost Ltd,
  P.O. Box 78, FOLKESTONE, Kent, CT19 5LJ
  Tel: Folkestone (0303) 77170
Trianco Redfyre Ltd,
  Stewart House, Brook Way, Kingston Road, LEATHERHEAD, Surrey,
  KT22 7LY
  Tel: Leatherhead 76453
UA Engineering Ltd,
  Canal Street, SHEFFIELD, S4 7ZE
  Tel: Sheffield (0742) 21167
Unidare Engineering Ltd,
  Seagoe Works, Church Road, Portadown, CRAIGAVON, Co. Armagh,
  BT63 5HU
  Tel: Portadown (0762) 33131
Warmback Ltd,
  Manor Works, Short Road, Leytonstone, LONDON, E11 4RH
  Tel: 01-539 6601
Waterford Ironfounders Exports Ltd,
  New Cut Lane Industrial Estate, Woolston, WARRINGTON, Cheshire,
  WA1 4AQ
  Tel: Padgate (0925) 815717
A. J. Wells & Sons,
  Crocker Lane, Kingates, NITON, Isle of Wight, PO38 2NT
  Tel: Niton (0983) 730329

STORAGE HEATERS

E. Chidlow & Co. Ltd,
  82–83 Spring Gardens,SHREWSBURY, Salop., SY1 2SY
  Tel: Shrewsbury (0743) 55776
Dimplex Heating Ltd,
  Millbrook, SOUTHAMPTON, SO9 2DP
  Tel: Southampton (0703) 777117
Erskine Westayr Engineering Ltd,
  P.O. Box 16, IRVINE, Strathclyde, KA12 8JL
  Tel: Irvine (0294) 75211

HVE-Thermoduct Ltd,
  Greenholme Mills, Burley in Wharfedale, ILKLEY, West Yorkshire,
  LS29 7EA
  Tel: Burley in Wharfedale (09435) 3587
Heatovent Electric Ltd,
  27 Lomond Street, Possilpark, GLASGOW, G22 6JQ
  Tel: 041-336 8321
Stiebel Eltron Ltd,
  25 Lyveden Road, Brackmills, NORTHAMPTON, NN4 0ED
  Tel: Northampton (0604) 66421
Storad Ltd,
  Belfield, ROCHDALE, Lancashire, OL16 2UX
  Tel: Rochdale (0706) 38271
TI Creda Ltd,
  Creda Works, Blythe Bridge, STOKE ON TRENT, Staffordshire, ST11 9LJ
  Tel: Blythe Bridge (07818) 2281
Unidare Engineering Ltd,
  Surrey Street, GLOSSOP, Derbyshire, SK13 9BX
  Tel: Glossop (04574) 4815

STORAGE OF FUEL

Acalor International Ltd,
  Crompton Way, CRAWLEY, Sussex, RH10 2QR
  Tel: Crawley (0293) 23271
Armac Engineering Co. Ltd,
  2239 London Road, GLASGOW, G32 8XN
  Tel: 041-778 2338
Armour Engineering (UK) Ltd,
  Wetherby Road, Osmaston Park Industrial Estate, DERBY, DE2 8HL
  Tel: Derby (0332) 363112
Banbury Buildings Ltd,
  P.O. Box 11, Ironstone Works, BANBURY, Oxfordshire, OX17 3NS
  Tel: Banbury (0295) 52500
Boatman Plastics (Staffs.) Ltd,
  Mount Industrial Estate, Mount Road, STONE, Staffordshire, ST15 8LL
  Tel: Stone (078583) 2435
Booth Concrete Products Ltd,
  Mill Road, BALLYCLARE, Co. Antrim BT39 9DX
  Tel: Ballyclare 2741
Brand & Rae Ltd,
  Russell Mill, Springfield, CUPAR, Fife KY15 5QX
  Tel: Cupar (0334) 2828/9

J. Redpath Buchanan & Co. Ltd,
   15 Dunstans Grove, LONDON SE22 0HJ
   Tel: 01-693 1107
Clifford Engineering Ltd,
   West Quay Road, SOUTHAMPTON, Hampshire, SO9 5GQ
   Tel: Southampton (0703) 25547/8
Colbar Engineers Ltd,
   Collingdon Road, CARDIFF, CF1 5EU
   Tel: Cardiff (0222) 41331
Compton Buildings Ltd,
   Station Works, Fenny Compton, LEAMINGTON SPA, Warwickshire,
   CV33 0XH
   Tel: Fenny Compton (029577) 291
Cookson & Zinn Ltd,
   Station Road Works, Hadleigh, IPSWICH, Suffolk, IP7 5PN
   Tel: Hadleigh (0473) 823061
Davies Brothers & Co. Ltd,
   P.O. Box 11, Cross Street North, WOLVERHAMPTON, WV1 1PR
   Tel: Wolverhampton (0902) 54122
Evans Bros (Concrete) Ltd,
   Riddings, DERBY, DE55 4FW
   Tel: Leabrooks (077384) 2301
M. J. Fry Ltd,
   1 Allens Lane, Hamworthy, POOLE, Dorset
   Tel: Lytchett Minster (0202) 622863
John Henning (Enginnering) Ltd,
   Unicorn Works, Waringstown, CRAIGAVON, Co. Armagh, BT66 7QB
   Tel: Waringstown (076288) 346
Heswall Engineering Ltd,
   Five Ways House, Liverpool Road, NESTON, Wirral, Cheshire,
   L64 3TL
   Tel: 051-336 3934
IVO Engineering & Construction Co. Ltd,
   Scrubs Lane, LONDON, NW10 6RH
   Tel: 01-969 7515/8
Luda Concrete Products Ltd,
   Scawby Station, BRIGG, South Humberside, DN20 9DS
   Tel: Brigg (0652) 55161/3
Marley Buildings Ltd,
   Peasmarch, GUILDFORD, Surrey, GU3 1LS
   Tel: Guildford (0483) 69922
Polystructures (Contracts) Ltd,
   Botley Road, North Baddesley, SOUTHAMPTON, SO5 9DQ
   Tel: Rownhams (0703) 733933

Printers Equipment & Engineering Co. Ltd,
  Barry Docks, BARRY, South Glamorgan, CF6 6XD
  Tel: Barry (0446) 737417
Reinforced Concrete Construction Co. Ltd,
  Delph Road, BRIERLEY HILL, West Midlands DY5 2RW
  Tel: (0384) 78079
Reliance Sheet Metal & Engineering Ltd,
  120 Montpelier Road, Dunkirk, NOTTINGHAM, NG7 2JZ
  Tel: Nottingham (0602) 703227/9
Rok-Crete Units Co. Ltd,
  112 Oxford Road, CLACTON ON SEA, Essex, CO15 3TN
  Tel: Clacton on Sea (0255) 24884
Solway Precast Products Ltd,
  Barholm Factory, Creetown, NEWTON STEWART, Dumfries and
  Galloway, DG3 7DD
  Tel: Creetown (067182) 391/5
Stepney Cast Stone Co. Ltd,
  Grovehill, BEVERLEY, North Humberside, HU17 0JN
  Tel: Hull (0482) 883255
Tanks & Drums Ltd,
  Bowling Iron Works, BRADFORD, West Yorkshire, BD4 8SX
  Tel: Bradford (0274) 28285
Thorpecrete Ltd,
  Thorpe Willoughby, SELBY, North Yorkshire, YO8 9LS
  Tel: Selby (0757) 706138
Thyssen (Great Britain) Group of Companies,
  Bynea, LLANELLI, Dyfed, SA14 9SU
  Tel: Llanelli (05542) 2244
Warecrete Products Ltd,
  London Road, WARE, Hertfordshire, SE12 9DD
  Tel: Ware (0920) 2468
C. Warrick & Son (Concrete) Ltd,
  Seagate Works, Long Sutton, SPALDING, Lincolnshire, PE12 9AD
  Tel: Long Sutton (0406) 362262
Western Welding & Engineering Co. Ltd,
  No. 1 Dock, BARRY, South Glamorgan, CF6 6XG
  Tel: Barry (0446) 733466
Whaley Welding Co. Ltd,
  Phoenix Sidings, STOCKTON ON TEES, Cleveland, TX19 0AD
  Tel: Stockton on Tees (0642) 62531

WARM AIR HEATERS

Afos Ltd,
  Manor Estate, Anlaby, HULL, North Humberside, HU10 6RL
  Tel: Hull (0482) 52152
Air Plants (Sales) Ltd,
  Batten Street, Aylestone Road, LEICESTER, LE2 7BC
  Tel: Leicester (0533) 833581
Armca Specialities Ltd,
  Armca House, 102 Beehive Lane, ILFORD, Essex, IG4 5EQ
  Tel: 01-551 0037
Boscombe Engineering Ltd,
  145 Sterte Road, POOLE, Dorset, BH15 2AF
  Tel: Poole (02013) 5141
Brake Shear Ltd,
  Bush House, Yattendon Road, HORLEY, Surrey, RH6 7BT
  Tel: Horley (02934) 5482
Chieftain Industries Ltd,
  Grange Road, Houstoun Estate, LIVINGSTON, Lothian, EH54 5DE
  Tel: Livingston (0589) 32223
Claudgen Ltd,
  South Way, Wembley Hill Estate, WEMBLEY, Middlesex, HA9 0DT
  Tel: 01-902 3682
Colt International Ltd,
  New Lane, HAVANT, Hampshire, PO9 2LY
  Tel: Havant (0705) 451111
Combat Engineering Ltd,
  Oxford Street, BILSTON, West Midlands, WV14 7EG
  Tel: Bilston (0902) 44425
Covrad Ltd,
  Sir Henry Parkes Road, Canley, COVENTRY, West Midlands, CV5 6BN
  Tel: Coventry (0203) 75544
Dragonair Ltd,
  Fitzherbert Road, Farlington, PORTSMOUTH, Hampshire, PO6 1SQ
  Tel: Cosham (0705) 376451
Dunham-Bush Ltd,
  Fitzherbert Road, Farlington, PORTSMOUTH, Hampshire, PO6 1RR
  Tel: Cosham (0705) 370961
Eltron (London) Ltd,
  Strathmore Road, CROYDON, Surrey, CR9 2NA
  Tel: 01-689 4341

Elvaco Ltd,
  Elvaco House, High Street, EGHAM, Surrey, TW20 9DN
  Tel: Egham 4400
Hanovia Lamps Ltd,
  480 Bath Road, SLOUGH, Berkshire, SL1 6BL
  Tel: Burnham (06286) 4041
Heatovent Electric Ltd,
  Lomond Street, Possilpark, GLASGOW, G22 6JQ
  Tel: 041-336 8321
Heatpak 77 Ltd,
  15 Raynham Road Industrial Estate, BISHOP'S STORTFORD,
  Hertfordshire, CM23 5PN
  Tel: Bishop's Stortford (0279) 55031
Hedin Ltd,
  Ravin Road, South Woodford, LONDON, E18 1HJ
  Tel: 01-504 6601
C. M. Hess Ltd,
  2 Westbourne Grove Mews, LONDON, W11 2RX
  Tel: 01-229 7666
Hi-Vee Heating Ltd,
  Carpenders Park, WATFORD, Hertfordshire, WD1 5BE
  Tel: 01-428 6221
HVE-Thermoduct Ltd,
  Greenholme Mills, Burley in Wharfedale, ILKLEY, West Yorkshire,
  LS29 7EA
  Tel: Burley in Wharfedale (09435) 3587
Interoven Ltd,
  70 – 72 Fearnley Street, WATFORD, Hertfordshire, WD1 7DE
  Tel: Watford (0923) 46761
ITT Reznor,
  Park Farm Road, FOLKESTONE, Kent, CT19 5DR
  Tel: Folkestone (0303) 59141
Johnson & Starley Ltd,
  8 Rothersthorpe Cresent, NORTHAMPTON, NN4 9JF
  Tel: Northampton (0604) 62881
Kaloric Heater Co. Ltd,
  31 Beethoven Street, LONDON, W10 4LJ
  Tel: 01-969 1367
Kiloheat Ltd,
  Vestry Estate, SEVENOAKS, Kent, TN14 5EL
  Tel: Sevenoaks (0732) 59224
Lennox Industries Ltd,
  P.O. Box 43, Lister Road, BASINGSTOKE, Hampshire,
  RG22 4AR
  Tel: Basingstoke (0256) 61261

Massrealm Ltd,
  Napier Way, CRAWLEY, West Sussex, RH10 2RA
  Tel: Crawley (0293) 21874
Mather & Platt Ltd,
  Park Works, Newton Heath, MANCHESTER, M10 6BA
  Tel: 061-205 2321
William May (Ashton) Ltd,
  Cavendish Street, ASHTON UNDER LYNE, Lancashire, OL6 7BR
  Tel: 061-330 3838
Nu-Way Benson Ltd,
  Temeside Works, LUDLOW, Salop., SY8 1JL
  Tel: Ludlow (0584) 3131
PJ Air Curtains Ltd,
  4A Hillingdon Parade, Uxbridge Road, HILLINGDON, Middlesex,
  UB10 0PE
  Tel: Uxbridge (0895) 39250
Powrmatic Ltd,
  Winterhay Lane, ILMINSTER, Somerset, TA19 9PQ
  Tel: Ilminster (04605) 3535
S & P Coil Products Ltd,
  Evington Valley Road, LEICESTER, LE5 5LU
  Tel: Leicester (0533) 730771
Sahara Products,
  St Mary's Road, Bowdon, ALTRINCHAM, Cheshire, WA14 2PL
  Tel: 061-928 9928
Stenor Ltd,
  Lydon Trading Estate, Mortlake Road, Kew, RICHMOND, Surrey,
  TW9 4AQ
  Tel: 01-876 0404
Taylor & Portway Ltd,
  Rosemary Lane, HALSTEAD, Essex, CO9 1HS
  Tel: Halstead (07874) 2551
Teleheaters Ltd,
  42–44 Waggon Road, AYR, Strathclyde, KA8 8BB
  Tel: Ayr (0292) 61635
Thermoscreens Ltd,
  Chandlers Ford Industrial Estate, EASTLEIGH, Hampshire, SO5 3DQ
  Tel: Chandlers Ford (04215) 4731
TI Creda Ltd,
  Creda Works, Blythe Bridge, STOKE ON TRENT, Staffordshire,
  ST11 9LJ
  Tel: Blythe Bridge (07818) 2281
Tomlinsons (Rochdale) Ltd,
  Newhey Road, Milnrow, ROCHDALE, Lancashire, OL6 3NR
  Tel: Rochdale (0706) 42411

Trailer Heaters Ltd,
  Springfield Way, Anlaby, HULL, North Humberside, HU10 6RL
  Tel: Hull (0482) 52152
Tronicair International Ltd,
  Spring Road, Kilsyth, GLASGOW, G65 0PT
  Tel: Kilsyth (0236) 821965
Unidare Engineering Ltd
  Surrey Street, GLOSSOP, Derbyshire, SK13 9BX
  Tel: Glossop (04574) 4815
Wanson Co. Ltd,
  7 Elstree Way, BOREHAMWOOD, Hertfordshire, WD6 1SA
  Tel: 01-953 6211
Warmco (Manchester) Ltd,
  Stamford Works, Manchester Road, MOSSLEY, Lancashire, OL5 9BJ
  Tel: Mossley (04575) 5511
Wynbourne-Sataba Equipment Ltd,
  82–86 City Road, LONDON, EC1Y 2BJ
  Tel: 01-251 4442
Wyndawaye Systems Ltd,
  London Road, BRANDON, Suffolk
  Tel: Thetford (0842) 810415

WATER HEATERS

Acoustics & Environmetrics Ltd,
  Winchester Road, WALTON ON THAMES, Surrey, KT12 2RP
  Tel: Walton on Thames 47644/6
Andrews Industrial Equipment Ltd,
  Dudley Road, WOLVERHAMPTON, West Midlands, WV2 3DB
  Tel: Wolverhampton (0902) 58111
Aquatron Showers Ltd,
  Radway Road, Shirley, SOLIHULL, West Midlands, B90 4NR
  Tel: 021-704 4193
Brefco (Northern) Ltd,
  P.O. Box 16, Brookhouse, Peel Green, Eccles, MANCHESTER
  Tel: 061-789 8111
Calomax (Engineers) Ltd,
  Lupton Avenue, LEEDS, West Yorkshire, LS9 7DD
  Tel: Leeds (0532) 49668
Chaffoteux Ltd,
  Concord House, Brighton Road, Salfords, REDHILL, Surrey,
  RH1 5DX
  Tel: Horley (02934) 72744

Cosybug Ltd,
  2 Napier Road, BROMLEY, Kent, BR2 9JA
  Tel: 01-464 3263
CTC Heat (London) Ltd (Rossfor Associates) Ltd,
  296b Station Road, HARROW, Middlesex, HA1 2DX
  Tel: 01-427 8454
W. H. Dean & Son Ltd,
  Accrington Road, BURNLEY, Lancashire, BB11 5DS
  Tel: Burnley (0282) 25901
Dolphin Showers Ltd,
  Bromwich Road, WORCESTER, WR2 4BD
  Tel: Worcester (0905) 42287
Doulton Heating Systems Ltd,
  School Lane, Knowsley, PRESCOTT, Merseyside, L34 9HT
  Tel: 051-546 8225/7
Dunsley Heating Appliance Co. Ltd,
  Fearnought, Holmfirth, HUDDERSFIELD, West Yorkshire, HD7 2TU
  Tel: Holmfirth (048489) 2635
Eclipse Cooper Co. of York Ltd,
  Midgley House, James Street, YORK, YO1 3DS
  Tel: York (0904) 412541
Eltron (London) Ltd,
  Accrington Works, Strathmore Road, CROYDON, Surrey, CR9 2NA
  Tel: 01-689 4341
Gainsborough Electrical Ltd,
  Shefford Road, Aston, BIRMINGHAM, B6 4PL
  Tel: 021-359 5631
Gardom & Lock Ltd,
  Aflow House, Soho Hill, Handsworth, BIRMINGHAM, B19 1AP
  Tel: 021-523 3311
Hamworthy Engineering Ltd,
  Fleets Corner, POOLE, Dorset, BH17 7LA
  Tel: Poole (02013) 5123
Heatrae-Sadia Heating Ltd,
  Hurricane Way, Norwich Airport, NORWICH, NR6 6EA
  Tel: Norwich (0603) 44144
Hedin Ltd,
  4 Raven Road, South Woodford, LONDON, E18 1HJ
  Tel: 01-504 6601
H. D. Howden Ltd,
  73 Coronation Road, New Stevenston, MOTHERWELL, Strathclyde,
  ML1 4JF
  Tel: Holytown (0698) 732303

HVE Thermoduct Ltd,
  Greenholme Mills, Burley in Wharfedale, ILKLEY, West Yorkshire, LS29 7EA
  Tel: Burley in Wharfedale (09435) 3587
IMI Santon Ltd,
  Somerton Works, NEWPORT, Gwent, NPT 0XU
  Tel: Newport (0633) 277711
Instaflow Ltd,
  Dellbow Road, Central Way, FELTHAM, Middlesex, TW14 0SQ
  Tel: 01-751 3117
Johnson & Starley Ltd,
  Rothersthorpe Crescent, NORTHAMPTON, NN4 9JF
  Tel: Northampton (0604) 62881
Lennox Industries Ltd,
  P.O. Box 43, Lister Road, BASINGSTOKE, Hampshire, RG22 4AR
  Tel: Basingstoke (0256) 61261
Lounsdale Electric Ltd,
  Lounsdale Works, Lounsdale Road, Meikleriggs, PAISLEY, Strathclyde, PA2 9DT
  Tel: 041-887 7511
Main Gas Appliances Ltd,
  Angel Road, Edmonton, LONDON, N18 3HL
  Tel: 01-807 3030
Prometheus Gas Appliances Ltd,
  Priory Buildings, Church Hill, ORPINGTON, Kent, BR6 0HE
  Tel: Orpington 34512
Radiation-Ascot Ltd,
  Nottingham Road, BELPER, Derbyshire, DE5 0PQ
  Tel: Belper (077382) 4141
Redring Electric Ltd,
  Reading Works, PETERBOROUGH, PE2 9JJ
  Tel: Peterborough (0733) 60431
Remploy Ltd,
  Remploy House, 415 Edgware Road, Cricklewood, LONDON, W2 6LR
  Tel: 01-452 8020
Sheathed Heating Elements Ltd,
  Wardley Industrial Estate, North Worsley, MANCHESTER, M28 5NJ
  Tel: 061-794 6122
W. H. Smith (Eziot) Ltd,
  32 Mansfield Street, Church Gate, LEICESTER, LE1 3DG
  Tel: Leicester (0533) 22514
Stiebel Eltron Ltd,
  25 Lyveden Road, Brackmills, NORTHAMPTON, NN4 0ED
  Tel: Northampton (0604) 66421

TI Creda Ltd,
  Creda Works, Blythe Bridge, STOKE ON TRENT, Staffordshire,
  ST11 9LJ
  Tel: Blythe Bridge (07818) 2281
Triton Aquatherm Ltd,
  Triton House, Weddington Terrace, Weddington Industrial Estate,
  NUNEATON, Warwickshire, CV10 0AG
  Tel: Nuneaton (0682) 325908
Walker Crossweller & Co. Ltd,
  Whaddon Works, CHELTENHAM, Gloucestershire, GL52 5EP
  Tel: Cheltenham (0242) 516317
A. K. Waugh Ltd,
  120 Kelvinhaugh Street, GLASGOW, G3 8PS
  Tel: 041-221 0325

Author Index

755

Subject Index

757